Managing and De
for Conservation

Conservation Science and Practice Series

Published in association with the Zoological Society of London

Blackwell Publishing and the Zoological Society of London are proud to present our new *Conservation Science and Practice* volume series. Each book in the series reviews a key issue in conservation today. We are particularly keen to publish books that address the multidisciplinary aspects of conservation, looking at how biological scientists and ecologists are interacting with social scientists to effect long-term, sustainable conservation measures.

Books in the series can be single or multi-authored and proposals should be sent to:

Ward Cooper, Senior Commissioning Editor, Blackwell Publishing Ltd, 9600 Garsington Road, Oxford OX4 2DQ, UK. Email: ward.cooper@oxon.blackwellpublishing.com

Each book proposal will be assessed by independent academic referees, as well as our Series Editorial Panel. Members of the Panel include:

Conservation Science and Practice Series

Managing and Designing Landscapes for Conservation: Moving from Perspectives to Principles

Edited by

David B. Lindenmayer
Richard J. Hobbs

ZSL
LIVING CONSERVATION

Blackwell
Publishing

BLACKWELL PUBLISHING
350 Main Street, Malden, MA 02148-5020, USA
9600 Garsington Road, Oxford OX4 2DQ, UK
550 Swanston Street, Carlton, Victoria 3053, Australia

First published 2007 by Blackwell Publishing Ltd

1 2007

Library of Congress Cataloging-in-Publication Data

Landscapes for conservation: moving from perspectives to principles / edited by David B. Lindenmayer, Richard J. Hobbs.
p. cm. – (Conservation science and practice series)
"Published in association with the Zoological Society of London."
Includes bibliographical references and index.
ISBN 978-1-4051-5914-2 (pbk. : alk. paper) 1. Ecological landscape design. 2. Landscape protection. I. Lindenmayer, David. II. Hobbs, R. J. (Richard J.) III. Zoological Society of London.

SB472.45.L363 2008
712–dc22

2007016681

A catalogue record for this title is available from the British Library.

Set in 10.5/12.4 Minion
by Newgen Imaging Systems, Chennai.
Printed and bound in Singapore
by Fabulous Printers Pte Ltd

The publisher's policy is to use permanent paper from mills that operate a sustainable forestry policy, and which has been manufactured from pulp processed using acid-free and elementary chlorine-free practices. Furthermore, the publisher ensures that the text paper and cover board used have met acceptable environmental accreditation standards.

For further information on
Blackwell Publishing, visit our website:
www.blackwellpublishing.com

Contents

Foreword: Landscapes in Peril

Paul R. Ehrlich

While many environmental scientists and some politicians are focused on climate change as a major threat to both biodiversity and the supply of crucial ecosystem services to humanity, a connected and likely equally daunting challenge is runaway land-use change. Indeed, the future of biodiversity greatly depends on the future of agricultural production for food and, increasingly, fuel. Ecologists know this, and it has spurred the development of fields called landscape ecology, conservation biology, countryside biogeography, and restoration ecology, which deal with how human activities are changing both terrestrial and aquatic ecosystems, what the consequences are for both biodiversity and ecosystem services, and what can be done to ameliorate deleterious changes. Accelerating anthropogenic climate change adds a dynamic urgency to all of these disciplines (e.g., Parmesan, 2006).

Sadly, despite the supreme importance of these fields, precious few generalizations have emerged from these scientific efforts, beyond the knowledge that spatial issues are extremely important to conservation, and that human-dominated landscapes can be managed in ways that will enhance their hospitality to biodiversity and their ability to deliver services. For example, even areas that were once tropical forest but are now mixed agricultural countrysides can, under certain conditions, maintain substantial portions of the original forest biota (e.g., Daily *et al.* 2003; Sekercioglu *et al.* 2007). In addition, there are guidelines for action that can be of immediate and direct use to managers.

Managing and Designing Landscapes for Conservation is a first attempt at identifying further principles from what environmental scientists have learned about these issues – revealing a rich lode of distilled knowledge. But in extracting the sorts of principles that can serve as actual guides for managers, it is clear there are no easy answers. So much is contingent on the nature and history of the ecosystem, the organisms or ecosystem processes to be sustained, and the socio–economic–political context that finding the equivalent of a Boyle's Law for landscape ecology seems highly unlikely.

For example, the ecological value of corridors – a classic topic in conservation science – turns out to be fiendishly complex and context-dependent.

Corridors often appear necessary to provide connectivity for animals with large home ranges, and to provide flexibility in the face of climate change. But they can also facilitate the spread of diseases such as the cancer now decimating Tasmanian devils (Bostanci, 2005) or the pernicious advance of myriad invasive organisms now threatening biodiversity in general (for overview see Baskin, 2002). Corridors can even create problems with native fauna, such as allowing the abundant and extremely aggressive edge-inhabiting passerine bird, the Noisy Miner, to reach habitat patches in which endangered birds thrive in the Miners' absence (Hastings and Beattie, 2006).

It is also clear that limiting the intensification of agriculture can benefit biodiversity and enhance ecosystem services (Ricketts *et al.* 2004). Sparing some areas of a Costa Rican coffee plantation by leaving them as forest patches actually increased coffee production above what would be achieved by clearing and cultivating those patches, because the actions of forest-dwelling pollinators resulted in a larger – and higher-quality – coffee harvest. But when coffee was replaced by pineapple, the forests no longer delivered the agricultural pollination service that had yielded the farm ~$US60,000 per year (Chan *et al.* 2007). Other, less easily monetized, ecosystem services were also supplied by the forest patches, but that example and the general complexity of bee distributions over the Costa Rican countryside (e.g., Brosi *et al.* 2007) demonstrate some of the problems of countryside management in a world where both biophysical and socioeconomic factors are continually changing the landscape (Chan *et al.* 2007, Fischer *et al.* 2007).

In the face of this complexity, what is a manager to do? The first thing is to read *Managing and Designing Landscapes for Conservation*, both to understand the complexity and to pick up, from the diversity of views presented, useful insights into the unique management situation he or she confronts. Another is to join with other scientists and managers to promote large-scale rewilding efforts (e.g., Soulé 1999, Soulé *et al.* 2004) such as the Wildlands Project, whose goal is to reconnect vast areas of land to create areas that will maintain viable populations of the large animals (especially top predators) that are critical to ecosystem functioning. The project hopes to do this on a continental scale, creating what are called "MegaLinkages." For instance, the absence of wolves in Yellowstone National Park in the United States for most of the twentieth century had severe impacts on the local ecosystem. Without the wolves, elk populations boomed and reduced stands of cottonwoods, the aspens and willows (e.g., Beschta, 2003). Lacking aspens to make lodges and dams, and willows to eat, the beavers disappeared, and with them the entire wetland ecosystems that the beavers created. Now that wolves have been reintroduced into the Park, some of these ecological patterns have begun to return to their former dynamics (Smith *et al.* 2003).

The goal of the Wildlands project and other rewilding projects is to create connected areas where large animals can run free and help maintain ecosystems in less-developed areas that people value for wilderness and recreational uses. One technique is to create wildlife bridges and underpasses ("ecoducts") across and beneath highways to keep the highways from dividing areas into habitat fragments too small to maintain large animals. Connectivity of areas is the key Wildlands principle – which, of course, can lead to the same sorts of problems as corridors between smaller habitat patches.

Notwithstanding these issues, lack of connectivity can produce quite general problems for conservation efforts. For example, in the San Diego area of California, patches of chaparral that are too isolated for coyotes to reach them have fewer birds than those with coyotes. The reason is that coyotes suppress foxes, raccoons, and especially cats, which are deadly predators of birds (Crooks and Soule, 1999). Areas of Australia with more dingoes also have more substantial populations of small native marsupials that are endangered elsewhere, because dingoes reduce the populations of feral cats, which are deadly to the marsupials (e.g., Johnson *et al.* 2006). The degree to which these wild dogs are excluded by lack of connectivity from areas where they would be ecologically beneficial is not clear.

In the above cases, predators such as wolves, coyotes, and dingoes can be considered "keystone species," ones whose impacts on their communities or ecosystems are disproportionately large relative to their abundance. Large mammalian herbivores, where they still exist, can play similar functional roles (Pringle *et al.* 2007), and they often require similar conditions to persist. For these reasons, creating expansive and highly connected habitats that support large keystone species is an important element of the Wildlands campaign.

Another promising project for managers to become familiar with is the *Natural Capital Project* (http://www.naturalcapitalproject.org/about.html). It was launched by an agreement among Stanford University, The Nature Conservancy, the World Wildlife Fund, and members of financial communities, with the hope of bringing in many other institutions and groups. The goal is to develop new scientific methods, new financial instruments, and new corporate and governmental policies to make preservation of natural capital as conventional as preservation of human and human-made capital is now. It recognizes, for example, that some of the goals of landscape management (e.g., preservation of biodiversity and of ecosystem services) may be partially incompatible (e.g., Chan et al, 2006, 2007). The project is taking the sensible scientific approach of sampling nature, focusing its initial efforts on three model systems: the Upper Yangtze Basin of China, the Eastern Arc Mountains of Tanzania, and the Sierra Nevada of California. In each case, efforts will be centered on developing new tools to incorporate ecosystem service values into landscape decision making. Elements of natural capital and the services flowing from them will be mapped, with the twin goals of motivating and enabling the public and policy makers to value and preserve them. Approaches, including the development of markets, will be explored to finance the necessary operations. The model systems, and subsidiary ones, will be used as test beds to see if and how natural capital and ecosystem services can be made a standard feature of land-use and investment decisions. All will involve great efforts to directly involve decision makers and managers, ranging from local people to financial professionals and government officials. This project is a grand experiment. It may not work, but it seems to have incredibly promising prospects for slowing or stopping the steady destruction of humanity's life-support systems.

The ultimate issue in managing landscapes is dealing with the drivers that are rapidly making conservation increasingly difficult. A good current example is the widespread and environmentally dangerous notion that the agricultural enterprise can be duplicated for the production of biofuels. The drivers, of course, are the Three Horseman of the **I = PAT** equation – human **P**opulation size and growth, **A**ffluence (i.e., overconsumption by the rich), and the use of environmentally faulty **T**echnologies and socio–political–economic arrangements (Ehrlich and Ehrlich, 2005). The latter are in large part caused by a maldistribution of political and economic power. No matter how many principles of landscape ecology, conservation biology, countryside biogeography, and restoration ecology are developed by environmental scientists, their efforts will be fruitless in the end if those critical drivers are not addressed effectively, rapidly, and with much attention paid to the needs of poor people who depend directly and heavily on the landscapes of concern. It therefore behooves all environmental scientists and environmental managers to devote some portion of their time to educating the public and generating political action on those issues. Otherwise, they are just trying to bail back a tsunami with thimbles.

But while we are necessarily wielding our thimbles, *Managing and Designing Landscapes for Conservation* can make an important contribution to enlarging their capacity and teaching us to bail more efficiently. Consider, for example, my colleague Harold Mooney's guidelines (Chapter 31) and the conclusion with which he accompanies them:

> Landscape Principles
>
> Protect what works since not only is it difficult to replace but also considerable economic losses can be incurred from its destruction as well as in attempts at restoration.
>
> Don't sell low when trading ecosystem services in development, for example, shrimp for coastal protection.
>
> Quantify, as completely as possible, all of the benefits derived from the services provided by any given ecosystem for which alterations are being proposed.
>
> When rehabilitating, incorporate an ecosystem approach in design that incorporates the ecosystem service paradigm and considers the range of services that can be restored, their value and their comparative costs for rehabilitation. Some services are very costly and difficult to replace whereas other valuable services may be less expensive and more likely to become self-sustaining.
>
> In either conservation or restoration or rehabilitation, prepare for a very uncertain future:
>
> A world of rapid change in climate and vegetation
> A world of increasing extreme events
> A world of weeds and diseases
> A world of regime shifts
> A world of rising seas and an acidifying ocean
> A world that has been diced and replumbed
> A world of increasing nitrogen and phosphorus redistribution

To which I would add "a world of fewer people leading lives of the sort to which the poorest third of today's population is condemned."

Lindenmayer and Hobbs are to be congratulated on this excellent effort to make a key area of conservation science available to other environmental scientists and accessible to managers. Both the breadth and the depth of the coverage are extraordinary, and time will certainly show the great value of their enterprise.

References

Baskin, Y. (2002) *A Plague of Rats and Rubber Vines: The Growing Threat of Species Invasions*. Island Press, Washington, DC.

Beschta, R.L. (2003) Cottonwoods, elk, and wolves in the Lamar Valley of Yellowstone National Park. *Ecological Applications* **13**, 1295–1309.

Bostanci, A. (2005) A devil of a disease. *Science* **307**, 1035.

Brosi, B.J., Daily, G.C., & Ehrlich, P.R. (2007) Bee community shifts with landscape context in a tropical countryside. *Ecological Applications*, in press.

Chan, K.M.A., Shaw, M.R., Cameron, D.R., Underwood, E.C. & Daily, G.C. (2006) Conservation planning for ecosystem services. *PLoS Biology* **4**, 2138–2152.

Chan, K.M.A., Pringle, R.M., Ranganathan, J., Boggs, C.L., Chan, Y.L., Ehrlich, P.R., Haff, P.K., Heller, N.E., Al-Khafaji, K. & MacMynowski, D.P. When agendas collide: Human welfare and biological conservation. *Conservation Biology*, in press.

Crooks, K.R. & Soule, M.E. (1999) Mesopredator release and avifaunal extinctions in a fragmented system. *Nature* **400**, 563–566.

Daily, G.C., Ceballos, G., Pacheco, J., Suzán, G. & Sánchez-Azofeifa, G.A. (2003) Countryside biogeography of Neotropical mammals: Conservation opportunities in agricultural landscapes of Costa Rica. *Conservation Biology* **17**, 1814–1826.

Ehrlich, P.R. & Ehrlich, A.H. (2005) *One with Nineveh: Politics, Consumption, and the Human Future (with new afterword)*. Island Press, Washington, DC.

Fischer, J., Brosi, B., Daily, G.C., Ehrlich, P.R., Goldman, R., Goldstein, J., Manning, A. D., Mooney, H.A., Pejchar, L., Ranganathan, J. & Tallis, H. (2007) Should agricultural policies encourage land-sparing or wildlife-friendly farming? Submitted to Science.

Hastings, R.A. & Beattie, A.J. (2006) Stop the bullying in the corridors: Can including shrubs make your revegetation more Noisy Miner free? *Ecological Management and Restoration* **7**, 105–112.

Johnson *et al.* 2006, *Proc.Roy.Soc.B*

Parmesan, C. (2006) Ecological and evolutionary responses to recent climate change. *Annual Review of Ecology, Evolution*, and Systematics, **37**, 637–669.

Pringle, R.M., Young, T.P., Rubenstein, D.I. & McCauley, D.J. (2007) Herbivore-initiated interaction cascades and their modulation by productivity in an African savanna. *Proceedings of the National Academy of Sciences of the United States of America*, in press.

Ricketts, T.H., Daily, G.C., Ehrlich, P.R. & Michener, C.D. (2004) Economic value of tropical forest to coffee production. *Proceedings of the National Academy of Sciences of the United States of America* **101**, 12579–12582.

Sekercioglu, Ç.H., Loarie, S.R., Ruiz-Gutierrez, V., Oviedo Brenes, F., Daily, G.C. & Ehrlich, P.R. (2007) Persistence of forest birds in tropical countryside. *Conservation Biology*, in press.

Smith, D.W., Peterson, R.O. & Houston, D.B. (1996) Yellowstone after Wolves. *BioScience* **53**, 330–340.

Soulé, M.E. (1999) An unflinching vision: Networks of people for networks of wildlands. *Wildlands* **9**, 38–46.

Soulé, M.E., Mackey, G., Recher, H.F., Williams, J.E., Woinarski, J.C.Z., Driscoll, D., Dennison, W.C. & Jones, M. (2004) Continental connectivity: Its role in Australian conservation. *Pacific Conservation Biology* **10**,266–279.

1

Introduction

David B. Lindenmayer and Richard J. Hobbs

Landscape ecology, conservation biology and restoration ecology are fields of study whose history of academic pursuit stretches back several decades or more. They are truly applied disciplines in that they attempt to generate insights that will promote better management of natural resources and biodiversity. All also are characterized by a large and rapidly expanding published literature. But are there any general principles that can be gleaned from the body of work to date? What can we say that is useful to guide the management of landscapes for conservation, and beyond that the ecological design (or redesign) of landscapes to achieve conservation goals? A meeting held at Craigieburn Resort, Bowral, in southeastern Australia in March 2006 attempted to determine if this was the case. Specifically, a group of leading and emerging researchers in landscape ecology and conservation biology was asked to appraise a given topic in landscape research and identify 5–10 general principles. This was by no means an easy task, especially because each essay had a ~4,000 word limit. The meeting covered 10 major themes, with an additional final theme of 'Bringing it all together'. Each theme was tackled by three authors working independently. We sought to gather together a highly diverse group of people to explore how truly general some of the proposed 'general principles' really are. Some authors have expertise in forest ecosystems, others in woodlands or grasslands. Many have worked in numerous different ecosystems. We made a deliberate effort to include aquatic ecosystems and streamscapes in this book on landscapes. This was because terrestrial and aquatic systems have for too long been treated as independent of one another, when clearly they strongly interact and have many similar properties (patchiness, connectivity, etc.). We also attempted to bring together a mix of people mostly concerned with traditional conservation biology issues such as species and habitat management and others more involved with ecosystem processes: too often these different groups do not interact.

To explore further the generality of principles, we selected participants from different continents and who have worked in different countries, states, regions and ecosystems within those continents. Of course there is a distinct Australian bias among the authors, but this was simply because of budget constraints and a desire to keep the size of the workshop small enough to foster insightful discussion. However, wherever possible, we pushed all authors to 'internationalize' their essays and consider the application of general principles to ecosystems beyond those in which they have worked. Finally, perspectives on general principles can be strongly shaped by research and management experience and world views. Therefore, the authors in each theme are a mix of 'old farts' and 'young turks' – those with prolonged research experience and those with potentially different perspectives through having relatively recently commenced their research careers. We leave the readers to decide who is which.

We make a final observation on the number and composition of the participants at the meeting. Time and logistical constraints limited the number of people that could attend the workshop. We acknowledge that it was not possible to invite all the leading workers in landscape ecology and conservation biology, all of whom we know could have made a major contribution to the meeting and to this book. At the same time, we could not afford to sponsor their participation and keep the dynamics of the group manageable and productive. One of us (DBL) received a barrage of email traffic from colleagues angered by their omission from the guest list. We sincerely apologize for any angst this may have caused and hope that one day we might be forgiven. Of course we fully understand that we may well not be invited to other people's workshops.

Each author completed a draft version of their essay before the meeting. They were then asked to make a short (five minute) presentation at Bowral based on the content of their essay. The three presentations on each topic then provided the basis for workshop discussions that aimed to pull together the generalities arising from the three talks. This proved a challenging enterprise, and some discussion sessions 'flew' while others 'flagged'. The chair of each discussion session was provided with 'axe' and 'barrow' cards. If any participant was perceived to be grinding his or her particular axe or pushing his or her particular barrow, they could be issued with the relevant card. Collection of two cards in any one session excluded the recipient from the workshop wine supply for that evening: fortunately, no participant had to be subjected to this fate. Nevertheless, towards the end of the workshop we perceived that the discussion sessions were not achieving the goal of deriving generalizable principles, and we reorganized the timing so that participants had the opportunity for some independent study and reflection time, from which they were asked to provide their top five principles from what they had heard in the workshop as a whole. This process produced a wealth of thoughtful and incisive material from which we pulled together a more prosaic summary for an extended all-of-workshop discussion. The final chapter of the book represents our distillation and interpretation of this material.

Authors were then given six weeks after the close of the meeting to revise their essays based on initial review and the extensive group discussions at Bowral. These revised essays were again reviewed by the editors. This volume of edited chapters is the product of that process. The book is set out in the 11 key sections reflecting the major themes of the meeting. We have written a short synthesis to conclude each theme. The synthesis is not simply a regurgitation of the material presented in the essays. Rather, we have attempted not only to extract the key points from the three essays, but also to supplement it with insights from group discussions conducted after the presentations on each theme. Of course we have also added our own perspectives and biases.

An initial external review of the book proposal highlighted concerns about overlap and duplication among the two to three essays on each topic. We had similar early concerns, but they proved to be ill founded. There was remarkable dissimilarity in the material covered by each of the authors in almost all of the key themes. An example is the theme on 'Connectivity, Corridors and Stepping Stones'. The three essays were not only highly insightful, but also extraordinarily diverse. The first by Fiona Schmiegelow highlighted the importance of matrix management for connectivity and how it can surpass the establishment of physical corridors in contributing to connectivity. The second (from Reed Noss) highlighted how in heavily cleared landscapes, physical corridors rather than matrix management can be critical in providing connectivity for particular species. Noss's chapter outlines what kinds of species might be the best ones to use as target taxa around which to design particular management actions that promote connectivity. The third by Denis Saunders was an instructive discussion about the objectives of establishing corridors and which landscape strategies might best contribute to connectivity.

The diversity of perspectives that characterizes the theme of 'Connectivity, Corridors and Stepping Stones' is replicated in all the other themes covered in this book. Indeed, we believe that the diversity of perspectives on each topic is a strength of this book. An appreciation of such a diversity of perspectives is critical for researchers, students and resource managers alike and an important antidote to the dangers of introspection. To this end, as the organizers of the Bowral meeting, we have not subjected any of the essays to heavy editing for style or content (apart from aiming for consistency of format and adherence to length restrictions). Rather, we have left them as thought-provoking (and sometimes quite provocative) explorations of key topics – unshackled by the highly sanitizing and homogenizing constraints that typify the publication of modern journal articles. This has left us with writing styles and approaches to topics as diverse as the authors and their different perspectives on particular issues. Some will perceive this as laziness on our part, but we believe that opportunities for considered 'thought pieces' are all too rare in modern scientific publishing and this book provided an important opportunity to counter this. Rather than squeeze the different individual approaches into a straight-jacketed formula, we have retained the individual quirkiness and tried to counterbalance that with the synthesis pieces in which we aimed for a degree of consistency of approach.

Of course in any meeting (and subsequent book of this type), there are important topics that could not be covered. There are limits to the size of a workshop consistent with sensible and productive dialogue and interaction. Also, there are budget constraints. Some important topics simply had to be left out for these reasons. For example, we did not touch upon such key topics as landscape genetics, spatial statistics or the socioeconomic dimensions of landscape management. We also did not cover marine ecosystems. Some will see these as major oversights, but there are limits to the length of a book.

As workshop coordinators, we also annoyed several participants from the outset by requesting that we restrict discussions to the scientific aspects of the problem, rather than bringing in socioeconomic considerations. We are fully aware, as were most of the workshop participants, that these issues are often of overriding importance in terms of what actually happens on the ground and in policy and planning. However, we felt that there is still plenty to be ironed out in the ecological arena and that the workshop should focus on this rather than dilute its efforts trying to cover the whole 'shooting match' of issues relating to landscape management and conservation. Various chapters in this book inevitably include discussion of some of the social, policy and economic issues that provide the broader context in which these ecological issues sit: however, a comprehensive treatment is again beyond the scope of this book.

Finally, we are of course also fully aware that attempts to identify 'general principles' can only be 'work in progress'. It is a massive challenge to find general principles that are not so general that they are truisms that make no advance, but yet are appropriate for conversion to on-ground prescriptions. The title of the book reflects our attempt to move from the individual perspectives provided by the diverse array of people attending the workshop towards a set of principles that may then be useful in practice. Notably, we did not attempt to apply any principles to any real-life landscapes at this meeting: that is a task for the next iteration of this process. Hence 'when it is all said and done there will be much more to be said and done'. Nevertheless, we hope that readers will find useful insights in the chapters in this book.

Acknowledgements

The workshop at Bowral (southern New South Wales, southeastern Australia) was funded by grants from Land and Water Australia and the Australian Research Council Kendall Foundation. The workshop was excellently organized by Dr Rebecca Montague-Drake. Notes from the Discussion sessions at the meeting in Bowral were expertly minuted by Mr Chris MacGregor and Dr Rebecca Montague-Drake. The staff at Craigieburn ensured that the workshop logistics and catering were smoothly and effectively taken care of. The Bowral meeting and the book concept were initially the brainchild of Jason Alexander, formerly of Land and Water Australia. Professor Will Steffen (Director of the Centre for Resource and Environmental Studies at the Australian National University) and Andrew Campbell (Land and Water Australia) were strong supporters of this project and made particular efforts to take part in discussions at Bowral. We thank Dr Martin Fortescue, Dr Nick Dexter (Department of Environment and Heritage) and Chris MacGregor and Darren Brown for coordinating a wonderful trip to Booderee National Park. Finally, we greatly appreciate the provision of heavily subsidized but nevertheless excellent wine by Jim and Diana Viggers that undoubtedly contributed to many fascinating and informative discussions!

Section 1
Classification of Landscapes and Terminology

2

The Whole Elephant: Classification and Terminology as Tools for Achieving Generality in Landscape Ecology

S. McIntyre

Abstract

The development of principles to guide land uses for conservation on landscapes requires us to negotiate between the hazards of oversimplification and too much complexity. The terminology and classification that we use to describe and investigate landscape processes are crucial to this process. I argue that we need to maintain breadth in the range of biophysical situations, land uses and organisms that are considered by landscape ecologists. Only then will it be possible to identify the important elements required to achieve management outcomes. These points are illustrated through a history of research in Australian eucalypt woodland landscapes. Guidelines for the development of landscape design principles are proposed. Concepts that are crucial to conservation at landscape scales include the consideration of all land uses that occur on landscapes. These land uses need to be linked to their effects on organisms through their disturbance characteristics and whether these represent exogenous and/or endogenous disturbances to the biotic communities. Through this understanding, it will be possible to develop principles regarding the types, amount and location of land uses that are compatible with the persistence of organisms on landscapes.

Keywords: endogenous disturbance; exogenous disturbance; land use; research models.

Introduction

It is a truth universally acknowledged that a landscape ecologist in possession of some good insights must be in want of general design principles. It is also true that as ecologists, we struggle with the constant tension between the quest for simple theoretical and management principles, and making sense of the overwhelming complexity of ecosystems. The way that individual researchers resolve this is as much a product of their personality, as it is a product of the scientific method. The tension is apparent among taxonomists, among whom the terms 'splitter' and 'lumper' have been coined to discriminate between people who prefer to highlight nature's complexity, and those who would rather allocate it to very large boxes. I possess the tendencies of a 'lumper', which has resulted in a display of general principles (McIntyre *et al.* 2002a). However, it is the purpose of this chapter to temper generalist tendencies appropriately with sufficient attention to variation within and between landscapes. How can we optimize the degree of detail in our depiction of landscapes while communicating with other landscape ecologists and a wider audience?

I argue that there is a need to consider the full range of environments, human actions and the biota that are associated with our landscapes. In doing so it is necessary to classify these elements and therefore identify the variables that will be most important for achieving desired management outcomes. This approach to the conceptualization of landscapes will therefore result in identification of the key hypothesized factors and thus creates useful simplifications that can subsequently be tested. I use the example of Australia's eucalypt grassy woodlands to illustrate some of these points. This ecosystem has been of interest to Australian landscape ecologists for some decades, and from it we are slowly learning to put our assumptions aside and consider the entire landscape, all of its land uses and the range of organisms inhabiting it.

Viewing the range of landscapes

There is a long and rich tradition of landscape classification, drawing on a range of biophysical and human-related variables by a range of disciplines, for a range of purposes, but most notably land use planning (e.g. Bastian 2000; Pressey *et al.* 2000; Mora & Iverson 2002; Mitchell *et al.* 2004). Less often, classifications are proposed with the explicit purpose of improving experimental models for the investigation of landscape processes (Collinge & Forman 1998; McIntyre & Hobbs 1999; Watson 2002; Hobbs & McIntyre 2005). The level of detail contained in the latter is generally less than the former, reflecting the need to identify a small number of testable hypotheses. However, detailed classifications are also important because they define the environmental domain and provide an array of potentially important variables. The task of the landscape ecologist is to select variables and partition levels in such a way as to highlight the limits of generality of various landscape design or management principles (e.g. see Hutchinson *et al.* 2005). Our research models are represented by the amount of detail and the terminology that we use to describe landscapes and their responses. Too often we do this unconsciously and with many unexamined assumptions.

As landscape ecologists we can also be deficient in the breadth of our perspective. This has become apparent from my experiences working as an ecologist in agricultural landscapes and within an agriculturally dominated research culture. In both ecology and agriculture, attention has been confined to elements of the landscape that are subject to different land uses, and these elements have been investigated in a largely independent manner. An apt metaphor is that of the elephant and the blind men, a parable that features in many cultures. When shown an elephant, a group of blind men felt different parts of the animal and each one concluded that it was a completely different thing, depending on whether they had grasped the tail, leg, trunk and so on. The consequences of ignoring important parts of the 'elephant' can be serious. The many problems in Australian rural landscapes – soil salinity, deterioration of water quality and flow, losses of biodiversity – have resulted from a consideration of the management of production areas without considering the functioning of the wider landscape. Landscape ecologists have also been blinkered in their view of the landscape, tending to focus on conservation reserves or habitats with particular physical characteristics. This has not so much resulted in environmental degradation, but rather hampered our understanding of landscape processes. It has also meant that integration of conservation and production land uses has been ignored. However, we need to be careful, particularly because there will be risks to the environment if, or when, our principles are translated into actual landscape management.

Viewing the range of human actions

The divide between ecology and agriculture, described above, is less applicable to Europe, where agriculture has been practised for sufficient time for the two to be culturally and ecologically embedded. Prior to European invasion, Australian landscapes were also cultural landscapes (Gott 2005), but the legacy of indigenous land management and the way it coevolved with the biophysical landscape is knowledge that had been lost for large parts of the continent. Present-day land managers generally operate in ignorance of all but the sketchiest details of this important cultural and biological context.

However, even under industrial agriculture, the biophysical features of the landscape determine the functions, goods and services that can be exploited. There is thus a close interrelationship between the biophysical landscape, the types of human land use that are imposed and the effects of this land use on the condition of the landscape. This strongly influences the patterns and typologies of land uses and conditions both within and between landscapes (Hobbs & McIntyre 2005; Huston 2005; Lunt & Spooner 2005). However, the biophysical attributes in themselves will determine the susceptibility to dysfunction and the forms of dysfunction that are likely to develop on a landscape.

Owing to the rate of change of human society in the 20th and 21st centuries, landscape ecologists have attributed to the human species a separateness from the biophysical landscape, and have viewed our influences as a superimposed layer, generally of greater disturbance and harmfulness to ecosystems. Even in cultural landscapes, this model can be appropriate when land use practices change faster than the rate of adaptation of the biota. It is also appropriate where the question at hand is: how are people to live on landscapes sustainably? It is the layer of human influence that needs to be modified, not the inherent features of the landscape. Nonetheless, the problem I see with this view is that it can lead to assumptions about the effects of humans on ecosystems, and a failure to see that our presence is not always a negative force. Lavish evidence of fauna and flora exploiting human activities and human-created habitats is presented by Low (2003). By ignoring this, and viewing humans as a dark force, it is quite possible that we will be blind to some of the solutions for coexistence with the biota that we so urgently seek.

Although human impacts are not essentially different from changes due to non-human factors, this does not mean that 'anything goes' (Haila 1999). In other words, we cannot afford to manage in an unthinking, human-centred manner. There needs to be recognition of a divide between the endogenous disturbance regime (that which the ecosystems have been subjected to over an evolutionary time frame) and the endogenous disturbance regime (novel human activities that disrupt ecosystem functions and that are the target of behavioural change). Endogenous disturbance regimes will include human activities in cultural landscapes, and it may be necessary to reinstate them to restore ecosystem function. However, to achieve this, it is necessary to assess dispassionately how the range of human actions actually influence the biota.

Viewing the entire landscape: A case study

There is evidence that we have often overlooked the presence of fauna and flora in sections of the landscape, because of our assumptions of habitat suitability and therefore our failure to look. These assumptions generally concern the perceived negative effects of human activities and have tended to be reinforced by the highly influential principles adapted from island biogeography. The concept of habitat fragmentation arose from here, in which landscape elements are perceived as either habitat or non-habitat depending on assumptions about the effects of human activities on vegetation (McIntyre & Hobbs 1999). While the translation of island biogeographical principles across to continental landscapes has had some uses in the development of conservation biology (Haila 2002), it has also left a legacy of misinterpretation of the landscape (McIntyre & Barrett 1992; Haila 2002). One of the reasons is that human-modified landscapes, which frequently form the matrix, are not always analogous to the sea (Watson 2002), but examination of the matrix for habitat suitability is rarely undertaken (Haila 2002).

The example of grassy eucalypt woodlands

The grassy ecosystems of eastern Australia have been of ongoing interest to ecologists since the 1970s due to the considerable impact that European settlement has had in applying agricultural technologies. This biome is an interesting case study regarding the evolution of research models and assumptions made about the landscape. Some of the earliest observations, by botanists, were of the impacts of agriculture on the natural grasslands (Willis 1964). The loss of plant species with livestock grazing was recorded (Stuwe & Parsons 1977), but the larger effects of cultivation and the use of fertilizers, which were assumed to result in complete replacement of the native community, were not documented by botanists or agronomists, who were primarily interested in the replacement species not the remnant native species. During the 1980s and 1990s attention was drawn to the wider areas of eucalypt grassy woodland from the tropics to the temperate regions. Note that this vegetation is termed 'grassy woodland', partly as a reference to the assumed pre-European condition, and ignoring the fact that the structure is frequently that of a forest (Fig. 2.1a,b).

The fragmentation model formed the paradigm for grassy woodlands, particularly for faunal ecologists: stands of trees above a certain size were considered to be habitat and the rest non-habitat. This model grouped forest and woodland as 'habitat' (Fig. 2.1a,b), and open woodland and scattered trees as 'non-habitat' (Fig. 2.1e,f). This model did provide some insights into the effects of landscape change on fauna but, in applying it, ecologists were forced to exclude large parts of some landscapes from experimental investigations. Recognition of intermediate levels of habitat modification was proposed and this enabled processes such as grazing and partial clearing to be considered and elements such as single trees to be highlighted (Fig. 2.1d–f). This was important for conservation, because opportunities for reserve creation were limited and the landscapes are highly dependent on conservation actions undertaken by landholders working primarily in an agricultural production context.

Taking account of the matrix in human-modified landscapes is now high on the agenda of landscape ecologists both in Australia and elsewhere (Murphy & Lovett-Doust 2004; Fischer *et al.* 2005) and represents a critically important paradigm shift. Variation of land use in the matrix has been observed to modify responses to fragmentation; for example, woodland birds were most sensitive to patch isolation in an agricultural matrix compared with urban and periurban matrices (Watson *et al.* 2005). Landscape context affected the composition of the bird assemblage in subtropical woodlands (Martin *et al.* 2006) and temperate forests (Yamaura *et al.* 2005). Highly modified habitats such as isolated trees in pastures have been recognized as significant for birds (Manning *et al.* 2004). Assumptions about the effects of livestock grazing are also being challenged in grassy woodlands in Australia, with high densities of native plant species occurring under commercial levels of livestock grazing (McIntyre *et al.* 2002b) and woodland birds persisting in grazed woodlands (Martin & McIntyre 2007).

Viewing the range of organisms: A lesson from woodlands

The recognition of only structural modification of grassy woodlands (e.g. tree clearing and grazing), while an improvement, in my view fails to recognize one

(a)

(b)

Figure. 2.1 **Ways of seeing the landscape. Grassy eucalypt ecosystems of temperate Australia in the Yass River valley, New South Wales. Although eucalypt forest (a) and woodland (b) are both associated with this landscape, they are generally lumped by landscape ecologists as 'eucalypt woodlands'. Work on habitat fragmentation in these landscapes initially focused on the contrasts of habitat and non-habitat such as that illustrated in (c), which shows eucalypt forest contrasting with cleared, fertilized pastures dominated by exotic species. (d–f) Intermediate**

(c)

(d)

Figure. 2.1 **(Continued) tree densities that do not fit into the patch-matrix model but form the dominant structure in many regions. Landscapes (e) and (f) are structurally similar, but to make further progress in landscape ecology it will become necessary to recognize that the understorey of (e [overleaf]) is of relatively low nutrient status and dominated by native herbaceous vegetation whereas (f [overleaf]) is dominated by exotic species as a result of fertilization and sowing 'improved' pasture species. (Photos by permission of Sue McIntyre.)**

(e)

(f)

Figure. 2.1 **(Continued)**

of the most important functional transformations of eucalypt grassy woodlands. Figures 2.1e and 2.1f are structurally similar and may currently support a similar woodland bird assemblage. However, Fig. 2.1f has been fertilized sufficiently to support an understorey dominated by exotic pasture grasses, and the functioning of the landscape has thus been transformed. Grassland fertilization dramatically reduces the diversity of native plants (Dorrough *et al.* 2006), can adversely affect eucalypt health (Wylie *et al.* 1993) and disrupt the regeneration processes of trees and shrubs (Semple & Koen 2003). Over time, these functionally different landscapes are most likely to translate into structurally different ones, in which the bird communities may well look entirely different. Conversely, a derived native grassland in which trees have been removed, but the understorey remains intact, may offer excellent prospects for woodland restoration. The introduction of trees is a relatively simple matter and regeneration processes are likely to proceed readily, owing to the low fertility status of the ground layer. If the immediate needs of birds only are considered, a poorer outcome for landscape restoration might be experienced than if landscape function and condition is assessed for plants in addition to fauna. This is but one important example that points to the need to keep our eyes open to the work of researchers who study groups other than our own.

Guidelines for the development of landscape design principles

I am proposing a set of guidelines to assist the development and communication of landscape design principles. If applied, these guidelines will help us to examine our assumptions, and identify the scope and generality of an observation or set of observations.

Design principles apply to the entire landscape, and so the following must be considered when developing them:

1. The biophysical nature of the landscape may give it particular susceptibilities or resistance to losses of ecological function or biodiversity. This may necessitate setting more or less stringent limits to land-use intensification, or developing a specific principle for a particular landscape type.
2. Endogenous disturbance regimes may no longer be operating on a particular landscape. Endogenous disturbances are those that have operated

on a landscape over evolutionary time frames and may be necessary to maintain biota or ecological function.

3 Human activities may be acting as exogenous disturbances on landscapes, disrupting ecosystem function and affecting the persistence of biota. In cultural landscapes, some human activities may represent endogenous disturbances. However, non-traditional management often acts as exogenous disturbance, which presents a threat to landscape biota and ecosystem functions.

4 Consider all land uses and land tenures that are present in the landscape. Assumptions should not be made that any particular land use will not support elements of biodiversity in some way.

5 Assess the functions and biodiversity associated with land uses and land tenures for a range of organisms. No single management factor or land use will be optimal for all functions and organisms.

6 Break down land uses into single factors and consider their effects on ecosystems both separately and in combination. Land uses tend to represent confounded factors and it may be necessary to determine experimentally the effects of single management factors.

7 Management factors that may be adverse when associated with a particular land use may partially substitute for endogenous disturbances when applied in isolation or in a different context. For example, livestock grazing can impact on plant diversity and habitat structure when used in association with pasture fertilization, but it may partially substitute for native mammal grazing (an endogenous disturbance) and be used to maintain plant diversity in low-fertility grasslands.

8 Design the overlay of human activities in such a way as to maintain or mimic endogenous disturbance regimes and limit the impacts of exogenous disturbance regimes. In multiple use landscapes, the principle is to optimize different land uses to allow for the persistence of organisms by considering the location, extent and impact of disturbances.

Before translating a set of observations into a general principle, observe the 'whole elephant' and consider:

9 The range of biophysical environments to which the generalization may or may not apply.

10 Whether the current state or condition of the landscape affects the application of the principle.

11 The range of organisms to which the generalization may or may not apply.

References

Bastian, O. (2000) Landscape classification in Saxony (Germany)—a tool for holistic regional planning. *Landscape and Urban Planning* **50**, 145–155.

Collinge, S.K. & Forman, R.T.T. (1998) A conceptual model of land conversion processes: predictions and evidence from a microlandscape experiment with grassland insects. *Oikos* **82**, 66–84.

Dorrough, J., Moxham, C. Turner, V. & Sutter, G. (2006) Soil phosphorus and tree cover modify the effects of livestock grazing on plant species richness in Australian grassy woodland. *Biological Conservation* **130**, 394–405.

Fischer, J., Fazey, I., Briese, R. & Lindenmayer D.B. (2005) Making the matrix matter: challenges in Australian grazing landscapes. *Biodiversity and Conservation* **14**, 561–578.

Gott, B. (2005) Aboriginal fire management in south-eastern Australia: aims and frequency. *Journal of Biogeography* **32**, 1203–1208.

Haila, Y. (1999) Biodiversity and the divide between culture and nature. *Biodiversity and Conservation* **8**, 165–181.

Haila, Y. (2002) A conceptual genealogy of fragmentation research: from island biogeography to landscape ecology. *Ecological Applications* **12**, 231–334.

Hobbs, R.J. & McIntyre, S. (2005) Categorizing Australian landscapes as an aid to assessing the generality of landscape management guidelines. *Global Ecology and Biogeography* **14**, 1–15.

Huston, M.A. (2005) The three phases of land-use change: implications for biodiversity. *Ecological Applications* **15**, 1864–1878.

Hutchinson, M.F., McIntyre, S., Hobbs, R.J. *et al.* (2005) Integrating a global agro-climatic classification with bioregional boundaries in Australia. *Global Ecology and Biogeography* **14**, 197–212.

Low, T. (2003) *The New Nature*. Penguin, Melbourne.

Lunt, I.D. & Spooner, P.G. (2005) Using historical ecology to understand patterns of biodiversity in fragmented agricultural landscapes. *Journal of Biogeography* **32**, 1–15.

Manning, A.D., Lindenmayer, D.B. & Barry, S.C. (2004) The conservation implications of bird reproduction in the agricultural 'matrix': a case study of the vulnerable superb parrot of south-eastern Australia. *Biological Conservation* **120**, 363–374.

Martin, T.G. & McIntyre, S. (2006) Livestock grazing and tree clearing: impacts on birds of woodland, riparian and native pasture habitats. *Conservation Biology* (in press).

Martin, T.G., McIntyre, S., Catterall, C.P. & Possingham, H.P. (2006) Is landscape context important for riparian conservation? Birds in grassy woodland. *Biological Conservation* **127**, 201–214.

McIntyre, S. & Barrett G.W. (1992) Habitat variegation, an alternative to fragmentation. *Conservation Biology* **6**, 146–147.

McIntyre, S. & Hobbs R.J. (1999) A framework for conceptualizing human impacts on landscapes and its relevance to management and research models. *Conservation Biology* **13**, 1282–1292.

McIntyre, S., McIvor, J.G. & Heard, K.M. (eds.) (2002a) *Managing and Conserving Grassy Woodlands*. CSIRO Publishing, Melbourne.

McIntyre, S., Heard, K.M. & Martin, T.G. (2002b) How grassland plants are distributed over five human-created habitats typical of eucalypt woodlands in a variegated landscape. *Pacific Conservation Biology* **7**, 274–285.

Mitchell, N., Espie, P. & Hankin R. (2004) Rational landscape decision-making: the use of meso-scale climatic analysis to promote sustainable land management. *Landscape and Urban Planning* **67**, 131–140.

Mora, F. & Iverson, L. (2002) A spatially constrained ecological classification: rationale, methodology and implementation. *Plant Ecology* **158**, 153–169.

Murphy, H.T. & Lovett-Doust, J. (2004) Context and connectivity in plant metapopulations and landscape mosaics: does the matrix matter? *Oikos* **105**, 3–14.

Pressey, R.L., Hager, T.C., Ryan, K.M. *et al.* (2000) Using abiotic data for conservation assessments over extensive regions: quantitative methods applied across New South Wales, Australia. *Biological Conservation* **96**, 55–82.

Semple, W.S. & Koen, T.B. (2003) Effect of pasture type on regeneration of eucalypts in the woodland zone of south-eastern Australia. *Cunninghamia* **8**, 76–84.

Stuwe, J. & Parsons, R.F. (1977) *Themeda australis* grasslands on the Basalt Plains, Victoria: floristics and management effects. *Australian Journal of Ecology* **2**, 467–476.

Watson, D.M. (2002) A conceptual framework for studying species composition in fragments, islands and other patchy ecosystems. *Journal of Biogeography* **29**, 823–834.

Watson, J.E.M., Whittaker, R.J. & Freudenberger, D. (2005) Bird community responses to habitat fragmentation: how consistent are they across landscapes? *Journal of Biogeography* **32**, 1353–1370.

Willis, J.H. (1964) Vegetation of the basalt plains in western Victoria. *Proceedings of the Royal Society of Victoria* **77**, 397–405.

Wylie, F.R., Johnston, P.J.M. & Eisemann, R.L. (1993) *A Survey of Native Tree Dieback in Queensland*. Queensland Department of Primary Industries, Forest Research Institute.

Yamaura, Y., Katoh, K., Fujita, G. & Higuchi, H. (2005) The effect of landscape contexts on wintering bird communities in rural Japan. *Forest Ecology and Management* **216**, 187–200.

3

Enacting Landscape Design: from Specific Cases to General Principles

Yrjö Haila

Abstract

My basic argument in this essay is that realizing landscape design is an enactment process in which the manager in charge and elements of the landscape are in intimate interaction: the manager influences the landscape and the landscape influences the manager. I discuss measures needed to start a process of enactment. Firstly, the landscape manager has to understand the main features of the landscape that she or he is working with. What is relevant varies depending on the perspective, but the manager has to consider the following basic aspects of the landscape: its spatial extensions; the scale of its main components; the natural dynamics of its components; and its relations to other landscape types in the vicinity. General concepts derived from landscape ecology and conservation biology help in forming a pre-image of the landscape, but enactment is essentially a coadaptive learning process. I regard the recruitment of human allies as a major challenge for the manager: every landscape has inherent potential for increasing its ecological value, but this is possible only under condition that all main actors participate in the process. For this to happen, the manager has to take into account symbolic meanings carried by landscape elements and concomitant subsistence practices.

Keywords: landscaper management dynamic conservation; enactment; process ontology; similarity vs. difference; trust.

Introduction: Assessing similarities and differences across landscapes

For classifying landscapes we need criteria for identifying critical similarities and differences. Specifying the criteria is not straightforward: every landscape has innumerable unique features that distinguish it from every other landscape. The landscape does not in itself, as a physical entity, specify the criteria that ought to be considered. 'Landscape' is a perspectival notion. The criteria are defined by whoever defines the landscape.

The term itself originated as a technical term in painting around the turn of the 17th century (*land-scape* or *landskip*, Bate 2000). Landscape was defined as an object of vision, and it was framed to separate it from the rest of the surroundings of the viewer. The term entered scientific discourses during the 19th century, finally to become a foundational concept in the ecological subdiscipline of landscape ecology. As defined within landscape ecology, landscape is a 'scientific kind', to use Ian Hacking's (1999) terms. Landscape ecology defines a systematic set of criteria for analysing and classifying landscapes. But does landscape ecology provide criteria that would be similarly important for any landscape anywhere? Instead of striving for universality at the outset, we would be wise to subject this question to critical scrutiny.

A suspicion about universality gets support from two broad arguments. First, landscapes are descriptively complex (Wimsatt 1974): when alternative criteria are used to carve a particular landscape into components, the resulting structures and patterns differ widely. From an ecological perspective, the main alternatives include: (i) flows and rates of various substances across the landscape; (ii) the composition and structure of ecosystems and communities in different parts of the landscape; and (iii) the favourability of habitats in the landscape for various species. Such criteria are relational; that is, the role of a particular landscape element depends on what is in its immediate surroundings and in what sort of configuration. This relationality is what makes 'landscape' an important ecological concept in the first place, but it also emphasizes the uniqueness of every landscape.

Second, landscape structures offer explanatory resources for understanding ecological processes, but the relevance of any particular variable depends on the type of process or population analysed. Whenever an explanatory framework and unambiguous variables are fixed, a quantitative assessment of similarities and differences is of course possible. However, the quantities that are measured do not by themselves, give guarantees for their meaningfulness. The use of area as the main variable in the context of island biogeography offers examples of misleading comparisons: a hectare on a small island and a hectare on a large island are not ecologically equivalent (Haila 1990).

Another layer of historical contextuality is added when considering landscape design. Design is site specific, and realizing a design requires a historical subject, a manager. Etymologically, the verb to manage is derived from the Italian verb *maneggiare*, 'to handle, and especially to handle and train a horse' (Williams 1983). A manager is not in control as if from the outside, instead she or he works together with what is managed. The subject-object relationship dominating the modern connotation of management is not adequate. In landscape management, the manager influences the landscape and the landscape influences the manager, analogously with what happens when someone rears a horse. This meaning can be captured with the term 'enactment'. In enactment, activities take place, but it is not clear who precisely is an actor: 'in practices, objects are enacted' (Mol 2002, p. 33).

In studies of complex and historically contextual situations, case study methodology shows good promise (Ragin & Becker 1992). Haila and Dyke (2006) propose the 'analogue model' as a tool derived from case studies: the purpose of an analogue model is to identify dynamically important features of the system under study. Dynamic features can potentially be generalized across situations of similar type irrespective of the actual material substance. Case studies and analogue models provide the only feasible methodological perspective in situations in which the idea of a 'representative sample' lacks meaning and generalizing from samples to background populations in the ordinary quantitative fashion is out of the question. This is largely true of landscapes, particularly when human actions are included in the picture. However, when we use the methodological principle of analogue models, other types of important generalizations become possible. I return to this issue in the last section.

The ontology of landscape design

The phrase 'enacting a landscape' gets concrete meaning through an understanding of what kind of an object a landscape is. Landscapes are made of real things. To begin with, I propose the following three historical facts.

1 The elements that make up a landscape are always already there. There is no initial point from where to begin. Some of the elements are inside the landscape and constitute its important features, whereas some of the elements make up borders of the landscape. Which are which depends on the perspective, but once the perspective is fixed the distinction is far from arbitrary.
2 There is human presence in every landscape. Humans, similar to all organisms, make their living from landscapes, and in so doing modify the elements of the landscape. This could be called **subsistence landscape**: the geographic setting within which the mode of life of human communities has historically evolved.
3 The elements of a landscape are dynamic, undergoing continuous change, due to both internal and external factors. As usual, it is a matter of definition where the boundary between internal and external is drawn (Levins 1998). Furthermore, as human-induced changes in the material world are based on the same physical and chemical processes as all other kinds of changes, human-induced and naturally occurring processes are inseparably merged together.

These facts, taken together, imply that a landscape is a process rather than an entity, and human activity is an integral part of this process. Landscape designs are conceived against this process-ontology. The activities of the manager who is in charge merge together with the processes already taking place. It is on this level that the enactment, the mutual coadaptation, takes place. The first duty of the manager is to understand what kind of processes can be influenced and modified by active intervention, and what kind of processes the manager has to learn to understand and respect. A landscape manager is constituted as a historical subject through such a context-specific learning process.

To succeed, a manager has to recruit allies, both human and non-human. By so doing, the manager can modify the range of options that are at her or his disposal in the long run. Recruiting allies is a social and political process. Crucial human allies include, above all, those whose subsistence happens in the landscape of concern. It is useful to distinguish between two dimensions in this regard. One is the historically shaped **subsistence landscape**, which usually also gives nourishment for strong, collectively shared emotional feelings. On the other hand, every person who lives in a particular landscape creates for herself or himself a **life-world landscape**—a concrete environment marked by elements and locations in which important observations and experiences have taken place in the past. In former times, the life-world landscape of any particular person and the subsistence landscape of her or his community tended to correspond to one another; in modern times they have tended to diverge. This has created important subcategories such as 'recreational landscape', 'aesthetic landscape', 'cultural landscape' and so on (Macnaghten & Urry 1998).

Non-human allies include a vast range of animals and plants and microorganisms, and also biophysical elements and processes such as flows of wind and water. The landscape manager forms alliances with non-human actors by getting familiar with them and acquiring understanding about the extent to which they can be modified in the course of enactment. This sort of familiarity can only result from genuinely interactive relationships. The analogy with rearing a horse offers a guideline.

What level of detail is necessary?

The landscape manager needs to understand specific features of the landscape that she or he is working with. What is relevant varies depending on the perspective, but some basic questions need to be asked about every landscape.

Figure 3.1 **An unmanaged corner of a summer residence, Loppi, southern Finland, 21 December 2005. Photo by Yrjö Haila.**

First, what are the spatial extensions of the landscape; that is, where is the boundary, what are the distinguishing criteria of the boundary—why is it where it is?—and what sort of influences cross the boundary?

Second, what is the scale of the elements that have to be recognized within the landscape? What are the criteria used to identify the elements? As regards this question, landscape ecology diverges into several research traditions that emphasize different principles of classification (McIntyre & Hobbs 1999). In an island-based model, the emphasis is put on the difference between patches of a particular habitat and their surroundings, usually called 'matrix'. This has been aptly called the patch-centred landscape concept. An alternative is to emphasize gradual change in various environmental characteristics across area: the continuum-based landscape concept (Manning *et al.* 2004).

A different question concerns the size, or grain, of the detail that has to be recognized within the landscape. The finer the scale, the more structural detail needs to be included, such as the layered structure of the vegetation cover and variation in soil quality, perhaps even fine-scaled variation in decomposer microbial communities and fungal-plant symbiotic systems (Whittaker & Levin 1977).

It is obviously not possible to focus on everything at the same time. A hierarchical perspective on the structure of landscapes is nowadays widely accepted (see several articles in Wiens & Moss 2005; for background see O'Neill *et al.* 1986). The whole pattern of variation across several orders of magnitude in scale can be studied using appropriate statistical devices such as the fractal dimension (Mandelbrot 1982).

Third, what sort of patchwork does the landscape consist of in a longer temporal perspective? In some forested biomes the notion of quasi-equilibrium landscape is adequate (Shugart 1998). Quasi-equilibrium landscape can be modelled either on the level of individual trees or on the level of disturbances such as wind-throws and forest fires.

Finally, large-scale interactions between different landscape elements have to be considered; for instance, the relations between upland and lowland habitats, hills and valleys, terra firma and wetlands, and so on.

Decisions on the level of detail are tied to the perspective adopted. This does not mean that a landscape would not be an objective entity. Quite the contrary: the enactment relationship that brings about the landscape has an objective existence. Enactment is a real material interaction. However, once criteria corresponding to a particular perspective are adopted, other types of criteria may be excluded. The experiences produced by the enactment of a particular landscape cannot be generalized to other landscapes straight away.

With these considerations, one more circle is closed: the conceptual perspective on a landscape becomes a material fact. Concepts do not become real by a relation of 'representation' or 'correspondence', they become real by shaping human action. This is a variant of the double hermeneutic that is a pertinent feature of the sciences of humans: if concepts are integrated into the self-image of human actors, they shape action and become materialized (e.g. Dyke 1981, Taylor 1985). As human actions are collectively shaped, the materialization of concepts attains a social and political dimension. Metaphors such as resonance, entrainment, synchronization and orchestration are useful for describing this process (Dyke 1999, 2006).

The dynamic challenge: The Siberian flying squirrel as an example

Landscapes undergo change in many temporal scales simultaneously. Defining relevant scales requires focal surrogates. As an example, I use the Siberian flying squirrel (*Pteromys volans*), which has become an emblem of conservation controversies in Finland in recent years. This is the result of its being included in the list of strictly protected species in the Habitats Directive of the European Union. The directive is a relatively direct adaptation from the Endangered Species Act of the USA (Haila *et al.* 2007).

Specific surrogates are continually needed in conservation policy, and endangered species retain a special status by being unambiguous at the limit: they either survive or go extinct (Margules & Pressey 2000). Nowadays, the legal obligations of species protection extend to cover the environments of the species. The critical ruling in the Habitats Directive is in its Article 12.1(d), as follows: 'Member States shall take the requisite measures to establish a system of strict protection for the animal species listed in Annex IV(a) . . . prohibiting . . . all forms of . . . *deterioration or destruction of breeding sites or resting places*.' (12.1(d) Italics as per original).

This requirement has created problems because the flying squirrel is in fact a relatively common and widespread resident of spruce-dominated mature forest in Finland. According to a recent estimate the total population comprises some 147,000 breeding females (Ilpo Hanski, personal communication, March 2006). Literal implementation of Article 12.1(d) inevitably triggers land-use conflicts. Ironically, the conflicts are centred around growing cities all over the southern half of Finland. Flying squirrels do not shy away from residential areas. They are numerous in suburban forests, which are typically excluded from commercial timber production but also serve as reserve land for housing or industrial development.

Recurring controversies over the protection of the flying squirrel are counterproductive for conservation policy. The controversies take up a lot of time and energy of both planners and conservationists, and other types of valuable habitat tend to be neglected, for instance cultural habitats that have great biodiversity potential at the outskirts of growing urban centres. Furthermore, the proliferation of what are perceived as variants of the same conflict undermines the credibility of species protection in general.

Flexibility should be created for preserving suitable flying squirrel habitats in the context of land-use planning by long-term arrangements that create several alternative ways to reach conservation goals (Haila *et al.* 2007). Squirrel habitats can be created proactively, but only under the condition that the characteristic temporal horizons of forest management and suburban land use on the one hand and forest succession producing favourable habitats for the squirrel on the other hand match each other. This is the challenge of **dynamic conservation**. Automatic protection of every site where the flying squirrel has been recorded contradicts this dynamic aim.

The identification of adequate scales is crucial for dynamic conservation. These should be operational scales in the sense that planners and landscape managers work on similar spatial units and within similar temporal horizons in which the viability of local squirrel populations is determined. On a preliminary basis, the following scales can be identified:

1 Suitable habitats for the survival and reproduction of the squirrels on the individual level.
2 Connectivity between suitable patches, necessary for juvenile dispersal.
3 The amount of suitable habitat at the regional scale. The temporal trajectory of habitat availability can be modelled, conditional upon forest management plans (Hurme *et al.* 2005).

Regulatory arrangements, knowledge and trust

Dynamic conservation is a political challenge. It sets new demands on existing institutional arrangements. Political scientist Maarten Hajer (2003) has analysed the demand by using the notion of 'institutional void'. Institutional void arises as a response to the dysfunctional nature of what has been characterized as 'classical-modernist statecraft', that is, hierarchical decision-making processes, strict separation between decisions and implementation, and trust in expertise that is detached from the implementation process. In contrast, Hajer (2003) defends a deliberative view of the policy process: acknowledging that decisions are made interactively on several levels, with multiple stakeholders being involved ('multi-level governance'), recognizing implementation as political and contextual and crafting knowledge to the situations in which it is used.

Four decision-making levels are important in species protection in Europe (and, *mutatis mutandis*, elsewhere as well): (i) the local context in which protection is implemented; (ii) the regional level on which the viability of the focal population is assessed; (iii) the national level on which general administrative rules are set; and (iv) the level of European Union legislation.

The success or failure of dynamic conservation depends on a range of questions: Is there room for flexibility and compensatory arrangements in the implementation of species protection on the local level? How binding are the arrangements agreed upon on the local level? Can the temporal horizon be extended in a reliable and trustworthy fashion far enough toward the future? Do all stakeholders actually accept the arrangements, and keep the promises that the arrangements require?

On a general level, the need to protect endangered species is widely acknowledged. What is lacking is a political process that turns this into reality. Mutual trust of all stakeholders is essential for creating such a process. Shared knowledge that all parties recognize as adequate is a prerequisite of trust. Relevant knowledge takes different forms. Following the work of Michel Foucault (1972), a useful distinction can be drawn between knowledge of **matters of fact**, and a **system of knowledge** that gives meaning to particular matters of fact.

Knowledge of the requirements of the target, that is, 'matters of fact', is particularly important on the local level. This can include also such specific and locally idiosyncratic knowledge that is not easily integrated into systematic science.

Knowledge on population viability is critical on the regional level. 'Systems of knowledge', established research traditions of conservation biology and landscape ecology offer resources for assessing viability.

On the national and EU level, the task is to build up a reflexive, appropriately flexible regulatory scheme that does not compromise the basic goal of species protection. The basic rulings of the Habitats Directive or the Endangered Species Act need not to be changed. Rather, a dynamic approach is needed in the implementation stage.

Getting enactment started

My basic argument in this essay is that realizing landscape design is an enactment process in which the manager in charge and elements of the landscape, including other human actors, are in intimate interaction. This process does not, however, run all by itself. Quite the contrary, purposeful measures are needed to get the process of enactment started.

Firstly, a perspective has to be defined that helps to set goals for the particular landscape that is at issue. If the manager is familiar with the landscape, she or he can start in a procedural mode by identifying the main features and developmental trends of the landscape. She or he can first sum up her/his understanding of the main process, features and developmental trends of the landscape. Another option is to select a focal species, such as the flying squirrel I discussed above, and specify goals with a restricted focus. As Lindenmayer *et al.* (2002) note, the focal species approach may be unnecessarily narrow and hide important potentialities of the landscape from sight. However, starting the work initializes a learning process.

Second, after the goals are set, the manager has to evaluate what is realistic with respect to the goals. In other words, the manager has to learn to understand the natural dynamics. Somewhat metaphorically speaking, she or he has to recruit the main dynamic elements of the landscape as her or his allies (see Latour 2004); this is the crux of the coadaptive enactment process. Insights derived from conservation biology and landscape ecology help, but the explanatory and hence predictive power of any particular idea has to be evaluated separately. Main developmental trends of different landscapes give grounds for classifications. For instance, quasi-equilibrium is a good model for spatial dynamics in many forested systems.

Then the manager has to consider what human allies need to be recruited. This is particularly important for dynamic conservation, which requires binding agreements between different actors on the local level, and trust that allows such agreements to be made in the first place. Recruiting human allies is a political process. Increasing familiarity with the features of the process creates potential for generalization. For instance, a comparison of the conservation histories of two species that are strictly protected according to the European Union Habitats Directive, the flying squirrel in Finland and the loggerhead turtle (*Caretta caretta*) in Greece, unveils amazing similarity although the species (a small arboreal mammal vs. a big sea turtle) could hardly be more different ecologically (Haila *et al.* 2007).

Finally, the manager has to understand how cultural views of this particular landscape have been constituted and what values are attached to its main elements. This is really a historical issue. The manager needs to ask what are the material practices like that gave rise to the landscape, effective in appropriate scales? Scale is relevant both in terms of habitat configuration and of texture. Intensive land use such as forestry and grazing (not to mention agriculture) change the texture of the exploited habitats.

However, symbolic meanings carried by the landscape are crucial. In rural areas, age-old subsistence practices are laden with deep symbolic significance for local inhabitants; conservationists neglect this at their peril (for a detailed ethnographic analysis, see Theodossopoulos 2003). Everyday use of particular environments by ordinary people may lead to 'sacralization', namely, a process that results in particular environmental types being shifted out of the reach of developers (Haila 1997). Natural monuments have been sacralized since the 19th century, and nowadays also endangered species provide the potential of sacralization, provided they are made familiar to ordinary people. Changes in the public perception of predators and wetland birds such as swans and cranes in recent decades give testimony to such a potential. Knowledge that is shared between the manager and the public at large is essential in this respect (Haila *et al.* 2007).

Principles

1 Define explicit goals for the management design, drawing upon a general understanding of the characteristics and trends of the landscape.
2 Identify the main dynamic elements of the landscape and get them into positive resonance with the spatial and temporal horizons of the management design.
3 Make sure all main human actors are involved in the management process.
4 Pay special attention to the symbolic dimensions of local subsistence and avoid confrontations with deeply held local values.
5 Promote conservation as a new and increasingly important source of symbolic meanings that become attached to particular elements of the landscape.

References

Bate, J. (2000) *The Song of the Earth*. Picador, London.
Dyke, C. (1981) *Philosophy of Economics*. Prentice-Hall, Englewood Cliffs, NJ.

Dyke, C. (1999) Bourdieuean dynamics: the American middle-class self-constructs. In: Shusterman, R. (ed.) *Bourdieu. A Critical Reader*, pp. 192–213. Blackwell, New York.

Dyke, C. (2006) Primer: on thinking dynamically about the human ecological condition. In: Haila, Y. & Dyke, C. (eds.) *How Nature Speaks. The Dynamics of the Human Ecological Condition*, pp. 279–301. Duke University Press, Durham, NC.

Foucault, M. (1972) *The Archaeology of Knowledge*. Tavistock, London.

Hacking, I. (1999) *The Social Construction of What?* Harvard University Press, Cambridge, MA.

Haila, Y. (1990) Toward an ecological definition of an island: a northwest European perspective. *Journal of Biogeography* **17**, 561–568.

Haila, Y. (1997) 'Wilderness' and the multiple layers of environmental thought. *Environment and History* **3**, 129–147.

Haila, Y. & Dyke, C. (2006) What to say about nature's 'speech'? In: Haila, Y. & Dyke, C. (eds.) *How Nature Speaks. The Dynamics of the Human Ecological Condition*, pp. 1–48. Duke University Press, Durham, NC.

Haila, Y. & Levins, R. (1992) *Humanity and Nature. Ecology, Science and Society*. Pluto Press, London.

Haila, Y., Kousis, M., Jokinen, A., Nygren, N. & Psarikidou, K. (2007) *Building Trust through Public Participation: Learning from Conflicts over the Implementation of the Habitats Directive. Final Report of WP4 of the PAGANINI project* (the 6th EU Framework Programme). To be published on www.paganini-project.net.

Hajer, M. (2003) Policy without a polity: Policy analysis and the institutional void. *Policy Sciences* **36**, 175–195.

Hurme, E., Mönkkönen, M., Nikula, A. *et al.* (2005) Building and evaluating predictive occupancy models for the Siberian flying squirrel using forest planning data. *Forest Ecology and Management* **216**, 241–256.

Latour, B. (2004) *Politics of Nature. How to Bring the Sciences into Democracy*. Harvard University Press, Cambridge, MA.

Levins, R. (1998) The internal and external in explanatory theories. *Science as Culture* **7**, 557–582.

Lindenmayer, D.B., Manning, A.D., Smith, P.L. *et al.* (2002) The focal-species approach and landscape restoration: a critique. *Conservation Biology* **16**, 338–345.

McIntyre, S. & Hobbs, R.J. (1999) A framework for conceptualizing human impacts on landscapes and its relevance to management and research models. *Conservation Biology* **13**, 1282–1292.

Macnaghten, P. & Urry, J. (1998) *Contested Natures*. Sage, London.

Mandelbrot, B.B. (1982) *The Fractal Geometry of Nature*. Freeman, New York.

Manning, A.D., Lindenmayer, D.B. & Nix, H.A. (2004) Continua and Umwelt: novel perspectives on viewing landscapes. *Oikos* **104**, 621–628.

Margules, C.R. & Pressey, R.L. (2000) Systematic conservation planning. *Nature* **405**, 243–253.

Mol, A. (2002) *The Body Multiple. Ontology in Medical Practice.* Duke University Press, Durham, NC.

O'Neill, R.V., DeAngelis, D.L., Waide, J.B. & Allen, T.F.H. (1986) *A Hierarchical Concept of Ecosystems.* Princeton University Press, Princeton, NJ.

Ragin, S.C. & Becker, H.S. (1992) *What is a Case? Exploring the Foundations of Social Inquiry.* Cambridge University Press, Cambridge.

Shugart, H.H. (1998) *Terrestrial Ecosystems in Changing Environments.* Cambridge University Press, Cambridge.

Taylor, C. (1985) Self-interpreting animals. In: Taylor, C. (ed.) *Philosophical Papers*, Vol. 1, pp. 45–76. Cambridge University Press, Cambridge.

Theodossopoulos, D. (2003) *Troubles with Turtles. Cultural Understandings of the Environment on a Greek Island.* Berghahn Books, Oxford.

Whittaker, R.H. & Levin, S.A. (1977) The role of mosaic phenomena in natural communities. *Theoretical Population Biology* **12**, 117–139.

Wiens, J. & Moss, M. (eds.) (2005) *Issues and Perspectives in Landscape Ecology.* Cambridge University Press, Cambridge.

Williams, R. (1983) *Keywords. A Vocabulary of Culture and Society*, 2nd edn. Fontana, London.

Wimsatt, W.C. (1974) Complexity and organization. In: Schaffner, K.F. & Cohen, S. (eds.) *Proceedings of the Meetings of the Philosophy of Science Association, 1972*, pp. 67–86. Reidel, Dordrecht, Holland.

4

Landscape Models for Use in Studies of Landscape Change and Habitat Fragmentation

David B. Lindenmayer and J. Fischer

Abstract

A range of conceptual models can be used to characterize landscapes. The type of model used is important because it can have a strong influence on the understanding of biotic responses to landscape change and on the conservation recommendations that might be made. Conceptual landscape models vary from single-species ones that are based on a species perspective of a landscape (e.g. the landscape contour model) to those based on a human perspective of a landscape. Examples of this second group of landscape models include extremely well-known and widely applied ones such as the island model, the patch-matrix-corridor model and the variegation model. The different models have different strengths and limitations. However, these are rarely considered in landscape and conservation planning and many workers appear to be captive to a particular conceptual framework and particular model (especially the island and patch-matrix-corridor models).

Keywords: conservation planning; island model; landscape models; landscape change and impacts on biota; landscape contour model; patch-matrix-corridor model; variegation model.

Introduction

Effective biodiversity conservation depends partly on determining and understanding how landscape change affects organisms. Given such an understanding, conservation biologists are often then called upon to make practical decisions about landscape and conservation planning (e.g. Lambeck 1999). However, the landscape model they use to characterize landscapes can have a significant influence on the practical conservation recommendations that are made. In this essay we present a brief overview of some of the landscape models used to conceptualize the effects of landscape change on biodiversity. A landscape model can be loosely defined as a conceptual tool that provides terminology and a visual representation that can be used to communicate and study how organisms are distributed through space.

In theory, landscape models could be applied at many organizational levels – from genes to ecosystems. In practice, the species is the most widely accepted organizational unit in both a scientific and land management context (Gaston & Spicer 2004). For this reason, we begin this essay by presenting a landscape model for individual species. Because it is impossible to know everything about the spatial distribution of all individual species, we then discuss landscape models that focus on the relationship between landscape pattern and aggregate measures of species occurrence (e.g. species richness and species composition).

A landscape model for individual species

Background

Much research on how landscape alteration impacts on organisms is strongly related to assessments of how species are distributed in relation to human-defined vegetation patches. However, different species occur in many different places, and for many different reasons (Elton 1927) – and the distribution pattern of some species may not be closely related to human-defined patches (Ingham & Samways 1996; Manning *et al.* 2004). An alternative is to consider the various requirements of a species that need to be met for it to complete its life cycle. For example, a species can be limited in its spatial distribution by climatic conditions, availability of food or shelter, insufficient space, or the presence of competitors, predators and mutualists (Krebs 1978; Mackey & Lindenmayer 2001).

Because different species have different requirements, they can coexist in a given location (Schoener 1974). In addition to some species co-occurring at a single location, the unique requirements of different species often mean that they occur at very different locations. Gleason (1939) discussed his 'individualistic concept of plant association' and argued that different plant species have unique biophysical requirements and their spatial co-occurrence is largely a function of these requirements being met. The idea of species responding uniquely to their biophysical environment was further elaborated by advocates of the continuum concept (Austin & Smith 1989), which recognizes that species composition changes gradually 'along environmental gradients, with each species having an individualistic and independent distribution' (Austin 1999, p. 171). The 'Gleasonian' school of thought further highlights that a human-defined vegetation patch may not necessarily equate to suitable habitat for all species; for example, for some species it may lack suitable locations to find shelter, or an area may be climatically unsuitable. On this basis, it is possible that all species in a given landscape respond differently to landscape change (Robinson *et al.*, 1992), and that each species may have its own unique spatial distribution.

The landscape contour model

The way animals perceive a landscape can be very different from how humans perceive it (Manning *et al.* 2004). This led Fischer *et al.* (2004) to develop the landscape contour model, which incorporates multiple species and their unique habitat requirements, and can characterize gradual habitat change across multiple spatial scales. The conceptual foundation comes from Wiens (1995), who suggested viewing landscapes as 'cost-benefit contours', and Lindenmayer *et al.* (1995), who highlighted how spatially explicit habitat models are similar to contour maps.

Contour maps provide a familiar graphical representation of complex spatial information. The landscape contour model represents a landscape as a map of habitat suitability contours overlaid for different species (Fig. 4.1).

Limitations of contour model and single species approaches

Fundamental to the landscape contour model is the recognition that different species occupy different habitats, and that no two species will respond to landscape change in precisely the same way. While strictly speaking this may be

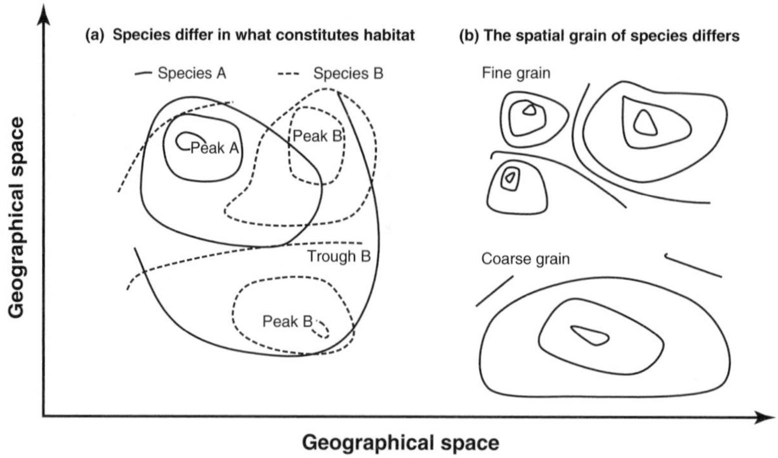

Figure 4.1 **Graphical representation of a contour-based conceptual landscape model. The model recognizes gradual changes in habitat suitability through space. Habitat contours are the emergent spatial pattern resulting from a myriad of ecological processes, including availability of suitable food, shelter and climate and sufficient space, as well as competition, predation and other interspecific processes.**

true, it poses a difficult challenge to land management (Simberloff 1998). First, the identity of all species in a given landscape is often not known (Lindenmayer & Burgman 2005). Second, even for those species that are known, it is often not known what constitutes habitat for most of them (Morrison *et al.* 1992). Hence, in the foreseeable future, detailed studies of the processes affecting individual species' distribution patterns probably will be impossible for the vast majority of species, and will most likely occur only for common or charismatic species, or species of particular conservation concern. This recognition means that some level of generalization across multiple species is often necessary.

One way to generalize across species is to consider situations where different species have broadly similar habitat requirements, and then focus management efforts on these species. Such groups of species may include 'woodland birds' (Watson *et al.* 2005), 'arboreal marsupials' (Lindenmayer 1997) or 'forest fungi' (Berglund & Jonsson 2003). The distribution patterns of such groups can be analysed as aggregate measures of species occurrence (such as species richness), for instance in relation to the size and spatial juxtaposition of forest remnants in a cleared landscape.

Pattern-based landscape models

An alternative to describing landscapes in relation to the habitat requirements of selected species is to describe landscapes from the perspective of humans. Investigating the relationship between human-defined landscape patterns and species occurrence patterns has been a popular, albeit controversial, research area in applied ecology and conservation biology (Haila 2002). Three landscape models are often used (both implicitly and explicitly): the island model, the patch-matrix-corridor model and, to a lesser extent, the variegation model.

The island model

The underlying premise of the island model is that fragments of original vegetation surrounded by cleared or highly modified land are analogous to oceanic islands in an 'inhospitable sea' of unsuitable habitat (reviewed by Haila 2002). Three broad assumptions are made under the island model: (i) islands (or vegetation patches) can be defined in a meaningful way for all species of concern; (ii) clear patch boundaries can be defined that distinguish patches from the surrounding landscape; and (iii) environmental, habitat and other conditions are relatively homogeneous within an island or patch.

Much empirical work has shown that large areas support more species than smaller ones, and the number of species often can be predicted (albeit crudely) from species-area functions (e.g. Arrhenius 1921; Rosenzweig 1995). The theory of island biogeography (MacArthur & Wilson 1967) was developed to explain species-area phenomena for island biotas. Part of this theory considers aggregate species richness on islands of varying size and isolation from a mainland source of colonists (Shafer 1990).

The literature on the theory of island biogeography is immense and well beyond the scope of this essay (e.g. see Simberloff 1988; Shafer 1990; Whittaker 1998). Much of it deals with the adaptation of island biogeography theory to reserve design (see Doak & Mills 1994) – another topic that is beyond the scope of this essay. However, as a model of landscape dynamics, the island model can fail (e.g. Zimmerman & Bierregaard 1986; Estades & Temple 1999; Gascon *et al.*, 1999) because: (i) areas between patches of remnant vegetation are rarely non-habitat for all species (Daily 2001); and (ii) important interactions between vegetation remnants and the landscapes surrounding them are not accounted for (Whittaker 1998; Manning *et al.* 2004).

Despite the inherent problems of the island model, the broad notion of islands in an inhospitable sea has spawned the development of many related theories and concepts (see Rosenzweig 1995; Whittaker 1998) that are used widely in work on landscape modification (Pulliam *et al.* 1992; Doak & Mills 1994). These include among many others: wildlife corridors (Bennett 1998), nested subset theory (Patterson & Atmar 1986), prioritization and reserve selection (Margules & Pressey 2000) and the notion of vegetation cover thresholds (Andrén 1994).

The patch-matrix-corridor model

Rather than conceptualizing landscapes as 'islands in a sea of non-habitat', Forman (1995) developed a model in which landscapes are conceived as mosaics of three components: patches, corridors and a matrix. The focus is not so much on aggregate species richness, but rather the geographical composition of landscapes with the different components having different characteristics, shapes and functions. Forman (1995) defined the three components of his conceptual model as follows:

- Patches are relatively homogeneous non-linear areas that differ from their surroundings.
- Corridors are strips of a particular patch type that differ from the adjacent land on both sides and connect two or more patches.
- The matrix is the dominant and most extensive patch type in a landscape. It is characterized by extensive cover and a major control over dynamics.

The name patch-matrix-corridor model is often used for Forman's (1995) conceptualization of landscape cover patterns. In the patch-matrix-corridor model, the matrix is intersected by corridors or perforated by smaller patches. Patches and corridors are readily distinguished from the background matrix (Forman & Godron 1986). Forman (1995) noted that every point in a landscape was either within a patch, corridor or the background matrix and that the matrix could be extensive to limited, continuous to perforated, and variegated to nearly homogeneous.

The patch-matrix-corridor model has been widely adopted in conservation biology. It helps land managers and researchers to translate their ideas into a spatial context. It is an extension of the island model, which often oversimplifies landscapes into areas of habitat and non-habitat. However, the patch-matrix-corridor model also makes some simplifying assumptions. First, it does not generally deal with spatial continua (apart from edge effects; see Laurance *et al.* 1997; Fischer *et al.* 2004). Second, it implicitly assumes that a single classification of landscape pattern can work for all species. This can be an important limitation because many organisms do not perceive landscapes in the same way as humans (Bunnell 1999; Lindenmayer *et al.* 2003). Therefore, patterns of landscape cover seen from a human perspective may not always provide a useful framework for interpreting biotic response to landscape conditions (Manning *et al.* 2004).

The variegation model

Some authors have objected to the sharp boundaries and discrete classes prevalent in the island model and patch-matrix-corridor model (Harrison 1991). In response, McIntyre and Barrett (1992) developed the variegation model (see also McIntyre & Hobbs 1999). In many landscapes, the boundaries between patch types are diffuse, and differentiating them from the background matrix may not be straightforward. The term 'variegated landscape' was coined to incorporate gradual spatial changes, or gradients, in vegetation cover (McIntyre & Barrett 1992).

The variegation model originally was proposed for semicleared grazing and cropping landscapes in rural eastern Australia. These landscapes are characterized by small patches of woodland and relatively isolated native trees scattered throughout grazing lands (McIntyre & Barrett 1992; McIntyre *et al.* 1996). Here, from a human perspective, 'patches' and 'corridors' are difficult to identify among the loosely organized and spatially dispersed trees and ecological communities. For instance, numerous trees scattered across a landscape collectively provide habitat for some species (e.g. for some woodland birds; Barrett *et al.* 1994; Fischer & Lindenmayer 2002). The variegation model takes account of small habitat elements that might otherwise be classified as 'unsuitable habitat' in the background matrix.

Limitations of pattern-based landscape models

One of the common goals of pattern-based landscape models is to reduce the complexity created by having to analyse every single species in its own right, which would pose insurmountable challenges to landscape management. Some level of simplification of reality in landscape models is necessary, and in fact desirable (Burgman *et al.* 2005). However, irrespective of the original intentions of the architects of pattern-based landscape models, these models are sometimes used uncritically and can oversimplify ecological patterns. Sometimes, biologists represent landscapes as universally suitable 'habitat patches' contrasting markedly with remaining areas of non-habitat, without carefully assessing if it is appropriate to aggregate multiple species in this way. Mapping tools (like Geographical Information Systems) are sometimes used to define 'habitat patches', assuming that species perceive 'patches' in the same way and at the same scale as humans (Bunnell 1999). Often, approaches based on landscape pattern do not consider the habitat requirements, movement patterns and other important ecological attributes of the organisms of interest.

In some cases, a possible solution to ensure that ecological complexity is not oversimplified is to forgo pattern-based landscape models and instead apply concepts such as the landscape contour model. In other cases, pattern-based landscape models can be improved by carefully assessing how species should be aggregated. For example, some workers have attempted to overcome the problem of species-specific responses to landscape conditions by classifying species according to their use of a given modified landscape – terms frequently used include 'forest-interior species' (Tang & Gustafson 1997; Villard 1998), 'edge species' (Bender *et al.* 1998; Euskirchen *et al.* 2001) and 'generalist species' (Andrén 1994; Williams & Hero 2001). Others have classified patches by the extent to which species use them or by their propensity to provide dispersers (the source–sink concept; Pulliam *et al.* 1992). Some concepts discriminate original vegetation cover from relictual vegetation cover (McIntyre & Hobbs 1999) or describe the matrix in more detail (Gascon *et al.* 1999). All these approaches are potentially useful refinements to what might otherwise be an overly simplistic or inappropriate anthropocentric classification of landscape pattern.

The link between single species and multiple species landscape models

The effects of landscape change can be assessed for a single species or for multiple species simultaneously. Single species investigations tend to be more detailed, and tend to have a reasonable grasp of the ecological processes that limit the distribution and abundance of a given species. In contrast, investigations on multiple species often need to aggregate species into groups and may need to make several assumptions about how landscape patterns are related to a given group of species. A common, but problematic, assumption is that human-defined patches correspond to habitat for a group of species. For example, Watson *et al.* (2001) implicitly assumed that human-defined patches of native vegetation provided essential habitat for woodland birds in the Australian Capital Territory. Later, Watson *et al.* (2005) demonstrated that the nature of the matrix (i.e. the area between woodland vegetation patches) actually had a major influence on woodland birds, most probably because some species could use the matrix for foraging or breeding, or could readily move through it.

When is it necessary to study every single species, and when is it reasonable to aggregate species and focus on landscape patterns? This dilemma is (often implicitly) a fundamental source of many debates in conservation biology – for example about indicator species (Landres *et al.* 1988; Lindenmayer *et al.* 2000), umbrella species (Lambeck 1997; Roberge & Angelstam 2004), small reserves (e.g. Gilpin & Diamond 1980; Simberloff & Abele 1982) and wildlife corridors (Noss 1987; Simberloff *et al.* 1992). The short answer to this dilemma is that there is no simple solution. Rather, it is important to recognize that detailed studies on single species (which often elucidate ecological processes) are complementary to studies that relate human-defined landscape patterns to aggregate measures like species richness or species composition. Neither approach is inherently superior because both approaches have limitations. Single-species approaches are limited mainly in their practical utility because it will be impossible to study every taxon. Multispecies approaches are limited because aggregation usually relies on simplifying assumptions about species co-occurrence patterns.

Landscape design principles

1 **Effective biodiversity conservation depends partly on determining and understanding how landscape change affects organisms.** Developing an understanding of biotic response to landscape change can, in turn, be strongly shaped by the underlying landscape model that is used to characterize a given landscape. Landscape models used to conceptualize the effects of landscape change on biodiversity can: (i) attempt to consider the perspective of a given species, or (ii) take a human perspective of landscapes.

2 **The landscape model used to characterize landscapes can make a significant difference to the practical conservation recommendations that are made.** Therefore, a careful appraisal of models is needed to underpin landscape and conservation planning – although this is rarely done, and either the island model or the patch-matrix-corridor model is almost always used.
3 **Considering the effects of landscape change from the perspective of a single species is a useful starting point to understand key ecological processes.** The landscape contour model is a potentially useful tool to conceptualize the distribution of individual species in landscapes.
4 **Given that it is impossible to study in detail every single species and every associated landscape change process that impacts on that species, it can be useful to focus on landscape patterns and multispecies responses.** Three landscape models that investigate the impacts of landscape change on biodiversity from a human perspective are the island model, the patch-matrix-corridor model and the variegation model.
5 A focus on single species and the ecological processes affecting them is complementary to a focus on landscape pattern in relation to assemblages of species.
6 Many workers appear to be captive to a particular conceptual framework such as a given model of landscape cover (e.g. the 'island model') and this constrains problem-solving and the development of robust conservation management strategies (Haila 2002).

Acknowledgements

This work was sponsored through grants from Land and Water Australia, the Kendall Foundation and the Australian Research Council. The authors have benefited greatly from past collaborative work with Dr Adrian Manning and Professor Ross Cunningham.

References

Andrén, H. (1994) Effects of habitat fragmentation on birds and mammals in landscapes with different proportions of suitable habitat: a review. *Oikos* **71**, 355–366.

Arrhenius, O. (1921) Species and area. *Journal of Ecology* **9**, 95–99.

Austin, M.P. (1999) A silent clash of paradigms: some inconsistencies in community ecology. *Oikos* **86**, 170–178.

Austin, M.P. & Smith, T.M. (1989) A new model for the continuum concept. *Vegetatio* **83**, 35–47.

Barrett, G.W., Ford, H.A. & Recher. H.F. (1994) Conservation of woodland birds in a fragmented rural landscape. *Pacific Conservation Biology* **1**, 245–256.

Bender, D.J., Contreras, T.A. & Fahrig, L. (1998) Habitat loss and population decline: a meta-analysis of the patch size effect. *Ecology* **79**, 517–529.

Bennett, A.F. (1998) *Linkages in the Landscape: The Role of Corridors and Connectivity in Wildlife Conservation*, 1st edn. IUCN, Gland, Switzerland.

Berglund, H. & Jonsson, B.G. (2003) Nested plant and fungal communities; the importance of area and habitat quality in maximizing species capture in boreal old-growth forests. *Biological Conservation* **112**, 319–328.

Bunnell, F. (1999) What habitat is an island? In: Rochelle, J., Lehmann, L.A. & Wisniewski, J. (eds.) *Forest Wildlife and Fragmentation. Management Implications*, 1st edn, pp. 1–31. Brill, Leiden, Germany.

Burgman, M.A., Lindenmayer, D.B. & Elith, J. (2005) Managing landscapes for conservation under uncertainty. *Ecology* **86**, 2007–2017.

Daily, G.C. (2001) Ecological forecasts. *Nature* **411**, 245.

Doak, D. & Mills, L.S. (1994) A useful role for theory in conservation. *Ecology* **75**, 615–626.

Elton, C.S. (1927) *Animal Ecology*, 1st edn. Methuen, London.

Estades, C.F. & Temple, S.A. (1999) Deciduous-forest bird communities in a fragmented landscape dominated by exotic pine plantations. *Ecological Applications* **9**, 573–585.

Euskirchen, E.S., Chen, J.Q. & Bi, R.C. (2001) Effects of edges on plant communities in a managed landscape in northern Wisconsin. *Forest Ecology and Management* **148**, 93–108.

Fischer, J. & Lindenmayer, D.B. (2002) The conservation value of paddock trees for birds in a variegated landscape in southern New South Wales. I. Species composition and site occupancy patterns. *Biodiversity and Conservation* **11**, 807–832.

Fischer, J., Lindenmayer, D.B. & Fazey, I. (2004) Appreciating ecological complexity: habitat contours as a conceptual model. *Conservation Biology* **18**, 1245–1253.

Forman, R.T. (1995) *Land Mosaics. The Ecology of Landscapes and Regions*, 1st edn. Cambridge University Press, New York.

Forman, R.T. & Godron, M. (1986) *Landscape Ecology*, 1st edn. John Wiley & Sons, New York.

Gascon, C., Lovejoy, T.E., Bierregaard, R.O.J. *et al.* (1999) Matrix habitat and species richness in tropical forest remnants. *Biological Conservation* **91**, 223–229.

Gaston, K.J. & Spicer, J.I. (2004) *Biodiversity: An Introduction*, 2nd edn. Blackwell Publishing, Oxford.

Gilpin, M.E. & Diamond, J.M. (1980) Subdivision of nature reserves and the maintenance of species diversity. *Nature* **285**, 567–568.

Gleason, H.A. (1939) The individualistic concept of plant association. *American Midland Naturalist* **21**, 92–110.

Haila, Y. (2002) A conceptual genealogy of fragmentation research from island biogeography to landscape ecology. *Ecological Applications* **12**, 321–334.

Harrison, S. (1991) Local extinction in a metapopulation context: an empirical evaluation. *Biological Journal of the Linnean Society* **42**, 73–88.

Ingham, D.S. & Samways, M.J. (1996) Application of fragmentation and variegation models to epigaeic invertebrates in South Africa. *Conservation Biology* **10**, 1353–1358.

Krebs, C.J. (1978) *Ecology: The Experimental Analysis of Distribution and Abundance*, 2nd edn. Harper International, New York.

Lambeck, R.J. (1997) Focal species: a multi-species umbrella for nature conservation. *Conservation Biology* **11**, 849–856.

Lambeck, R.J. (1999) *Landscape Planning for Biodiversity Conservation in Agricultural Regions. A Case Study from the Wheatbelt of Western Australia.* Biodiversity Technical Paper No. 2, pp. 1–96. Environment Australia, Canberra, Australia.

Landres, P.B., Verner, J. & Thomas, J.W. (1988) Ecological uses of vertebrate indicator species: a critique. *Conservation Biology* **2**, 316–328.

Laurance, W.F., Bierregaard, R.O., Gascon, C. *et al.* (1997) Tropical forest fragmentation: synthesis of a diverse and dynamic discipline. In: Laurance, W.F. & Bierregaard, R.O. (eds.) *Tropical Forest Remnants. Ecology, Management and Conservation of Fragmented Communities*, 1st edn, pp. 502–525. University of Chicago Press, Chicago.

Lindenmayer, D.B. (1997) Differences in the biology and ecology of arboreal marsupials in forests of southeastern Australia. *Journal of Mammalogy* **78**, 1117–1127.

Lindenmayer, D.B. & Burgman, M.A. (2005) *Practical Conservation Biology*, 1st edn. CSIRO Publishing, Melbourne, Australia.

Lindenmayer, D.B., Ritman, K., Cunningham, R.B., Smith, J.D.B. & Horvath, D. (1995) A method for predicting the spatial distribution of arboreal marsupials. *Wildlife Research* **22**, 445–456.

Lindenmayer, D.B., Margules, C.R. & Botkin, D. (2000) Indicators of forest sustainability biodiversity: the selection of forest indicator species. *Conservation Biology* **14**, 941–950.

Lindenmayer, D.B., McIntyre, S. & Fischer, J. (2003) Birds in eucalypt and pine forests: landscape alteration and its implications for research models of faunal habitat use. *Biological Conservation* **110**, 45–53.

MacArthur, R.H. & Wilson, E.O. (1967) *The Theory of Island Biogeography*, 1st edn. Princeton University Press, Princeton.

McIntyre, S. & Barrett, G.W. (1992) Habitat variegation, an alternative to fragmentation. *Conservation Biology* **6**, 146–147.

McIntyre, S. & Hobbs, R. (1999) A framework for conceptualizing human effects on landscapes and its relevance to management and research models. *Conservation Biology* **13**, 1282–1292.

McIntyre, S., Barrett, G.W. & Ford, H.A. (1996) Communities and ecosystems. In: Spellerberg, I.F. (ed.) *Conservation Biology*, 1st edn, pp. 154–170. Longman, Harlow, UK.

Mackey, B.G. & Lindenmayer, D.B. (2001) Towards a hierarchical framework for modelling the spatial distribution of animals. *Journal of Biogeography* **28**, 1147–1166.

Manning, A.D., Lindenmayer, D.B. & Nix, H.A. (2004) Continua and umwelt: alternative ways of viewing landscapes. *Oikos* **104**, 621–628.

Margules, C.R. & Pressey, R.L. (2000) Systematic conservation planning. *Nature* **405**, 243–253.

Morrison, M.L., Marcot, B.G. & Mannan, R.W. (1992) *Wildlife Habitat Relationships: Concepts and Applications*, 1st edn. University of Wisconsin Press, Madision, WI.

Noss, R.F. (1987) Corridors in real landscapes: a reply to Simberloff and Cox. *Conservation Biology* **1**, 159–164.

Patterson, B.D. & Atmar, W. (1986) Nested subsets and the structure of insular mammalian faunas and archipelagos. *Biological Journal of the Linnean Society* **28**, 65–82.

Pulliam, H.R., Dunning, J.B. & Liu, J. (1992) Population dynamics in complex landscapes: a case study. *Ecological Applications* **2**, 165–177.

Roberge, J.-M. & Angelstam, P. (2004) Usefulness of the umbrella species concept as a conservation tool. *Conservation Biology* **18**, 76–85.

Robinson, G.R., Holt, R.D., Gaines, M.S. *et al.* (1992) Diverse and contrasting effects of habitat fragmentation. *Science* **257**, 524–526.

Rosenzweig, M.L. (1995) *Species Diversity in Space and Time*, 1st edn. Cambridge University Press, Cambridge.

Schoener, T.W. (1974) Resource partitioning in ecological communities. *Science* **185**, 27–39.

Shafer, C.L. (1990) *Nature Reserves: Island Theory and Conservation Practice*, 1st edn. Smithsonian Institution Press, Washington, DC.

Simberloff, D.A. (1988) The contribution of population and community biology to conservation science. *Annual Review of Ecology and Systematics* **19**, 473–511.

Simberloff, D.A. (1998) Flagships, umbrellas, and keystones: is single-species management passé in the landscape era. *Biological Conservation* **83**, 247–257.

Simberloff, D. & Abele, L.G. (1982) Refuge design and island geographic theory: effects of fragmentation. *American Naturalist* **120**, 41–45.

Simberloff, D.A., Farr, J.A., Cox, J. & Mehlman, D.W. (1992) Movement corridors: conservation bargains or poor investments? *Conservation Biology* **6**, 493–504.

Tang, S.M. & Gustafson, E.J. (1997) Perception of scale in forest management planning: challenges and implications. *Landscape and Urban Planning* **39**, 1–9.

Villard, M.A. (1998) On forest-interior species, edge avoidance, area sensitivity, and dogmas in avian conservation. *The Auk* **115**, 801–805.

Watson, J., Freudenberger, D. & Paull, D. (2001) An assessment of the focal-species approach for conserving birds in variegated landscapes in southeastern Australia. *Conservation Biology* **15**, 1364–1373.

Watson, J.E., Whittaker, R.J. & Freudenberger, D. (2005) Bird community responses to habitat fragmentation: how consistent are they across landscapes? *Journal of Biogeography* **32**, 1353–1370.

Whittaker, R.J. (1998) *Island Biogeography. Ecology, Evolution and Conservation*, 1st edn. Oxford University Press, Oxford.

Wiens, J. (1995) Landscape mosaics and ecological theory. In: Hansson, L., Fahrig, L. & Merriam, G. (eds.) *Landscape Mosaics and Ecological Processes*, 1st edn., pp. 1–26. Chapman & Hall, London.

Williams, S.E. & Hero, J.M. (2001) Multiple determinants of Australian tropical frog biodiversity. *Biological Conservation* **98**, 1–10.

Zimmerman, B.L. & Bierregaard, R.O. (1986) Relevance of equilibrium theory of island biogeography and species-area relations to conservation with a case study from Amazonia. *Journal of Biogeography* **13**, 133–143.

5

Synthesis: Landscape Classification

David B. Lindenmayer and Richard J. Hobbs

Landscape classification involves the use of an underlying conceptual model to characterize a landscape. The translation of the model into a classification is usually (although not always) a representation in the form of a map. Landscape classification is the first theme in this book for good reason. It is a crucial task because how a landscape is defined, characterized and then classified can have significant effect on a wide range of management decisions. Landscape classification will affect what management, conservation and other investments are made and where. Landscape classification also can expand or constrain our thinking. Issues of landscape classification and the definition of key elements and processes rightly permeate much of this book. As just one example, Calhoun (Chapter 37) discusses the importance of wetland classification systems and how it impacts markedly on legislation, land-use planning and conservation and management actions.

Despite the importance of landscape classification, it remains a tough task. This is because

- There are different models for conceptualizing landscapes.
- Every landscape is complex.
- Every landscape is different.
- No landscape is static.
- Different elements of the biota perceive the same landscape differently.
- Landscapes embody both patterns and processes and these are multiscaled entities. As Walker (Chapter 34) notes, all workers should be concerned about at least three scales: the one of inherent interest plus a scale either side of that (smaller/larger; longer/shorter, etc.).
- Many landscape patterns and processes are continuous entities or gradients rather than discrete entities marked by sharp boundaries.

- Different problems, objectives and goals may require different classifications. For example, a classification suitable to guide a research program may well be markedly different from one needed to address the needs of a landscape manager. This, for example, is evident in the landscape classification used in the case study presented by Gibbons *et al.* (Chapter 19). Their aim was to develop a classification and subsequent map to limit land clearing.

Attempts to classify landscapes are also subject to the same human subjectivity as other forms of taxonomy – there are 'lumpers' (who favour generality) and 'splitters' (whose focus is on complexity; see McIntyre, Chapter 2). Hence, there are many ways of perceiving the same landscape. This is not just among humans – landscapes also will be perceived differently by different organisms. Thus, unlike Linnean taxonomy, there is no one broad system of landscape classification, but rather multiple classification systems. Given this, one of the key aspects of successful landscape classification is to select from the suite of available different underlying conceptual models for the task at hand and use one that not only has sufficient generality to reduce complexity and allow meaningful application, but also has sufficient flexibility to capture complexity when it is needed.

Many workers remain unaware that there is a range of models that can be used to classify landscapes. The island model has typically been the default classification used (often unwittingly) by researchers and land managers, particularly in landscapes subject to human modification (e.g. in the majority of 'fragmentation' studies). Simple models like the island model portray landscapes in a largely binomial context – 'habitat' and 'non-habitat'. This simplicity is often then readily transferred to simple vegetation maps that can be quickly generated through tools such as Geographic Information Systems. All the authors in the Landscape Classification theme highlighted the limitations of such a simple model (and the maps created through using it). Franklin (in the section on Structure and Degradation – see Chapter 12) and Miller (in the section on Habitat Loss – see Chapter 8) also highlight the limitations of many simple models and note that no single model can deal adequately with all aspects of landscapes. Lindenmayer and Fischer (Chapter 4) touched upon other models such as the landscape contour model (for individual species) and 'pattern' models for physical attributes of vegetation cover like the variegation and continuum models. It is important to be aware of the existence of a range of models for perceiving and classifying landscapes other than the island model (and its derivatives like Richard Forman's patch-corridor-matrix model). This does not mean that the island and patch-corridor-matrix models are 'dead'. Rather, different models may need to be applied to tackle different problems or achieve particular management outcomes, even in the same landscape. Moreover, models for landscape classification should not be treated as 'black boxes'. Rather, there is a need to be familiar with the assumptions and limitations associated with each of them.

An additional issue fundamental to landscape classification is that a landscape is not only an entity with compositional attributes but also a process with flows of water, nutrients, etc. Maps are usually the medium through which landscape classifications are represented, and while these can capture compositional attributes reasonably well, they do not readily capture processes. Hence, in some cases landscape classifications cannot equate to maps.

As humans are a major factor influencing these processes, they also must be a crucial part of landscape classification. McIntyre (Chapter 2) elegantly highlights the problems that arise when the impacts of human activities are segregated in landscape classification, particularly for the maintenance of key ecological processes. She concludes that humans need to be considered as part of the solution and not only as the underlying problem.

Finally, and perhaps most importantly, the complexity of successful landscape classification makes it essential to be clear and explicit about problem definition and objective setting. Scientists and resource managers have been notoriously bad at this. Often the scientific process is done backwards – 'These are the data we have gathered, now what can we make from them?' A better way forward is to articulate: What are the goals? What are the problems being addressed? What are the appropriate scales of concern? What are the priorities? These will influence both the underlying conceptual model that is used and landscape classification. Haila (Chapter 3) notes that once these questions are appropriately framed and well articulated, well-informed landscape classification(s) and subsequent landscape management decisions will be best guided by close interactions between landscape managers and the landscape they are managing.

Section 2
Habitat, Habitat Loss and Patch Sizes

6

Remnant Geometry, Landscape Morphology, and Principles and Procedures for Landscape Design

Ralph Mac Nally

Abstract

Landscape design must be contingency based – it makes little sense to be prescriptive. Goals must be firmly established to lead to continuous ecological improvement, agreed upon by all stakeholders and narrated as a time-course to provide rational expectations. The current state of the landscape, its regional context and the levels of investment need to be known or articulated before designs are contemplated. Piecemeal tinkering with patches or vegetation ribbons has little chance of producing continuous ecological improvement. Multiple alternative designs need to be considered and assessed for their likelihood of success and for their vulnerability to calamity (choose 'failsafe' designs). Some hard choices need to be confronted, one of which is the reacquisition from production of some of the most profitable parts of landscapes to improve biodiversity and give other ecological benefits. Making do with impoverished, low-value parts of many landscapes probably will doom most designs to long-term failure.

Keywords: change trajectory; contingent design; continuous ecological improvement; failsafe design; tactical vs. strategic design; whole-landscape measurement.

Introduction

Any researcher is captive to his or her research experiences. Much of my thinking is coloured by a long history of studying birds in terrestrial landscapes, with birds collectively being the most mobile group. This mobility makes the researcher think beyond the patch, linkages (e.g. riparian and roadside vegetation) and even the landscape itself to the regional and inter-regional dynamics of the organisms (Mac Nally 2005). I also believe in the overarching significance of concentrating on assemblage-level processes and analysis rather than on single or small sets of species. I feel that the latter is a little like spot-fire management when a general conflagration is imminent. While one is strongly sympathetic to the plight of the superb parrot (*Polytelis swainsonii*) and the regent honeyeater (*Xanthomyza phrygia*) in southern Australia, where I work, I have come to a view that the 'near-threatened' (Garnett *et al.* 2003) and 'decliner' (Barrett *et al.* 2003) species ought to be capturing our attention right now lest they soon reach the dire situation of the parrot and the honeyeater.

Results from my own work suggest that patch characteristics (other than area *per se*) and landscape context of a patch make relatively little difference overall to birds in their ongoing occupation of heavily modified landscapes (Mac Nally & Horrocks 2002; Harwood & Mac Nally 2005; see also Bender *et al.* 1998). Temporal variation may well overwhelm spatial considerations at a patch scale (Mac Nally & Horrocks 2000; Maron *et al.* 2005). I believe that focusing on single species and tinkering with small-scale attributes of patches within landscapes (e.g. Lambeck 1997), or linking them with mere ribbons of vegetation, will lead to ongoing declines in ecological condition, especially of native biodiversity.

The overall amount of native vegetation in a landscape is likely to be a surrogate for the general ecological condition of a landscape. This condition may be affected substantially by ecological threats not directly related to cover *per se*, especially invasive plants and animals, but many ecological processes and biodiversity will be broadly related to the amount and status ('health') of native vegetation in the landscape. Work conducted at explicitly landscape scales shows that the total amount of vegetation in a landscape is by far the most significant explanatory variable for native species richness, and for occurrence probabilities of most species (Radford *et al.* 2005; Bennett *et al.* 2006). Therefore, patch geometry and landscape context of patches (including corridors/linkages) should not greatly control or guide reconstruction designs for landscapes.

Principles for design and implementation of landscape reconstruction

My recommended principles form a sequence or procedure by which landscape design can be tackled. I feel that one cannot be prescriptive about designing landscapes because this ignores the many contingencies surrounding the decisions about how to deal with the ecological problems of a given landscape. For comparison, one would not set exact prescriptions for constructing the footings for a building without detailed knowledge of the substratum, the slope of the land, the amount of 'flashiness' (temporal variability) of rainfall and prevalence of strong winds, and the nature of surrounding constructions – these contingencies will dictate the kinds of design (and implementation) that are appropriate. Therefore, my principles are general in character and are listed below with commentary.

1. Set goal(s) for continuous ecological improvement

Many will think that this is a 'given', but making it a formal principle means that there will need to be due consideration about what continuous ecological improvement means for a landscape. There also will need to be an explicit prioritization of alternative objectives, which needs to be rationally discussed and, most importantly, agreed upon at the outset. One of the major problems in landscape management for ecological outcomes is that there are always many possible objectives that one might attempt to meet. Prioritization, and concurrent specification of a timeframe within which different levels of 'achievement' would occur if the actions were working as expected, also should be articulated. There also is an implicit **trajectory** built into this principle rather than a stopping point beyond which no further actions are needed.

2. *Measurements of response and effector variables are made at whole-landscape scales*

Very few studies have been conducted at a true landscape scale (Bennett *et al.* in press). This is different from measuring lots of 'objects' in a landscape and aggregating data to make inferences about whole-landscape responses and processes. It is also very different from partitioning sources of variation along the classical α-, β- and γ-diversity lines (e.g. Legendre *et al.* 2005) or at different spatial scales (e.g. Ricotta 2005). What is needed are measurements that characterize biodiversity or ecological processes over the entire landscape, which probably entails moving away from point-based surveys/measurements to more flexible schemes, perhaps by using stopping rules (Christen & Nakamura 2003; Watson 2003). If one is thinking biodiversity, then the kinds of information needed relate to viability rather than to just presence in a landscape, but viability analyses are notoriously difficult to conduct at landscape scales. However, it is possible to monitor evidence of reproductive activity, or at least measure patterns of relative abundance within assemblages, which may provide superior information beyond presence or reporting rates.

One also needs to measure effector variables that describe the spatial state of the landscape in ways that are meaningful to the chosen response variable. These may or may not be well described by many of the measures available in software packages like FRAGSTATS (McGarigal & Marks 1995).

The basic questions in principle 2 are (i) how is the whole landscape 'doing' ecologically, and (ii) how is the landscape structured, organized or functioning (physically) at that same scale? An important shift of emphasis will be to thinking about the landscape in gradient rather than binary terms, especially for biodiversity management (Bennett *et al.* 2006). Parts of the landscape effectively may be black or white from the perspectives of many organisms, but a more comprehensive and usable view is gradient-based. One also should think about 'keystone areas' in landscapes, which are parts of the landscape that happen to be much more ecologically significant than would be expected from their spatial extent. Many have suggested that wetlands and riparian zones may be in this group (see Lake, Chapter 38), but I suggest that many crucial areas of high fertility have already been lost, and without some reacquisition of those areas (see Vesk & Mac Nally 2006) there may be little chance of achieving continuous ecological improvement.

One final comment on whole-landscape thinking is that I believe there has been far too much focus on reserve-system design, which has been a profitable area for academic research (>400 papers; Pressey *et al.* 1993 has been cited >350 times to June 2006) but one that maintains a 'patches-in-a-matrix' mode of thought. I think this is antithetical to producing continuous ecological improvement at landscape scales and fools one into thinking the problem is being addressed (Allison *et al.* 1998).

3. Establish where you are on a gradient of ecological dysfunction

Use of the term 'ecological dysfunction' may seem pejorative. However, it reflects the reason why we are interested in design as an avenue to landscape reconstruction. Options available, and investment decisions among different landscapes at catchment (watershed), state or federal scales, must be conditional upon the state of the landscape. Hobbs *et al.* (2003) have written about 'triage' in a restoration context and this seems a reasonable general principle for decisions about resource allocation. The Victorian River Health Strategy also effectively incorporates triage as a principle (NRE 2002). While decisions may be overwhelmed by local political decisions that distort the relative rankings among landscapes, the principle is sound. One measurement that probably needs to be made is the level of ecological dysfunction as assessed in a regional context (is the landscape of moderate, high or extreme dysfunction?). One also might consider the regional context of a landscape in determining which actions, if any, to undertake. For example, Landscapes 3, 18 and 22 (all 100 km^2) in the study of Radford *et al.* (2005) are similar structurally to one another (all very low amounts of remnant vegetation) but differ in regional context. Landscape #3 is set within a regional context of large amounts of remnant vegetation, Landscape #18 is near a lowland floodplain with relatively large amounts of extensive riverine vegetation, while Landscape #22 is in a region with little remnant vegetation in its vicinity. Therefore, notwithstanding their similarities, the regional contexts of the three landscapes will lead to differing expectations vis-à-vis their respective rates of recolonization from nearby landscapes. Consequently, priorities and design will differ as a function of regional context.

One must also consider the nature of the dysfunctionality. Comparisons of landscapes in northern and southern Australia, especially the production belts of southwestern and southeastern regions, would provide different kinds of dysfunctionality and hence different forms of action (i.e. largely feral animals/weeds and fire management in the north, general loss of natural vegetation and encroaching salinity/soil quality characteristics in the south) (Hobbs & McIntyre 2005).

4. Establish investment resources available or necessary for effecting continuous ecological improvement

One of the major difficulties facing us when attempting to effect continuous ecological improvement in landscapes is the lack of clear communication regarding the magnitude of the ecological problems, and the costs involved in implementing designs.

There need to be clear statements to managers, politicians and stakeholders in general that there is a gradient from small-scale, tactical ('rinky dink') design elements and large-scale, strategic design for landscapes. Tactical designs cost relatively little, but have little ecological benefit that will accrue at best very slowly and have a high probability of failing to achieve continuous ecological improvement in the landscape. High investment associated with strategic-scale designs are much more likely to produce continuous ecological improvement much more quickly. It seems to me that conservation ecologists have locked themselves into a cost-minimization mindset rather than a goal-achievement way of thinking. I suspect a business analyst assessing most tactical plans would recommend not pursuing them. The upshot is that design depends upon knowledge of willingness to invest in implementation, and the likelihood of continuous ecological improvement is contingent on investment level.

5. Evaluate and compare multiple plausible design scenarios

One should not be prescriptive about a specific design because it is desirable to find landscape solutions that are robust against shocks and nascent threats ('insurance factors', Allison *et al.* 2003). Evaluation of multiple scenarios should be undertaken, and decisions made on the basis of

- the probability of achieving continuous ecological improvement;
- the consequences of failing to achieve continuous ecological improvement; and
- the feasibility and staging of implementation.

It seems prudent to adopt a minimax approach to selection of designs – a 'failsafe' design, with an emphasis on avoiding catastrophic outcomes at the expense of seeking the long-term 'best' effects in continuous ecological improvement, which may be more risky and potentially disastrous if certain exogenous factors occur. I re-emphasize that a time-course of change should be made explicit to shape the expectations of all stakeholders and to provide criteria for selecting strategies.

Two general criteria

From a biodiversity perspective, landscapes should be designed for landscape-responsive species, typically those that 'experience' the landscape at management scales (e.g. most birds, bats, mobile mammals, some mobile insects such as butterflies). Species whose lifetimes or viable populations are confined to small habitat patches relative to the landscape should have little bearing on whole-landscape design. *In situ* management of habitat quality is significant for these species.

Where there are low-nutrient soils and moisture availability is limited, as in southern Australia, replanting on the poorest farming sites in the landscape probably will fail to deliver vegetation with the structure and ecological resources needed to support the full complement of species sufficiently rapidly to arrest serious declines in biodiversity. The corollary is that some (unclear how much) prime agricultural land may need to be acquired publicly, or 'clawed back', to achieve this outcome (Allison *et al.* 2003).

Principles

1 Set goal(s) for continuous ecological improvement.
2 Measurements of response and effector variables should be made at whole-landscape scales.
3 Establish where you are on a gradient of ecological dysfunction.
4 Establish investment resources available or necessary for effecting continuous ecological improvement.
5 Evaluate and compare multiple plausible designs.
6 Design landscapes for species that 'experience' landscapes.
7 Given the urgency of the need for reconstruction, and that ecological resources take decades to centuries to become available, finding ways to accelerate maturation of vegetation is a high priority.

Acknowledgements

I thank David Lindenmayer and Richard Hobbs for the opportunity to contribute to the project, and other workshop participants in helping to (re-)forge my thinking. Sam Lake, Erica Fleishman and Jim Thomson commented on a previous version of the manuscript. Much of my experience has been derived from projects funded by the Australian Research Council, Land & Water Australia, the Victorian Departments of Sustainability and Environment and Primary Industry, the Nevada Biodiversity Research and Conservation Initiative and the Joint Fire Sciences Program through the Rocky Mountain Research Station, US Department of Agriculture Forest Service. This is publication number 97 from the Australian Centre for Biodiversity: Analysis, Policy and Management at Monash University.

References

Allison, G.W., Lubchenco, J. & Carr, M.H. (1998) Marine reserves are necessary but not sufficient for marine conservation. *Ecological Applications* **8**, S79–S92.

Allison, G.W., Gaines, S.D., Lubchenco, J. & Possingham, H.P. (2003) Ensuring persistence of marine reserves: catastrophes require adopting an insurance factor. *Ecological Applications* **13**, S8–S24.

Barrett, G., Silcocks, A., Barry, S., Cunningham, R. & Poulter, R. (2003) *The New Atlas of Australian Birds*. Birds Australia (Royal Australasian Ornithologists Union), Melbourne.

Bender, D.J., Contreras, T.A. & Fahrig, L. (1998) Habitat loss and population decline: A meta-analysis of the patch size effect. *Ecology* **79**, 517–533.

Bennett, A.F., Radford, J.Q. & Haslem, A. (2006) Properties of land mosaics: implications for nature conservation in agricultural landscapes. *Biological Conservation* **133**, 250–264.

Christen, J.A. & Nakamura, M. (2003) Sequential stopping rules for species accumulation. *Journal of Agricultural Biological and Environmental Statistics* **8**, 184–195.

Garnett, S., Crowley, G. & Balmford, A. (2003) The costs and effectiveness of funding the conservation of Australian threatened birds. *Bioscience* **53**, 658–665.

Harwood, W. & Mac Nally, R. (2005) Geometry of large woodland remnants and its influence on avifaunal distributions. *Landscape Ecology* **20**, 401–416.

Hobbs, R.J. & McIntyre, S. (2005) Categorizing Australian landscapes as an aid to assessing the generality of landscape management guidelines. *Global Ecology and Biogeography* **14**, 1–15.

Hobbs, R.J., Cramer, V.A. & Kristjanson, L.J. (2003) What happens if we cannot fix it? Triage, palliative care and setting priorities in salinising landscapes. *Australian Journal of Botany* **51**, 647–653.

Lambeck, R.J. (1997) Focal species: a multi-species umbrella for nature conservation. *Conservation Biology* **11**, 849–856.

Legendre, P., Borcard, D. & Peres-Neto, P.R. (2005) Analyzing beta diversity: partitioning the spatial variation of community composition data. *Ecological Monographs* **75**, 435–450.

McGarigal, K. & Marks, B. (1995) FRAGSTATS: spatial pattern analysis program for quantifying landscape structure, Rep. No. Gen. Tech. Report PNW-GTR-351. USDA Forest Service, Pacific Northwest Research Station, Portland, OR.

Mac Nally, R. (2005) Scale and an organism-centric focus for studying interspecific interactions in landscapes. In: Wiens, J.A. & Moss, M.R. (eds.) *Issues in Landscape Ecology*, pp. 52–69. Cambridge University Press, New York.

Mac Nally, R. & Horrocks, G. (2000) Landscape-scale conservation of an endangered migrant: the Swift Parrot *Lathamus discolor* in its winter range. *Biological Conservation* **92**, 335–343.

Mac Nally, R. & Horrocks, G. (2002) Relative influences of site, landscape and historical factors on birds in a fragmented landscape. *Journal of Biogeography* **29**, 395–410.

Maron, M., Lill, A., Watson, D.M. & Mac Nally, R. (2005) Temporal variation in bird assemblages: How representative is a one-year snapshot? *Austral Ecology* **30**, 383–394.

NRE (2002) Healthy rivers, healthy communities and regional growth. The Victorian River Health Strategy. Department of Natural Resources and Environment, Melbourne, Australia.

Pressey, R.L., Humphries, C.J., Margules, C.R., Vanewright, R.I. & Williams, P.H. (1993) Beyond opportunism – key principles for systematic reserve selection. *Trends in Ecology and Evolution* **8**, 124–128.

Radford, J.Q., Bennett, A.F. & Cheers, G.J. (2005) Landscape-level thresholds of habitat cover for woodland-dependent birds. *Biological Conservation* **117**, 375–391.

Ricotta, C. (2005) On hierarchical diversity decomposition. *Journal of Vegetation Science* **16**, 223–226.

Vesk, P. & Mac Nally, R. (2006) Changes in vegetation structure and distribution in rural landscapes: implications for biodiversity and ecosystem processes. *Agriculture, Ecosystems and Environment* **112**, 356–366.

Watson, D.M. (2003) The 'standardized search': an improved way to conduct bird surveys. *Austral Ecology* **28**, 515–525.

7

Estimating Minimum Habitat for Population Persistence

Lenore Fahrig

Abstract

A central problem for ecological landscape design is to determine how much habitat is needed to ensure persistence of a wildlife population. Various attempts at solving this problem have appeared in the ecological literature over the past six decades. These attempts include patch-scale and landscape-scale ideas, and have incorporated several different processes relating habitat amount to population persistence. In this essay I review these ideas, emphasizing the processes whose rates change with declining habitat amount. I argue that the currently available generic PVA (population viability analysis) models should not be used to estimate minimum habitat for population persistence, because they omit an important process, namely increasing per capita emigration rate and dispersal mortality rate with decreasing habitat amount. I discuss alternative modelling frameworks for estimating minimum habitat for population persistence, and I conclude with some principles for ecological landscape design.

Keywords: dispersal mortality; emigration rate; habitat loss; population viability analysis; Skellam's process.

Introduction: Factors relating habitat amount to population persistence

Population persistence cannot be measured directly, so estimating minimum habitat for population persistence requires a model. There are at least five factors that must be included in any such model. The first two are largely responsible for differences between species in minimum habitat requirements: species need more habitat for population persistence if (i) their individuals have larger area requirements, and (ii) they have lower reproductive rates (Pimm *et al.* 1988; Casagrandi & Gatto 1999; With & King 1999; Fahrig 2001; Vance *et al.* 2003; Holland *et al.* 2005). The other three factors are processes whose rates actually change with changing habitat amount: (i) per capita emigration rate and associated dispersal mortality rate increase with decreasing habitat amount; (ii) the influence of demographic and environmental stochasticity on local extinction rate increases with decreasing habitat amount; and (iii) immigration rate and colonization rate decrease with decreasing habitat amount.

The main reason that the minimum habitat amount for population persistence is difficult to estimate is that the rates associated with these last three factors change with changing habitat amount. In other words, one might estimate emigration rate and dispersal mortality rate, rate of local extinction and immigration and/or colonization rates for a particular species in a particular landscape. However, these values could not be used to estimate minimum habitat requirements for that species, because the rates themselves change with changing habitat amount. In this section I discuss these three processes in more detail.

Increasing per capita emigration rate and dispersal mortality rate with decreasing habitat amount

Possibly the earliest work on the problem of estimating the amount of habitat needed for population persistence is Skellam's (1951) classic paper on dispersal. Skellam argued that there is a **critical patch size**, or a minimum patch size, for population persistence, which is determined by the population's intrinsic growth rate and its dispersal function. Imagine a population in a single, isolated patch (no immigration), with random dispersal. The combination of the dispersal distance and the patch size will determine the probability of an individual dispersing outside the boundary of the patch (i.e. the **per capita emigration rate**). Because the patch is isolated, these emigrants are lost to **dispersal mortality**. The smaller the patch, the larger the per capita emigration rate, because the edge: area ratio of the patch increases with decreasing patch size, which means that a larger proportion of the population is within dispersal range of the edge of a small patch than a large patch. Therefore, there is a minimum patch size (the critical patch size) below which losses through emigration are not balanced by reproduction in the patch, and the population in the patch goes extinct (Fig. 7.1). Note that the existence of a critical patch size is a necessary consequence of geometry; the probability of encountering the patch edge increases with decreasing patch size, no matter what the particular movement behaviour of the organism. Since Skellam's (1951) paper, several studies have evaluated various factors influencing the critical patch size. As summarized by Holmes *et al.* (1994), 'factors that increase movement out of a patch lead to larger critical patch sizes, while factors that decrease movement out of the patch lead to smaller critical patch sizes.'

Landscape-scale models have also demonstrated that emigration rate and dispersal mortality can have a large effect on the amount of habitat needed for

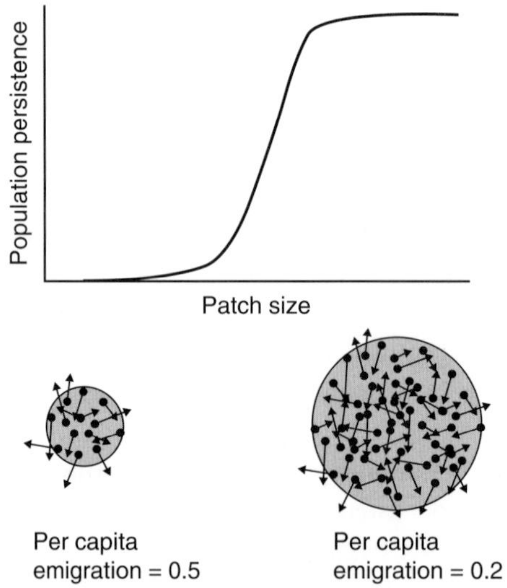

Figure 7.1 **Illustration of the critical patch size concept (Skellam 1951). The per capita emigration rate increases with decreasing patch size, leading to a critical patch size below which reproduction in the patch cannot balance losses from emigration.**

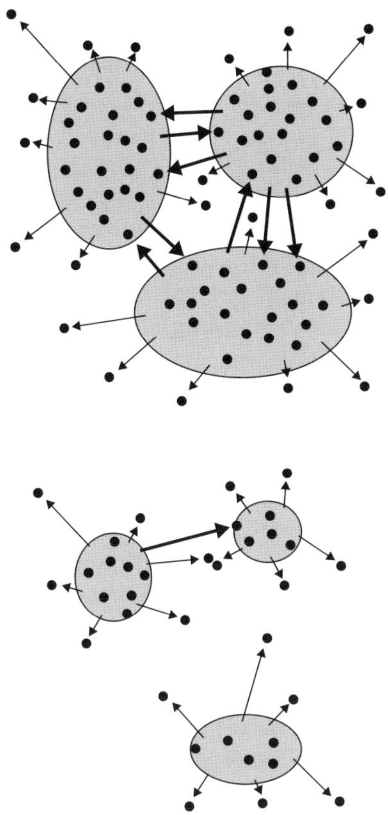

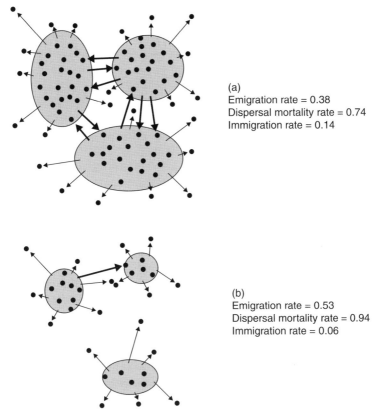

Figure 7.2 **Illustration of the effects of habitat loss on emigration rate, dispersal mortality rate and immigration rate. Ovals are habitat patches. Arrows represent emigration events of individuals between the previous time step and the present. Emigration rate is the proportion of all individuals present in the previous time step that left their patch: (a) 31/82; (b) 18/34. Dispersal mortality rate is the proportion of all emigrants that died without reaching another patch: (a) 23/31; (b) 17/18. Immigration rate is the proportion of all individuals currently present in patches that were in a different patch in the previous time step: (a) 8/59; (b) 1/18.**

population persistence (Casagrandi & Gatto 1999; Fahrig 2001; Flather & Bevers 2002). The rate of loss to the population from dispersal mortality increases with decreasing amount of habitat in the landscape, for the same geometric reasons that create the critical patch size (Fig. 7.2). The less habitat there is, the higher the per capita probability of individuals emigrating from habitat because the edge: area ratio in the landscape generally increases with declining habitat. Also, the less habitat there is, the larger the proportion of these emigrants that die in the matrix without reaching habitat, as shown experimentally by Smith and Batzli (2006). The increase in per capita dispersal mortality rate with decreasing habitat amount affects the amount of habitat needed for population persistence, and needs to be included in any model aimed at estimating minimum habitat for population persistence.

Increasing rate of stochastic extinctions with decreasing habitat amount

Soulé and Simberloff (1986) introduced a very different patch-scale approach to the minimum habitat question, based on the **minimum viable population** (MVP) concept (Shaffer 1981). MVP is based on the idea that population persistence is most strongly related to population size, because extinction is essentially a stochastic event, and the effects of demographic and environmental stochasticity increase with decreasing population size. The MVP is defined as the smallest population that ensures a high probability of persistence (say 95%) over a long time period (say 100 or 1000 generations). Soulé and Simberloff (1986) asked, 'what is the smallest nature reserve that will ensure persistence of a population of interest?' They suggested that one could first estimate the MVP, and then multiply the area requirement of each individual by the number of individuals in the MVP to give the minimum reserve area for MVP. This concept has been used to estimate minimum reserve size for individual species (e.g. Reed *et al.* 1988; Wielgus 2002; Tisdell *et al.* 2005) and groups of species (Pereira *et al.* 2004).

The critical patch size concept and the minimum habitat area for MVP concept are based on very different assumptions about the processes that create a relationship between habitat amount and population persistence. The critical patch size is mainly the result of an increase in the per capita rate of loss of individuals through emigration out of the patch (a net loss from the population) with decreasing patch size. The minimum habitat area for MVP actually makes no assumption about loss of individuals through dispersal mortality; rather, the MVP results from an increase in the effects of demographic and environmental stochasticity with decreasing population size. Both processes are important factors in determining the relationship between habitat amount and persistence of any population.

Decreasing rates of immigration and recolonization with decreasing habitat amount

For at least the past 35 years, ecologists have recognized that populations typically do not occupy single, completely isolated patches. Habitat patches occur within landscapes containing other patches of similar habitat, which may also contain the species of interest. Extinction of the population in a particular patch may be only temporary if immigrants arrive from other nearby patches and recolonize the empty patch. This implies that in many situations the calculation of minimum habitat area for population persistence should be done on a landscape scale, not a patch scale.

The first attempts to determine minimum habitat amount on a landscape scale were derived from the metapopulation model (Levins 1970), and are reviewed in Hanski *et al.* (1996). Here, the question was how many patches are needed for population persistence? Hanski *et al.* (1996) referred to this as the 'minimum viable metapopulation'. All patches were assumed to be the same size. The rate of colonization of empty patches was an increasing function of the number of occupied patches in the metapopulation, so the proportion of occupied patches decreased with decreasing number of patches in the landscape. This led to a minimum number of patches below which the proportion of occupied patches was zero, that is, the population was extinct. This number of patches is the **minimum viable metapopulation**.

The original minimum viable metapopulation formulation assumed all patches were the same size, so it did not include the processes relating patch size to population persistence discussed above. Later formulations do allow for patch size effects on both extinction and colonization (Hanski & Ovaskainen 2003): larger patches are assumed to have higher colonization rates and lower extinction probabilities. The reduced extinction probability in larger patches is assumed to be due to reduced effects of demographic and environmental stochasticity with increasing patch size (and presumably increasing population size). More recent metapopulation models (e.g. Drechsler *et al.* 2003) recognize that the probabilities of both extinction and recolonization of a patch population should also depend quite strongly on the rate of immigration to the patch. Immigration rate should depend on the size of the recipient patch as well as the number of patches that are within the dispersal range of the recipient patch, and the population sizes in them, which are assumed to be positively related to the sizes of these patches. Therefore, in these models, both extinction and colonization probabilities of a patch are functions of the number and sizes of other patches within dispersal range of the focal patch.

However, there is an important process missing from these metapopulation models, namely, the increase in losses through emigration and dispersal mortality with decreasing amount of habitat, that is, the process introduced by Skellam in 1951 (here called 'Skellam's process'). This process is also missing from current generic PVA tools, which are based, to varying degrees, on metapopulation models. In the following section I discuss this omission and its implications for using PVA tools for estimating minimum habitat for population persistence.

Misuse of current PVA tools for estimating minimum habitat for population persistence

Several generic computer programs have been developed for estimating population viability at a landscape scale; examples include RAMAS GIS (Akçakaya 2000), VORTEX (Lacy 2000), ALEX (Possingham & Davies 1995) and META-X (Grimm *et al.* 2004). These models are all patch-based, so the landscape is described in terms of the sizes and relative locations of habitat patches. The models allow the user to estimate the viability of a particular species in a particular landscape, by supplying parameter values appropriate for that species. Input parameters include: the maximum number of individuals in each patch, which is a function of patch size and individual area requirements of the species; reproductive rate; mortality rate; and immigration rate to each patch, which is a function of patch size and the number and sizes of other patches within dispersal range of the patch. All of these parameters can be subject to stochastic variation, resulting in demographic stochasticity, and the models all include environmental stochasticity.

Once a PVA model has been parameterized for a particular species in a particular landscape, it is tempting to use that model to evaluate viability of hypothetical populations in different landscape scenarios, representing either different landscapes, or the same landscape in a future where habitat amount and pattern have been altered (e.g. Akçakaya *et al.* 2004; Schtickzelle *et al.* 2005). In principle, one could use such scenarios to estimate minimum habitat requirements for population persistence, by evaluating the viability of the population in hypothetical model landscapes containing different amounts of habitat. The minimum habitat in which the modelled population has a high probability of persistence is then the estimated minimum habitat amount for population persistence.

Is this a valid use of PVA models? The first step in answering this is to ask whether these tools contain at least the minimum set of five factors discussed above, which are necessary components of any model aimed at estimating minimum habitat for population persistence. On inspection of the PVA tools, one finds that all of the models lack one important factor: Skellam's process, that is, the increase in per capita emigration rate and dispersal mortality with decreasing habitat amount. The most recent model, META-X, actually predicts a monotonic increase in population persistence with increasing 'emigration' from patches (Grimm *et al.* 2004) (Fig. 7.3a). This is opposite to the predictions of models that explicitly include emigration from patches (Casagrandi & Gatto 1999; Fahrig 2001; León-Cortés *et al.* 2003) (Fig. 7.3b–d). This difference is due to the fact that in META-X 'emigration' is only used to calculate the immigration or colonization resulting from emigration. Emigration itself, with its associated losses to dispersal mortality, does not actually occur in this model. Even though emigration from patches is necessary for immigration to and colonization of other patches, the overall effect of emigration on population persistence can easily be negative when these factors are weighed against dispersal mortality (Casagrandi & Gatto 1999; Fahrig 2001; León-Cortés *et al.* 2003). Because the generic PVA tools do not include one of the processes responsible for the negative effects of habitat loss on population persistence, they must underestimate the minimum habitat needed for population persistence, and they should therefore not be used for this purpose.

It has been suggested that this is a moot point because PVA models really should not be used to provide quantitative estimates of population viability at all; they should only be used qualitatively to compare different management options (landscape scenarios) for their relative impacts on viability (D.B. Lindenmayer, personal communication; L. Tischendorf, personal communication; H.R. Akçakaya, personal communication). The argument is that there must be huge uncertainty associated with any population viability estimate, because estimating viability entails long-term extrapolation based on current parameter values. Small errors in current parameter estimates are propagated and compounded over time, and true parameter values can change over time. Both of these lead to great uncertainty in viability predictions. While this is true, it does not obviate the need for models that can be used to estimate minimum habitat for population persistence. Managers need to know how much habitat will ensure population persistence and, because viability cannot be measured directly, models are the only option. We need models that are unbiased and that contain reasonable estimates of uncertainty and error propagation, so that careful decisions on landscape design can be taken. Current PVA

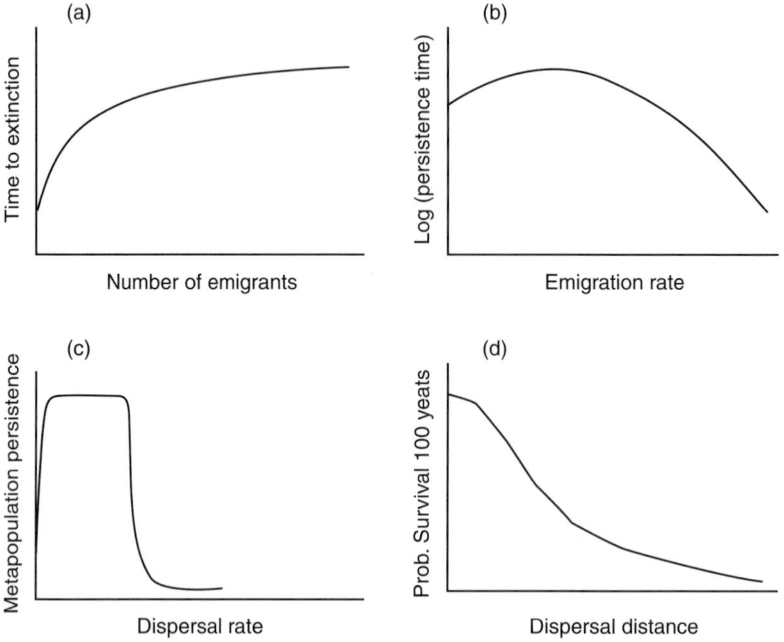

Figure 7.3 **Relationship between emigration and population persistence from four models. (a) In META-X, population persistence is a monotonically increasing function of emigration, due to increasing colonization. Redrawn from Grimm *et al.* 2004, with permission. (b) Population persistence generally decreases with increasing emigration rate, due to increasing dispersal mortality, except at very low emigration rates (Fahrig 2001). Redrawn from Fahrig (in press), with permission. (c) Metapopulation persists only at low to moderate dispersal rates, and not at higher or very low dispersal rates. Curve inferred from Figure 1a, at intrinsic rate of increase = 0.4, in Casagrandi and Gatto (1999). (d) Metapopulation persistence decreases with increasing dispersal distance; in this model dispersal rate from patches is an increasing function of dispersal distance. Redrawn from León-Cortés *et al.* 2003, with permission.**

tools do make some attempt to estimate uncertainty in model predictions; however, they are all biased because they omit Skellam's process.

Because of this bias, I suggest that current PVA tools should not even be used for qualitative comparisons of different landscape scenarios with different habitat amounts. Users of the models are generally unaware of the bias, because there is no mention of it in the documentation for any of the PVA tools. In comparing different scenarios, managers naturally favour those whose PVA runs indicate long-term persistence (e.g. Larson *et al.* 2004; Wintle *et al.* 2005). This may create a sense of false confidence in the effectiveness of the proposed management. If managers knew that the true expected population persistence for the scenarios under consideration was much lower than estimated by the PVA tools, possibly even outside the range of estimates provided, they might look for other options.

Alternative modelling approaches for estimating minimum habitat for population persistence

How would one build a model to estimate minimum habitat for population persistence? A possible starting point would be to take an existing PVA model and add to it the missing pieces, namely, increasing per capita emigration rate with decreasing patch size, and increasing dispersal mortality rate with decreasing habitat amount. The available PVA tools are patch-based. This means that the user provides input parameter values for each patch. Because patches are not all equivalent, that is, they differ in size and location, one needs to have information or make assumptions about how patch size and location affect the parameter values, and then use this information to adjust the values depending on the actual patch sizes and locations. For example, one may assume a particular functional relationship between immigration rate to a patch and the size of that patch, and the distances to and sizes of potential donor patches. These assumptions would be used to estimate the immigration rate for each patch. Similarly, one would need a functional relationship between patch size and emigration rate. The most appropriate way to measure patch size – for example, as its area, or perimeter or linear dimension – would depend on the dispersal behaviour of the organism (Hambäck & Englund 2005). The point here is that in such a patch-based model, the relationships between the parameter values and patch sizes and locations must be included as input assumptions in the model, so they must be known *a priori* by the researcher.

A completely different modelling approach for estimating minimum habitat for population persistence is the individual-based, grid-based 'landscape population model' (e.g. Fahrig 2001; reviewed in Fahrig, in press). These models represent the landscape as a spatial grid containing habitat and non-habitat cells, over which the population is distributed. Individuals are tracked through time as they move, give birth and die. The probabilities of an individual giving birth or dying depend on whether the individual happens to be in a habitat cell or a non-habitat cell. As for patch-based models, variation in habitat quality can be included by assigning cell-specific carrying capacity, reproduction and mortality values. An individual's movement probability, direction and distance also depend on whether it is in a habitat or non-habitat cell, and perhaps on the distribution of habitat and non-habitat cells within some neighbourhood. None of these probabilities or movement rules is a function of the size or location of the patch that a cell happens to be in. Therefore, the relationship between emigration and immigration and patch sizes and locations are not included as input assumptions of the models, but they emerge as model output. Even the simplest movement algorithms – for instance, movement occurs with a given probability in a random direction to a random distance – produces as output: (i) the negative relationship between patch size and per capita emigration rate from patches; (ii) increasing dispersal mortality with decreasing amount of habitat on the landscape; (iii) the positive relationship between patch size and per patch immigration rate; and (iv) higher immigration rates to patches that have more large patches nearby (i.e. are less isolated).

Whether patch-based models or individual-based, grid-based models will be more useful for estimating minimum habitat amount for persistence of a real species will depend on which type of model is easier to parameterize for that species. The basic requirements for parameterizing these two model types are very different and are summarized in Table 7.1. In general, patch-based models require functional relationships between population parameters (rates of reproduction, mortality, emigration, immigration) and patch characteristics (patch size, shape) and the amount and distribution of habitat in the landscape. Variances in these parameter values are also needed, so that the effects of stochastic variation can be included. Individual-based, grid-based models require relationships between probabilities (probability of reproduction, mortality, movement, movement directionality) and the individual's location (in habitat, in matrix, at a habitat–matrix boundary). It seems likely that individual-based models will be easier to parameterize for species such as large animals that can be followed individually, for example using radiotelemetry, to determine movement behaviour in matrix, in habitat and at habitat–matrix boundaries, and to determine mortality rates in matrix. This could also be done for some small animals where experimentation to determine parameter

Table 7.1 **Comparison of the parameters needed in a simple patch-based model and a simple individual-based, grid-based model, adequately to represent the processes that determine the relationship between population persistence and amount of habitat on a landscape.**

Process	**Parameters needed for patch-based population model**	**Parameters needed for individual-based, grid-based model**
Reproduction and mortality in habitat, and individual area requirements	Per capita reproduction and mortality rates or intrinsic population growth rate, and associated variance(s); carrying capacity as a function of patch size	Expected number of offspring per individual per time step and probability of dying per time step; carrying capacity per habitat cell
Emigration (should increase with decreasing amount of habitat)	Proportion of individuals emigrating from each patch, as a function of patch size and shape, and associated variance	Movement rules in habitat – probability of moving per time step, directionality of movement, distribution of movement distances – and probability that an individual leaves habitat when it encounters the edge
Dispersal Mortality (should increase with decreasing amount of habitat)	Proportion of dispersers that die, as a function of the amount of habitat available within dispersal range of the donor patch	Movement rules in matrix – probability of moving per time step, directionality of movement, distribution of movement distances – and probability of dying per time step in matrix
Immigration (should decrease with decreasing amount of habitat)	Proportion of surviving emigrants from all other patches arriving in the recipient patch, as a function of the distances to and sizes of potential donor patches and the size and shape of the recipient patch, and associated variance	Probability of individuals entering a patch on encountering its edge

values is possible (e.g. Goodwin & Fahrig 2002). Patch-based models may be more appropriate for species such as birds, where individuals are hard to follow and large-scale experimentation is not practical. To estimate the parameter values for a patch-based model one would need measures of emigration and immigration for patches of different sizes and shapes and situated in landscapes containing different amounts of habitat. Of course, the best (though most data intensive) choice is to build both patch-based and individual-based, grid-based models; decisions based on predictions from several models with very different structures are more robust (Burgman *et al.* 2005).

Finally, it is important to note that the factors discussed here are only the minimum set of necessary ingredients in any model aimed at estimating minimum habitat for population persistence. There are many additional factors that will be important in some situations, depending on the biology of the species of interest. For example, space use by the animal may change in various ways with decreasing habitat amount, due to conspecific attraction (e.g. Gautier *et al.* 2006; Nocera *et al.* 2006), increasing territory sizes (e.g. Lindenmayer *et al.* 2001) or altered emigration rates depending on the type of matrix that is adjacent to the habitat (Collinge & Palmer 2002). Species interactions may also be affected by habitat loss (Melian & Bascompte 2002): predation rates can either increase or decrease (Rushton *et al.* 2000; Schneider 2001: Swihart *et al.* 2001; Ryall & Fahrig 2006), and the strength of competitive interactions can either increase or decrease (Tilman *et al.* 1994; Klausmeier 2001). Because such factors can alter population viability (Lindenmayer *et al.* 2003), models for estimating minimum habitat for population persistence need to be flexible enough to allow incorporation of such effects when needed.

Principles

1 An important question in ecological landscape design is how much habitat is needed for population persistence? Virtually any ecological management plan includes as one of its objectives the assurance that particular species of interest and/or importance will not disappear from the landscape. Because habitat availability is the most important determinant of population persistence, to meet this objective managers need to know how much habitat is needed for persistence.

2 Answering this question (How much habitat is needed for population persistence?) requires a model, because population persistence cannot be measured directly.

3 Any model aimed at estimating minimum habitat for population persistence must include at least the following five factors: (i) individual

area requirements; (ii) reproductive rate; (iii) increasing per capita emigration rate and dispersal mortality rate with decreasing habitat amount (Skellam's process); (iv) increasing effects of demographic and environmental stochasticity with decreasing habitat amount; and (v) decreasing colonization and immigration rates with decreasing habitat amount.

4 Current population viability analysis (PVA) tools should not be used to estimate minimum habitat for population persistence, because they omit Skellam's process, that is, increasing per capita emigration rate and dispersal mortality rate with decreasing habitat amount. Use of current PVA tools for estimating minimum habitat would result in underestimates of the amount of habitat needed for population persistence.

Acknowledgements

I thank Dan Bert, Julie Brennan, Felix Eigenbrod, Adam Ford, Mark Harvey, Richard Hobbs, Sara Gagné, David Lindenmayer, Leif Olson, Adam Smith, Lutz Tischendorf, Rebecca Tittler, Trina Rytwinski, Lisa Venier and Ruth Waldick for comments on earlier drafts. This work was supported by the Natural Sciences and Engineering Research Council of Canada (NSERC).

References

Akçakaya, H.R. (2000) Viability analysis with habitat-based metapopulation models. *Population Ecology* **42**, 45–53.

Akçakaya, H.R., Radeloff, V.C., Mlandenoff, D.J. & He, H.S. (2004) Integrating landscape and metapopulation modeling approaches: viability of the sharp-tailed grouse in a dynamic landscape. *Conservation Biology* **18**, 526–537.

Burgman M.A., Lindenmayer, D.B. & Elith, J. (2005) Managing landscapes for conservation under uncertainty. *Ecology* **86**, 2007–2017.

Casagrandi, R. & Gatto, M. (1999) A mesoscale approach to extinction risk in fragmented habitats. *Nature* **400**, 560–562.

Collinge, S.K. & Palmer, T.M. (2002) The influences of patch shape and boundary contrast on insect response to fragmentation in California grasslands. *Landscape Ecology* **17**, 647–656.

Drechsler, M., Frank, K., Hanski, I., O'Hara, R.B. & Wissel, C. (2003) Ranking metapopulation extinction risk: from patterns in data to conservation management decisions. *Ecological Applications* **13**, 990–998.

Fahrig, L (2001) How much habitat is enough? *Biological Conservation* **100**, 65–74.

Fahrig, L. (2007) Landscape heterogeneity and metapopulation dynamics. In: Wu, J.& Hobbs, R.J. (eds) *Key Topics and Perspectives in Landscape Ecology*. Cambridge University Press, Cambridge, UK.

Flather, C.H. & Bevers, M. (2002) Patchy reaction-diffusion and population abundance: The relative importance of habitat amount and arrangement. *American Naturalist* **159**, 40–56.

Gautier, P., Olgun, K., Uzum, N. & Miaud, C. (2006) Gregarious behaviour in a salamander: attraction to conspecific chemical cues in burrow choice. *Behavioral Ecology and Sociobiology* **59**, 836–841.

Goodwin, B.J. & Fahrig, L. (2002) Effect of landscape structure on the movement behaviour of a specialized goldenrod beetle, *Trirhabda borealis*. *Canadian Journal of Zoology* **80**, 24–35.

Grimm, V., Lorek, H., Finke, J. *et al.* (2004) META-X: generic software for metapopulation viability analysis. *Biodiversity and Conservation* **13**, 165–188.

Hambäck, P.A. & Englund, G. (2005) Patch area, population density and the scaling of migration rates: the resource concentration hypothesis revisited. *Ecology Letters* **8**, 1057–1065.

Hanski, I. & Ovaskainen, O. (2003) Metapopulation theory for fragmented landscapes. *Theoretical Population Biology* **64**, 119–127.

Hanski, I., Moilanen, A. & Gyllenberg, M. (1996) Minimum viable metapopulation size. *American Naturalist* **147**, 527–541.

Holland, J.D., Fahrig, L. & Cappuccino, N. (2005) Fecundity determines the extinction threshold in a Canadian assemblage of longhorned beetles (Coleoptera: Cerambycidae). *Journal of Insect Conservation* **9**, 109–119.

Holmes, E.E., Lewis, M.A., Banks, J.E. & Veit, R.R. (1994) Partial-differential equations in ecology – spatial interactions and population-dynamics. *Ecology* **75**, 17–29.

Klausmeier, C.A. (2001) Habitat destruction and extinction in competitive and mutualistic metacommunities. *Ecology Letters* **4**, 57–63.

Lacy, R.C. (2000) Structure of the VORTEX simulation model for population viability analysis. *Ecological Bulletin* **48**, 191–203.

Larson M.A., Thompson, F.R., Millspaugh, J.J., Dijak, W.D. & Shifley, S.R. (2004) Linking population viability, habitat suitability, and landscape simulation models for conservation planning. *Ecological Modelling* **180**, 103–118.

León-Cortés, J.L., Lennon, J.J. & Thomas, C.D. (2003) Ecological dynamics of extinct species in empty habitat networks. 1. The role of habitat pattern and quantity, stochasticity and dispersal. *Oikos* **102**, 449–464.

Levins, R. (1970) Extinction. In: Gerstenhaber, M. (ed.) *Lecture Notes on Mathematics in the Life Sciences 2*, pp. 77–107. American Mathematics Society, Providence, RI.

Lindenmayer, D.B., McCarthy, M.A., Possingham, H.P. & Legge, S. (2001) A simple landscape-scale test of a spatially explicit population model: patch occupancy in fragmented south-eastern Australian forests. *Oikos* **92**, 445–458.

Lindenmayer, D.B., Possingham, H.P., Lacy, R.C., McCarthy, M.A. & Pope, M.L. (2003) How accurate are population models? Lessons from landscape-scale tests in a fragmented system. *Ecology Letters* **6**, 41–47.

Melian C.J. & Bascompte, J. (2002) Food web structure and habitat loss. *Ecology Letters* **5**, 37–46.

Nocera, J.J., Forbes, G.J. & Giraldeau, L.A. (2006) Inadvertent social information in breeding site selection of natal dispersing birds. *Proceedings of the Royal Society B – Biological Sciences* **273**, 349–355.

Pereira, H.M., Daily, G.C. & Roughgarden, J. (2004) A framework for assessing the relative vulnerability of species to land-use change. *Ecological Applications* **14**, 730–742.

Pimm, S.L., Jones, H.L. & Diamond, J. (1988) On the risk of extinction. *American Naturalist* **132**, 757–785.

Possingham, H.P. & Davies, I. (1995) ALEX: A model for the viability analysis of spatially structured populations. *Biological Conservation* **73**, 143–150.

Reed, J.M., Doerr, P.D. & Walters, J.R. (1988) Minimum viable population size of the red-cockaded woodpecker. *Journal of Wildlife Management* **52**, 385–391.

Rushton, S.P., Barreto, G.W., Cormack, R.M., Macdonald, D.W. & Fuller, R. (2000) Modelling the effects of mink and habitat fragmentation on the water vole. *Journal of Applied Ecology* **37**, 475–490.

Ryall, K.L. & Fahrig, L. (2006) Response of predators to loss and fragmentation of prey habitat: a review of theory. *Ecology* **87**, 1086–1093.

Schneider, M.F. (2001) Habitat loss, fragmentation and predator impact: spatial implications for prey conservation. *Journal of Applied Ecology* **38**, 720–735.

Schtickzelle, N., Choutt, J., Goffart, P., Fichefet, V. & Baguette, M. (2005) Metapopulation dynamics and conservation of the marsh fritillary butterfly: population viability analysis and management options for a critically endangered species in Western Europe. *Biological Conservation* **126**, 569–581.

Shaffer, M.L. (1981) Minimum population sizes for species conservation. *BioScience* **31**, 131–134.

Skellam, J.G. (1951) Random dispersal in theoretical populations. *Biometrika* **38**, 196–218.

Smith, J.E. & Batzli, G.O. (2006) Dispersal and mortality of prairie voles (*Microtus ochrogaster*) in fragmented landscapes: a field experiment. *Oikos* **112**, 209–217.

Soulé, M.E. & Simberloff, D. (1986) What do genetics and ecology tell us about the design of nature reserves? *Biological Conservation* **35**, 19–40.

Swihart, R.K., Feng, Z., Slade, N.A., Mason, D.M. & Gehring, T.M. (2001) Effects of habitat destruction and resource supplementation in a predator–prey metapopulation model. *Journal of Theoretical Biology* **210**, 287–303.

Tilman, D., May, R.M., Lehman, C.L. & Nowak, M.S. (1994) Habitat destruction and the extinction debt. *Nature* **371**, 65–66.

Tisdell, C., Wilson, C. & Swarna Nantha, H. (2005) Policies for saving a rare Australian glider: economics and ecology. *Biological Conservation* **123**, 237–248.

Vance, M.D., Fahrig, L. & Flather, C.H. (2003) Relationship between minimum habitat requirements and annual reproductive rates in forest breeding birds. *Ecology* **84**, 2643–2653.

Wielgus, R.B. (2002) Minimum viable population and reserve sizes for naturally regulated grizzly bears in British Columbia. *Biological Conservation* **106**, 381–388.

Wintle B.A., Bekessy, S.A., Venier, L.A., Pearce, J.L. & Chisholm, R.A. (2005) Utility of dynamic-landscape metapopulation models for sustainable forest management. *Conservation Biology* **19**, 1930–1943.

With, K.A. & King, A.W. (1999) Extinction thresholds for species in fractal landscapes. *Conservation Biology* **13**, 314–326.

8

Habitat and Landscape Design: Concepts, Constraints and Opportunities

James R. Miller

Abstract

Fundamental to conservation design is the amount, arrangement and quality of habitat in a landscape. Habitat is typically characterized in one of two ways, one being organism-specific and the other land-based. Whereas the latter may be useful, particularly at broad scales, an organism-based understanding is necessary to assess habitat quality and to determine appropriate restoration goals. Because existing reserves are inadequate to provide for the full complement of biodiversity, it is necessary to acquire additional land with habitat value or restoration potential and also to achieve conservation goals on private land. In targeting habitat for acquisition, a salient factor in many regions where the need for conservation is greatest is cost, including the price of restoration and ongoing maintenance. Because costs are often high and available land is limited in human-dominated landscapes, it is essential to consider carefully the merits of all available properties, including relatively small parcels. Habitat conservation on private land will require that regulatory approaches are complemented with economic incentives, and a mix of top-down and grassroots organizational structures. Greater emphasis must also be placed on the benefits accruing to landowners through habitat conservation on their properties.

Keywords: cultural sustainability; fitness surface; private land; restoration costs.

Introduction

Design is the first signal of intent – the intent to develop a course of action for changing existing conditions to preferred ones (Simon 1969). Here, the focus is landscape-scale conservation design, and the intent is to devise courses of action for modifying landscapes that have been altered by human activities so as to accommodate better the needs of native species. Fundamental to this endeavour is consideration of habitat – where it currently exists on a landscape and in what amount, its quality, and where there needs to be more of it. In this chapter, I explore two issues that bear directly on the ability to make such assessments. The first has to do with the way 'habitat' is typically depicted in landscape ecology and conservation biology, a matter of some contention in recent years. The second issue has to do with the practice of conservation in areas that not only have been altered by humans but also are currently occupied by them, often in substantial numbers. Both issues have a number of implications for the way we go about designing, or redesigning, landscapes to enhance conditions for biodiversity. Although it could be argued that there are few if any areas in the world that have not been modified by human activities (Vitousek *et al.* 1997), I emphasize the portion of the land-use gradient that extends from production landscapes (e.g. agricultural areas) to those dominated by human settlements.

Characterizing habitat

'Habitat' can be thought of in two distinct ways. Originally, it was conceived as a species-specific property embodying a functional relationship between an organism and its environment (Corsi *et al.* 2000). In this spirit, Hall *et al.* (1997, 175) defined habitat as 'the resources and conditions present in an area that produce occupancy'. A second, more prevalent, usage essentially equates habitat with land cover, as in 'riparian habitat' or 'old-growth habitat'. In the latter half of the 20th century, this concept of habitat types gave natural resource managers a convenient tool for mapping wildlife distributions over large areas (Corsi *et al.* 2000), often using remote-sensing products developed for some other purpose.

The notion of habitat types is operationally not very different from the patch-based definition of habitat favoured by landscape ecologists. Patch theory has long served as the conceptual and analytical foundation of landscape ecology, possibly because it was already familiar from its use in optimal foraging (Charnov 1976) and because it essentially amounted to an extension of traditional population dynamics models into the spatial domain (Wiens 1995; Mitchell & Powell 2003). Forman (1995) asserts that every point on a landscape is within either a patch, a corridor (a narrow linear patch) or the background matrix (the dominant patch type). In theory, the point itself could be considered a patch (Dramstad *et al.* 1996). In practice, however, patches are typically much larger and in many cases are defined by the algorithms used in classifying remote-sensing imagery or by pre-existing data layers from a geographical information system (GIS). Like habitat types, habitat patches are portrayed as distinct polygons that are internally homogeneous and are often defined on the basis of vegetation structure and composition.

The idea that vegetation is an effective surrogate for other resources may be a convenient yet tenuous assumption for many species (Hall *et al.* 1997; Morrison 2001). When an insecticide is applied to an agricultural field and overspray envelops an adjacent grassland, the vegetation remains but insectivorous birds will probably have to seek foraging opportunities elsewhere. More generally, Mitchell and Powell (2003) caution that any definition of habitat embodies a set of assumptions, explicit or not, about the relationship between an animal and its environment. They point out that defining habitat as patches assumes a correlation between resources measured at fine scales and coarse-scale landscape data, and that those resources are uniformly distributed within well-defined polygons and therefore increase in proportion to patch area. Garshelis (2000) underscores the weakness in this logic with a simple analogy, noting that doubling the size of a kitchen in a house does not necessarily mean that the room's resources (stove, refrigerator, etc.) will double or that the house's occupants will spend twice as much time there.

One way around some of the drawbacks of a typological, patch-based definition of habitat is to model the relationship between organisms and critical resources directly. Mitchell and Powell (2003) propose the use of fitness surfaces, or maps of continuous values depicting the locations of resources weighted by their importance to a given species. Thus, the contribution of resources to fitness is not abstracted, but rather modelled directly based on known or hypothesized relationships. The authors contend that their framework not only avoids the pitfalls inherent in typological definitions of habitat, but also provides an *a priori* hypothesis that is amenable to further testing and refinement. Of course, patch-based approaches can also generate testable hypotheses, but as Mitchell and Powell observe, the emphasis in landscape ecology has too often been on generating predictions rather than testing them. This idea of a fitness surface is conceptually similar to Wiens' (1997) proposal that landscapes be characterized in terms of the costs and benefits (and hence fitness) conferred to an organism (also see Chapter 4).

Fitness surfaces and typological definitions of habitat need not be mutually exclusive. A patch-based characterization is justified in some instances and at broad spatial scales, where logistical constraints and constraints imposed by available technology may render this the only option. Such an approach may provide accurate predictions of the occurrence of some species if there is a mechanistic relationship between habitat use and landscape features that can be captured via remote sensing (Miller *et al.* 2004a) or if the correlation between resources and landscape data is consistently strong. Whereas fitness surfaces derived from resource distributions may be useful in resolving some issues of scale, as Mitchell and Powell (2003) suggest, a patch-based landscape depiction may still be needed to identify broad-scale constraints on local processes (Johnson 1980; Bissonette *et al.* 1997; Donovan *et al.* 1997; Horn *et al.* 2005).

The strength of defining habitat as a fitness surface is that the definition is based on biological first principles, whereas patch-based approaches are one step removed. For this reason, the added financial costs of quantifying resources critical to the target species over relatively large areas may well be worth it in the long term. Explicitly considering the spatial distribution of resources and the costs incurred by an animal in accessing them could go a long way towards devising more effective landscape designs, or in facilitating comparisons of the merits inherent in competing alternatives. At the very least, ecologists should be aware of the assumptions inherent in a patch-based characterization of habitat and be able to justify them (see also Chapter 19).

Conservation design in human-dominated landscapes

The need to design landscapes that include sufficient habitat for native species has never been greater. The world's existing reserve network is inadequate to conserve the full complement of biodiversity because there is not enough area in reserves (Rosenzweig 1995, 2001); taxonomic coverage and representation of the world's biomes in reserves is grossly uneven (Brooks *et al.* 2004); and the land that is protected tends to be characterized by relatively low productivity

(Scott *et al.* 2001; Hansen & Rotella 2002; Huston 2005). To complicate matters further, many large protected areas that were once considered remote are experiencing pressures from human population growth just outside their borders (Miller & Hobbs 2002; Sanderson *et al.* 2002; Huston 2005). It follows that as we consider landscape designs aimed at protecting biodiversity, we must not only seek opportunities to acquire additional land with habitat value or restoration potential, but also work to achieve conservation goals on private land.

Protecting habitat through acquisition

Acquiring property in landscapes undergoing urban or exurban development can be an expensive proposition, yet land costs have not received much consideration in designs aimed at expanding reserve networks (Newburn *et al.* 2005). By way of example, in Cook County, Illinois (the heart of the city of Chicago), undeveloped land may cost several million dollars per hectare. To the west in neighbouring Kane County, land on Chicago's urban/rural fringe that has not yet been converted to subdivisions and shopping malls is priced at approximately $100,000/ha. In the western portion of Kane Co., where agriculture is still the dominant form of land use, property values are <$24,000/ha (S.A. Snyder *et al.*, unpublished). In any of these locations, habitat restoration costs will probably have to be added to these prices if acquired lands are to serve as habitat for native species of concern. The price tag for restoring areas currently in row-crop agriculture to some semblance of a tallgrass prairie is estimated to be just over $4,000/ha; restoring pastures or hayfields to tallgrass will cost approximately $2,000/ha. Of course, these prairies will then have to be maintained with fire, mowing or mechanical removal of woody vegetation on a regular basis, incurring additional costs that are ongoing. There is a sense of purpose among Kane County's staff of natural resource managers and planners in their efforts to expand the existing reserve network there, and an urgency driven by the fact that the county's population is growing at the rate of 11% annually (http://www.nipc.org/forecasting/cnty2004.html). Although the majority of citizens are generally supportive of these efforts, the costs involved can strain the budgets of local government agencies or non-governmental organizations and impose further constraints on already dwindling options. Difficult decisions must be made and land managers and planners urgently need input from conservation scientists to help evaluate the trade-offs involved.

Given the cost of acquiring additional habitat in places where the need is often greatest, such as urbanizing or agricultural regions, careful consideration must be given to all available options. In this regard, Schwartz and van Mantgem (1997) have pointed out some of the limitations imposed by the bigger-is-better paradigm that has dominated conservation biology since the early 1990s. This paradigm has been reinforced by the patch-based view of habitat and the underlying assumption noted earlier that resources increase with area. Although large tracts of habitat clearly have advantages over small ones and are a necessity for some species, small parcels may be the only option in some places (Schwartz & van Mantgem 1997; Miller & Hobbs 2002). It is therefore important to emphasize that small parcels can serve as valuable components of landscape-level conservation networks (Shafer 1995; Schwartz 1999) when considered in light of biological first principles rather than just size. Highlighting instances in which small tracts of habitat may help to meet conservation objectives would serve to broaden the set of available options for decision-makers. One example along these lines has to do with the size of wetlands.

Several workers have noted the weak relationship between wetland area and amphibian diversity (Lehtinen *et al.* 1999; Snodgrass *et al.* 2000; Oertli 2002; Babbitt 2005; see also Chapter 37). In fact, many naturally occurring wetlands are small, yet support a large number of amphibian species and serve as sources of juvenile recruits (Semlitsch & Bodie 1998). Factors that tend to be much more important than area *per se* include connectivity among wetlands (Lehtinen *et al.* 1999; Oertli 2002; Semlitsch 2002), wetland perimeter (Lehtinen *et al.* 1999), presence of a shallow littoral zone (Porej & Hetherington 2005), the absence of predatory fish (Ficetola & De Bernardi 2004; Porej & Hetherington 2005), hydroperiod (Semlitsch 2000) and a sufficient amount of upland habitat (Semlitsch & Bodie 2003; Trenham & Shaffer 2005).

Habitat quality also appears to be more important than area as a determinant of occupancy for some native butterfly species (Thomas *et al.* 2001), and there is evidence to suggest that it may be possible to compensate for reductions in habitat area by enhancing habitat quality (Summerville & Crist 2001). As with amphibians, connectivity among habitats is another key consideration for butterflies, suggesting that a single-patch management focus needs to be complemented by a landscape perspective (Thomas *et al.* 2001; Bergman *et al.* 2004; Koh & Sodhi 2004).

Perhaps no other taxon has received more attention with regard to the importance of habitat area than birds, and recommendations for habitat acquisition and restoration are often based on patch size (e.g. Robinson *et al.* 1995). Once again, however, small habitat remnants have been shown to play a valuable role as components of conservation networks in a variety of landscape contexts and should not be disregarded (Friesen *et al.* 1995; Schieck *et al.* 1995; Miller & Cale 2000; Fischer & Lindenmayer 2002; Lindenmayer *et al.* 2002; Skagen *et al.* 2005; Brawn 2006). Patch size may not be a major determinant in habitat selection for species associated with habitats that were naturally fragmented or variable in size prior to European settlement (Miller *et al.* 2004b; Brawn 2006). There is a widely held belief that birds breeding in smaller patches necessarily suffer higher rates of nest predation (Martin 1988). Yet, accumulating evidence indicates that this relationship is not so simple and that rates of nest loss are a function of multiple interacting factors, such as the amount of suitable habitat in the surrounding landscape and regional differences among predator communities (Cavitt & Martin 2002; Chalfoun *et al.* 2002; Horn *et al.* 2005). It is also worth noting that even if habitats do have limited value for breeding birds, they may serve crucial functions at other times of the year, such as stopover sites for migratory species (Hutto 1998; Rodewald & Brittingham 2004; Mehlman *et al.* 2005).

One particularly vexing problem in landscapes where conservation areas are intermixed with human land use is the potential for seemingly 'good' habitat to actually operate as an ecological trap (Schlaepfer *et al.* 2002; Battin 2004). In environments that have been rapidly altered by humans, animals may be 'trapped' by their evolutionary response to formerly reliable cues regarding habitat suitability (Schlaepfer *et al.* 2002). This phenomenon may go undetected in some instances as a function of some measures typically used to assess habitat quality. Van Horne (1983) warned us over two decades ago that the number of individuals present in a habitat may or may not correlate with its suitability. This highlights the need for a systematic monitoring programme that includes not only counting animals but also quantifying demographic processes. Although this may be expensive and logistically difficult, to do otherwise runs the risk that habitat conservation or restoration may actually do more harm than good for the target species.

Habitat on private lands

Protecting habitat on private lands is a challenge that must be met for at least two reasons. First, the majority of the world's biodiversity occurs on lands in private ownership, and second, the primary determinant of our ability to maintain viable populations of native species on reserved lands is the condition of the matrix surrounding them (Franklin 1993). For conservation goals to be attained on private lands, traditional regulatory approaches must be complemented by strategies that provide economic incentives to landowners

(Bean & Wilcove 1997; Knight 1999). In addition to compensation, most landowners will require some form of guidance regarding the necessary management steps that need to be taken to maintain 'good' habitat.

There are a growing number of programmes that provide private landowners with incentives, advice or both, with the goal of fostering good stewardship. Indeed, in some regions the array of programmes is overwhelming. The challenge is to coordinate these programmes so as to achieve a synergistic effect at landscape scales, whether that takes the form of buffering existing reserves from external influences, enhancing connectivity or coalescing habitats on several parcels to form a larger contiguous tract. For example, the State of Minnesota in the north-central USA contains approximately 6.6 million ha of forest, of which about a third is in private ownership. There are currently dozens of programmes sponsored by the federal government, state or county agencies or non-governmental organizations that provide economic incentives or guidance on conservation practices to interested landowners (Ann Pierce, Minnesota Department of Natural Resources, personal communication). A given landowner might apply to one programme and his neighbour to another. However well intentioned these programmes may be, their objectives vary widely and some are quite narrowly focused. The sole objective of several programmes is improving habitat for white-tailed deer (*Odocoileus virginianus*) – a species that is already at historically high densities and is causing profound changes in vegetation structure and composition (Rooney 2001; Rooney & Waller 2002). Even when the goals are framed more comprehensively, it is unlikely that any real conservation benefits can emerge at broader scales from uncoordinated management on a parcel-by-parcel basis (see Chapter 18). What is needed is some degree of top-down coordination to guide grassroots efforts. There are promising examples to show the way, such as burning cooperatives in Texas in which such an organizational structure has resulted in hundreds of landowners assisting one another in applying fire to restore native grasslands (Taylor 2005).

Many in the conservation community are gradually shifting their stance from one of confrontation with private landowners to one of cooperation in order to achieve mutual goals. To be successful, it is important that conservationists invest greater effort in understanding the motivations, attitudes and perceptions of landowners across the land-use gradient. As one example, the key to maintaining large expanses of grassland habitat in the central USA is keeping locally owned cattle operations on the landscape. Obviously, good stewardship on the part of the ranchers will be central to accommodating species that depend on grassland habitats, but to many ranchers, good stewardship means closely cropped fields of Eurasian grasses. A recently developed fire-grazing model (Fuhlendorf & Engle 2004) holds much promise for restoring the heterogeneity that characterized these landscapes historically and that is necessary to provide suitable conditions for a broad suite of grassland-obligate species (Roper 2003; Fuhlendorf *et al.* 2006). For this to work, ranchers will need to appreciate the value of heterogeneity and native species, and see some benefit in applying fire to their lands. As to the latter, one result of immense benefit to these landowners is the control or elimination of invasive woody vegetation that can greatly reduce the suitability of their lands for grazing operations.

At the other end of the urban-rural gradient, gaining support for maintaining habitats among people who live nearby is of crucial importance. Some habitats are more easily appreciated than others by the non-ecologist. For example, most people would agree that a mountain view is aesthetically pleasing, but many see wetlands as unattractive and forbidding (Nassauer 2004). However, there are some relatively simple steps that can be taken to enhance the public's appreciation for such habitats, involving a mix of education and slight modifications that will convey the sense that the area is being managed and cared for (Nassuer 2004). As with the fire-grazing example mentioned above, public awareness of the benefits accruing from habitat conservation on or near private property is a necessary but too-often ignored ingredient in achieving cultural sustainability for conservation in human-dominated landscapes (Miller 2005).

Principles

I have explored two issues that bear directly on our ability to design landscapes that achieve conservation goals. The first of these has to do with the way that habitat is defined and the key points can be summarized as follows:

1 A species-based definition of 'habitat' necessitates consideration of a set of conditions, both biotic and abiotic, and resources for a given organism. Vegetation alone may not be a useful surrogate for these conditions and resources.
2 Conceiving of habitat as patches or internally homogeneous polygons with sharp boundaries limits our ability to make accurate predictions

about the viability of species or even the occurrence of species in many cases.

3 Cost-benefit or fitness surfaces that depict the spatial distribution of resources based on biological first principles provide a more direct way of characterizing species-habitat relationships.
4 Patch-based definitions of habitat have merit in some cases and may be the only option in others, but it is important at least to consider the assumptions inherent in defining habitat this way.

The second issue has to do with several aspects of habitat conservation or restoration in areas dominated by human activities; here, the key points are:

1 Because conservation is competing with other land uses in human-dominated landscapes, the full costs of acquiring and restoring habitat must be carefully considered during the design phase of reserve networks.
2 Spatially extensive tracts of habitat are clearly desirable and necessary for some species, but a sole focus on large areas is unnecessarily restrictive and will cause some viable options to be undervalued or overlooked completely. Small patches may also play key roles as components of conservation reserve networks.
3 Human-dominated landscapes are often characterized by rapid change and novel combinations of land uses compared with more natural settings, increasing the likelihood that apparently suitable habitat will actually function as an ecological trap. This highlights the need for monitoring schemes that include demographic processes.
4 Habitat conservation on private lands will require not only incentive programmes and technical guidance, but also some degree of top-down coordination in order to realize a synergistic effect at landscape scales.
5 The extent to which habitats in human-dominated landscapes are ecologically sustainable will depend on the extent to which they are culturally sustainable. Therefore, conservationists must invest greater effort in understanding the motivations and perceptions of private landowners and in defining the benefits that accrue to landowners by maintaining these habitats.

References

Babbitt, K.J. (2005) The relative importance of wetland size and hydroperiod for amphibians in southern New Hampshire, USA. *Wetlands Ecology and Management* **13**, 269–280.

Battin, J. (2004) When good animals love bad habitats: ecological traps and the conservation of animal populations. *Conservation Biology* **18**, 1482–1491.

Bean, M.J. & Wilcove, D.S. (1997) The private land problem. *Conservation Biology* **11**, 1–2.

Bergman, K-O., Askling, J., Ekberg, O., Ignell, H., Wahlman, H. & Milberg, P. (2004) Landscape effects on butterfly assemblages in an agricultural region. *Ecography* **27**, 629.

Bissonette, J.A., Harrison, D.J., Hargis, C.D. & Chapin, T.G. (1997) The influence of spatial scale and scale-sensitive properties on habitat selection by American marten. In: Bissonette, J.A. (ed.) *Wildlife and Landscape Ecology: Effects of Pattern and Scale*, pp. 368–385. Springer-Verlag, New York.

Brawn, J.D. (2006) Effects of restoring oak savannas on bird communities and populations. *Conservation Biology* **20**, 460–469.

Brooks, T.M., Bakarr, M.I., Boucher, T. *et al.* (2004) Coverage provided by the global protected-area system: is it enough? *BioScience* **54**, 1081–1091.

Cavitt, J.F. & Martin, T.E. (2002) Effects of forest fragmentation on brood parasitism and nest predation in eastern and western landscapes. *Studies in Avian Biology* **25**, 73–80.

Chalfoun, A.D., Thompson, F.R. & Ratnaswamy, M.J. (2002) Nest predators and fragmentation: a review and meta-analysis. *Conservation Biology* **16**, 306–318.

Charnov, E.L. (1976) Optimal foraging, the marginal value theorem. *Theoretical Population Biology* **9**, 129–136.

Corsi, F., de Leeuw, J. & Skidmore, A. (2000). Modeling species distribution with GIS. In: Boitani, L. & Fuller, T.K. (eds.) *Research Techniques in Animal Ecology: Controversies and Consequences*, pp. 389–424. Columbia University Press, New York.

Donovan, T.M., Jones, P.W., Annand, E.M. & Thompson, F.R.I. (1997) Variation in local-scale edge effects: mechanisms and landscape context. *Ecology* **78**, 2064–2075.

Dramstad, W.E., Olson, J.D. & Forman, R.T. (1996) *Landscape Ecology Principles in Landscape Architecture and Land-Use Planning*. Island Press, Washington, DC.

Ficetola, G.F. & De Bernardi, F. (2004) Amphibians in a human-dominated landscape: the community structure is related to habitat features and isolation. *Biological Conservation* **119**, 219–230.

Fischer, J. & Lindenmayer, D.B. (2002) Small patches can be valuable for biodiversity conservation: two case studies on birds in southeastern Australia. *Biological Conservation* **106**, 129–136.

Forman, R.T.T. (1995) *Land Mosaics*. Cambridge University Press, Cambridge.
Franklin, J.F. (1993) Preserving biodiversity: species, ecosystems, or landscapes. *Ecological Applications* **3**, 202–205.
Friesen, L.E., Eagles, P.F.J. & MacKay, R.J. (1995) Effects of residential development on forest-dwelling neotropical migrant songbirds. *Conservation Biology* **9**, 1408–1414.
Fuhlendorf, S.D. & Engle, D.M. (2004) Application of the fire-grazing interaction to restore a shifting mosaic on tallgrass prairie. *Journal of Applied Ecology* **41**, 604–614.
Fuhlendorf, S.D., Harrell, W.C., Engle, D.M., Hamilton, R.G., Davis, C.A. & Leslie, D.M., Jr. (2006) Should heterogeneity be the basis for conservation? Grassland bird response to fire and grazing. *Ecological Applications* **16**(5), 1706–1716.
Garshelis, D.L. (2000). Delusions in habitat evaluation: measuring use, selection, and importance. In: Boitani, L. & Fuller, T.K. (eds.) *Research Techniques in Animal Ecology: Controversies and Consequences*, pp. 111–164. Columbia University Press, New York.
Hall, L.S., Krausman, P.R. & Morrison, M.L. (1997) The habitat concept and a plea for standard terminology. *Wildlife Society Bulletin* **25**, 173–182.
Hansen, A.J. & Rotella, J.J. (2002) Biophysical factors, land use, and species viability in and around nature reserves. *Conservation Biology* **16**, 1112–1122.
Horn, D.J., Phillips, M.L., Koford, R.R., Clark, W.R., Sovada, M.A. & Greenwood, R.J. (2005) Landscape composition, patch size, and distance to edges: interactions affecting duck reproductive success. *Ecological Applications* **15**, 1367–1376.
Huston, M.A. (2005) The three phases of land-use change: implications for biodiversity. *Ecological Applications* **15**, 1864–1878.
Hutto, R.L. (1998) On the importance of stopover sites to migrating birds. *Auk* **115**, 823–825.
Johnson, D.H. (1980) The comparison of usage and availability measurements for evaluating resource preference. *Ecology* **61**, 65–71.
Knight, R.L. (1999) Private lands: the neglected geography. *Conservation Biology* **13**, 223–224.
Koh, L.P. & Sodhi, N.S. (2004) Importance of reserves, fragments, and parks for butterfly conservation in a tropical urban landscape. *Ecological Applications* **14**, 1695–1708.
Lehtinen, R.M., Galatowitsch, S.M. & Tester, J.R. (1999) Consequences of habitat loss and fragmentation for wetland amphibian assemblages. *Wetlands* **19**, 1–12.
Lindenmayer, D.B., Cunningham, R.B., Donnelly, C.F., Nix, H. & Lindenmayer, B.D. (2002) Effects of forest fragmentation on bird assemblages in a novel landscape context. *Ecological Monographs* **72**, 1–18.
Martin, T.E. (1988) Habitat and area effects on forest bird assemblages: is nest predation an influence? *Ecology* **69**, 74–84.

Mehlman, D.W., Mabey, S.E., Ewert, D.N. *et al.* (2005) Conserving stopover sites for forest-dwelling migratory landbirds. *Auk* **122**, 1281–1290.
Miller, J.R. (2005) Biodiversity conservation and the extinction of experience. *Trends in Ecology and Evolution* **20**, 430–434.
Miller, J.R. & Cale, P. (2000) Behavioral mechanisms and habitat use by birds in a fragmented agricultural landscape. *Ecological Applications* **10**, 1732–1748.
Miller, J.R. & Hobbs, R.J. (2002) Conservation where people live and work. *Conservation Biology* **16**, 330–337.
Miller, J.R., Turner, M.G., Smithwick, E.A.H., Dent, L.C. & Stanley, E.H. (2004a) Spatial extrapolation: the science of predicting ecological patterns and processes. *Bioscience* **54**, 310–320.
Miller, J.R., Dixon, M.D. & Turner, M.G. (2004b) Response of avian communities in large-river floodplains to environmental variation at multiple scales. *Ecological Applications* **14**, 1394–1410.
Mitchell, M.S. & Powell, R.A. (2003). Linking fitness landscapes with the behavior and distribution of animals. In: Bissonette, J.A. & Storch, I. (eds.) *Landscape Ecology and Resource Management. Linking Theory With Practice*, pp. 93–124. Island Press, Washington, DC.
Morrison, M.L. (2001) A proposed research emphasis to overcome the limits of wildlife-habitat relationship studies. *Journal of Wildlife Management* **65**, 228–240.
Nassauer, J.I. (2004) Monitoring the success of metropolitan wetland restorations: cultural sustainability and ecological function. *Wetlands* **24**, 756–776.
Newburn, D., Reed, S., Berck, P. & Merenlender, A. (2005) Economics and land-use change in prioritizing private land conservation. *Conservation Biology* **19**, 1411–1420.
Oertli, B. (2002) Does size matter? The relationship between pond area and biodiversity. *Biological Conservation* **104**, 59–70.
Porej, D. & Hetherington, T.E. (2005) Designing wetlands for amphibians: the importance of predatory fish and shallow littoral zones in structuring of amphibian communities. *Wetlands Ecology and Management* **13**, 445–456.
Robinson, S.K., Thompson, F.R., III, Donovan, T.M., Whitehead, D.R. & Faaborg, J. (1995) Regional forest fragmentation and the nesting success of migratory birds. *Science* **267**, 1987–1990.
Rodewald, P.G. & Brittingham, M.C. (2004) Stopover habitats of landbirds during fall: use of edge-dominated and early-successional forests. *Auk* **121**, 1040–1055.
Rooney, T.P. (2001) Deer impacts on forest ecosystems: a North American perspective. *Forestry* **74**, 201–208.
Rooney, T.P. & Waller, D.M. (2002) Direct and indirect effects of white-tailed deer in forest ecosystems. *Forest Ecology and Management* **181**, 165–176.
Roper, A. (2003) Invertebrate response to patch-burning. MS Thesis, Oklahoma State University, Stillwater, OK.

Rosenzweig, M.L. (1995) *Species Diversity in Space and Time*. Cambridge University Press, Cambridge.

Rosenzweig, M.L. (2001) Loss of speciation rate will impoverish future diversity. *Proceedings of the National Academy of Sciences of the USA*. **98**, 5403–5410.

Sanderson, E.W., Jaiteh, M., Levy, M.A., Redford, K.H., Wannebo, A.V. & Woolmer, G. (2002) The human footprint and the last of the wild. *Bioscience* **52**, 891–904.

Schieck, J., Lertzman, K., Nyberg, B. & Page, R. (1995) Effects of patch size on birds in old-growth montane forests. *Conservation Biology* **9**, 1072–1084.

Schlaepfer, M.A., Runge, M.C. & Sherman, P.W. (2002) Ecological and evolutionary traps. *Trends in Ecology and Evolution* **17**, 474–480.

Schwartz, M.W. (1999) Choosing the appropriate scale of reserves for conservation. *Annual Review of Ecology and Systematics* **30**, 83–108.

Schwartz, M.W. & van Mantgem, P.J. (1997) The value of small preserves in chronically fragmented landscapes. In: Schwartz, M.W. (ed.) *Conservation in Highly Fragmented Landscapes*, pp. 379–394. Chapman & Hall, New York.

Scott, J.M., Davis, F.W., McGhie, R.G., Wright, R.G., Groves, C. & Estes, J. (2001) Nature reserves: do they capture the full range of America's biological diversity? *Ecological Applications* **11**, 999–1007.

Semlitsch, R.D. (2000) Principles for management of aquatic-breeding amphibians. *Journal of Wildlife Management* **64**, 615–631.

Semlitsch, R.D. (2002) Critical elements for biologically based recovery plans of aquatic-breeding amphibians. *Conservation Biology* **16**, 619–629.

Semlitsch, R.D. & Bodie, J.R. (1998) Are small, isolated wetlands expendable? *Conservation Biology* **12**, 1129–1133.

Semlitsch, R.D. & Bodie, J.R. (2003) Biological criteria for buffer zones around wetlands and riparian habitats for amphibians and reptiles. *Conservation Biology* **17**, 1219–1228.

Shafer, C.L. (1995) Values and shortcomings of small reserves. *Bioscience* **45**, 80–88.

Simon, H. (1969) *The Sciences of the Artificial*. MIT Press, Cambridge, MA.

Skagen, S.K., Adams, A.A.Y. & Adams, R.D. (2005) Nest survival relative to patch size in a highly fragmented shortgrass prairie landscape. *Wilson Bulletin* **117**, 23–34.

Snodgrass, J.W., Komoroski, M.J., Bryan, A.L., Jr. & Burger, J. (2000) Relationships among isolated wetland size, hydroperiod, and amphibian species richness: implications for wetland regulations. *Conservation Biology* **14**, 414–419.

Summerville, K.S. & Crist, T.O. (2001) Effects of experimental habitat fragmentation on patch use by butterflies and skippers (Lepidoptera). *Ecology* **82**, 1360–1370.

Taylor, C.A., Jr. (2005) Prescribed burning cooperatives: empowering and equipping ranchers to manage rangelands. *Rangelands* February, pp. 18–23.

Thomas, J.A., Bourn, N.A.D., Clarke, R.T. *et al.* (2001) The quality and isolation of habitat patches both determine where butterflies persist in fragmented landscapes. *Proceedings of the Royal Society of London* **269**, 1791–1796.

Trenham, P.C. & Shaffer, H.B. (2005) Amphibian upland habitat use and its consequences for population viability. *Ecological Applications* **15**, 1158–1168.

Van Horne, B. (1983) Density as a misleading indicator of habitat quality. *Journal of Wildlife Management* **47**, 893–901.

Vitousek, P.M., Mooney, H.A., Lubchenco, J. & Melillo, J.M. (1997) Human domination of earth's ecosystems. *Science* **277**, 494–499.

Wiens, J.A. (1995). Landscape mosaics and ecological theory. In: Hansson, L., Fahrig, L. & Merriam, G. (eds.) *Mosaic Landscapes and Ecological Processes*, pp. 1–26. Chapman & Hall, London.

Wiens, J.A. (1997). The emerging role of patchiness in conservation biology. In: Pickett, S.T.A. (ed.) *The Ecological Basis of Conservation: Heterogenity, Ecosystems, and Biodiversity*, pp. 93–107. Chapman & Hall, New York.

9

Synthesis: Habitat, Habitat Loss and Patch Sizes

David B. Lindenmayer and Richard J. Hobbs

Habitat loss is one of the major factors governing species loss and population decline worldwide. Given the importance of habitat loss as a threatening process, a key question for landscape managers and conservation biologists is: How much habitat is needed? The series of chapters in this section have explored a range of issues associated with habitat.

Perhaps the key starting point for this synthesis section is to define **habitat**. The term is used loosely in conservation biology and landscape ecology – a problem that has led to much confusion. The concept is used in two ways. One is the specific sense. That is, habitat is a species-specific entity – the environment and other conditions suitable for a particular taxon. The other usage of the term habitat is more general and typically refers to the amount of native vegetation cover (Miller, Chapter 8; see also Schmiegelow, Chapter 22). These differences may seem semantic, but they can be crucial for conservation and management. Indeed, the problem with the second usage is that the amount of native vegetation cover in a landscape or region may not reflect the amount of suitable habitat available for a particular species. As an example, a landscape may support 5,000 ha of forest (native vegetation cover), but only the 500 ha of old-growth forest is potentially suitable habitat for a particular old-growth-dependent species. Within that only 350 ha of old growth with certain features (e.g. large logs with thick mats of moss) will provide actual habitat for that species (see Schmiegelow, Chapter 22).

A general consensus across the essays in this section (Chapters 6–8) together with the group discussions from the workshop is that the amount of native vegetation cover is a significant factor governing the occurrence and abundance of the majority of species in a landscape. It is considered more important than the configuration of patches of vegetation cover and hence influences other things like connectivity (see Chapters 22–25). Mac Nally (Chapter 6) argues that the past focus of researchers on patch content, patch geometry and the landscape context of patches has failed to reveal the importance of increasing the overall amount of native vegetation in landscapes. Many of these arguments are made in the light of the need for larger-scaled and more ambitious vegetation restoration programmes and the potential limitations of current small-scaled local efforts. Mac Nally (Chapter 6) also believes that the scale of past research work has been inappropriate, and that work at a true landscape scale (rather than at the patch scale) is imperative to make progress. There is some congruence here with the arguments of Bennett and Radford (Chapter 18) on the need to look at the responses of species to landscape mosaics or ensembles of patches. However, issues and significant challenges remain about how to gather reliable data on biota at these larger scales. In addition, there are arguments made in many of the chapters in this book that landscapes are multiscaled entities, and factors at many spatial scales influence patterns of species distribution and abundance and hence need to be carefully considered and quantified (e.g. Walker, Chapter 34).

Miller (Chapter 8) notes that while vegetation quantity is important, so is quality – in some cases it can be more important. Small areas also can have considerable value for conservation, such as providing stepping stones or stopover sites for mobile species. This leads to his recommendations about careful consideration of quality and size in land acquisition for conservation – acquisition should not be focused solely on large areas. As in the case of the section on Landscape Classification (see Chapters 2–5), Miller (Chapter 8) recognizes the importance of efforts to conserve habitats (and native vegetation cover) in human-dominated landscapes. Private lands can be valuable, particularly because they are often the more productive parts of landscapes that are essential to the survival of some species. For example, they may be where growth rates of trees and other vegetation are fastest and restoration programmes more effective.

A very different perspective on habitat is offered by Fahrig (Chapter 7). Some landscape managers may use models like those embedded in population viability analysis (PVA) in an attempt to simulate how much habitat is required to ensure the persistence of a species in a landscape. She outlines some of the potential problems inherent in this approach, notably from omitting key processes such as changes in emigration rates and mortality rates that are associated with habitat losses. This is an important consideration, and one that has largely been overlooked to date. However, the underpinning conceptual landscape models used to simulate such changes in modified landscapes, including most metapopulation models, are still highly simplified binomial habitat/non-habitat models (see Chapters 2–5 on Landscape Classification); moreover, the potential habitat and connectivity roles of the matrix around 'patches' could be important (see Schmiegelow, Chapter 22). Herculean efforts will be required to parameterize and validate models through gathering the empirical data on changes in dispersal mortality and emigration rates coupled with changing amounts of vegetation cover. Perhaps rather than using PVA models in an attempt to estimate accurately the total amount of habitat needed for a given species, a more effective use of such models might be to assess the **relative** merit of different management options for landscape management. This issue is representative of an array of important conceptual issues not currently captured effectively in models and hence unable to be accounted for in practical applications, but which would take considerable time and effort to parameterize effectively. It highlights the ongoing dilemma for much of conservation biology: the need to move forward with pressing conservation decisions versus the desirability of waiting for sound information and improved understanding. In many parts of the world this issue is apparent at a much more basic level – for instance, the acquisition of basic life-history characteristics or habitat requirements of species of conservation concern.

Section 3
Structure, Degradation and Condition

10

Nature's Infinite Variety: Conservation Choice and Management for Dynamic Ecological Systems

J.C.Z. Woinarski

Abstract

Any parcel of land has some conservation value, but this value is relative, influenced by the spatial context of the parcel, and variable over time. To a large extent, landscape managers and conservation planners have considered spatial context reasonably effectively, but have been far less successful in incorporating temporal variability. The ecological structure, composition and function of a site or landscape will vary over time. To an increasing degree, such change is directional, as human influence simplifies and regulates natural systems; but ecological simplification may also be an important stage within regular or disorderly natural perturbations. There is a burgeoning ecological discipline involved in measurement of such divergence of sites from an assumed norm ('benchmark'), but current approaches are typically simplistic and subjective, and inadequately consider the system's necessary dynamism. Rather, condition assessment needs to be contextualized within a conservation management system that recognizes the array of states possible at a site and with reference to the current and future array of environments across spatial scales in the region. With such contextualization and an objective measurement of conservation value, it is feasible to identify (sets of) sites in preferred condition and those in less preferred condition. Across all spatial scales, the former – intact natural landscapes – should be the linchpins of biodiversity conservation. Informed management may enhance the ecological value of degraded systems, but the costs of such intervention may represent an inefficient distortion of limited conservation resources.

Keywords: benchmark; condition; conservation planning; degradation; dynamism.

Introduction: key issues

Any particular site may support a characteristic ecological assemblage. But that assemblage is not fixed. Its composition and ecological structure and interactions will vary in response to disturbance factors operating at local, regional or even global scales, at timescales varying from the bewilderingly rapid to the almost imperceptibly long, and in temporal patterns that may be regular, gradational and predictable or chaotic and unpredictable. Dynamism is a natural and necessary attribute of ecological systems; and all environmental appearances are illusions of variable transience.

Some sites may be predisposed to more rapid and substantial change than others. There may be far fewer realizable options for sites with extreme and relatively fixed climates (such as the most arid deserts or hot-water springs) or at sites with peculiar chemical characteristics (such as sulphur lakes) than for sites that better fit our concepts of benign environments. Arguably, some ecological assemblages may be relatively invadible and others more impregnable. For example, the baroque ecological assemblages of Hawaii proved eggshell fragile once the isolation that sheltered them from most external influences was broken.

Measuring and interpreting change in the ecological assemblage of a site is not necessarily straightforward. Apparently trivial changes in the composition or relative abundance of species within the assemblage may cause profound changes in the array of ecological interactions and fabric, especially so where the community structure may be moulded largely by a keystone species, such as a top-order predator (Estes 1996; Terborgh *et al.* 1999) or a major herbivore that structures vegetation patterning (Zimov *et al.* 1995). Conversely, in some cases, broad-scale changes in the species composition of the assemblage may have little impact on the ecological functioning of the system.

Human influences have substantially altered the dynamic rhythms and prevailing manifestations at most sites across the globe. In most cases, such influence has resulted in trends towards more simplified systems, more orderly disturbance regimes, an increased incidence of allochthonous elements (exotic weeds and pests) and a reduced incidence of species that are larger, more specialized, longer-lived and/or less disturbance tolerant.

But, conversely, some sites exposed to relatively little human intervention may (at least temporarily) display ecological assemblages that are highly simplified relative to the potential ecological complexity supportable at that site. Sites denuded by recent volcanic activity, tsunami, extreme weather events or outbreaks of herbivorous insects are such examples. In contrast, some sites exposed to substantial human intervention may support assemblages at maximum complexity.

At any site, there is a tendency for increased architectural complexity to foster increased species richness, and more dimension to the array of ecological guilds present. Rivers with snags tend to provide better reproductive habitat for fish and other aquatic fauna than those in which the snags have been removed (Larson *et al.* 2001). Forests and woodlands with more complex foliage profiles tend to support more bird species (MacArthur *et al.* 1966); those with fallen logs and other woody debris tend to support specialized fungi, plants and animals, and hence more biodiversity (Mac Nally *et al.* 2002; Woldendorp & Keenan 2005), and those with old trees bearing hollows are necessary to support obligate hollow-nesting fauna (Lindenmayer *et al.* 1990). Strip these environments of such architectural structures, and the biodiversity they support is much diminished (Scotts 1991).

Accepting that there is a range of ecological assemblages and systems possible at any site, it is conceivable to order these alternatives in terms of desirability; and thence to attempt to manipulate factors to increase the persistence or extent (or to increase the likelihood of reappearance) of the preferred manifestation. With variable success, humans have been engaged in this ecosystem design practice since the development of agriculture more than 5,000 years ago, and somewhat less precisely since the introduction of landscape burning at least 20,000 years ago. Variably robust and complex descriptors for the manifestations possible within any system have been coined by agricultural ecologists, based largely on the utility (agricultural productivity) of a site's possible environmental states; and also by ecologists measuring environmental impacts, based largely on the correlations between the amount of gunk we add to a system and that system's consequential ecological patterning. Ecologists more generally have inherited and (often largely uncritically) applied the lexicon (notably through the now widespread use of such terms as 'condition', 'state', 'degradation') and theory (such as state-transition). Sometimes usefully (and sometimes misleadingly), some ecologists have also adopted descriptors of landscapes based on analogies to medical health (Costanza *et al.* 1992; Suter 1993; Wicklum & Davies 1995; Hobbs & Kristjanson 2003).

Challenges

Conservation biology is a *parvenu* player in this field. There are some uncomfortable and awkward considerations that fall well outside our scientific provenance.

1. If we accept fluidity and impermanence as the definitional lot of ecological systems, there may be little rational basis for describing an ideal or benchmark ecological assemblage (or for that matter an undesirable state) for any given site. The issue becomes particularly murky when states are attributed as 'good', 'bad', 'disturbed', 'pristine', 'healthy', etc. Such terms carry the whiff of a moral system that can distort ecological science and conservation outcomes. But if we are overly egalitarian and fail to set an explicit, preferred and realistic target state (or set of states) for a given site, it is impossible to harness ecological expertise to develop and implement an effective management strategy for that site.
2. Differing ecological perspectives may produce contrasting 'preferred' states. For example, it may be preferable to maintain a forest at old age for the benefit of a particular threatened species, at intermediate age for total plant species richness, or at a relatively young age for carbon sequestration. These different currencies may not be readily convertible or blendable. Further the 'worth' of any state at a given site is context dependent. If all other representations of woodlands dominated by mallee (multistemmed eucalypts in semiarid Australia) across a region are less than 20 years post-fire, then it may be especially desirable to maintain one site far longer unburnt. Grasslands invaded by the shrub *Acacia nilotica* may be more structurally complex and support more and different species than non-invaded grasslands, but if native grasslands in a region are a diminishing conservation resource, then the invaded state may be undesirable. Some ecologists have argued that we should maintain or restore environments to the state they may have had before the influence of modern development (Hopkins 1999); others have argued that such states are too conjectural and their re-achievement impractical (Oliver *et al.* 2002).
3. Even if conservation biologists could agree on a preferred state for any given site, other interest groups use different value systems. 'Old-growth' forests have exemplified this most pointedly, with different sectors attributing them near-religious significance or describing them as senescent wastelands. Given rapidly expanding human populations, such competing value systems will be increasingly ranged against residual natural

environments, even within our headline conservation reserves (e.g. Robinson *et al.* 2005). Why, where and when should an (at least quasi-) objective biodiversity conservation system prevail? And how can we persuade others that it should?

4 Assuming that we want to and can articulate an ideal state for any given site, how do we measure departure from that state? Now this can be so dog-obvious that expertise seems unnecessary. A lake stifled by an outbreak of algal bloom, a floodplain invaded by the exotic woody weed *Mimosa*, a grassland denuded by overstocking – these seem self-evidently bad, unhealthy, degraded. But, in part, even these are slippery and subjective assessments. Productivity may have increased in at least some of these 'undesirable' states, and there will be a set of species that benefit in each case. The enumeration of condition becomes far more complex when changes are less extreme, and the recent proliferation of condition scoring systems includes many examples where indices are highly idiosyncratic, artificial, and only dubiously related to ecological functioning or defensible conservation outcomes. This is not to say that such scoring is useless. I'd rather add to our reserve system a lush never-grazed tract of native grassland than a bare over-grazed paddock, and I'd probably need some measurements to convince my administrators of the grounds for such choice.

5 If the condition of any site is suboptimal, how do we improve it? This is generally a more straightforward problem, challenged more by the limitations in our technical competence (and resources) than by the weakness of our philosophical foundations. But our capability is highly limited. Notwithstanding our best intentions, most ecological systems are probably in decline (Sattler & Creighton 2002), and the current rate of species loss is without recent precedent. But some recent heroic advances deserve celebrating: the active use of fire for considered management goals in conservation reserves (Dombeck *et al.* 2004; Saab & Powell 2005), and the reintroduction of top-order predators (Fritts *et al.* 1997; Terborgh *et al.* 1999; Smith *et al.* 2003), have each restored ecological dynamism to systems that had hitherto been held in an artificial stasis. In rare cases (Bond & Lake 2003; Hobbs 2003), these restorative system changes have been carefully measured, to enable ongoing management refinement, more explicit cost-benefit assessments and measurement of the success of ecological outcomes (Edwards *et al.* 2003). Faced with a 'degraded' system, we need to be able to predict the cost for the menu of possible restoration actions, the likelihood of their success, their collateral benefits and detriments and to develop and measure explicit indicators of progress towards restoration.

Restoration is a worthy and necessary part of landscape conservation. But it is expensive, can take a very long time, and in some cases may never succeed in reproducing the predisturbance state. It is a truism worth repeating that conservation will be most effective and efficient where representative landscapes are maintained in well-functioning condition rather than when we attempt to recreate such natural systems once all that is left of them is a mess of broken pieces. Ecological systems are far easier to dismantle than to reconstruct (Bond & Lake 2003).

For biodiversity conservation, landscape management will be optimized when (i) the array of possible states available to any site is described, and attributed with values; (ii) the ecological underpinnings (the disturbances and processes that shape the system) are understood; (iii) a desired and practical array of the system's ecological manifestations is designed across space and time; (iv) managers are armed with the knowledge and resources to achieve such an array; and (v) the system is monitored sufficiently to validate (and improve) the design and management prescriptions.

To some extent, recent technological advances – notably in geographical information systems (GIS), modelling and remote sensing – have greatly aided this process (particularly such as for the maintenance of mosaic burning and representation of different seral stages across large conservation reserves: Richards *et al.* 1999; Gill *et al.* 2000). But conservation managers confront an increasingly formidable set of obstacles to orderly practice, and may be all too readily waylaid by short-term expediency or unpredicted events. In some cases, the seemingly capricious occurrence of natural disturbance events may obstruct our attempts at orderly restoration, a reminder of our limitations as ecosystem designers and of the necessary chaos instigated by natural disturbance events as the kaleidoscope turns.

Key landscape conservation principles

1 Dynamism is a characteristic and essential feature of all natural systems. Some disturbances are necessary to promote and maintain that dynamism, and hence to allow for the diversity of states possible within any system.
2 There will be an array of ecological manifestations (states) possible at any site: these need to be identified, defined and attributed with values. There is then a need to articulate which state (or set of states) is most

desirable, and hence is managed for. This demands some objective relative valuation of each possible state. Such valuation should include a measure of conservation worth (such as the occurrence of threatened, highly specialized or range-restricted species, richness, functional value), fragility and management requirements.

3 Any such valuation is contextual, requiring information on the relative occurrence of alternative states in the surrounding region, and over time.
4 Any such valuation by conservation biologists may differ from that of other sectors of the human community. If unresolved, this difference is likely to lead to management conflict and failure to meet conservation goals.
5 Human influences have distorted the natural range and relative occurrence of possible states at any site, typically by altering the disturbance regime, increasing the incidence of exotic species, simplifying ecological structure and habitat complexity and reducing the incidence of specialized, long-lived and/or large species.
6 It is feasible to measure this distortion, and there is an obvious functional utility in providing some index of such condition (in part in order to help rank conservation choices). However, such indices have proven slippery – partly because biodiversity is not a conveniently unitary concept and partly because the ingredients of any such index may need to be realigned for each individual environment. Nonetheless, ecologists should be able to design reasonably robust measures of condition that describe the deviation of any site from a defined benchmark or target state. Ingredients for such a condition scoring scheme should include: the proportion (and ecological 'health') of autochthonous species (and particularly those that are specialized, keystone, long-lived and disturbance-sensitive) in the assemblage; habitat complexity; deviation in the prevailing disturbance regime from that existing before recent human intervention; and the extent of intact functioning ecological interactions.
7 A sound knowledge of ecological functioning (particularly the interactions among species and responses to disturbance) is needed by managers to maintain (or attempt to resurrect) the 'optimal' state or mix of states. Some individual species and processes may be particularly significant for the maintenance of ecological structure and function at a site, and such critical species and processes should be targets for management intervention.

8 At local, regional and national scales, intact well-functioning ecological systems are the greatest assets for biodiversity conservation and should be the linchpins for conservation design. In general they will be relatively inexpensive to manage and are likely to provide greatest security to most species.

9 Judicious and cost-effective conservation planning needs to balance the extent to which degraded sites may represent the only vestiges of a particular ecological system against the costs and likelihood of rehabilitation for such degraded sites (Hobbs & Kristjanson 2003). At many (probably most) sites, key ecological attributes have been degraded and such sites offer less for conservation design. In some cases, such degraded residues may provide a necessary complement to the core conservation system, such as where they represent the only remnants of some formerly more widespread assemblages, or where they provide a means for linking otherwise isolated cores. But degraded sites may also be a liability, an ecological and financial sink – resource-hungry to rehabilitate and with sometimes a low probability of returning to a preferred state.

References

Bond, N.R. & Lake, P.S. (2003) Local habitat restoration in streams: constraints on the effectiveness of restoration for stream biota. *Ecological Management and Restoration* **4**, 193–198.

Costanza, R., Norton, B. & Haskell, B.J. (eds.) (1992) *Ecosystem Health: New Goals for Ecosystem Management*. Island Press, Washington, DC.

Dombeck, M.P., Williams, J.E. & Wood, C.A. (2004) Wildfire policy and public lands: integrating scientific understanding with social concerns across landscapes. *Conservation Biology* **18**, 883–889.

Edwards, A., Kennett, R., Price, O., Russell-Smith, J., Spiers, G. & Woinarski, J. (2003) Monitoring the impacts of fire regimes on biodiversity in northern Australia: an example from Kakadu National Park. *International Journal of Wildland Fire* **12**, 427–440.

Estes, J.A. (1996) Predators and ecosystem management. *Wildlife Society Bulletin* **24**, 390–396.

Fritts, S.H., Bangs, E.E., Fontaine, J.A. *et al.* (1997) Planning and implementing a reintroduction of wolves to Yellowstone National Park and central Idaho. *Restoration Ecology* **5**, 7–27.

Gill, A.M., Ryan, P.G., Moore, P.H.R. & Gibson, M. (2000) Fire regimes of World Heritage Kakadu National Park, Australia. *Australian Journal of Ecology* **25**, 616–625.

Hobbs, R.J. (2003) Ecological management and restoration: assessment, setting goals and measuring success. *Ecological Management and Restoration* **4S**, 2–3.

Hobbs, R.J. & Kristjanson, L.J. (2003) Triage: how do we prioritize health care for landscapes? *Ecological Management and Restoration* **4S**, 39–45.

Hopkins, A. (1999) *National Land and Water Resources Audit Vegetation Theme Condition of Vegetation*. Discussion paper. WA Department of Conservation and Land Management, Perth.

Larson, M.G., Booth, D.B. & Morley, S.A. (2001) Effectiveness of large woody debris in stream rehabilitation projects in urban basins. *Ecological Engineering* **18**, 211–226.

Lindenmayer, D.B., Cunningham, R.B., Tanton, M.T. & Smith, A.P. (1990) The conservation of arboreal marsupials in the montane ash forests of the Central Highlands of Victoria, South East Australia. II. The loss of trees with hollows and its implications for the conservation of Leadbeater's possum *Gymnobelideus leadbeateri* McCoy (Marsupialia: Petauridae). *Biological Conservation* **54**, 133–145.

MacArthur, R.J., Recher, H.F. & Cody, M.L. (1966) On the relation between habitat selection and bird species diversity. *American Naturalist* **100**, 319–322.

Mac Nally, R., Horrocks, G. & Pettifer, L. (2002) Experimental evidence for potential beneficial effects of fallen timber in forests. *Ecological Applications* **12**, 1588–1594.

Oliver, I., Smith, P.L., Lunt, I. & Parkes, D. (2002) Pre-1750 vegetation, naturalness and vegetation condition: what are the implications for biodiversity conservation? *Ecological Management and Restoration* **3**, 176–178.

Richards, S.A., Possingham, H.P. & Tizard, J. (1999) Optimal fire management for maintaining community diversity. *Ecological Applications* **9**, 880–892.

Robinson, C.J., Smyth, D. & Whitehead, P.J. (2005) Bush tucker, bush pets, and bush threats: cooperative management of feral animals in Australia's Kakadu National Park. *Conservation Biology* **19**, 1385–1391.

Saab, V.A. & Powell, H.D.W. (2005) Fire and avian ecology in North America: process influencing pattern. *Studies in Avian Biology* **30**, 1–13.

Sattler, P. & Creighton, C. (2002) *Australian Terrestrial Biodiversity Assessment 2002*. National Land & Water Resources Audit, Canberra.

Scotts, D.J. (1991) Old-growth forests: their ecological characteristics and value to forest-dependent vertebrate fauna of south-east Australia. In: Lunney, D. (ed.) *Conservation of Australia's Forest Fauna*, pp. 147–159. Royal Zoological Society of New South Wales, Mosman.

Smith, D.W., Peterson, R.O. & Houston, D.B. (2003) Yellowstone after wolves. *Bioscience* **53**, 330–340.

Suter, G.W. (1993) A critique of ecosystem health concepts and indexes. *Environmental Toxicology and Chemistry* **12**, 1533–1539.

Terborgh, J., Estes, J.A., Paquet, P. *et al.* (1999) The role of top carnivores in regulating terrestrial ecosystems. In: Soule, M.E. & Terborgh, J. (eds.) *Continental Conservation: Scientific Foundations of Regional Reserve Networks*, pp. 39–64. Island Press, Washington, DC.

Wicklum, D. & Davies, R.W. (1995) Ecosystem health and integrity? *Canadian Journal of Botany* **73**, 997–1000.

Woldendorp, G. & Keenan, R.J. (2005) Coarse woody debris in Australian forest ecosystems: a review. *Austral Ecology* **30**, 834–843.

Zimov, S.A., Churprynin, V.I., Oreshko, A.P., Chapin, F.S., Reynolds, J.F. & Chapin, M.C. (1995) Steppe-tundra transition: a herbivore-driven biome shift at the end of the Pleistocene. *American Naturalist* **146**, 765–794.

11

The Diverse Impacts of Grazing, Fire and Weeds: How Ecological Theory Can Inform Conservation Management

Don A. Driscoll

Abstract

Disturbances, such as fire, livestock grazing and introduced weeds, all have profound effects on vegetation structure and the composition of remnant vegetation. Native fauna within remnants frequently shows contrasting responses to vegetation changes; some species increase and others decline. Changes in vegetation structure can alter animal abundance through loss of shelter or foraging areas. However, changes in plant species composition are also influential, because specialist animal species are dependent on particular plants. Predicting the extent of specialization would enable better management by highlighting the components of the vegetation that could be retained or recovered to benefit the most species. Disturbance mosaics seem a logical solution to the diverse response of species to disturbance. However, evidence that tests the intermediate disturbance hypothesis, a model that incorporates the mosaics concept, suggests that mosaics rarely sustain the highest number of species. Given these lessons for management from ecological theory, there is an ongoing need to integrate theory and management. Furthermore, the lack of congruence between the expected benefits of mosaics and the results of intermediate disturbance experiments emphasizes the need for further landscape-scale experiments. Implementing management in the form of designed experiments will be an important part of the solution.

Keywords: ecological theory; fire; grazing; invasive weeds.

Introduction

Spatial ecological theory is often applied in the context of habitat fragmentation, with isolation and patch size the main focus. For example, classic metapopulation theory (Levins 1970) and patch-dynamic metacommunity theory (Leibold & Mikkelson 2002) assume habitat patches are identical, with colonization rates playing a key role in how the systems function. However, in many, perhaps most, examples of metapopulations, extinction results from deterministic processes associated with habitat quality (Thomas 1994). Furthermore, patch condition can have a stronger influence on the risk of extinction than spatial arrangement (Harrison & Bruna 1999) and so both aspects need to be considered (Hobbs 2003). Empirical research in fragmented landscapes supports this position, demonstrating that both spatial and habitat variables explain species distributions in fragmented landscapes (Diaz *et al.* 2000; Schweiger *et al.* 2000; Brotons & Herrando 2001; Davis 2004; Driscoll 2004; Driscoll & Weir 2005). How habitat quality influences extinction risk is important for developing ecological theory, but also for making appropriate management decisions. In this essay I will review what I think are three of the most important and widespread processes that influence habitat condition for wildlife, namely grazing, fire and invasive plant species. I will focus on the way these processes affect vegetation structure and composition, and the consequences of those changes for animals. I acknowledge that there are other processes that can lead to degradation of native vegetation, particularly those related to physical and chemical changes in the environment (e.g. salinity, pollution, changed hydrological regimes: (Hobbs *et al.* 1993; Toth *et al.* 1995; Letnic & Fox 1997), but I focus on these three processes because they act extensively throughout much of the globe, and are amenable to management manipulation. I will then examine the relevance of ecological theory for understanding and predicting the impacts of vegetation change on fauna, which will highlight the gaps in our knowledge and the research directions that are now critical.

How land management alters the vegetation of native communities

Grazing

Grazing by livestock is a widespread management practice that alters vegetation structure and composition. The implications for resident biota are frequently negative (Bennett 1993; Bromham *et al.* 1999; Hobbs 2001; Luck 2003; Driscoll 2004), but not always. James *et al.* (1999) observed that many reptile and bird species were less abundant at high grazing levels, but one gecko species and some birds were more abundant in intensively grazed sites. Similar patterns were observed by Martin *et al.* (2005) and Soderstrom *et al.* (2001), where one group of bird species was more common in grazed sites, with different species preferring less disturbed sites. In their review, James *et al.* (1999) pointed out that highest species richness usually occurs at low or medium grazing levels, and that at high grazing intensity only a few resistant species remained. These diverse responses to grazing in woodlands and grasslands imply that maximizing biodiversity conservation is likely to involve a diversity of grazing regimes, but the avoidance of very high grazing intensities.

Limited research into the mechanisms of biodiversity decline resulting from grazing highlight a range of ways in which grazing affects biodiversity. Grazing-related changes to the understorey of grassy woodlands can alter bird foraging modes, reducing foraging efficiency (Maron & Lill 2005). Impacts of grazing on reptiles are usually inferred to arise through changes in vegetation structure (James *et al.* 1999). Structure also may be important for birds, with Martin and Possingham (2005) correctly predicting the direction of response to grazing in 80% of bird species, based on whether or not they forage or nest in the shrub layer, the layer most affected by grazing. Large grazing animals such as Asian elephant (*Elephas maximus*), gaur (*Bos gaurus*) and chital (*Axis axis*) may be influenced by direct competition with livestock for food (Madhusudan 2004). These examples show that the effect on native animals of grazing can be mediated by changes to native vegetation, but that the mechanism could include loss of shelter, altered foraging behaviour or loss of food resources.

The influences exerted by plant composition and structure on animal species are not easily separated, but when detailed biological information about the requirements of an animal species is known, it is possible to recognize the key role of particular plant species. Gouldian finches (*Erythrura gouldiae*) show a negative correlation with grazing, like other declining granivorous birds from tropical Australian woodlands (Franklin *et al.* 2005). Although grazing does affect the structure of grasslands, Woinarski *et al.* (2005) noted that the finches were dependent on seeds from one grass species, *Alloteropsis semialata*, and that cattle grazing reduced the availability of this grass. So although grazing may lead to substantial changes in the vegetation structure, it was the reduction of one grass species that impacted on Gouldian finches. With knowledge of relationships between plant species and animals, it may be possible to provide resources for a proportion of the native biota despite exploitative human activities.

Fire

Fire is a major cause of vegetation change. Fire suppression has been a common management strategy that inadvertently allows post-fire succession to proceed to a late stage, often to the detriment of early successional species (Gibb & Hochuli 2002; Hobbs 2003). Fire exclusion promotes a range of vegetation changes that are specific to particular ecological communities, such as invasion of woody plants into grasslands (Blake & Schuette 2000; Davis *et al.* 2000) and the expansion of rainforest into sclerophyll communities (Harrington & Sanderson 1994). At the other extreme, very frequent fire is often employed with the aim of reducing fire hazard, and can also lead to vegetation change, including the extinction of fire-sensitive species (Morrison *et al.* 1996). Substantial progress has been made in understanding the response of animal species to plant successional change after fire (Andersen 1991; Friend 1993; Nelle *et al.* 2000; Friend & Wayne 2003). Fox *et al.* (2003) demonstrated experimentally that the effects of fire on small mammals can be mimicked by physically altering the vegetation structure, emphasizing the central role of structure in mediating post-fire succession. Vegetation structure can also influence bird communities (Skowno & Bond 2003) through the provision of nesting sites (Hamel 2003), or protection from nest predators (Larison *et al.* 2001). Biological legacies such as dead trees in burnt forest stands (Lindenmayer & McCarthy 2002) also attest to the key importance of structural features in mediating the response of fauna to fire.

Although studies on bird and mammal succession after fire emphasize that vegetation structure has a strong influence, the plant species that are present may also be important. For example, in tropical Australian woodlands,

Woinarski *et al.* (2005) demonstrated that black-footed rats (*Mesembriomys gouldii*) were especially dependent on *Cycas* species for food during one part of the year. The loss of any one of five out of 14 food plant species, which could occur through late dry season fires, could result in population extinction, emphasizing that plant species rather than structure can determine the impact of fire (Woinarski *et al.* 2005).

Given the strong successional responses to fire, and the dependence of some species on multi-aged habitats (Bradstock *et al.* 2005; Hayward 2005), it is likely that a mosaic of successional ages is needed in the landscape to maintain biodiversity (Briani *et al.* 2004). If successional responses are known, it may be possible to optimize fire management to maintain appropriate areas of each successional stage in the landscape (Richards *et al.* 1999). Richards *et al.* (1999) provide a model for choosing between competing management strategies such as suppressing fires or conducting fuel-reduction burns. Their model (and the later model of McCarthy *et al.* 2001) did not consider temporal or spatial proximity of successional stages. Proximity may be a critical issue for species that must disperse between areas at a suitable successional stage (Richards *et al.* 1999), but currently we know very little about the capacity of species to disperse from unburnt to burnt patches.

The optimal spatial arrangement and frequency of fires is beginning to be examined for particular species. Bradstock *et al.* (2005) used a spatially explicit model to explore the impact of different fire regimes on mallee fowl (*Leipoa ocellata*) in Australia. Taking dispersal capacity and known responses to fire into account, Bradstock *et al.* (2005) were able to distinguish among competing management options. To manage entire communities, models like that for the mallee fowl are needed for a broad range of species. To do that, there is an urgent need to understand the response of species to fire, and to obtain good estimates of the dispersal capacity of species that occur in only a limited range of successional stages.

Invasive plant species

Invasion of introduced plant species causes enormous changes to vegetation structure and composition and has large impacts on native fauna (Hobbs 2001; Pavey 2004; Ferdinands *et al.* 2005; Maerz *et al.* 2005; Simberloff 2005). Recognized weeds continue to expand their range (LeMaitre *et al.* 1996; Lindenmayer & McCarthy 2001; Setterfield *et al.* 2005) and new weeds continue to be introduced (Hacker & Waite 2001; Bortolussi *et al.* 2005), even though a critical part of the solution is to stop releasing non-native species (McIntyre *et al.* 2005). Weed invasion reduces plant species diversity (Galatowitsch & Richardson 2005) and can lead to a bigger reduction in species richness than other causes of vegetation change such as grazing or rainfall variation (Clarke *et al.* 2005).

Despite the overwhelmingly negative effects of invasive weeds, some invasive weeds can benefit native species. In North America, leafy spurge (*Euphorbia esula*) invasion of the Great Plains led to the decline of two grassland birds but increased nesting success in a third species (Scheiman *et al.* 2003). In Australia, the endangered southern brown bandicoot (*Isoodon obesulus*) is often found among the dense clumps of invasive blackberry (*Rubus* spp.), where it may be protected from introduced foxes and cats (Alessio 2000). The structural importance of weeds was emphasized by Fleishman *et al.* (2003), who studied a North American desert bird community. In riparian communities invaded by salt-cedar (*Tamarix ramosissima*), Fleishman *et al.* (2003) found that bird species richness was maintained because the weed provided essential habitat structure. For optimal conservation management, a vegetation management plan will need to weigh up the benefits of the structure provided by weeds against the likelihood that native species may be displaced.

Weed invasion and interactions with fire and grazing

A range of mechanisms promote weed invasion. Some weeds invade by outcompeting natives in undisturbed environments (Clarke *et al.* 2005). More frequently, however, invasive plant species are favoured under changed environmental conditions such as increased light intensity and nutrient availability (Green & Galatowitsch 2002; Daehler 2003). These environmental changes may result from human activities and are often associated with fire and grazing (Briese 1996; Duggin & Gentle 1998; Hobbs 2001; Keeley *et al.* 2003; Kimball & Schiffman 2003). Wide dispersal of seeds, either on animals (Couvreur *et al.* 2005; McIntyre *et al.* 2005) or aerially has the potential to maintain substantial densities of invasive species, even when the species are unable to sustain those densities through *in situ* reproduction (Brandt & Rickard 1994). This last example emphasizes that aspects of the surrounding landscape may impact on vegetation structure and composition within remnants (Hobbs 2001).

Invasive plants may not just be promoted by human disturbance; there may be a positive feedback loop where the initial invasion changes local conditions to favour ongoing invasion and plant community change (Milberg & Lamont 1995; Brooks *et al.* 2004). This process frequently involves invasive grasses in the grass/fire cycle (D'Antonio & Vitousek 1992). Examples include Buffel grass (*Cenchrus ciliaris*) in semiarid Australia (Clarke *et al.* 2005) and Gamba grass (*Andropogon gayanus*) in tropical Australia (Setterfield *et al.* 2005), the latter of which increases fire fuel loads by a factor of seven and burns with eight times the intensity of fires in native grasslands (Rossiter *et al.* 2003).

Although both fire and grazing can promote weed invasion, there are situations where use of fire or grazing may reduce the abundance or spread of invasive plants (Grice 1997; Safford & Harrison 2001; Bellingham & Coomes 2003; Maron & Lill 2005). For example, Emery and Gross (2005) suggested that annual summer burning could be used to control spotted knapweed (*Centaurea maculosa*) in North American prairies. Hayes and Holl (2003) point out that grazing benefits different components of the flora, increasing some exotics and some natives, thereby demanding a careful assessment of the overall biodiversity benefit. It is likely that a mosaic of management practices should be implemented across the landscape, given the contrasting responses of flora and fauna (Hayes & Holl 2003).

Discussion

Two strong themes emerge from this brief review. First, to manage native vegetation in a way that will conserve most biodiversity, a range of management regimes is required. However, the fire-impacts literature shows that we cannot yet determine the spatial scale of habitat mosaics because so little is known about dispersal through mosaic landscapes (see Chapter 18). Second, aspects of both vegetation structure and vegetation composition can influence the resident fauna, but the mechanisms are not well understood. In the following discussion, I explore how ecological theory may help to resolve these gaps in knowledge, or at least to provide some context for doing so.

Patterns of specialization: niche theory

How many species are likely to be affected by the removal or addition of a particular element of vegetation? Niche theory has the potential to help answer this question. There are many studies demonstrating that even very closely related species are ecologically differentiated and in some cases the specialization may be linked to habitat structure (Winemiller & Pianka 1990; Gignac 1992; Landmann & Winding 1993). Ecologists often classify species as generalists or specialists, implying that they occupy different niche breadths (Harrison 1999; Niemela 2001). Understanding the degree and nature of niche specialization is important for predicting community responses to vegetation change. For example, in Australian semiarid woodlands, the loss of the grass *Triodia* spp. through fire or grazing can lead to population crashes of several *Triodia*-specialist reptile species (Cogger 1984; Caughley 1985; Driscoll 2004). When vegetation structure is simplified or plant species are eliminated, generalist species will face an increased risk of extinction because a proportion of their resources is removed. However, certain specialist species may be entirely eliminated because all of their resources are lost. Predicting the nature of specialization is central to understanding the consequences of loss of particular elements of native habitat, and therefore, which elements are most important to preserve or re-establish.

There is the potential to use niche theory (Winemiller *et al.* 2001) and related keystone theory (Paine 1969; Hurlbert 1997; Tews *et al.* 2004) to predict the degree of specialization in communities of a certain complexity. Perhaps the most promising approach is community viability analysis (Ebenman *et al.* 2004; Ebenman & Jonsson 2005), in which niche breadth is measured by the number and strength of food-web links. With knowledge of food-web structure, species interaction strengths and population growth rates, it is possible to model the cascading effects through a community following a species' extinction (Ebenman & Jonsson 2005). Interaction strength is difficult to establish, but can involve estimates of resource use from observation (Wootton 1997). In addition to the trophic strength of interactions, the structural importance of species interactions needs to be incorporated into community viability analysis, given the importance of structure emerging from this review. Although, like population viability analysis, accurate predictions of outcomes in complex systems are unlikely (Lindenmayer *et al.* 2003), it may be possible to differentiate among competing management options (McCarthy *et al.* 2003).

Habitat heterogeneity and intermediate disturbance

Given that mosaics of vegetation condition may be an important management approach, the concept of habitat heterogeneity seems especially salient. One explanation for the species-area relationship is the habitat-heterogeneity model, in which larger areas include a greater diversity of habitats, and therefore more species (MacArthur & Wilson 1967; Connor & McCoy 1979). However, creating habitat mosaics differs from the habitat-heterogeneity model because the latter assumes increasing area, whereas the former does not. Creating habitat mosaics may not be qualitatively different from habitat loss and fragmentation for habitat specialists. Two processes are of concern: reduction in area and increased fragmentation of favoured habitat as additional habitats are added to a finite landscape (Fahrig 2003). In addition to improving our understanding of appropriate spatial scales for creating mosaics, we need to discover when the extremes of habitat heterogeneity become habitat fragmentation. This problem is related to the idea of niche specialization discussed above, in which the extent and degree of specialization determine the likely importance of habitat heterogeneity.

The intermediate disturbance hypothesis (Connell 1978) is conceptually linked to the habitat heterogeneity hypothesis. According to the intermediate disturbance hypothesis, disturbance produces habitat heterogeneity, but disturbance may occur within a patch, between patches or be temporally spaced (Roxburgh *et al.* 2004). Greatest species richness occurs at intermediate levels of disturbance because the latter generally provide conditions that suit the largest number of species. The model has been extensively tested, and is supported in some empirical studies (Molino & Sabatier 2001; Shea *et al.* 2004), but not by most studies (Mackey & Currie 2001). Mackey and Currie (2001) reported that evidence for a humped species distribution along a disturbance gradient was observed in only 16% of 244 cases examined. Importantly in the current context, when disturbances were caused by humans, an even smaller proportion of studies showed a humped relationship (Mackey & Currie 2001). Disturbances in spatial mosaics or within patches were equally unlikely to produce maximum species richness (Mackey & Currie 2001). This is rather discouraging given the apparent importance of mosaics of disturbance levels that I inferred previously. Shea *et al.* (2004) outlined research protocols to explore why intermediate disturbance levels sometimes lead to greater diversity. The same recommendations would also enable study of habitat mosaics needed to conserve species when there are contrasting responses to disturbance such as fire, grazing or weed invasion. Shea *et al.* (2004) used grazing disturbance as their example and emphasized that studies need to consider all aspects of the timing of the disturbance, as well as the spatial scale of disturbance and dispersal capacity of the species within the community. This is the key direction in which research needs to go, for both practical conservation management and to test and refine ecological theory.

Metacommunity theory

Metacommunity theory examines how dispersal influences local community composition and may be particularly valuable in predicting the outcome of

Figure 11.1 **Livestock grazing in remnant vegetation can have a profound effect on the understorey (top right), and contrasting effects on different animal species. Retention of the prickly spinifex grass (*Triodia scariosa*) (top left) in grazed sites can enable spinifex specialists to survive grazing. Photos by Don Driscoll, with permission.**

habitat loss and fragmentation (Holyoak *et al.* 2005). Although the spatial arrangement of habitat is a critical component of metacommunity theory, the condition of habitat may also influence metacommunity processes. For example, under the species-sorting model, community differentiation is based on either differential survival in patches of particular condition, or patch selection (Cottenie & De Meester 2004). The mass-effects model (Leibold & Mikkelson 2002) also assumes that patches differ in quality, but high dispersal rates between patches mean that local communities can be more a product of immigration than internal recruitment. For example, mass effects may help explain the abundance of some weeds in remnant vegetation (Brandt & Rickard 1994). If mass effects or species sorting dominate community development, then given a variety of grazing regimes or fire regimes in a landscape within the dispersal capacity of most species, few extinctions are likely.

Discovering the optimal spatial scale of habitat mosaics for taxa with diverse dispersal abilities remains critical.

Other metacommunity theories are not so helpful for understanding species' responses to vegetation structure and condition. Neutral metacommunity theory implies that species should respond in the same direction to disturbance because they are ecologically equivalent (Bell 2001; Hubbell 2001), and nested subset theory (Patterson & Atmar 1986; Fischer & Lindenmayer 2005) implies there will be predictable, nested responses across a disturbance gradient. However, given the general observation I made above that species frequently show opposite responses to vegetation change, nestedness and the neutral models are unlikely to be helpful guides to biodiversity conservation. A fourth metacommunity model, the patch dynamic model (Leibold & Mikkelson 2002), considers scenarios where patch condition is uniform, and focuses on the spatial arrangement of patches.

Conclusions

Although the establishment of habitat mosaics seemed like a general management principle from the empirical studies, research addressing the intermediate disturbance hypothesis raises doubts about the likely effectiveness of this strategy. The principle emerging is that **habitat/disturbance mosaics may be a valuable approach to biodiversity management, but they should be implemented in an experimental context and thoroughly tested**. The general research model defined by Shea *et al.* (2004) provides a valuable approach to collecting data that will facilitate design of effective habitat mosaics, as well as providing data to test and refine ecological theory.

There is ample evidence to conclude that animal species depend on particular plant species for specific resources, and that the structural features of plants are also essential for animal species to survive. These observations give rise to a second general principle that **both plant structure and plant species composition need to be considered to manage the faunal community composition effectively**. Given the importance of vegetation structure, it will be necessary for community viability models to expand to include all aspects of species interactions, not just trophic interactions, if they are reasonably to predict species extinctions.

Modelling optimal fire regimes (Richards *et al.* 1999; Bradstock *et al.* 2005) is a good example of the research direction that needs to be developed to enable landscape-scale management. Current examples show how it is possible to choose between competing fire management strategies for one species, taking dispersal into account (Bradstock *et al.* 2005). These kinds of models need to be extended to incorporate more management options, such as grazing or weed control, and more species. Incorporating the developing field of community viability analysis into the spatial models that have so far been applied to fire management is a potentially fruitful avenue to follow, allowing management decisions that take a broad spectrum of the biota into account. These models would also help to answer questions like 'how many species will be threatened if we allow grazing in this reserve and remove the shrub layer?', or 'how much biodiversity can we restore if we replant a particular understorey species in certain locations?' My argument here is underpinned by a third principle: **conservation management of landscapes must take into account a broad range of the biota, and not be based on models of limited taxonomic scope**.

Restoring and managing ecological communities is complex and our knowledge is incomplete (Young *et al.* 2005). The way forward will involve continuing to test and refine ecological theory while exploring practical solutions to management problems. By addressing those practical management issues in the context of theoretical ecology, there is the potential to develop more useful general rules and tools that would mean we don't have to approach every situation on a case-by-case basis. My fourth principle is therefore that **research into managing habitat quality should be undertaken in the context of ecological theory**. To accomplish this research at an appropriate spatial scale, **land management should be implemented as a research programme, following the rigours of experimental design and statistical testing**, my fifth principle. Implementing management as an experiment to answer specific questions originated as 'adaptive management' (Walters 1986; Morghan *et al.* 2006). However, the term 'adaptive management' is often taken to mean implementing management and observing the outcome without using contrasts, controls or replicates, and so managers never really know why a particular outcome occurred (Walters 1997). Implementing adaptive management (Walters 1986), perhaps better named experimental management (Walters & Green 1997), would reflect our state of knowledge of species' responses to vegetation change, and it would help to improve our ability to make better management decisions in future.

Acknowledgements

I thank Jenny Lau, David Lindenmayer and Richard Hobbs for commenting on an earlier draft. This review was completed while I was a lecturer in the School of Biological Sciences, Flinders University, South Australia.

Principles

1 Habitat/disturbance mosaics may be a valuable approach to biodiversity management, but they should be implemented in an experimental context.
2 Both plant structure and plant species composition need to be considered to manage fauna community composition effectively.
3 Conservation management of landscapes must take into account a broad spectrum of the biota, and not be based on models of limited taxonomic scope.
4 Research into managing habitat quality should be undertaken in the context of ecological theory.
5 Land management should be implemented as a research programme (experimental management).

References

Alessio, J. (2000) The habitat preference and diet of the southern brown bandicoot (*Isoodon obesulus obesulus*), within Scott Creek Conservation Park, South Australia. Honours Thesis, Flinders University, Adelaide.

Andersen, A.N. (1991) Responses of ground-foraging ant communities to 3 experimental fire regimes in a savanna forest of tropical Australia. *Biotropica* **23**, 575–585.

Bell, G. (2001) Neutral macroecology. *Science* **293**, 2413–2418.

Bellingham, P.J. & Coomes, D.A. (2003) Grazing and community structure as determinants of invasion success by Scotch broom in a New Zealand montane shrubland. *Diversity and Distributions* **9**, 19–28.

Bennett, A.F. (1993) Microhabitat use by the long-nosed potoroo, *Potorous tridactylus*, and other small mammals in remnant forest vegetation of south-western Victoria. *Wildlife Research* **20**, 267–285.

Blake, J.G. & Schuette, B. (2000) Restoration of an oak forest in east-central Missouri – Early effects of prescribed burning on woody vegetation. *Forest Ecology and Management* **139**, 109–126.

Bortolussi, G., McIvor, J.G., Hodgkinson, J.J., Coffey, S.G. & Holmes, C.R. (2005) The northern Australian beef industry, a snapshot. 5. Land and pasture development practices. *Australian Journal of Experimental Agriculture* **45**, 1121–1129.

Bradstock, R.A., Bedward, M., Gill, A.M. & Cohn, J.S. (2005) Which mosaic? A landscape ecological approach for evaluating interactions between fire regimes, habitat and animals. *Wildlife Research* **32**, 409–423.

Brandt, C.A. & Rickard, W.H. (1994) Alien taxa in the North-American shrub-steppe 4 decades after cessation of livestock grazing and cultivation agriculture. *Biological Conservation* **68**, 95–105.

Briani, D.C., Palma, A.R.T., Vieira, E.M. & Henriques, R.P.B. (2004) Post-fire succession of small mammals in the Cerrado of central Brazil. *Biodiversity and Conservation* **13**, 1023–1037.

Briese, D.T. (1996) Biological control of weeds and fire management in protected natural areas: Are they compatible strategies? *Biological Conservation* **77**, 135–141.

Bromham, L., Cardillo, M., Bennett, A.F. & Elgar, M.A. (1999) Effects of stock grazing on the ground invertebrate fauna of woodland remnants. *Australian Journal of Ecology* **24**, 199–207.

Brooks, M.L., D'Antonio, C.M., Richardson, D.M. *et al.* (2004) Effects of invasive alien plants on fire regimes. *Bioscience* **54**, 677–688.

Brotons, L. & Herrando, S. (2001) Factors affecting bird communities in fragments of secondary pine forests in the north-western Mediterranean basin. *Acta Oecologica-International Journal of Ecology* **22**, 21–31.

Caughley, J. (1985) Effect of fire on the reptile fauna of mallee. In: Grigg, G., Shine, R. & Ehmann, H. (eds.) *Biology of Australasian frogs and reptiles*, pp. 31–34. Royal Zoological Society of NSW and Surrey Beatty & Sons, Chipping Norton, NSW.

Clarke, P.J., Latz, P.K. & Albrecht, D.E. (2005) Long-term changes in semi-arid vegetation: Invasion of an exotic perennial grass has larger effects than rainfall variability. *Journal of Vegetation Science* **16**, 237–248.

Cogger, H.G. (1984) Reptiles in the Australian arid zone. In: Cogger, H.G. & Cameron, E.E. (eds.) *Arid Australia*, pp. 235–252. Surrey Beatty & Sons, Chipping Norton, NSW.

Connell, J.H. (1978) Diversity in tropical rain forests and coral reefs. *Science* **199**, 1302–1310.

Connor, E.F. & McCoy, E.D. (1979) The statistics and biology of the species-area relationship. *American Naturalist* **113**, 791–833.

Cottenie, K. & De Meester, L. (2004) Metacommunity structure: Synergy of biotic interactions as selective agents and dispersal as fuel. *Ecology* **85**, 114–119.

Couvreur, M., Verheyen, K. & Hermy, M. (2005) Experimental assessment of plant seed retention times in fur of cattle and horse. *Flora* **200**, 136–147.

Daehler, C.C. (2003) Performance comparisons of co-occurring native and alien invasive plants: Implications for conservation and restoration. *Annual Review of Ecology, Evolution and Systematics* **34**, 183–211.

D'Antonio, C.M. & Vitousek, P.M. (1992) Biological invasions by exotic grasses, the grass fire cycle, and global change. *Annual Review of Ecology and Systematics* **23**, 63–87.

Davis, M.A., Peterson, D.W., Reich, P.B. *et al.* (2000) Restoring savanna using fire: Impact on the breeding bird community. *Restoration Ecology* **8**, 30–40.

Davis, S.K. (2004) Area sensitivity in grassland passerines: Effects of patch size, patch shape, and vegetation structure on bird abundance and occurrence in southern Saskatchewan. *Auk* **121**, 1130–1145.

Diaz, J.A., Carbonell, R., Virgos, E., Santos, T. & Telleria, J.L. (2000) Effects of forest fragmentation on the distribution of the lizard *Psammodromus algirus*. *Animal Conservation* **3**, 235–240.

Driscoll, D.A. (2004) Extinction and outbreaks accompany fragmentation of a reptile community. *Ecological Applications* **14**, 220–240.

Driscoll, D.A. & Weir, T. (2005) Beetle responses to habitat fragmentation depend on ecological traits, habitat condition, and remnant size. *Conservation Biology* **19**, 182–194.

Duggin, J.A. & Gentle, C.B. (1998) Experimental evidence on the importance of disturbance intensity for invasion of *Lantana camara* L. in dry rainforest-open forest ecotones in north-eastern NSW, Australia. *Forest Ecology and Management* **109**, 279–292.

Ebenman, B. & Jonsson, T. (2005) Using community viability analysis to identify fragile systems and keystone species. *Trends in Ecology and Evolution* **20**, 568–575.

Ebenman, B., Law, R. & Borrvall, C. (2004) Community viability analysis: The response of ecological communities to species loss. *Ecology* **85**, 2591–2600.

Emery, S.M. & Gross, K.L. (2005) Effects of timing of prescribed fire on the demography of an invasive plant, spotted knapweed *Centaurea maculosa*. *Journal of Applied Ecology* **42**, 60–69.

Fahrig, L. (2003) Effects of habitat fragmentation on biodiversity. *Annual Review of Ecology, Evolution and Systematics* **34**, 487–515.

Ferdinands, K., Beggs, K. & Whitehead, P. (2005) Biodiversity and invasive grass species: multiple-use or monoculture? *Wildlife Research* **32**, 447–457.

Fischer, J. & Lindenmayer, D.B. (2005) Nestedness in fragmented landscapes: a case study on birds, arboreal marsupials and lizards. *Journal of Biogeography* **32**, 1737–1750.

Fleishman, E., McDonal, N., Mac Nally, R., Murphy, D.D., Walters, J. & Floyd, T. (2003) Effects of floristics, physiognomy and non-native vegetation on riparian bird communities in a Mojave Desert watershed. *Journal of Animal Ecology* **72**, 484–490.

Fox, B.J., Taylor, J.E. & Thompson, P.T. (2003) Experimental manipulation of habitat structure: a retrogression of the small mammal succession. *Journal of Animal Ecology* **72**, 927–940.

Franklin, D.C., Whitehead, P.J., Pardon, G., Matthews, J., McMahon, P. & McIntyre, D. (2005) Geographic patterns and correlates of the decline of granivorous birds in northern Australia. *Wildlife Research* **32**, 399–408.

Friend, G. & Wayne, A. (2003) Relationships between mammals and fire in south-west Western Australian ecosystems: what we know and what we need to know. In: Abbott, I. & Burrows, N. (eds.) *Fire in Ecosystems of South-West Western Australia*, pp. 363–380. Backhuys, Leiden.

Friend, G.R. (1993) Impact of fire on small vertebrates in mallee woodlands and heathlands of temperate Australia – a review. *Biological Conservation* **65**, 99–114.

Galatowitsch, S. & Richardson, D.M. (2005) Riparian scrub recovery after clearing of invasive alien trees in headwater streams of the Western Cape, South Africa. *Biological Conservation* **122**, 509–521.

Gibb, H. & Hochuli, D.F. (2002) Habitat fragmentation in an urban environment: large and small fragments support different arthropod assemblages. *Biological Conservation* **106**, 91–100.

Gignac, L.D. (1992) Niche structure, resource partitioning, and species interactions of mire bryophytes relative to climatic and ecological gradients in western Canada. *Bryologist* **95**, 406–418.

Green, E.K. & Galatowitsch, S.M. (2002) Effects of *Phalaris arundinacea* and nitrate-N addition on the establishment of wetland plant communities. *Journal of Applied Ecology* **39**, 134–144.

Grice, A.C. (1997) Post-fire regrowth and survival of the invasive tropical shrubs *Cryptostegia grandiflora* and *Ziziphus mauritiana*. *Australian Journal of Ecology* **22**, 49–55.

Hacker, J.B. & Waite, R.B. (2001) Selecting buffel grass (*Cenchrus ciliaris*) with improved spring yield in subtropical Australia. *Tropical Grasslands* **35**, 205–210.

Hamel, P.B. (2003) Winter bird community differences among methods of bottomland hardwood forest restoration: results after seven growing seasons. *Forestry* **76**, 189–197.

Harrington, G.N. & Sanderson, K.D. (1994) Recent contraction of wet sclerophyll forests in the wet tropics of Queensland due to invasion by rainforest. *Pacific Conservation Biology* **1**, 319–327.

Harrison, S. (1999) Local and regional diversity in a patchy landscape: Native, alien, and endemic herbs on serpentine. *Ecology* **80**, 70–80.

Harrison, S. & Bruna, E. (1999) Habitat fragmentation and large-scale conservation: what do we know for sure? *Ecography* **22**, 225–232.

Hayes, G.F. & Holl, K.D. (2003) Cattle grazing impacts on annual forbs and vegetation composition of mesic grasslands in California. *Conservation Biology* **17**, 1694–1702.

Hayward, M.W. (2005) Diet of the quokka (*Setonix brachyurus*) (Macropodidae: Marsupialia) in the northern jarrah forest of Western Australia. *Wildlife Research* **32**, 15–22.

Hobbs, R.J. (2001) Synergisms among habitat fragmentation, livestock grazing, and biotic invasions in southwestern Australia. *Conservation Biology* **15**, 1522–1528.

Hobbs, R.J. (2003) How fire regimes interact with other forms of ecosystem disturbance and modification. In: Abbott, I. & Burrows, N. (eds.) *Fire in Ecosystems of South-West Western Australia*, pp. 421–436. Backhuys, Leiden.

Hobbs, R.J., Saunders, D.A., Lobry de Brun, L.A. & Main, A.R. (1993) Changes in biota. In: Hobbs, R.J. & Saunders, D.A. (eds.) *Reintegrating Fragmented Landscapes: Towards Sustainable Production and Nature Conservation*, pp. 65–106. Springer-Verlag, New York.

Holyoak, M., Leibold, M.A., Mouquet, N.M., Holt, R.D. & Hoopes, M.F. (2005) Metacommunities: a framework for large-scale community ecology. In: Holyoak, M., Leibold, M.A. & Holt, R.D. (eds.) *Metacommunities. Spatial Dynamics and Ecological Communities*, pp. 1–31. University of Chicago Press, Chicago, IL.

Hubbell, S.P. (2001) *The Unified Neutral Theory of Biodiversity and Biogeography.* Princeton University Press, Princeton, NJ.

Hurlbert, S.H. (1997) Functional importance vs keystoneness: reformulating some questions in theoretical biocenology. *Australian Journal of Ecology* **22**, 369–382.

James, C.D., Landsberg, J. & Morton, S.R. (1999) Provision of watering points in the Australian arid zone: a review of effects on biota. *Journal of Arid Environments* **41**, 87–121.

Keeley, J.E., Lubin, D. & Fotheringham, C.J. (2003) Fire and grazing impacts on plant diversity and alien plant invasions in the southern Sierra Nevada. *Ecological Applications* **13**, 1355–1374.

Kimball, S. & Schiffman, P.M. (2003) Differing effects of cattle grazing on native and alien plants. *Conservation Biology* **17**, 1681–1693.

Landmann, A. & Winding, N. (1993) Niche segregation in high-altitude Himalayan chats (Aves, Turdidae) – does morphology match ecology? *Oecologia* **95**, 506–519.

Larison, B., Laymon, S.A., Williams, P.L. & Smith, T.B. (2001) Avian responses to restoration: nest-site selection and reproductive success in song sparrows. *Auk* **118**, 432–442.

Leibold, M.A. & Mikkelson, G.M. (2002) Coherence, species turnover, and boundary clumping: elements of meta-community structure. *Oikos* **97**, 237–250.

LeMaitre, D.C., VanWilgen, B.W., Chapman, R.A. & McKelly, D.H. (1996) Invasive plants and water resources in the Western Cape Province, South Africa: Modelling the consequences of a lack of management. *Journal of Applied Ecology* **33**, 161–172.

Letnic, M. & Fox, B.J. (1997) The impact of industrial fluoride fallout on faunal succession following sand-mining of dry sclerophyll forest at Tomago, NSW. 2. Myobatrachid frog recolonization. *Biological Conservation* **82**, 137–146.

Levins, R. (1970) Extinction. In: Gerstenhaber, M. (ed.) *Some Mathematical Questions in Biology. Lectures on Mathematics in Life Sciences 2*, pp. 77–107. American Mathematical Society, Providence, RI.

Lindenmayer, D.B. & McCarthy, M.A. (2001) The spatial distribution of non-native plant invaders in a pine-eucalypt landscape mosaic in south-eastern Australia. *Biological Conservation* **102**, 77–87.

Lindenmayer, D. & McCarthy, M.A. (2002) Congruence between natural and human forest disturbance: a case study from Australian montane ash forests. *Forest Ecology and Management* **155**, 319–335.

Lindenmayer, D.B., Possingham, H.P., Lacy, R.C., McCarthy, M.A. & Pope, M.L. (2003) How accurate are population models? Lessons from landscape-scale tests in a fragmented system. *Ecology Letters* **6**, 41–47.

Luck, G.W. (2003) Differences in the reproductive success and survival of the rufous treecreeper (*Climacteris rufa*) between a fragmented and unfragmented landscape. *Biological Conservation* **109**, 1–14.

MacArthur, R.H. & Wilson, E.O. (1967) *The Theory of Island Biogeography*. Princeton University Press, Princeton, NJ.

McCarthy, M.A., Andelman, S.J. & Possingham, H.P. (2003) Reliability of relative predictions in population viability analysis. *Conservation Biology* **17**, 982–989.

McCarthy, M.A., Possingham, H.P. & Gill, A.M. (2001) Using stochastic dynamic programming to determine optimal fire management for *Banksia ornata*. *Journal of Applied Ecology* **38**, 585–592.

McIntyre, S., Martin, T.G., Heard, K.M. & Kinloch, J. (2005) Plant traits predict impact of invading species: an analysis of herbaceous vegetation in the subtropics. *Australian Journal of Botany* **53**, 757–770.

Mackey, R.L. & Currie, D.J. (2001) The diversity-disturbance relationship: Is it generally strong and peaked? *Ecology* **82**, 3479–3492.

Madhusudan, M.D. (2004) Recovery of wild large herbivores following livestock decline in a tropical Indian wildlife reserve. *Journal of Applied Ecology* **41**, 858–869.

Maerz, J.C., Blossey, B. & Nuzzo, V. (2005) Green frogs show reduced foraging success in habitats invaded by Japanese knotweed. *Biodiversity and Conservation* **14**, 2901–2911.

Maron, M. & Lill, A. (2005) The influence of livestock grazing and weed invasion on habitat use by birds in grassy woodland remnants. *Biological Conservation* **124**, 439–450.

Martin, T.G. & Possingham, H.P. (2005) Predicting the impact of livestock grazing on birds using foraging height data. *Journal of Applied Ecology* **42**, 400–408.

Martin, T.G., Kuhnert, P.M., Mengersen, K. & Possingham, H.P. (2005) The power of expert opinion in ecological models using Bayesian methods: Impact of grazing on birds. *Ecological Applications* **15**, 266–280.

Milberg, P. & Lamont, B.B. (1995) Fire enhances weed invasion of roadside vegetation in southwestern Australia. *Biological Conservation* **73**, 45–49.

Molino, J.F. & Sabatier, D. (2001) Tree diversity in tropical rain forests: A validation of the intermediate disturbance hypothesis. *Science* **294**, 1702–1704.

Morghan, K.J.R., Sheley, R.L. & Svejcar, T.J. (2006) Successful adaptive management – the integration of research and management. *Rangeland Ecology & Management* **59**, 216–219.

Morrison, D.A., Buckney, R.T., Bewick, B.J. & Cary, G.J. (1996) Conservation conflicts over burning bush in south-eastern Australia. *Biological Conservation* **76**, 167–175.

Nelle, P.J., Reese, K.P. & Connelly, J.W. (2000) Long-term effects of fire on sage grouse habitat. *Journal of Range Management* **53**, 586–591.

Niemela, J. (2001) Carabid beetles (Coleoptera: Carabidae) and habitat fragmentation: a review. *European Journal of Entomology* **98**, 127–132.

Paine, R.T. (1969) A note on trophic complexity and community stability. *American Naturalist* **103**, 91–93.

Patterson, B.D. & Atmar, W. (1986) Nested subsets and the structure of insular mammalian faunas and archipeligos. In: Heaney, L.R. & Patterson, B.D. (eds.) *Island Biogeography of Mammals*, pp. 65–82. Academic Press, London.

Pavey, C. (2004) *Recovery Plan for Slater's Skink,* Egernia slateri, *2005–2010*. Northern Territory Department of Infrastructure, Planning and Environment, Darwin.

Richards, S.A., Possingham, H.P. & Tizard, J. (1999) Optimal fire management for maintaining community diversity. *Ecological Applications* **9**, 880–892.

Rossiter, N.A., Setterfield, S.A., Douglas, M.M. & Hutley, L.B. (2003) Testing the grass-fire cycle: alien grass invasion in the tropical savannas of northern Australia. *Diversity and Distributions* **9**, 169–176.

Roxburgh, S.H., Shea, K. & Wilson, J.B. (2004) The intermediate disturbance hypothesis: Patch dynamics and mechanisms of species coexistence. *Ecology* **85**, 359–371.

Safford, H.D. & Harrison, S.P. (2001) Grazing and substrate interact to affect native vs. exotic diversity in roadside grasslands. *Ecological Applications* **11**, 1112–1122.

Scheiman, D.M., Bollinger, E.K. & Johnson, D.H. (2003) Effects of leafy spurge infestation on grassland birds. *Journal of Wildlife Management* **67**, 115–121.

Schweiger, E.W., Diffendorfer, J.E., Holt, R.D., Pierotti, R. & Gaines, M.S. (2000) The interaction of habitat fragmentation, plant, and small mammal succession in an old field. *Ecological Monographs* **70**, 383–400.

Setterfield, S.A., Douglas, M.M., Hutley, L.B. & Welch, M.A. (2005) Effects of canopy cover and ground disturbance on establishment of an invasive grass in an Australia savanna. *Biotropica* **37**, 25–31.

Shea, K., Roxburgh, S.H. & Rauschert, E.S.J. (2004) Moving from pattern to process: coexistence mechanisms under intermediate disturbance regimes. *Ecology Letters* **7**, 491–508.

Simberloff, D. (2005) Non-native species do threaten the natural environment! *Journal of Agricultural & Environmental Ethics* **18**, 595–607.

Skowno, A.L. & Bond, W.J. (2003) Bird community composition in an actively managed savanna reserve, importance of vegetation structure and vegetation composition. *Biodiversity and Conservation* **12**, 2279–2294.

Soderstrom, B., Part, T. & Linnarsson, E. (2001) Grazing effects on between-year variation of farmland bird communities. *Ecological Applications* **11**, 1141–1150.

Tews, J., Brose, U., Grimm, V. *et al.* (2004) Animal species diversity driven by habitat heterogeneity/diversity: the importance of keystone structures. *Journal of Biogeography* **31**, 79–92.

Thomas, C.D. (1994) Extinction, colonization and metapopulations: environmental tracking by rare species. *Conservation Biology* **8**, 373–378.

Toth, L.A., Arrington, D.A., Brady, M.A. & Muszick, D.A. (1995) Conceptual evaluation of factors potentially affecting restoration of habitat structure within the channelized Kissimmee River ecosystem. *Restoration Ecology* **3**, 160–180.

Walters, C. (1986) *Adaptive Management of Renewable Resources*. Macmillan, New York.

Walters, C.J. (1997) Challenges in adaptive management of riparian and coastal ecosystems. *Conservation Ecology* [online] **1**, 1 (http://www.consecol.org/vol1/iss2/art1/).

Walters, C.J. & Green, R. (1997) Valuation of experimental management options for ecological systems. *Journal of Wildlife Management* **61**, 987–1006.

Winemiller, K.O. & Pianka, E.R. (1990) Organization in natural assemblages of desert lizards and tropical fishes. *Ecological Monographs* **60**, 27–55.

Winemiller, K.O., Pianka, E.R., Vitt, L.J. & Joern, A. (2001) Food web laws or niche theory? Six independent empirical tests. *American Naturalist* **158**, 193–199.

Woinarski, J.C.Z., Williams, R.J., Price, O. & Rankmore, B. (2005) Landscapes without boundaries: wildlife and their environments in northern Australia. *Wildlife Research* **32**, 377–388.

Wootton, J.T. (1997) Estimates and tests of per capita interaction strength: diet, abundance, and impact of intertidally foraging birds. *Ecological Monographs* **67**, 45–64.

Young, T.P., Petersen, D.A. & Clary, J.J. (2005) The ecology of restoration: historical links, emerging issues and unexplored realms. *Ecology Letters* **8**, 662–673.

12

Forest Landscape Structure, Degradation and Condition: Some Commentary and Fundamental Principles

Jerry F. Franklin and Mark E. Swanson

Abstract

Forest landscapes are characterized by structure, function and composition. At the landscape scale, functionality of the forest ecosystem is heavily influenced by landscape structure (pattern) and content (composition of patches or other elements). Degradation occurs when certain structures decline, or when landscape patterns are altered to the point where flows of materials, energy or organisms are influenced. Landscape design for ecological purposes is a complex and challenging field, but some general principles may be applied to landscape management. Although homogeneous landscapes can maximize particular functions and flows, forest landscape designs should incorporate complexity and heterogeneity at multiple spatial scales. Landscape design must reflect the influence of the geophysical template, utilize network and gradient (concepts) as well as patch paradigms, and take patch content into consideration. The matrix (the dominant patch type) and its content must be adequately considered. These themes merit consideration in a world where the output of multiple values, economic and ecological, is increasingly demanded by consumers and stakeholders.

Keywords: degradation; heterogeneity; landscape condition; landscape design; landscape ecology; structure.

Introduction

Structure, function and composition are the fundamental attributes of all ecosystems. Structure is of particular interest from both ecosystem and landscape perspectives because the attributes of function and composition can often be dealt with through the surrogate of structure. For example, although interest in managing forest stands and landscapes is often centred around composition (e.g. biodiversity) or ecological functions (e.g. productivity), these attributes most often are indexed by structure. Further, structure is what we typically manipulate and attempt to regulate.

In this essay we provide some perspectives on aspects of landscape structure, including the functionality of landscapes (i.e. issues related to condition and degradation) and aspects of internal patch structure that relate to landscape function. This discussion is organized around a diverse set of landscape design principles focused on the conservation of biodiversity and other landscape-level values.

Some definitions

In this essay **structure** refers primarily to physical attributes of patches and landscapes; in teaching we sometimes use the terms 'structure' and 'architecture' interchangeably. For example, at the stand level, forest structure includes various biological and physical features of the ecosystems, such as living trees of various sizes and conditions, standing dead trees, and boles and other coarse wood on the forest floor (e.g. Franklin & Spies 1991). In forest stands, as well as other vegetation types, there is also a component of structure related to spatial pattern – for example, structures may be homogeneously or heterogeneously distributed in the stand or occur in some mixed arrangement (e.g. Franklin & Van Pelt 2004; Franklin 2005).

Similarly, landscapes exhibit structure in the form of patches of different sizes, shapes and internal conditions as well as gradients of physical and biological attributes and well-defined networks (Forman 1995). Landscape function and composition will depend greatly upon the spatial arrangement of these patches, gradients and networks and also upon the internal structure of the patches.

Degradation is a term that implies that the capacity of a landscape to provide for an ecological function – whether habitat for a species or an ecosystem process – has been reduced below some previous capacity. Degradation is a much more subjective term than structure or condition in the sense that it implies an assessment about the functionality of a landscape. Subjective statements about ecosystems or landscapes can only be made with reference to specified management objectives in terms of a species, function or output of interest, as well as the relevant spatial and temporal dimensions. An exemplary objective might be provision of nesting, roosting and foraging habitat for Northern spotted owls (*Strix occidentalis caurina*) in a drainage basin in northwestern North America over the next century.

Degradation also may consist of departure from a generally agreed set of benchmark (or reference) conditions. Caution is warranted, because quantified conditions from a previous time may have been atypical in time. Ecosystems are spatiotemporally unique, and some may represent an extreme condition within the historic range of variability (Swetnam *et al.* 1999). Degradation should not be confused with ecosystem shifts within the historic range of variability.

Condition generally refers to a current state of an ecosystem or landscape. This may be assessed in terms of some objective set of parameters or, more subjectively, with reference to some specified management objective, such as provision of dispersal or nesting habitat for a species. An example of an objective characterization of a stand condition would be its description in terms of the density of hollow-bearing trees and the spatial arrangement of those trees (even, random or aggregated). Characterizing a landscape in terms of the average and range of patch sizes can be done objectively. However, assessments of the functionality of either a stand or a landscape as habitat can only be done with reference to the habitat requirements of a specific species (e.g. nesting or foraging habitat for a species of owl). As with the term degradation, condition is sometimes used with reference to some previous state or capacity.

Some landscape design principles

Scarcity of general principles for interpreting effects of specific landscape patterns on ecological functions, including habitat

Interpreting the effects of landscape structure on ecological function requires stipulating which ecosystem functions or species are of interest as well as the relevant spatial and temporal scale. It is not possible to interpret the functionality of a particular landscape design – good or bad, functional or dysfunctional – until that is done. Some generalizations about pattern and its effects are possible (and we will make some later) but the consequences of those patterns will differ with the process or organism of interest and the spatial and temporal dimensions of interest. Conditionality or limitation in application of general ecological theory is a common problem when attempting to use such theory in guiding decision-making (see e.g. Shrader-Frechette & McCoy 1993).

For example, heterogeneous landscapes (landscapes with a diversity of patch types and spatial arrangements of patches) can be expected to have: (i) more niches, which, in turn, will result in (ii) higher levels of species diversity and (iii) greater diversity in the types and rates of ecological processes present in the landscape. Specific consequences for a process, parameter or organism of interest will vary from extremely positive to extremely negative. However, few processes or species are likely to dominate overwhelmingly in all portions of the heterogeneous landscape. For example, the net annual wood accumulation or, more generally, the annual biomass increment, is maximized in landscapes dominated by uniform, densely stocked, young stands of trees. The net annual wood accumulation will be lower, sometimes significantly lower, in heterogeneous stands or landscapes. This is why humans often create simpler (homogeneous) landscapes in order to maximize some particular process, such as wood or food crop production, or a population of some particular organism, such as domestic cattle. In contrast, there are many organisms that are successful in heterogeneous landscapes. For example, whitetailed deer in North America (Geist 1998) and several kangaroo species in Australia (McAlpine *et al.* 1999) are most abundant where several patch types are well dispersed.

Patch scale in heterogeneous landscapes can be crucial with regards to functionality. For example, the US Forest Service selected a dispersed patch clearcutting system for harvesting old-growth forests in the northwestern USA following World War II (Franklin & Forman 1987). Clearcut patches measuring 15–25 ha were dispersed through a pristine forest landscape that was characterized by much larger patches created by large, stand-replacement fire events. Dispersed clearcutting also generated large amounts of high-contrast edge between residual old-growth forest patches and recent cutovers, which resulted in pervasive edge influences on the environment and biota of the residual forest patches (e.g. Chen *et al.* 1995). Effectively, this harvest system greatly reduced the amount of functional interior old-growth forest habitat in the cutover landscapes and eventually produced a landscape composed of patches that were too small to accommodate many old-growth-dependent species, including the

Northern spotted owl. The result of cutting was a more heterogeneous landscape but one that was no longer functional for many organisms and processes.

Forest landscape designs should incorporate complexity and heterogeneity at multiple spatial scales

Nature tends towards complexity and heterogeneity at all spatial scales (stand through landscape), whereas humans tend towards simplicity and homogeneity at all spatial scales in managed landscapes. Hence, where a primary management objective is to sustain biodiversity and ecological processes, forest landscape design should strive to incorporate more complexity and heterogeneity at appropriate scales. This necessitates the utilization of a diversity of management approaches as well as conservation of naturally existing landscape heterogeneity; hence the admonition not to 'do the same thing everywhere in the landscape'. This is especially true when ecologists or managers subscribe to a given management paradigm as the 'right approach', for example thinning mature stands or replanting burned areas.

At the stand level, this complexity takes the form of a greater variety in structural elements and in the spatial arrangement of these structures. Richness in individual structural elements includes variability in sizes and conditions of live trees, of standing dead trees, and of tree boles lying on the forest floor (Lindenmayer *et al.* 2000; Franklin *et al.* 2002). Heterogeneity in the spatial arrangement of the structures (e.g. trees, snags and logs) adds much to the structural complexity of natural forests (Franklin and Van Pelt 2004). The high level of diversity in species and ecological processes typically associated with old-growth forests is a consequence of spatial heterogeneity in structures in both the vertical (e.g. multiple or continuous canopy layers) and horizontal (e.g. structural gaps) dimensions (Franklin *et al.* 2002; Franklin & Van Pelt 2004).

Heterogeneity at the landscape level is represented by (among other things) a diversity of patch types and conditions. Some of the patches represent areas of special ecological significance, such as those associated with aquatic features or specialized habitats. Other patches may represent areas dedicated to commodity production; hopefully these patches also incorporate sufficient internal heterogeneity in structure and composition to allow them to contribute to habitat and ecological functions along with commodities (Lindenmayer & Franklin 2002) (see below). To accommodate the needs of organisms requiring a high diversity of vegetation types at a fine spatial scale or significant area away from edge effects ('interior area'), a diversity of patch sizes, from very small to very large, may be maintained in the landscape (Spies & Turner 1999). At the small end, this may be canopy gaps or aggregates of retained trees. At the large end, large reserves of late-successional or other special habitats will aid in maintenance of specialized organisms.

Homogeneous landscapes can maximize particular functions and flows

The maximization of function is why many domesticated landscapes, such as intensive plantation forestry or commercial tree fibre farming, tend to be homogeneous. A forest landscape of dense young stands maximizes net primary productivity in the form of harvestable wood (as well as return on investment), which is the objective of industrial forestry. However, such homogeneous landscapes generally are more susceptible to forest pests and pathogens as well as other disturbances (Larsen 1995; Baleshta *et al.* 2005). They also provide limited habitat for many native forest species and may result in suboptimal levels of other important ecological processes. Maintenance of biodiversity and conservation of other functions can be negatively impacted by conversion of a compositionally and structurally diverse forest landscape to a landscape dominated by short-rotation, intensive culture of exotic tree species or a single native species. Chile and New Zealand, for example, have replaced large tracts of native forest with radiata pine and hybrid *Eucalyptus* plantations (Lara & Veblen 1993). The southeastern USA has seen the conversion of its diverse forest types to plantations of a few species of quickly growing pine species (Sharitz *et al.* 1992).

Foresters do not always recognize the homogeneity of the regulated landscape that they attempt to create (Davis *et al.* 2001). For example, they may interpret a landscape composed of 30 different age classes of loblolly pine plantation as a heterogeneous or diverse landscape. Unfortunately, such 'diversity' makes little or no contribution to species diversity or variability in ecological processes, because all of the landscape is dominated by a single structural and compositional condition.

Landscape design must reflect the influence of the geophysical template for more effective incorporation of natural patterns and processes

The geophysical template influences or controls many biologically significant processes or conditions in landscapes and communities (Swanson *et al.* 1988), and is a primary driver of heterogeneity in ecosystems at the spatial scale of regions (Kruckeberg 2002). Many of the landscape components dealt with in this chapter owe their geospatial characteristics or even existence to attributes of the geophysical template. Topographical variables (elevation, aspect and slope) modify abiotic conditions, and geology and lithology are direct contributors to the physical and chemical environment with which the biota must interact. The distribution of vegetation communities is strongly organized by gradients created by topography and lithology, such as moisture, temperature and soil gradients (Ohmann & Spies 1998; McKenzie *et al.* 2003). The topology of the hydrological network is determined largely by geophysical and climatic features (Forman 1995; Montgomery & Buffington 1998). Complex networks, such as animal trails, are frequently constrained by topography (Forman 1995). Topographical extremes, such as cliffs, equator-facing ridges, pole-facing coves, waterfall spray zones and mountain peaks often host insular communities of specialized, locally endemic or geographically disjunct biota (Warshall 1994; Kruckeberg 2002; Lindenmayer & Franklin 2002). Management at the landscape scale will more effectively conserve biodiversity when it acknowledges the underlying sources of heterogeneity due to geophysical phenomena. For example, special geoedaphic substrates, such as those derived from ultramafic or calcareous rocks, merit recognition in management plans.

Another crucial element in landscape design related to the geophysical template is the need to accommodate disturbance regimes that are responsive to landform influences (Cissel *et al.* 1999). For example, topography can partially determine the intensity and extent of wildfire (Agee 1993; Hessburg *et al.* 2005; Schulte *et al.* 2005) and the amounts of live biological legacies persisting into the post- disturbance system (Keeton & Franklin 2004). Another important element is to mitigate landform-induced positive feedback mechanisms that may exacerbate the effects of management-related disturbance. The effects of wind vary across the landscape as a function of topography (Hannah *et al.* 1995; Kramer *et al.* 2001; Harcombe *et al.* 2004; Schulte *et al.* 2005), and forest harvest and silvicultural planning may be modified in order to avoid the unintended consequences of edge-mediated windthrow (Ruel 1995; Mitchell *et al.* 2001). Inappropriate placement or orientation of timber harvest unit edges may cause adjacent forest to blow down, thus creating more edge and a progressive intensification of windthrow effects (Franklin & Forman 1987; Perry 1994). The size and location of timber harvest units in montane areas with heavy snowfall must be modified to reduce the possibility of exacerbating snow avalanches and thus reducing the amount of forested area (Weir 2002). Large harvest units in snow avalanche-prone terrain lead to disturbance to downslope forest, creating the potential for more intense avalanches and thus causing more forest loss (Weir 2002). Channel migration in major higher-order rivers is also a topography-mediated process that has implications for landscape management (Gregory *et al.* 1991; Naiman *et al.* 2000), including road design, timber harvest and location of human settlements. Harvest of riparian forests in a river channel migration zone leads to lower inputs of large woody debris into the channel, increasing the velocity of storm flows. This increases the amount of forest disturbed by bank cutting, thus reducing the capacity of the riparian forest to generate large woody material (Collins *et al.* 2002; Montgomery & Abbe 2006). Monitoring of managed forest landscapes is important to detect and quantify these feedback mechanisms and their impact on landscape condition.

The planimetric view of landscapes (from above, with no sense of topographical relief) that is characteristic of most landscape analysis often imparts a skewed perspective of the forces that influence vegetation structure and animal dispersal. Recognition of geoedaphic and topographical influences in both modelling and planning offers a host of benefits, from the conservation of endemic biota to decreased resource extraction costs, but requires a departure from traditional ways of viewing landscapes.

Landscape designs must use network and gradient paradigms as well as patch paradigms

Landscape ecology has traditionally been dominated by a patch-based paradigm, sometimes called the 'patch-matrix-corridor' model (Forman 1995). The patchwork paradigm has been useful for facilitating the adoption of landscape ecology. However, neither this model nor any other single construct can deal adequately with all aspects of landscape ecology.

Networks are important features of forest landscapes that are receiving much greater attention (see e.g. Chapter 8 in Forman 1995). Network perspectives are critical in dealing with highly connected parts of the landscape, such as hydrological systems (e.g. stream/river networks) and transportation networks. Flow paths and rates and source and sink phenomena are important aspects of such networks. The patch-based paradigm is not well adapted to such analysis of networks and their interactions. One of the most important of those interactions in forest landscapes is the interaction between the natural dendritic network of streams and the artificial rectilinear network of roads or other transportation systems (Trombulak & Frissell 2000). Trails of large mammals represent another important network type that reflects a variety of landscape-scale factors, including topography (Forman 1995; Ganskopp *et al.* 2000) and resource availability (Blake & Inkamba-Nkulu 2004).

Networks of patches, or connectivity of certain types of patches, also merit consideration in management of landscapes. Mladenoff *et al.* (1993) demonstrate the loss of connectivity between certain stand types in a forested landscape affected by timber harvest when compared with a relatively undisturbed landscape dominated by old-growth.

Gradients also have paradigmatic significance in landscape design (McGarigal & Cushman 2005). Gradients take many forms in landscapes, including classical gradients such as those associated with climate and with landforms. For example, regional landscapes often have well-defined macroclimatic gradients that reflect interactions between dominant atmospheric flows (e.g. continental or marine air masses) and mountain ranges. On a more localized basis within mountainous topography, climatic gradients are often associated with variability in elevation, slope and aspect. As a consequence, systematic gradient-related variability within patches of a given type (e.g. old-growth forest) can be expected. Gradients also exist within patches as a consequence of high-contrast edges (Chen *et al.* 2006).

Patch content matters: viewing patches as 'shades-of-grey'

An unfortunate consequence of reliance on the patch-matrix-corridor paradigm has been that it often encourages adoption of a dichotomous view of the landscape – a division of the landscape into habitat and non-habitat. This results in part from conservation biologists merging the patch paradigm of landscape ecology with island biogeography theory. What typically emerges is a vision of undisturbed (and often reserved) patches ('islands') of suitable habitat within a matrix ('ocean') of unsuitable habitat. The landscape is effectively divided into black and white – suitable and unsuitable habitat.

In fact, patch functionality – including habitat suitability – typically represents a spectrum or continuum, rather than a condition of absolute suitability or unsuitability. Ecosystem scientists, with their focus on ecosystem functions – such as hydrological regulation or net primary productivity – intuitively recognize this continuum of functionality. Conservation biologists traditionally have not recognized habitat as a continuum, perhaps partially because of preferences for natural patches and antipathy towards human-manipulated patches.

Adopting a vision of patch functionality as a potential continuum – shades of grey, rather than black or white – allows managers to design landscapes that sustain organisms and processes over the entire landscape. Such an approach is essential, given the overwhelming importance of the matrix or working landscapes in most regions of the world (Lindenmayer & Franklin 2002). Patch content – the internal condition of patches – becomes a universally important concern once one accepts the premise that landscapes are not simply dichotomous patchworks of habitat and non-habitat, functional and dysfunctional. Sole reliance upon a 'black and white' perspective of landscape functionality will be inimical to achieving most conservation objectives. The use of continuous variables representing quantifiable patch content characteristics may often be more useful than a Boolean classification of habitat vs. non-habitat.

Landscape design must focus on the matrix and its content

The final landscape design principle follows from the preceding discussion, and it is the need for a holistic view of the landscape. This holistic landscape vision specifically includes conditions on lands that are managed for production of commodities, sometimes referred to as the **matrix** (Lindenmayer & Franklin 2002; Fischer *et al.* 2006).

Matrix has various definitions. In landscape ecology it is typically defined as the dominant and most extensive patch type (Forman 1995). In the conservation and forest planning literature, areas that are not dedicated primarily to nature conservation are often referred to as the matrix, the perspective being that ecological reserves are typically habitat islands within a managed landscape or matrix.

Lindenmayer and Franklin (2002) define the matrix as:

> landscape areas that are not designated primarily for conservation of natural ecosystems, ecological processes, and biodiversity, regardless of their current condition (i.e., whether developed or natural).

They then proceed to describe the critical ecological roles of the matrix in: (i) supporting populations of species; (ii) regulating the movement of organisms; (iii) buffering sensitive areas and reserves; and (iv) maintaining the integrity of aquatic ecosystems.

The landscape design principle is that the entire landscape has to be considered in developing a design that allows for integration of and provision for conservation of biological diversity and production of various commodities, such as wood, for human needs. This is done by approaching landscape design in a series of steps that includes (Lindenmayer & Franklin 2002; Lindenmayer *et al.*, 2006):

- establishment of large ecological reserves;
- identification and protection of important specialized habitats within the matrix, including aquatic ecosystems, specialized habitats and biological hotspots;
- conservation of important structural features, such as snags and logs, in managed forest patches;
- careful consideration of the spatial and temporal patterns of timber harvest, including internal structure and size of harvest units and rotation periods; and
- planning of transportation networks, particularly with regards to interactions with drainage networks.

The ultimate goal in such design is the creation of landscapes that provide for the full array of human objectives by integrating structural goals at all spatial scales, including both the scale of the overall landscape with its patchworks and networks of land allocations and the internal structure of the patches, including the patches allocated to commodity objectives.

Conclusions

Issues related to landscape structure, degradation and condition are complex, and defy strict characterization. However, some general principles may be applied to the management of landscapes towards the objective of conserving biodiversity and landscape function.

These themes may be profitably included in landscape management for simultaneous commodity production and conservation of biodiversity. In a world where forest lands are coming under increasing pressure to fulfil seemingly incongruous roles, such as commodity production and provision of ecological services, the concepts and principles presented in this essay will prove useful to the manager attempting to manage for a diverse set of products and values.

General principles for landscape design

1 General principles for interpreting effects of specific landscape patterns on ecological functions are scarce.
2 Forest landscape designs should incorporate complexity and heterogeneity at multiple spatial scales.

3 Homogeneous landscapes can maximize particular functions and flows.
4 Landscape design must reflect the influence of the geophysical template for more effective incorporation of natural patterns and processes.
5 Landscape designs must utilize network and gradient (concepts) as well as patch paradigms.
6 Patch content matters: viewing patches as 'shades of grey'.
7 Landscape design must focus on the matrix and its content.

References

Agee, J.K. (1993) *Fire Ecology of Pacific Northwest forests*. Island Press, Washington, DC.

Baleshta, K.E., Simard, S.W., Guy R.D. & Chanway C.P. (2005) Reducing paper birch density increases Douglas-fir growth rate and Armillaria root disease incidence in southern interior British Columbia. *Forest Ecology and Management* **208**, 1–13.

Blake, S. & Inkamba-Nkulu, C. (2004) Fruit, minerals, and forest elephant trails: do all roads lead to Rome? *Biotropica* **36**, 392–401.

Chen, J., Franklin, J.F. & Spies, T.A. (1995) Growing season macroclimatic gradients from clearcut edges into old-growth Douglas-fir forests. *Ecological Applications* **5**, 74–86.

Chen, J., Saunders, S.C., Brosofske, K.D. & Crow, T.R. (2006) *Ecology of Hierarchical Landscapes: From Theory to Application*. Nova Science Publishers, Hauppauge, NY.

Cissel, J.H., Swanson, F.J. & Weisberg, P.J. (1999) Landscape management using historical fire regimes: Blue River, Oregon. *Ecological Applications* **9**, 1217–1231.

Collins, B.D., Montgomery, D.R. & Haas, A.D. (2002) Historical changes in the distribution and functions of large wood in Puget Lowland rivers. *Canadian Journal of Fisheries and Aquatic Sciences* **59**, 66–76.

Davis, L.S., Johnson, K.N., Bettinger, P. & Howard, T.E. (eds.) (2001) *Forest Management to Sustain Ecological, Economic, and Social values*, 4th edn. Waveland Press, Long Grove, IL.

Fischer, J., Lindenmayer, D.L. & Manning, A.D. (2006) Biodiversity, ecosystem function, and resilience: ten guiding principles for commodity production landscapes. *Frontiers in Ecology and the Environment* **4**, 80–86.

Forman, R.T.T. (1995) *Land Mosaics: The Ecology of Landscapes and Regions*. Cambridge University Press, Cambridge, UK.

Franklin, J.F. (2005) Spatial pattern and ecosystem function: reflections on current knowledge and future directions. In: Lovett, G.M. *et al.* (eds.) *Ecosystem Functions in Heterogeneous Landscapes*, pp. 427–441. Springer-Verlag, New York.

Franklin, J.F. & Forman, R.T.T. (1987) Creating landscape patterns by forest cutting: ecological consequences and principles. *Landscape Ecology* **1**, 5–18.

Franklin, J.F. & Spies, T.A. (1991) Composition, function, and structure of old-growth Douglas-fir forests. In: Ruggiero, L.F., Aubry, K.B., Carey, A.B. & Huff, M.H. (eds.) *Wildlife and Vegetation of Unmanaged Douglas-fir Forests*, pp. 71–83. General Technical Report PNW-GTR-285, US Department of Agriculture Forest Service. Pacific Northwest Research Station, Portland, OR.

Franklin, J.F. & Van Pelt, R. (2004) Spatial aspects of structural complexity in old-growth forests. *Journal of Forestry* **102**, 22–27.

Franklin, J.F., Spies, T.A., Van Pelt, R. *et al.* (2002) Disturbances and structural development of natural forest ecosystems with silvicultural implications, using Douglas-fir forests as an example. *Forest Ecology & Management* **155**, 399–423.

Ganskopp, D., Cruz, R. & Johnson, D.E. (2000) Least-effort pathways? A GIS analysis of livestock trails in rugged terrain. *Applied Animal Behaviour Science* **68**, 179–190.

Geist, V. (1998) *Deer of the World: Their Evolution, Behaviour, and Ecology*. Stackpole Books, Mechanicsburg, PA.

Gregory, S.V., Swanson, F.J., McKee, W.A. & Cummins, K.W. (1991) An ecosystem perspective of riparian zones. *Bioscience* **41**, 540–551.

Hannah, P., Palutikof, J.P. & Quine, C.P. (1995) Predicting wind-speeds for forest areas in complex terrain. In: Coutts, M.P. & Grace, J. (eds.) *Wind and Trees*, pp. 113–129. Cambridge University Press, Cambridge.

Harcombe, P.A., Greene, S.E., Kramer, M.G. *et al.* (2004) The influence of fire and windthrow dynamics on a coastal spruce-hemlock forest in Oregon, USA, based on aerial photographs spanning 40 years. *Forest Ecology and Management* **194**, 71–82.

Hessburg, P.F., Agee, J.K. & Franklin, J.F. (2005) Dry forests and wildland fires of the inland Northwest USA: contrasting the landscape ecology of the pre-settlement and modern eras. *Forest Ecology and Management* **211**, 117–139.

Keeton, W.S. & Franklin, J.F. (2004) Fire-related landform associations of remnant old-growth trees in the southern Washington Cascade Range. *Canadian Journal of Forest Research* **34**, 2371–2381.

Kramer, M.G., Hansen, A.J., Taper, M.L. & Kissinger, E.J. (2001) Abiotic controls on long-term windthrow disturbance and temperate rain forest dynamics in southeast Alaska. *Ecology* **82**, 2749–2768.

Kruckeberg, A.R. (2002) *Geology and Plant Life*. University of Washington Press, Seattle, WA.

Lara, A. & Veblen, T.T. (1993) Forest plantations in Chile: a successful model? In: Mather, A. (ed.) *Afforestation, Policies, Planning, and Progress*, pp. 118–139. Belhaven Press, London.

Larsen, J.B. (1995). Ecological stability of forests and sustainable silviculture. *Forest Ecology and Management* **73**, 85–96.

Lindenmayer, D.B. & Franklin, J.F. (2002) *Conserving Forest Biodiversity: a Comprehensive Multiscaled Approach*. Island Press, Washington, DC.

Lindenmayer, D.B., Cunningham, R.B., Donnelly, C.F. & Franklin, J.F. (2000) Structural features of old-growth Australian montane ash forests. *Forest Ecology and Management* **134**, 189–204.

Lindenmayer, D.B., Franklin, J.F., & Fischer, J. (2006) General management principles and a checklist of strategies to guide forest biodiversity conservation. *Biological Conservation* **129**, 511–518.

McAlpine, C.A., Grigg, G.C., Mott, J.J. & Sharma, P. (1999) Influence of landscape structure on kangaroo abundance in a disturbed semi-arid woodland of Queensland. *Rangeland Journal* **21**, 104–134.

McGarigal, K. & Cushman, S.A. (2005). The gradiant concept of landscape structure. In: Wiens, J.A. and Moss, M.R. *Issues and Perspectives in Landscape Ecology*. 112–119. Cambridge University Press, Cambridge, UK.

McKenzie, D., Peterson, D.W., Peterson, D.L. & Thornton, P.E. (2003) Climatic and biophysical controls on conifer species distributions in mountain forests of Washington State, USA. *Journal of Biogeography* **30**, 1093–1108.

Mitchell, S.J., Hailemariam, T. & Kulis, Y. (2001) Empirical modeling of cutblock edge windthrow risk on Vancouver Island, Canada, using stand level information. *Forest Ecology and Management* **154**, 117–130.

Mladenoff, D.J., White, M.A., Pastor, J. & Crow, T.R. (1993) Comparing spatial pattern in unaltered old-growth and disturbed forest landscapes. *Ecological Applications* **3**, 294–306.

Montgomery, D.R. & Abbe, T.B. (2006) Influence of logjam-formed hard points on the formation of valley-bottom landforms in an old-growth forest valley, Queets River, Washington, USA. *Quaternary Research* **65**, 147–155.

Montgomery, D.R. & Buffington, J.M. (1998) Channel processes, classification, and response. In: Naiman, R.J. & Bilby, R.E. (eds.) *River Ecology and Management: Lessons From the Pacific Coastal Ecoregion*, pp. 13–42. Springer-Verlag, New York.

Naiman, R.J., Bilby, R.E. & Bisson, P.A. (2000) Riparian ecology and management in the Pacific coastal rain forest. *Bioscience* **50**, 996–1011.

Ohmann, J.L. & Spies, T.A. (1998) Regional gradient analysis and spatial pattern of woody plant communities of Oregon forests. *Ecological Monographs* **68**, 151–182.

Perry, D.A. (1994) *Forest Ecosystems*. The Johns Hopkins University Press, Baltimore, MD.

Ruel, J.C. (1995) Understanding windthrow: silvicultural implications. *Forestry Chronicle* **71**, 434–445.

Schulte, L.A., Mladenoff, D.J., Burrows, S.N. *et al.* (2005) Spatial controls of pre-Euro-American wind and fire disturbance in northern Wisconsin (USA) forest landscapes. *Ecosystems* **8**(1), 73–94.

Sharitz, R.R., Boring, L.R., Van Lear, D.H. & Pinder, J.E.I. (1992) Integrating ecological concepts with natural resource management of southern forests. *Ecological Applications* **2**, 226–237.

Shrader-Frechette, K.S. & McCoy, E.D. (1993) *Method in Ecology: Strategies For Conservation*. Cambridge University Press, Cambridge.

Spies, T.A. & Turner, M.E. (1999) Dynamic forest mosaics. In: Hunter, M.L. (ed.) *Maintaining Biodiversity in Forest Ecosystems*, 95–160. Cambridge University Press, Cambridge, UK.

Swanson, F.J., Kratz, T.J., Caine, N. & Woodmansee, R.G. (1988) Landform effects on ecosystem patterns and processes. *Bioscience* **38**, 92–98.

Swetnam, T.W., Allen, C.D. & Betancourt, J.L. (1999) Applied historical ecology: using the past to manage for the future. *Ecological Applications* **9**, 1189–1206.

Trombulak, S.C. & Frissell, C.A. (2000) Review of ecological effects of roads on terrestrial and aquatic communities. *Conservation Biology* **14**, 18–30.

Warshall, P. (1994) The Madrean sky island archipelago: a planetary overview. In: Debano, L.F., Gottfried, G.J., Hamre, R.H. *et al.* (tech. co-ordinators) *Biodiversity and Management of the Madrean Archipelago: the Sky Islands of Southwestern United States and Northwestern Mexico*, pp. 6–18. Rocky Mountain Forest and Range Experiment Station Technical Report No. RM-GTR-264. USDA Forest Service, Fort Collins, CO.

Weir, P. (2002) *Snow Avalanche: Management in Forested Terrain*. Land Management Handbook No. 55. Research Branch, British Columbia Ministry of Forests, Victoria, BC.

13

Synthesis: Structure, Degradation and Condition

David B. Lindenmayer and Richard J. Hobbs

The three essays in Chapters 10–12 – touch on the massive topic of Structure, Degradation and Condition. As with all of the sections in the book, the three essays are notable for the diverse perspectives that overlap little but are highly complementary in their content. In common with the other themes tackled at the meeting in Bowral, the group discussions on this topic were lively, often controversial and raised more questions than they answered.

In his chapter, Woinarski (Chapter 10) reminds us of the well-known relationships between structural complexity and species richness – an important consideration given that many forms of human land use (e.g. livestock grazing, forestry) can greatly simplify vegetation structure and significantly alter vegetation condition. A more difficult and far less tractable issue is that of appropriate benchmarks for assessing vegetation condition. The concept of benchmarks is complicated by various problems.

- Vegetation structure is dynamic. How do we determinine what is 'natural', and how appropriate is this concept in many landscapes long influenced by human activities.
- The existence of various alternative states of vegetation condition and associated other biotic assemblages. For example, some species will be associated with long-undisturbed vegetation whereas others will be early post-disturbance specialists.
- Difficulties in developing robust metrics for quantifying departure from benchmarks and the tenuous links between these metrics (in a generic form) and biota.
- Different human perspectives on what is appropriate (or suitable) vegetation structure and condition.

- Temporal issues, such as at what time in the past would it be appropriate to set a given benchmark condition for a given type of vegetation – pre-arrival of European people? pre-arrival of indigenous people? (see also Hunter Jr., Chapter 35).

In Chapter 28, Burgman *et al.* raise further issues concerning benchmarks, and in particular, the ways that metrics are calculated to assess vegetation condition. Each of these kinds of issues highlights the potential for differences in what the 'preferred state' of the vegetation might be at a site and, in turn, the need for careful consideration of the issues associated with labelling the condition of vegetation cover as 'good', 'bad', 'disturbed', 'healthy', etc. Here science and values intersect and explicit recognition of the dual features of these concepts is required. Despite such inherent complexity, these are nevertheless important issues to tackle, not least because appraisals of vegetation structure and condition are fundamental to assessments of the suitability of areas for: land clearing versus conservation (see Gibbons *et al.*, Chapter 19); targets for attempted restoration efforts; and using natural disturbance as a template to guide human disturbance (e.g. logging – see Hunter Jr., Chapter 35). One approach to tackling some of these problems is to limit attempts to manage towards a given benchmark or endpoint and embrace a philosophy of continuous improvement as part of a management trajectory.

In his chapter, Driscoll (Chapter 11) explores relationships between vegetation structure, patch condition and key human-derived mechanisms of change, and provides short summaries of the effects of grazing, fire and invasive species. He concludes that the array of species with different requirements in all ecosystems demands that many different parts of any given ecosystem be managed via a range of different management regimes – akin to the 'don't do the same thing everywhere' principle. Driscoll argues that mosaics of different patches (in different condition and characterized by different internal structure) will be important for conservation management (see also Bennett & Radford, Chapter 18), although what kinds of mosaics and how they should be arranged remains unclear. He laments both the failure of some ecological theories associated with vegetation condition and structure (e.g. the intermediate disturbance hypothesis) and the failure better to link theory and practice in assessing relationships between vegetation condition and biodiversity response. Driscoll suggests that an increased use of management-by-experiment approaches is an important way to test existing theories and improve knowledge to manage landscapes better. Unfortunately the paucity of adaptive management studies (see Simberloff, Chapter 26) and almost non-existent current commitments to monitoring in virtually every jurisdiction worldwide undermine the potential for management-by-experiment to generate the much-needed knowledge to improve resource management and conservation actions.

Franklin and Swanson (Chapter 12) cover remarkably little similar ground to that of the preceding authors, although they touch on issues, covered by Woinarski (Chapter 10), associated with determining departure from benchmarks and issues in identifying what is 'natural' in a historical context. These authors highlight the importance of the content of patches for landscape management and also the need for landscape heterogeneity – different kinds of patches across the landscape. Franklin and Swanson also stress the role of the matrix in contributing to ecosystem functions and the persistence of biodiversity. But through matrix management and its contribution to both structural complexity within patches and increased heterogeneity across landscapes, traditional binomial perspectives of habitat and non-habitat in landscapes become blurred (see also Chapters 2–5), and models like the island model and patch-corridor-matrix model that underpin landscape classifications become problematic.

Section 4
Edge Effects

14

Incorporating Edge Effects into Landscape Design and Management

Thomas D. Sisk

Abstract

Landscape design and management attempt to anticipate and mitigate the effects of human activities on the composition and structure of landscapes serving diverse and often conflicting purposes. Natural landscape heterogeneity is typically increased by anthropogenic activities, and one of the most pervasive results is the proliferation of habitat edges. A long history of research on edge effects has led to the general impression that they are frustratingly idiosyncratic and inconsistent. However, recent advances suggest that the direction – if not the magnitude – of edge effects can be predicted; and spatial models allow managers to explore the likely effects of edges on focal species and key ecological processes. These tools provide a new capacity for anticipating changes in animal abundance and ecological processes near edges, while exploring the consequences of alternative landscape designs. By combining species- and process-level understanding with spatial data describing real and hypothetical future landscapes, managers can integrate consideration of edge effects into their decisions and improve the likelihood that sensitive species and key ecosystem services will be conserved as society places increasing demands on a finite land base.

Keywords: edge response; effective area model; habitat heterogeneity; planning; spatial modelling.

Introduction

Land planning and management involve the identification of different landscape features, assessment of appropriate land uses, and allocation of land to different purposes to meet diverse and often conflicting objectives. This process almost inevitably leads to fragmentation of native habitats and increases the complexity of landscape mosaics. Concomitant with fragmentation of native habitats and the subsequent allocation of lands to different, continually evolving land uses, the diversity and amount of edge habitat increases. Because the ecological effects of edges are many, highly variable and difficult to quantify, we often give them less consideration in landscape design than we give to the effects of decreasing patch size or increasing isolation of habitat fragments. Yet edge effects are often pronounced, and they can affect habitat quality and ecosystem processes well into the interiors of adjoining habitat patches. Because the demands of expanding human populations are likely to result in ever more finely subdivided landscapes with increasingly intensive management, landscape designers cannot afford to treat edge effects as an afterthought. Edges are ubiquitous, highly influential elements of most landscapes, and they are among the landscape features most amenable to management.

Wherever two or more different habitat types abut, they form an edge, or ecotone, defined by ecological gradients created by the transition from one relatively homogeneous habitat to another. Human manipulation of terrestrial ecosystems has dramatically increased the abundance and altered the nature of these transitions, as ownership and management boundaries are imposed on native landscapes, fostering changes in land use that create anthropogenic habitats, increase fragmentation and transform gradual transitions between habitats into sharp edges. These transformations of the landscape mosaic lead to the proliferation of edges as larger patches are fragmented into multiple, smaller and irregularly shaped patches with high ratios of edge to interior habitat (Sisk & Margules 1993).

The ecological effects of edges have been studied for many decades, and the creation of edge habitat was a central objective of many wildlife managers through much of the 20th century, because it was thought that edges increased species diversity and abundance (Leopold 1933; Dasmann 1981). More recently, as ongoing habitat fragmentation has increased the amount of edge exponentially, the negative effects of edges on biodiversity have become well recognized (Murcia 1995; Robinson *et al.* 1995). This evolving perspective among landscape ecologists regarding edge effects highlights the complexity of edge effects and the challenges of integrating their consideration into landscape design efforts (Sisk & Haddad 2002).

To simplify the challenges associated with the study of edge effects, ecologists often concentrate on particular habitat types or 'focal patches', relegating the rest of the landscape to a category of non-habitat, often called the 'matrix'. Yet this matrix is usually composed of many different land uses and habitat types, each potentially exerting different effects via their edges with the 'focal patches'. This conceptual simplification of complex habitat mosaics to binary habitat–matrix landscapes has proven helpful in isolating particular questions, developing null hypotheses and making habitat fragmentation amenable to more focused research (Fahrig 2003). However, landscape planners are often unable to follow suit; they are tasked with integrating many diverse and often competing land uses into a unified plan for intelligent management. Landscape design, therefore, must address the entire landscape, in all its complexity, including the proliferation of different types of habitat edges and their influence on ecological processes and species abundances (Fig. 14.1). From an applied perspective, explicit consideration and incorporation of edge effects in landscape design efforts is essential.

While edge effects are among the most studied of ecological phenomena, the development of practical tools to predict the effects of fragmentation and to design appropriate mitigation efforts has progressed slowly (Saunders *et al.* 1991; Wiens 1995). Reviews of the hundreds of papers published on habitat edges have highlighted their highly variable effects, while ignoring or failing to identify the underlying mechanisms that generate them (Murcia 1995; Sisk & Battin 2002). Until recently, few general patterns had been identified, and landscape planners often lacked the understanding and predictive capacity required to inform landscape design efforts. The whole topic of edge effects seemed intractable and idiosyncratic (Ehrlich 1997).

The emerging mechanistic understanding of edge effects

In recent years, research advances in distinct but related fields have begun to generate a more unified perspective on edge effects and their role in landscape

(a) (b)

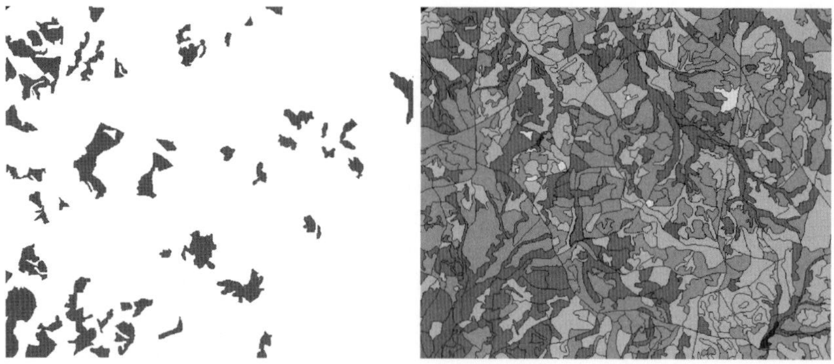

Figure 14.1 **Old-growth pine stands at Fort Benning, Georgia, USA, constitute important habitat for sensitive species in highly altered landscapes. (a) A simplified view of edges between habitat and non-habitat in a 'fragmented' landscape. (b) Edges are more complex when the same landscape is viewed in the context of a mosaic of different patches. Studies of habitat fragmentation and the resulting isolation of patches often assume a binary landscape with patches of habitat embedded in a matrix of non-habitat (a). In reality, the focal habitat patches are embedded in a complex mosaic of different habitats, each creating unique edge types and influences on the focal patches (b). Landscape design and management must deal with the complexity of real landscapes, because differing values and land uses must be addressed in an integrated manner that supports intelligent management. Reproduced from Ries *et al.* 2004, with permission.**

composition and function. By focusing on distinct classes of effects, rather than on the myriad variations described from particular field studies, Fagan *et al.* (1999) provided a theoretical framework for examining the mechanisms underlying edge effects. At the same time, recent empirical studies demonstrate emerging techniques for quantifying biophysical aspects of edge effects (e.g. Chen *et al.* 1999), and studying their effects on organisms (e.g. Meyer & Sisk 2001). These developments have, in turn, stimulated the formulation of general models of edge effects that organize disparate empirical observations around an emerging mechanistic understanding, and provide a capacity for predicting edge responses and their effects on a wide range of organisms and ecological processes (Ries and Sisk 2004).

Ries *et al.* (2004) provide an overarching model for edge responses that synthesizes from an extensive literature four fundamental influences on organismal abundance patterns near habitat edges (Fig. 14.2). These mechanistic

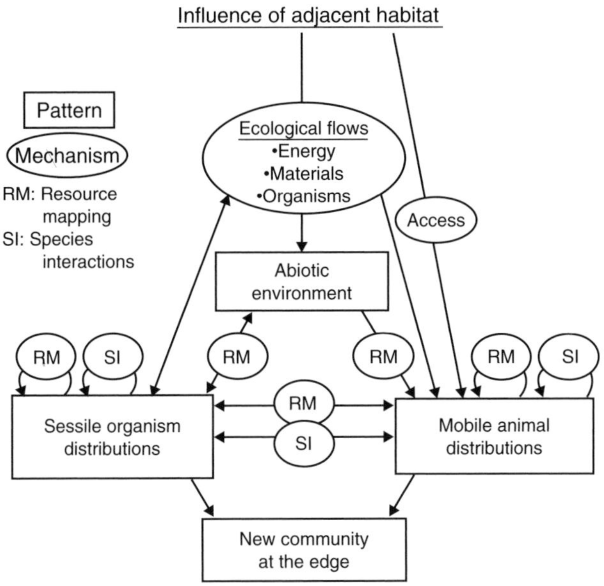

Figure 14.2 **A mechanistic model describing four pathways by which edge effects may arise. Patterns in the abiotic environment, the distribution of organisms and community structure (boxes) are affected by four fundamental mechanisms (ovals). Ecological flows of energy, material and organisms across the edge influence the abiotic environment as well as organismal distributions. Mobile organisms whose resources are spatially separated can gain better access by being near edges. Organisms map onto changes in the distribution of their resources. Changes in distributions near edges can lead to novel species interactions, further influencing patterns in abundance and distribution. And the aggregated effects of changes in distribution and abundances lead to altered community structure near edges. Reproduced from Ries *et al.* 2004, with permission.**

influences are (i) ecological flows of organisms or resources, which can be amplified, attenuated or reflected by edges (Strayer *et al.* 2003); (ii) access to spatially separated resources, which is highest near the edge; (iii) resource mapping, whereby the distribution and abundance of organisms track gradients in resources and abiotic conditions; and (iv) species interactions, where edges bring into proximity species that otherwise would not interact, sometimes with profound ecological consequences. Clarifying the state of our understanding of how these mechanisms influence the distribution of organisms and community structure near habitat edges provides a framework for addressing the potential consequences of particular landscape designs and management plans. Gradually, this integration of mechanistic explanations is replacing the seemingly idiosyncratic view of edge effects with a clearer conceptual model.

Predictive models of edge responses

A mechanistic understanding of edge effects provides some insight into where and under what conditions edge effects are likely to manifest, but it does not provide the predictive power needed to anticipate specific effects on particular species or ecological processes. Until recently, no models had been developed that would allow a planner to predict, for example, how the abundance of a particular species or changes in ambient temperature might be altered near habitat edges. Ries and Sisk (2004) developed and tested a simple model based on the relative habitat quality of adjacent patches and the distribution of resources within each patch. Species with strong habitat associations show the most consistent edge responses, but the nature of those responses varies, depending on how resources are distributed in the landscape. For example, two North American bird species show consistent edge responses. Ovenbirds (*Seiurus aurocapillus*) are dependent on forest habitat, and they consistently show negative responses near edges with less suitable habitat. Brown-headed cowbirds (*Moluthrus ater*) forage in open habitat, but lay their eggs in the nests of forest-dwelling songbirds. Key resources in the open habitat complement different resources in the forest habitat, so being near the boundary between these two habitats offers convenient access to both critical resources. As might be expected, cowbirds consistently show positive edge responses. In contrast, if two bordering habitats are both used by a species, but resources are available in both (so the resources in one habitat supplement the resources in the other), no edge response is expected, because being near the edge offers no advantage. Finally, in some cases, resources are known to be concentrated along the edge, and species dependent on those resources often show increased density near edges. Shrub-dependent birds, for example, often are found at increased densities near forest edges that develop a shrub layer where the latter is absent or rare in either bordering habitat. These four situations constitute the basis for a predictive model of edge effects. When little is known about edge responses at a particular edge type, this model can be used to anticipate the effects of edges under alternative landscape designs and management plans. To apply this model, the only information needed is the species' habitat associations in each bordering habitat type, and information about the distribution of principal resources.

Tests of the model, drawing on the results from over 260 published papers, indicate that the direction of change in species abundance was predictable for the majority of species in four taxa (birds, butterflies, mammals and plants), based on patterns in the distribution of key resources among adjacent habitat patches. More detailed models of edge responses, drawing on extensive life-history attributes for well-studied species, also show promise for moving our understanding of edge responses from a descriptive to a predictive footing (e.g. Luck *et al.* 1999; Brand *et al.* 2004, 2006).

Modelling edge effects at the landscape level

Knowledge of edge responses provides important insight into the effects of landscape structure on organisms and ecological processes, but use of this insight in planning requires tools for extrapolating from specific responses to complex edge configurations in real landscapes. Temple (1986) was perhaps the first to predict species' responses to landscape pattern based on edge effects. His early Core Area Model drew on empirical data concerning the penetration of nest predators into forest fragments to assess the impacts of changing land use on 'fragmentation-sensitive' species. Laurance and Yensen (1991) added realism to the Core Area approach by applying it to complex patch shapes and allowing variation in edge responses. Early Core Area Models focused on 'interior species' that were negatively associated with habitat edges and, therefore, sensitive to fragmentation. Expansion of the core-area approach to include any combination of habitats, and any species or environmental parameter of interest, is the objective of Effective Area Modelling approaches (Sisk & Margules 1993; Sisk *et al.* 1997). Like the core-area approach, Effective Area Models assume that the amount of suitable habitat, from the perspective of any species in the landscape, is generally different from the collective area of the patches of its habitat(s). In contrast to the concept of 'core area', the effective area may be larger (for edge exploiters) or smaller (for edge avoiders). Thus, the effective area of a particular patch may differ from species to species, depending on the degree to which proximity to edges enhances or degrades habitat quality. Similarly, other response variables (e.g. microclimatic edge effects) can be modelled, providing a way to generate predictions about the effects of landscape structure on the distributions of organisms, resources or environmental conditions in heterogeneous landscapes. The rapid development of geographical information systems (GIS) technology and the availability of remotely sensed data allow Effective Area Models to combine edge responses with digital maps of real and hypothetical landscapes, making it feasible to explore the implications of alternative landscape designs on a wide range of species and ecological processes (Sisk *et al.* 2002).

Current capabilities and remaining challenges

A better understanding of the mechanisms underlying edge effects, the ability to predict edge responses for many species and abiotic variables, and the development of tools for modelling edge effects across heterogeneous landscapes make it possible for landscape planners to incorporate edge effects into practical design efforts. While each of these developments is ongoing and in need of continuing scientific attention, it seems likely that edge effects need no longer be considered an idiosyncratic or intractable source of 'noise' in landscape models. Initial efforts to model edge effects have generated important insights into ecological dynamics relevant to landscape design, but many considerable challenges remain. A few of the areas where knowledge is currently rudimentary, and research investments are most likely to generate practical advances, are discussed below.

Improving mechanistic understanding

Despite over 1000 published studies of edge effects, our understanding of the mechanisms that cause edge effects is still incomplete. Only recently have reviewers begun to synthesize these studies and identify the causes of edge effects (Fagan *et al.* 1999; Sisk & Battin 2002; Ries *et al.* 2004). Even less is known about how edges affect community organization (Bender *et al.* 1998). Work is needed to identify under what circumstances, and for which species, the various mechanisms causing edge effects are likely to be important.

Effects of edge orientation

Edges often comprise structural features that make their effects sensitive to their orientation to abiotic factors, such as insolation and prevailing winds (Chen *et al.* 1999). Differences between east- and west-facing edges, for example, have been shown to affect arthropod movements (Meyer & Sisk 2001), and it is likely that edge orientation mediates the influence of other biotic and abiotic effects. Similarly, the review by Ries *et al.* (2004) found that edge effects tend to be stronger at south-facing vs. north-facing edges in the northern hemisphere; the opposite is likely to be true in the southern hemisphere, although data to test this hypothesis are lacking. Orientation of edges may emerge as a central issue in landscape design, but a better understanding is needed.

Effects of edge contrast

Some edges exhibit greater contrast in habitat structure than others. Some researchers have noted that high-contrast edges, such as those between forests and grasslands, tend to show more pronounced edge effects than low-contrast edges, such as those between different grasslands (e.g. Yahner *et al.* 1989). Empirical studies show variable results. Because edge contrast is potentially amenable to management, better understanding of its influences on the magnitude of edge effects is desirable.

Understanding interacting influences of multiple edges

Most of the factors described above affect the strength of edge effects, rather than the nature or direction (positive or negative) of their influence. Orientation or contrast may make edge effects stronger or weaker, but they are unlikely to transform a positive effect into a negative one. Likewise, when multiple edges interact to influence habitat quality in a particular location, each is likely to be manifest in a predictable way (Kolbe & Janzen 2002). But because the nature of their combined influence is poorly understood at present, the magnitude of their aggregated effect – although likely to be large – is harder to predict (Fletcher 2005) (Fig. 14.3). Because landscapes worldwide are becoming more fragmented and intensively managed, the issue of multiple edge effects deserves greater attention than it currently receives.

Meeting or taming the demands of data-hungry landscape modelling approaches

Even the simplest approaches for modelling edge effects in complex landscapes pose daunting challenges for parameter estimation, especially when

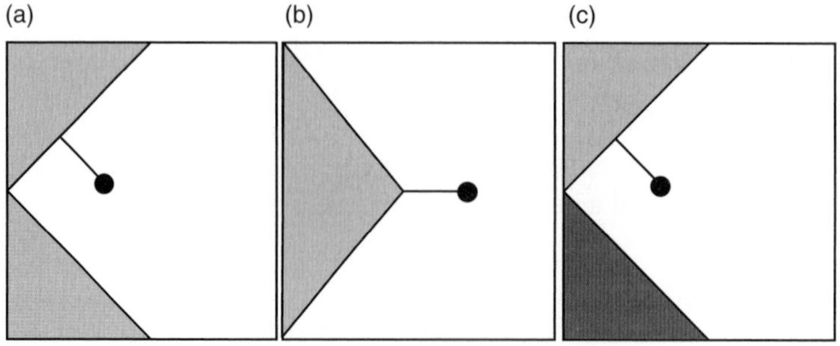

Figure 14.3 **Multiple edges often exert complex influences on habitat quality and biophysical processes (e.g. microclimatic conditions). Even when distance to the nearest edge is held constant, as in the three panels shown, the magnitude of the edge effect may differ depending on whether the orientation of the edges tends to amplify (a) or attenuate (b) the effects at a particular location. Where different edge types interact (c), not only the magnitude but the direction of the effect (positive or negative) may change (Fletcher & Koford 2003). Efforts to apply our limited knowledge of edge effects to landscape-level design problems require an improved understanding of the effects of multiple, interacting edges. Reproduced from Ries *et al.* 2004, with permission.**

multiple species are modelled in heterogeneous landscapes. Unless ecologists can streamline models and reduce the variance associated with estimation, propagation of error threatens to swamp insights from spatially explicit modelling approaches.

Predicting edge responses for poorly known species

Neither ecologists nor managers will be able to measure the edge responses of every species or process at each type of edge in a landscape. Thus, general rules are needed to allow the prediction of edge responses for many poorly studied phenomena. Species' edge responses, for example, are almost certainly related to aspects of life history, demography and/or behaviour. Some progress has been made in mobilizing life-history information to predict edge responses, but further advances are needed for managers to address edge effects efficiently, without unrealistic demands for detailed empirical data for each species at every conceivable edge. Advances in this area would facilitate the application of Core Area and Effective Area Models to landscapes that have not been the focus of previous research efforts.

Predicting effects of edges on community composition and ecosystem function

We currently know little about how the effects of edges on ecological processes and species abundance influence landscape-level patterns in community composition and ecosystem function. Most of the research and most landscape design efforts have focused on population size or the interactions between species. A major challenge for edge studies is to provide tools for scaling up from individual and populational responses to the effects of edges on community assembly and ecosystem functioning.

Principles for addressing edge effects in landscape design

Although tools for explicit consideration and quantification of edge effects in landscape design are in their infancy, research on edge responses and early applications of emerging spatial planning tools have resulted in some key insights of practical value for landscape planners.

While the principles outlined here fall short of a prescriptive set of guidelines for managing edges in heterogeneous landscapes, their consideration and thoughtful application can reduce complex edge effects to a level of 'manageable complexity' and provide practical guidance to managers. Given the pervasive influence of edges on many organisms and ecological processes, their effects on biological diversity should not be ignored in landscape design and planning efforts. While the variety of edge responses and possible future landscape configurations make the prediction of edge effects challenging, landscape modelling tools can help managers anticipate ecological changes near habitat edges and assess the cumulative effects of all edge types in landscapes undergoing rapid change in composition and structure.

Principles

1 **Edges are omnipresent features of managed landscapes,** and decades of research show that they often have strong effects on habitat quality, the distribution of organisms and a wide range of ecological processes. They should be considered explicitly in landscape design efforts, but even the limited research information that is currently available is underutilized.

2 **Edges are amenable to management;** they often are among the landscape features that can be directly manipulated by managers. While patch area and habitat types are often constrained by economic and logistical factors, the nature and position of edges are more easily altered and therefore provide design opportunities.
3 **Most species show variable responses to edges based on the characteristics of particular edge types.** The common perception that organisms conform to broad response categories, such as 'edge' and 'interior' species, is overly simplistic. Natural history and the history of empirical ecology provide useful, often detailed information to guide landscape design.
4 **The direction of many edge responses (i.e. positive, negative, neutral) is consistent, and to some extent predictable,** despite the great variability and uncertainty reported in the literature. Particular species show similar responses to similar edge types.
5 **Recent research has identified several approaches for predicting the edge responses of lesser-studied species.** For mobile species, the distribution of resources across the edge can explain considerable variation in responses; examination of life-history characteristics can increase predictive power considerably.
6 **Many mechanisms that generate edge effects have been identified and studied in isolation. Combined, they can generate complex responses that result in high uncertainty about the magnitude of response.** But as noted above, the direction of edge responses is much more consistent.
7 **Extrapolation from specific edge responses, which are manifested at the level of the organism, to landscape-level effects requires spatial models dependent on landscape maps.** While quantitative prediction of broadscale edge effects is fraught with uncertainty, current modelling tools are sufficient to compare alternative landscape designs and management scenarios, and their likely outcomes can be assessed and compared.

Acknowledgements

Sincere thanks to James Battin, Arriana Brand, Haydee Hampton, Barry Noon and especially Leslie Ries, all of whom have helped develop the ideas presented in this chapter. Much of the research that underlies the points presented here was supported by the Strategic Environmental Research and Development Program (SERDP), under projects CS 1100 and SI 1597.

References

Bender, D.J., Contreras, T.A. & Fahrig, L. (1998) Habitat loss and population decline: A meta-analysis of the patch size effect. *Ecology* **79**, 517–533.

Brand, L.A. (2004) Prediction and assessment of edge response and abundance for desert riparian birds in southeastern Arizona. PhD Thesis, Colorado State University, Fort Collins, CO.

Brand, L.A., Noon, B.R. & Sisk, T.D. (2006) Predicting abundance of desert riparian birds: validation and calibration of the effective area model. *Ecological Applications* **16**, 1090–1102.

Chen, J., Saunders, S.C., Crow, T.R. *et al.* (1999) Microclimate in forest ecosystem and landscape ecology. *BioScience* **49**, 288–297.

Dasmann, R. (1981) *Wildlife Biology*. John Wiley & Sons, New York.

Ehrlich, P.R. (1997) *A World of Wounds: Ecologists and the Human Dilemma*. Ecology Institute, Oldendorf/Luhe, Germany.

Fagan, W.E., Cantrell, R.S. & Cosner, C. (1999) How habitat edges change species interactions. *American Naturalist* **153**, 165–182.

Fahrig, L. (2003) Effects of habitat fragmentation on biodiversity. *Annual Review of Ecology and Systematics* **34**, 487–515.

Fletcher, R.J. & Koford, R.R. (2003) Spatial responses of bobolinks (Dolichonyx oryzivorus) near different types of edges in northern Iowa. Auk **120**, 799–810.

Fletcher, R.J. (2005) Multiple edge effects and their implications in fragmented landscapes. *Journal of Animal Ecology* **74**, 342–352.

Kolbe, J.J. & Janzen, F.J. (2002) Spatial and temporal dynamics of turtle nest predation: Edge effects. *Oikos* **99**, 538–544.

Laurance, W.F. & Yensen, E. (1991) Predicting the impacts of edge effects in fragmented habitats. *Biological Conservation* **55**, 77–97.

Leopold, A. (1933) *Game Management*. John Wiley & Sons, New York.

Luck, G.W., Possingham, H.P. & Paton, D.C. (1999) Bird responses at the inherent and induced edges in the Murray mallee, South Australia. 1. Differences in abundance and diversity. *Emu* **99**, 157–169.

Meyer, C. & Sisk, T.D. (2001) Butterfly responses to microclimatic changes following ponderosa pine restoration. *Restoration Ecology* **9**, 453–461.

Murcia, C. (1995) Edge effects in fragmented forests: Implications for conservation. *Trends in Ecology and Evolution* **10**, 58–62.

Ries, L., Fletcher, R.J., Jr., Battin, J. & Sisk, T.D. (2004) Ecological responses to habitat edges: mechanisms, models and variability explained. *Annual Review of Ecology, Evolution and Systematics* **35**, 491–522.

Ries, L. & Sisk, T.D. (2004) A predictive model of edge effects. *Ecology* **85**, 2917–2926.

Robinson, S.K., Thompson, F.R., III, Donovon, T.M., Whitehead, D.R. & Faaborg J. (1995) Regional forest fragmentation and the nesting success of migratory birds. *Science* **267**, 1987–1990.

Saunders, D.A., Hobbs, R.J. & Margules, C.R. (1991) Biological consequences of ecosystem fragmentation – a review. *Conservation Biology* **5**, 18–32.

Sisk, T.D. & Battin, J. (2002) Habitat edges and avian habitat: Geographic patterns and insights for western landscapes. *Studies in Avian Biology* **25**, 30–48.

Sisk, T.D. & Haddad, N.M. (2002) Incorporating the effects of habitat edges into landscape models: Effective area models for cross-boundary management. In: Liu, J. & Taylor, W. (eds.) *Integrating Landscape Ecology into Natural Resource Management*, Pp. 208–240. Cambridge University Press, Cambridge.

Sisk, T.D. & Margules, C.R. (1993) Habitat edges and restoration: Methods for quantifying edge effects and predicting the results of restoration efforts. In: Saunders, D.A., Hobbs, R.J. & Ehrlich, P.R. (eds.) *Nature Conservation 3: Reconstruction of Fragmented Ecosystems*, pp. 57–69. Surrey, Beatty & Sons, Sydney.

Sisk, T.D., Haddad, N.M. & Ehrlich, P.R. (1997) Bird assemblages in patchy woodlands: modeling the effects of edge and matrix habitats. *Ecological Applications* **7**, 1170–1180.

Sisk, T.D., Noon, B.R. & Hampton, H.M. (2002) Estimating the effective area of habitat patches in variable landscapes. In: Scott, M.J., Heglund, P., Morrison, M.L. *et al.* (eds.) *Predicting Species Occurrences: Issues of Accuracy and Scale*, Pp. 713–725. Island Press, Washington, DC.

Strayer, D.L., Power, M.E. & Fagan, W.F. (2003) A classification of ecological boundaries. *BioScience* **53**, 723–729.

Temple, S.A. (1986) Predicting impacts of habitat fragmentation on forest birds: A comparison of two models. In: Verner, J., Morrison, M.L. & Ralph, C.J. (eds.) *Wildlife 2000: Modeling Habitat Relationships of Terrestrial Vertebrates*, pp. 301–304. University of Wisconsin Press, Madison, WI.

Wiens, J.A. (1995) Habitat fragmentation: Island vs. landscape perspective on bird conservation. *Ibis* **137**, S97–S104.

Yahner, R.H., Morrell, T.E. & Rachel, V. (1989) Effects of edge contrast on depredation of artificial avian nests. *Journal of Wildlife Management* **53**, 1135–1138.

15

Edge Effects

David B. Lindenmayer and J. Fischer

Abstract

Edges can be a prominent part of landscapes that are subject to human alteration. They can have profound impacts and may be classified in various ways, including (among others): biotic/abiotic; soft/hard; and natural/human-derived. Biotic edge effects are characterized by marked variations in response between species, between vegetation types and between regions. It remains unclear how it will be possible accurately to predict which ecosystems, vegetation communities and individual vegetation species will be most susceptible to edge effects. Recent conceptual models offer promise but await further detailed testing. In the absence of a better understanding of edge effects, and of well-tested models, attempts to mitigate edge effects might be best tackled through traditional approaches such as buffers, the management of spatial patterns of vegetation cover (to limit the length of human-created boundaries) and targeted management of the matrix surrounding vegetation patches that are potentially susceptible to edge effects (e.g. to limit levels of structural and physical contrast between them).

Keywords: abiotic edge effects; biotic edge effects; buffering systems; edge effects; edges; edge sensitivity; models for edge effects.

Introduction

A common change in landscape pattern resulting from landscape alteration is the increase in the length of boundaries or edges between remaining patches of original vegetation and the surrounding matrix. Edge effects can occur at such boundaries and have profound impacts on assemblages of taxa (Laurance 2000; Siitonen *et al.* 2005). This essay discusses abiotic and biotic edge effects, some factors that are related to the variability of edge effects, and the relationship between edge effects and species extinctions. The concluding section focuses on landscape principles designed to limit edge effects.

Types of edges and edge effects

The edge of a vegetation patch can be broadly defined as a marginal zone of altered microclimatic and ecological conditions that contrasts with its interior (Matlack 1993). A range of edge types has been recognized. One way of classifying edges is by their origin – natural or human-derived (Luck *et al.* 1999). Edges usually occur naturally at the interface of two ecological communities. Such natural boundaries are sometimes termed ecotones, and can support distinctive plant and animal communities (Witham *et al.* 1991).

In modified landscapes, many edges are created by humans. Examples are the boundaries between recently clearcut forest and adjacent unlogged stands (Chen *et al.* 1990) or between cropped areas and remnant native vegetation (Sargent *et al.* 1998). Edges can be 'soft', where the transition between different types of patches is gradual; or 'hard', at boundaries with marked contrasts in vegetation structure and other features (e.g. clearfelled–unlogged forest; Forman 1995).

Edge effects refer to changes in biological and physical conditions that occur at an ecosystem boundary and within adjacent ecosystems (Wilcove 1985). The ecological edge that results from a disturbance is the result of interactions between the kind and intensity of the disturbance event and the ecological dynamics within the adjacent, undisturbed environment. In a major review of edge effects in forest landscapes, Harper *et al.* (2005) classified edge effects as either primary responses that arise directly from edge creation or secondary responses that arise indirectly as a result of edge creation. Primary responses include structural damage to the vegetation, disruption of the forest floor and soil layer, altered nutrient cycling and decomposition, changed evaporation, and altered pollen and seed dispersal (Harper *et al.* 2005). Secondary or indirect responses (that flow from primary responses) include patterns of plant growth, regeneration, reproduction and mortality, and are manifested as altered patterns of vegetation structure and species composition (Harper *et al.* 2005). The intensity of edge effects, or the area of a patch subject to significant edge influence depend on the parameter of interest (Fig. 15.1).

Another way of classifying edge effects is by the impacts they have on climate and abiotic processes (e.g. altered wind penetration) or on biota (e.g. altered levels of predation). These two broad kinds of effects, abiotic edge effects and biotic edge effects, are the subject of the following two sections.

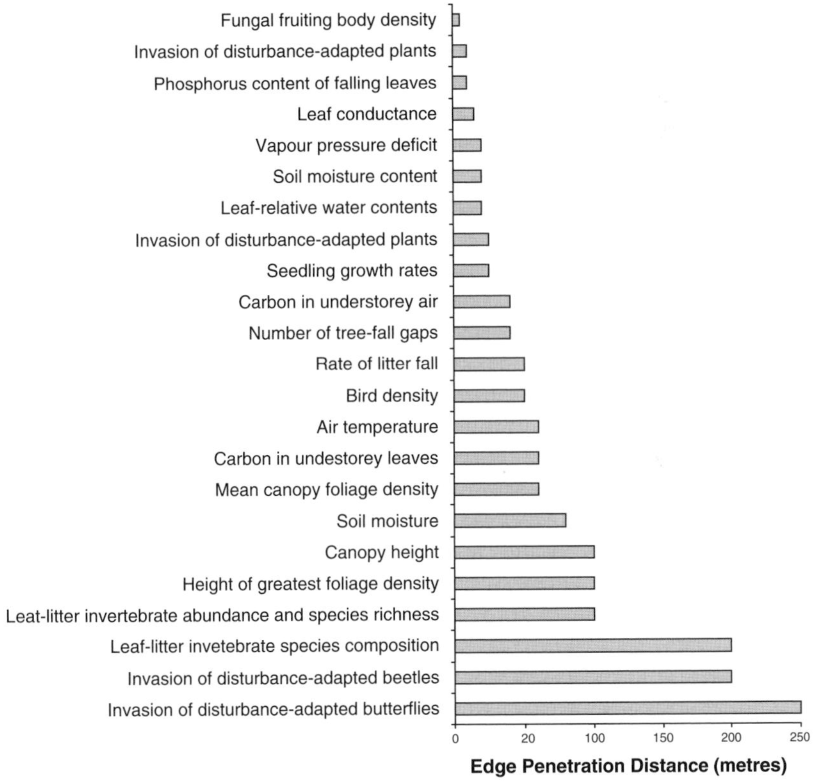

Figure 15.1 **Variation in edge penetration for a range of measures in the Biological Dynamics of Forest Fragments Project in Brazil. Modified and redrawn from Laurance *et al.* 1997.**

Abiotic edge effects

Edges may experience microclimatic changes such as increased temperature and light, or decreased humidity that extend tens or hundreds of metres from an edge, depending on the environmental variable, the physical nature of the edge and weather conditions (reviewed by Saunders *et al.* 1991). Table 15.1 lists a range of abiotic edge effects.

Biotic edge effects

Biological factors that affect ecological communities across a boundary, such as diseases, weeds and predators, may penetrate hundreds of metres into vegetation remnants (e.g. Wilcove *et al.* 1986). Thus, edge effects can significantly influence the distribution and abundance of assemblages of species that inhabit vegetation remnants (Fletcher 2005) (Table 15.2).

Changes in vegetation communities

Edge environments can affect reproduction, growth, seed dispersal and mortality in plants (Hobbs & Yates 2003). For example, Chen (1991) observed increased reproduction and growth of surviving mature trees in old-growth forests bordering recently clearfelled areas in the Pacific Northwest of the USA. Similarly, weed invasion is a major biotic edge effect in many heavily disturbed landscapes (Brothers & Spingarn 1992). Altered microclimatic conditions at

Table 15.1. **Examples of abiotic edge effects.**

Modification of wind speeds
Modification of temperature regimes and humidity
Modification of light fluxes
Changes in localized precipitation and frost intensity
Altered fire ignition probabilities
Altered levels of nutrients, chemicals
Accelerated levels of windthrow and disrupted landscape connectivity between vegetation remnants

Table 15.2 **Examples of biotic edge effects.**

Increased nest predation and brood parasitism in birds and reptiles
Impaired breeding success in birds and other animals
Altered patterns of animal and plant dispersal
Altered patterns of behaviour such as nest building and food gathering
Lowered rates of fledging success among birds
Reduced habitat quality for individual species
Increased numbers of browsing and/or game animals
Increased human hunting of large vertebrates
Altered invertebrate community composition
Altered levels of insect activity
Increased levels of weed invasion
Altered density, reproduction, growth and mortality in native plants
Altered composition of soil-borne bacterial and fungal populations

edges may make conditions particularly favourable for the growth of non-native plants (Honnay *et al.* 2002). Regenerating vegetation and patch edges often experience a seed rain of environmental weeds and other introduced plants that are frequently better adapted to exposed and disturbed environments (Janzen 1983).

Changes in animal communities

A range of studies have demonstrated the existence of altered animal community composition at edges (e.g. Hansson 1998) including assemblages of invertebrates (e.g. Magara 2002). Some vertebrates avoid forest edges and are classified as 'interior forest' species (Gates & Gysel 1978). Some forest birds fall into this category (Terborgh 1989), in part because arthropod densities may be lower near edges (Zanette & Jenkins 2000) or because foraging efficiency may be impaired by edge conditions (Huhta *et al.* 1999). Another common reason for species to avoid edges is that higher abundances of predators and parasites at edges can have negative impacts on prey and host species near edges (Chalfoun *et al.* 2002). For example, bird populations inhabiting edge environments can experience lower rates of successful pairings and impaired breeding success compared with populations of the same taxa occupying other parts of a landscape (Chalfoun *et al.* 2002). Frequently studied predators in this context include corvids, small mustelids and other small mammals (e.g. red squirrel *Tamiasciurus hudsonicus*) as well as foxes (*Vulpes* spp.) and the badger (*Meles meles*). The most frequently studied nest parasite is the brown-headed cowbird (*Molothrus ater*) in North America (Zanette *et al.* 2005).

In the northern hemisphere, some game species have strong positive preferences for edge environments where foraging and cover are close together (Matlack & Litvaitis 1999). As an example, large populations of white-tailed deer (*Odocoileus virginianus*) heavily graze food plants at forest edges (Johnson *et al.* 1995) including endangered taxa (Miller *et al.* 1992). This can limit stand regeneration (Tilghman 1989) and alter patterns of species diversity (McShea & Rappole 2000).

Variation in edge effects

Regional variation

Strong spatial variation in edge effects has been recognized by a range of authors (Rudnicky & Hunter 1993) and has even been used to hypothesize which ecosystems and vegetation communities are likely to be most prone to edge effects (Harper *et al.* 2005). Elevated nest predation and rates of nest parasitism show strong patterns of variation between locations (Batary & Baldi 2004). However, while such effects are commonly associated with edges, the patterns are far from universal (reviewed by Kremsater & Bunnell 1999). Forest edges on the boundary of clearcuts in Sweden have lower bird species diversity than continuous forest areas (Hansson 1983), but no such pattern has been detected at some clearcut forest edges in the northeastern USA (Rudnicky & Hunter 1993). Similarly, patterns of brood parasitism characteristic of edge environments in many temperate northern hemisphere landscapes appear to be rare in Australia (Piper *et al.* 2002) and are less pronounced in central Europe than they are in northwestern Europe (Batary & Baldi 2004). They are also not prominent in many tropical landscapes (Stratford & Robinson 2005). Even within North America, increased nest predation and brood parasitism as observed on the eastern side of the continent is less common in the northwest (Kremsater & Bunnell 1999). The reasons underpinning regional differences in edge effects are not clear. Regional differences in vegetation types might account for some of the observed differences.

The influence of the matrix

The magnitude of many types of edge effects is related to the strength of the contrast between the matrix and other landscape units; where the contrast is strong there will often be more intense interactions and edge effects (Laurance *et al.* 1997). The extent of the area supporting high-contrast conditions influences the magnitude of edge effects. For example, microclimate edge effects in forests may be greater where a large clearfelled area abuts a retained patch than where a cutover is small (Lindenmayer *et al.* 1997). Elevated nest predation is often observed in agricultural landscapes where there are strong contrasts between vegetation remnants and the surrounding environment (Andrén 1992).

Edge sensitivity and extinction proneness

Edge effects vary markedly between species (Schlaepfer & Gavin 2001), and some authors have argued that edge-sensitive species may be among those at particular risk in heavily modified environments (Lehtinen *et al.* 2003). However, relationships between extinction-proneness and edge responses have almost never been demonstrated empirically. A notable exception is the work of Lehtinen *et al.* (2003), who found that edge-avoiding species of reptiles and frogs in Madagascar were less likely to persist within forest remnants than taxa that were insensitive to boundaries with the surrounding modified landscape matrix.

Predicting edge effects

To date it appears there is limited predictive ability in attempts to forecast accurately which species will be edge-sensitive in landscapes subject to significant human modification (Murcia 1995). This appears to be a result of several factors. First, there are markedly different results from studies of edge effects from, for example, different ecosystem types (e.g. forests vs. woodland), similar ecosystem types in different regions (e.g. central and western North America; Kremsater and Bunnell 1999), the same type of ecosystem but where the surrounding matrix is different (Bayne & Hobson 1997) and different species in the same ecosystem (Lehtinen *et al.* 2003). In addition, even within the same local area, abiotic edge effects can vary between aspects

(Hylander 2005) and seasons (Schlaepfer & Gavin 2001), and the response of individual species can change markedly as a result (De Maynadier & Hunter 1998). Similarly, the shape of vegetation patches as well as the distance from the edge may influence the intensity of edge effects. Some authors have argued that ecologists often lack sufficient appreciation of the range of spatial and temporal complexities that characterize the type and magnitude of edge effects (e.g. Murcia 1995).

A second factor limiting predictive ability lies with the use of markedly different methods in different investigations. This leads to major problems in making rigorous meta-analytical comparisons across investigations (Murcia 1995). A third problem is that many factors influence the distribution and abundance of organisms, leading to high variability in datasets and resulting statistical relationships.

Although some authors have argued that there currently appears to be a limited ability to predict edge-sensitive ecosystems, vegetation communities and species, a recent review proposed an alternative view. Ries *et al.* (2004) introduced a conceptual model of edge effects that considered four mechanisms underlying ecological changes near edges: changed ecological flows; species accessing spatially separated resources; resource mapping; and species interactions. Using these four broad categories as a conceptual basis, Ries *et al.* (2004) argued that ecological responses to edges were far less idiosyncratic and unpredictable than often suggested. Rather, the correct prediction of edge effects depended on the careful definition of the types of processes under consideration (Ries *et al.* 2004).

Caveats

It is often forgotten in studies of edge effects that humans define patch boundaries (see Chapter 4). These 'boundaries' may or may not be relevant to particular species or processes. What actually defines a patch boundary for a particular species will be defined by that species. For example, edge effects might mean that a 60-ha area of native forest contains only 30 ha of suitable habitat for a species that avoids edge environments. In other cases, individuals of a species will move well beyond a patch boundary into the adjacent matrix to gather additional resources – sometimes termed landscape supplementation or a 'halo' effect (Tubelius *et al.* 2004).

There are other important considerations in the study and interpretation of edge effects. First, because the nature and strength of edge effects are species specific (Hansson 2002), not all elements of an assemblage will respond to boundaries in the same way. Some will be edge avoiders, others may be interior avoiders, and others will respond to boundaries in neither a positive nor a negative way (e.g. Noss 1991). Moreover, many kinds of abiotic edge effects are likely to display marked temporal variability. As climate regimes vary throughout the year, including at boundaries, biotic edge effects also may exhibit seasonal changes. For example, in Madagascar, Lehtinen *et al.* (2003) found that a suite of reptiles and amphibians that avoided edges during the dry season showed no response to boundaries in the wet season.

Finally, edge effects are sometimes regarded as univariate entities, but in fact many factors can influence the magnitude and nature of edge effects, including the existence of multiple edges in the same local area (Fletcher 2005). As an example, in Australia aggressive interactions at edges between small native honeyeaters and the noisy miner (*Manorina melanocephala*) are influenced by the body size of small honeyeaters, remnant patch size and the density of understorey vegetation within areas of remnant native vegetation (Piper *et al.* 2002).

Landscape design principles

1 **Human landscape modification often results in an increase in perimeter-to-area ratios of remaining patches of native vegetation.** The boundaries created between remaining vegetation patches and the surrounding matrix can be subject to edge effects. Two broad types of edge effects are recognized – abiotic and biotic edge effects. Abiotic edge effects are primarily concerned with altered microclimatic conditions (e.g. modification of wind speeds, light fluxes and temperature regimes). Biotic edge effects may refer to changes in ecological processes, community composition or species interactions. Examples of biotic edge effects include changes in the rate of organic matter decomposition, increased nest predation and brood parasitism, and lowered rates of fledging success among birds with territories located at edges. Not all species respond negatively to edges and some taxa can be more common at patch boundaries than elsewhere in a landscape.

2 **The magnitude of edge effects varies in response to a number of factors such as the type of ecosystem and the level of contrast between**

remaining patches and the surrounding matrix. There is no generic edge depth, and the intensity of edge effects will vary with the processes and species in question. Edge effects also vary between locations – some types of edge effects such as increased nest parasitism found in the mid-western USA and northern Europe appear to be less important elsewhere (e.g. western North America, Australia, central Europe and tropical environments).

3 **It remains unclear if it is possible to predict accurately which ecosystems, vegetation communities and individual species are most susceptible to edge effects.** However, recent conceptual models that attempt to capture the range of processes, responses and interspecies interactions that can occur across patch boundaries indicate that edge effects may be more predictable than many authors had earlier proposed.

4 **There are many examples that demonstrate that remnants of pre-modification ecosystems may not maintain their species composition because of edge effects** (reviewed by Kremsater and Bunnell 1999). Buffers are areas that surround and protect sensitive areas better to conserve the species within them (Hylander *et al.* 2004). Typical aims of buffers may be to limit the impacts of a disturbance regime on native ecosystems, or maximize native species richness within a protected area (Baker 1992). What comprises a suitable buffer will depend on many factors such as the type of process that generates an edge effect, and the taxa or other attributes that need protection. For example, the width of buffer strips to protect riparian and aquatic areas from pesticides and to reduce instream invertebrate mortality needs to exceed 50 m in Australian eucalypt plantations (Barton & Davies 1993). In contrast, buffers may need to be several hundred metres wide effectively to mitigate changed wind patterns (Harris 1984). Recommended buffer widths can vary significantly between regions, even for the same group of species (Hagar 1999).

5 **Edge effects typically arise as a consequence of human alterations in spatial patterns of vegetation cover.** It follows then that one approach to mitigating edge effects is to try to manage spatial patterns of landscape vegetation cover to limit the length of human-created boundaries between patch types. Such approaches are perhaps best developed in forested landscapes subject to logging and associated road building for the transport of wood. Planning the spatial arrangement of harvest

units so they are aggregated rather than dispersed has the potential to limit the extent and magnitude of negative edge effects (Franklin & Forman, 1987).

6 The magnitude of many kinds of edge effects is often strongly associated with the level of contrast in physical, structural and other conditions between vegetation remnants, reserves, etc. and the surrounding matrix. It follows then that management strategies in the matrix that attempt to limit the level of contrast between remnant vegetation and the surrounding landscape should reduce the magnitude of many (but not always all) of the kinds of abiotic and biotic edge effects that might otherwise develop (Rodriguez *et al.* 2001). In the case of commodity (forestry and agricultural) production landscapes, several studies have shown that land-use practices that produce structural characteristics in the matrix similar to those found in retained areas of vegetation will reduce the magnitude of edge effects (as well as other kinds of effects).

Acknowledgements

This work was sponsored through grants from the Kendall Foundation, Land and Water Australia and the Australian Research Council. The authors have benefited greatly from past collaborative work with Dr Adrian Manning and Professor Ross Cunningham.

References

Andrén, H. (1992) Corvid density and nest predation in relation to forest fragmentation: a landscape perspective. *Ecology* **73**, 794–804.

Baker, W.L. (1992) The landscape ecology of large disturbances in the design and management of nature reserves. *Landscape Ecology* **7**, 181–194.

Barton, J.L. & Davies, P.E. (1993) Buffer strips and streamwater contamination by atrazine and pyrethrenoids aerially applied to *Eucalyptus nitens* plantations. *Australian Forestry* **56**, 201–210.

Batary, P. & Baldi, A. (2004) Evidence of an edge effect on avian nest success. *Conservation Biology* **18**, 389–400.

Bayne, E.M. & Hobson, K.A. (1997) Comparing the effects of landscape fragmentation by forestry and agriculture on predation of artificial nests. *Conservation Biology* **11**, 1418–1429.

Brothers, T.S. & Spingarn, A. (1992) Forest fragmentation and alien plant invasion of central Indiana old-growth forests. *Conservation Biology* **6**, 91–100.
Chalfoun, A.D., Thompson, F.R. & Ratnaswamy, M.J. (2002) Nest predators and fragmentation: a review and meta-analysis. *Conservation Biology* **16**, 306–318.
Chen, J. (1991) Edge effects: microclimatic pattern and biological responses in old-growth Douglas fir forests. PhD Thesis, University of Washington, Seattle, WA.
Chen, J., Franklin, J.F. & Spies, T.A. (1990) Microclimatic pattern and basic biological responses at the clearcut edges of old-growth Douglas Fir stands. *Northwest Environmental Journal* **6**, 424–425.
De Maynadier, P. & Hunter, M. (1998) Effects of silvicultural edges on the distribution and abundance of amphibians in Maine. *Conservation Biology* **12**, 340–352.
Fletcher, R.J. (2005) Multiple edge effects and their implications in fragmented landscapes. *Journal of Animal Ecology* **74**, 342–352.
Forman, R.T. (1995) *Land Mosaics. The Ecology of Landscapes and Regions*, 1st edn. Cambridge University Press, New York.
Franklin, J.F. & Forman, R.T. (1987) Creating landscape patterns by forest cutting: ecological consequences and principles. *Landscape Ecology* **1**, 5–18.
Gates, J.E. & Gysel, L.W. (1978) Avian nest dispersion and fledging success in field-forest ecotones. *Ecology* **59**, 871–883.
Hagar, J.C. (1999) Influence of riparian buffer width on bird assemblages in Western Oregon. *Journal of Wildlife Management* **63**, 484–496.
Hansson, L. (1983) Bird numbers across edges between mature conifer and clearcuts in central Sweden. *Ornis Scandinavia* **14**, 97–103.
Hansson, L. (1998) Vertebrate distributions relative to clear-cut edges in a boreal forest landscape. *Landscape Ecology* **9**, 105–115.
Hansson, L. (2002) Mammal movements and foraging at remnant woodlands inside coniferous forest landscapes. *Forest Ecology and Management* **160**, 109–114.
Harper, K.A., Macdonald, S.E., Burton, P.J. *et al.* (2005) Edge influence on forest structure and composition in fragmented landscapes. *Conservation Biology* **19**, 768–782.
Harris, L.D. (1984) *The Fragmented Forest: Island Biogeography Theory and the Preservation of Biotic Diversity*, 1st edn. University of Chicago Press, Chicago.
Hobbs, R.J. & Yates, C.J. (2003) Impacts of ecosystem fragmentation on plant populations: generalising the idiosyncratic. *Australian Journal of Botany* **51**, 471–488.
Honnay, O., Verheyen, K. & Hermy, M. (2002) Permeability of ancient forest edges for weedy plant species invasion. *Forest Ecology and Management* **161**, 109–122.
Huhta, E., Jokimäki, J. & Rahko, P. (1999) Breeding success of Pied Flycatchers in artificial forest edges: the effect of a suboptimally shaped foraging area. *The Auk* **116**, 528–535.
Hylander, K. (2005) Aspect modifies the magnitude of edge effects on bryophyte growth in boreal forests. *Journal of Applied Ecology* **42**, 518–525.
Hylander, K., Nilsson, C. & Gőthner, T. (2004) Effects of buffer-strip retention and clearcutting on land snails in boreal riparian forests. *Conservation Biology* **18**, 1052–1062.

Janzen, D.H. (1983) No park is an island: increase in interference from outside as park size decreases. *Oikos* **41**, 402–410.

Johnson, A.S., Hale, P.E., Ford, W.M. *et al.* (1995) White-tailed deer in relation to successional stage, overstorey type and management of southern Appalachian forests. *American Midland Naturalist* **133**, 18–35.

Kremsater, L. & Bunnell, F.L. (1999) Edge effects: theory, evidence and implications to management of western North American forests. In: Rochelle, J., Lehmann, L.A. & Wisniewski, J. (eds.) *Forest Wildlife and Fragmentation: Management Implications*, 1st edn., pp. 117–153. Brill, Leiden, Germany.

Laurance, W.F. (2000) Do edge effects occur over large spatial scales? *Trends in Ecology and Evolution* **15**, 134–135.

Laurance, W.F., Bierregaard, R.O., Gascon, C. *et al.* (1997) Tropical forest fragmentation: synthesis of a diverse and dynamic discipline. In: Laurance, W.F. & Bierregaard, R.O. (eds.) *Tropical Forest Remnants. Ecology, Management and Conservation of Fragmented Communities*, 1st edn, pp. 502–525. University of Chicago Press, Chicago.

Lehtinen, R.M., Ramanamanjato, J-B. & Raveloarison, J.G. (2003) Edge effects and extinction proneness in a herpetofauna from Madagascar. *Biodiversity and Conservation* **12**, 1357–1370.

Lindenmayer, D.B., Cunningham, R.B. & Donnelly, C.F. (1997) Tree decline and collapse in Australian forests: implications for arboreal marsupials. *Ecological Applications* **7**, 625–641.

Luck, G.W., Possingham, H.P. & Paton, D.C. (1999) Bird responses at inherent and induced edges in the Murray mallee, South Australia. 1. Differences in abundance and diversity. *Emu* **99**, 157–169.

McShea, W.J. & Rappole, J.H. (2000) Managing the abundance and diversity of breeding bird populations through manipulation of deer populations. *Conservation Biology* **14**, 1161–1170.

Magara, T. (2002) Carabids and forest edge: spatial pattern and edge effect. *Forest Ecology and Management* **157**, 23–37.

Matlack, G.R. (1993) Microenvironment variation within and among forest edge sites in the eastern United States. *Biological Conservation* **66**, 185–194.

Matlack, G.R. & Litvaitis, J.A. (1999) Forest edges. In: Hunter, M., III (ed.) *Managing Biodiversity in Forest Ecosystems*, 1st edn., pp. 210–233. Cambridge University Press, Cambridge.

Miller, S.G., Bratton, S.P. & Hadidian, J. (1992) Impacts of white-tailed deer on endangered and threatened vascular plants. *Natural Areas Journal* **12**, 67–74.

Murcia, C. (1995) Edge effects on fragmented forests: implications for conservation. *Trends in Ecology and Evolution* **10**, 58–62.

Noss, R.F. (1991) Landscape connectivity: different functions at different scales. In: Hudson, W.E. (ed.) *Landscape Linkages and Biodiversity*, 1st edn., pp. 27–39. Island Press, Wahington, D.C.

Piper, S., Catterall, C.P. & Olsen, M. (2002) Does adjacent land use affect predation of artificial shrub-nests near eucalypt forest edges? *Wildlife Research* **29**, 127–133.

Ries, L., Fletcher, R.J., Battin, J. & Sisk, T.D. (2004) Ecological responses to habitat edges: mechanisms, models, and variability explained. *Annual Review of Ecology, Evolution and Systematics* **35**, 491–522.

Rodríguez, A., Andrén, H. & Jansson, G. (2001) Habitat-mediated predation risk and decision-making of small birds at forest edges. *Oikos* **95**, 383–396.

Rudnicky, T.C. & Hunter, M.L. (1993) Avian nest predation in clearcuts, forests, and edges in a forest-dominated landscape. *Journal of Wildlife Management* **57**, 358–364.

Sargent, R.A., Kilgo, J.C., Chapman, B.R. & Miller, K.V. (1998) Predation of artificial nests in hardwood fragments enclosed by pine and agricultural habitats. *Journal of Wildlife Management* **62**, 1438–1442.

Saunders, D.A., Hobbs, R.J. & Margules, C.R. (1991) Biological consequences of ecosystem fragmentation: a review. *Conservation Biology* **5**, 18–32.

Schlaepfer, M.A. & Gavin, T.A. (2001) Edge effects on lizards and frogs in tropical forest fragments. *Conservation Biology* **15**, 1079–1090.

Siitonen, P., Lehtinen, A. & Siitonen, M. (2005) Effects of forest edges on the distribution, abundance and regional persistence of wood-rotting fungi. *Conservation Biology* **19**, 250–260.

Stratford, J.A. & Robinson, W.D. (2005) Gulliver travel to the fragmented tropics: geographic variation in the mechanisms of avian extinction. *Frontiers in Ecology and Environment* **3**, 91–98.

Terborgh, J. (1989) *Where Have all the Birds Gone*?, 1st edn. Princeton University Press, Princeton, NJ.

Tilghman, N.G. (1989) Impacts of white–tailed deer on forest regeneration in northwestern Pennsylvania. *Journal of Wildlife Management* **53**, 524–532.

Tubelius, D.P., Lindenmayer, D.B., Saunders, D.A., Cowling, A. & Nix, H.A. (2004) Landscape supplementation provided by an exotic matrix: implications for bird conservation and forest management in a softwood plantation system in southeastern Australia *Oikos* **107**, 634–644.

Wilcove, D.S. (1985) Nest predation in forest tracts and the decline of migratory songbirds. *Ecology* **66**, 1211–1214.

Wilcove, D.S., McLellen, C.H. & Dobson, A.P. (1986) Habitat fragmentation in the temperate zone. In: Soulé, M.E. (ed.) *Conservation Biology: The Science of Scarcity and Diversity*, 1st edn., pp. 237–256. Sinauer, Sunderland, MA.

Witham, T.G., Morrow, P.A. & Potts, B.M. (1991) Conservation of hybrid plants. *Science* **254**, 779–780.

Zanette, L. & Jenkins, B. (2000) Nesting success and nest predators in forest fragments: a study using real and artificial nests. *The Auk* **117**, 445–454.

Zanette, L., MacDougall-Shakleton, E., Clinchy, M. & Smith, J.N. (2005) Brown-headed cowbirds skew host offspring sex ratios. *Ecology* **86**, 815–820.

16

Edges: Where Landscape Elements Meet

Gary W. Luck

Abstract

An edge can be defined as the place where landscape elements meet. Structural change in landscape elements may be accompanied by changes in ecosystem function, commonly referred to as edge effects. Edges have been a prominent topic of exploration in landscape ecology, particularly in the last two to three decades as research focused on the implications of landscape fragmentation. Species' responses to edges vary and may be taxa and context specific, although some researchers have been successful in predicting spatial patterns in edge response based on species–habitat relationships and resource abundance. Despite some success in this area, great uncertainty remains regarding the circumstances under which particular edge effects (which ultimately drive the patterns in distribution) manifest themselves. More work needs to be done in comparing the same variables across different edge types in the same landscape, with greater emphasis on temporal, as opposed to spatial, changes. These factors will influence the strength (although possibly not the direction) of edge responses and the penetration distance of associated edge effects. Owing to the complexity of ecological processes occurring across landscapes it is very difficult to identify widely applicable land management principles. When edges favour undesirable species, land managers may reduce the amount of edge through strategic revegetation or alter edge structure or permeability through land-use change. However, such management should not reduce the capability of desirable species to access key resources.

Keywords: edge; edge effect; edge responses; landscape fragmentation.

Introduction

Edges are a prominent feature of our world. They range from minute to immense scales (e.g. the boundary between microcosms in a Petri dish or where a continental landmass meets the ocean). Not surprisingly, edges are a common topic of exploration in landscape ecology. Over the last two to three decades, this has been influenced by the dramatic growth of research on the consequences of habitat fragmentation for species conservation. Edges are readily identifiable and commonly occurring attributes of fragmented landscapes. 'Landscape' in this sense refers to a particular spatial scale that is anthropocentric in nature (Fig. 16.1). This has largely driven our conceptualization of edges

Figure 16.1 **A landscape defined from an anthropocentric view showing edges between various landscape elements including native and pine forest, exotic grassland and roads. A myriad of less obvious edges occur at various spatial scales. Photo is from northeast Victoria, Australia, taken by T. Korodaj.**

and the way we go about studying them. We need to temper our human-centric approach and focus on explicitly linking changes in structure and function across edges to understand better their role in defining ecological process.

Nevertheless, edge theory (see below) has largely evolved within a scale-dependent, human-centred landscape context – that is, defining landscapes as heterogeneous areas of a certain size often encompassing elements such as fields, forest patches, waterways and roads (I refer to these as human-scale landscapes). Most current research deviates little from this perspective. For mobile species, research on edges is dominated by studies of birds (e.g. Gates & Gysel 1978; Kroodsma 1982; Rich *et al.* 1994; Driscoll & Donovan 2004) and to a lesser extent mammals (see Lidicker 1999 for review). These organisms generally operate at spatial scales that coincide with our perception of landscapes. Hence, our initial perception drives the questions we ask, and in many cases the focal taxa we choose to study reinforce our perceptions.

I begin this chapter with a brief discussion of the need to develop scale-independent definitions of 'edge' and 'edge effect'. I follow this with a short history of the evolution of edge theory and briefly review the current state of knowledge, gaps in our understanding and where we should focus our attention in the future. The chapter concludes with a set of key principles for landscape design based on the last two decades of edge-related studies. Because the literature is biased towards human-scale landscapes, the principles will focus on these landscapes and edges occurring between vegetation patches.

Defining landscapes, edges and edge effects

Before defining edges and edge effects, a quick definition of 'landscape' is warranted. Turner *et al.* (2001) define a landscape as an area that is spatially heterogeneous in at least one feature of interest. This definition is attractive because it is scale independent and applicable in a variety of circumstances. A useful definition of 'edge' should follow the same principles. An edge can be identified as the place where two landscape elements meet (Thomas *et al.* 1979). Once again this definition is scale independent and widely applicable. However, a more complete definition should elaborate on the notion of 'place' and acknowledge the role of ecological processes and the behaviour of organisms in identifying landscape boundaries.

An edge can be defined structurally by measuring changes in a particular characteristic(s) of one or both of the landscape elements. The zone of this structural change may or may not correspond with functional changes occurring across the edge. For example, Gates and Mosher (1981) demonstrated that the structural width of their forest–field edges ranged from 11 to 12 m based on measurements of the tree canopy, but the functional width of these edges, defined by the distribution of nests associated with edge-using birds, ranged from 12 to 64 m. It is therefore crucial to link edge-related structural and functional characteristics, particularly because it is the latter that landscape ecologists are generally more interested in. Indeed, 'edge effects' can be defined as changes in ecosystem function occurring across edges ('function' is used in the broadest sense and includes the behaviour of organisms that may ultimately lead to community-level differences across edges).

Edges can result from subtle changes in the structure and function of landscapes, whereas visually obvious changes may be inconsequential. For example, a transition from one soil moisture level to another across a heathland may influence the distribution and density of certain plant species, yet such a transition may not be readily identified without careful measurements of soil characteristics and plant distribution. The obvious structural change from savanna to woodland may have no influence on the distribution, abundance or behaviour of wide-ranging vertebrates. Moreover, spatial changes may constitute an edge at one point in time, but not at another (e.g. mammals establishing breeding territories). Therefore, to understand the role edges play in landscape ecology we must consider variation in time and space and the implications of this for our focal organism(s).

A brief history of edge theory

Clements (1907) is attributed as the first to identify the importance of edges by introducing the term 'ecotone'. Nearly three decades later, Leopold (1933, p. 132) defined his principle of edge which states: 'Abundance of non-mobile wild life [*sic*] requiring two or more [habitat] types, appears, in short, to depend on the degree of interspersion of those types . . .' Some ecologists interpreted this principle to mean that the diversity and abundance of animal species increased at the edges of vegetation patches when compared with the adjacent interior (e.g. Dasmann 1964; Odum 1971).

The acceptance of Leopold's principle was encouraged by the game management bias of early wildlife managers in North America, who focused on the use of edges by large mammals (Meffe & Carroll 1994). It was generally accepted that game animals exploit edges (Leopold 1933; Odum 1971) and this led to the assumption that the creation of edges would benefit a range of faunal species. Early studies of species distribution across edges provided empirical evidence to support this assumption (Johnston 1947; Johnston & Odum 1956). Explanations for these positive effects were based on the often greater vegetative complexity occurring at the boundaries of two patch types (compared with the relatively homogeneous interior of either type) and the ability of organisms to exploit more than one patch (Thomas *et al.* 1979; Harris 1988; Yahner 1988).

As research began to focus on the negative impacts of habitat fragmentation, some ecologists challenged the idea that edges benefit most wildlife. Gates and Gysel (1978) and Brittingham and Temple (1983) suggested that increased densities of nest predators and nest parasites near edges might be inhibiting reproduction in some bird species occupying vegetation patches. Ambuel and Temple (1983) found that invasive, generalist birds established themselves at edges, and later studies identified certain bird species that were restricted to patch interiors (e.g. Lynch & Whigham 1984; Freemark & Merriam 1986). Edges are currently perceived as having negative impacts on a number of species. Yet the structural and functional changes that occur across edges are extremely varied and likely to be context dependent.

What do we currently know?

Edge-related studies have generally focused on community-level patterns in the diversity and abundance of species (e.g. Chasko & Gates 1982; Lynch & Saunders 1991; Euskirchen *et al.* 2001; Dauber & Wolters 2004), detailed responses of individual species (e.g. Kremsater & Bunnell 1992; Mazerolle & Hobson 2003; Orrock & Danielson 2005) or investigations of mechanisms that may underlie observed patterns (e.g. Murcia 1995; Kingston & Morris 2000; Kristan *et al.* 2003). In this latter category is the large body of literature that has examined avian nest success and predation as an edge effect (see Paton 1994; Hartley & Hunter 1998; Lahti 2001; Chalfoun *et al.* 2002b for reviews). A few studies have directly tested the hypothesis that species diversity increases at edges (e.g. Luck *et al.* 1999) or that edges contain communities distinct from the adjacent patch types (e.g. Baker *et al.* 2002), yet there is little support in the literature for the general interpretation of Leopold's principle (Guthery & Bingham 1992; Murcia 1995).

Species responses at edges are highly variable making the identification of general patterns difficult (see Murcia 1995; Villard 1998; Ries *et al.* 2004, for reviews). It is reasonable to expect habitat generalists to adapt readily to edges because, by definition, these species exploit a range of patch types (e.g. Wilder & Meikle 2005), although Imbeau *et al.* (2003) found that 28 out of 30 edge-using birds were considered specialists of early successional habitat, which just happened to occur most commonly at patch edges. Species that require multiple patch types are favoured by edges between these types because this represents the optimum location for access to the spatially separated resources. This situation is likely to arise when the edge occurs between two patches that provide complementary resources (e.g. foraging and nesting habitat; Ries *et al.* 2004).

Some invasive species also may favour edges and their increased abundance may be detrimental to other species. Two prominent examples are the brown-headed cowbird (*Molothrus ater*), a nest parasite in North America (Donovan *et al.* 1997; Lloyd *et al.* 2005), and the noisy miner (*Manorina melanocephala*), a highly aggressive honeyeater in Australia that actively excludes smaller birds from otherwise suitable habitat (Piper & Catterall 2003). At the other extreme are species that avoid edges either because they are habitat specialists and their preferred habitat type does not occur at the edge, or they are excluded from the edge by invasive or aggressive species (e.g. Mills 1995; Burke & Nol 1998). Species that require a single habitat type for all life-history attributes are likely to exhibit a neutral or negative response at the edges of their preferred habitat.

Ecologists have known for a long time that the distribution of species largely reflects the distribution of the resources they use (with competition and predation as additional influences). Hence, patterns in species distribution at edges are likely to reflect changes in resource distribution. Ries and Sisk (2004) provide a predictive model of species' edge responses based on the quality of adjacent habitats and whether the resources contained in those habitats are supplementary (of the same type) or complementary (of different types). When both habitats contain required, complementary resources then a positive edge response is predicted. If the habitats are of similar quality and contain supplementary resources, then the edge response is predicted to be neutral. Tests of the model suggest substantial concordance with empirical data for a range of taxa (see Ries & Sisk 2004; Ries *et al.* 2004, for details). The responses of species to edges are largely predictable based on the distribution of resources across the edges.

Murcia (1995) documents a number of edge effects that have been reported in the literature, including changes in air temperature, light, tree density, seed dispersal and herbivory. For each effect, a penetration distance into one or both patch types is often noted. For example, Kapos (1989) showed that variation in light intensity was evident up to 20 m from the edges of tropical forests. Murcia (1995) argues that the penetration distance of such abiotic edge effects is probably influenced by edge orientation and physiognomy. For example, north-facing edges in the northern hemisphere may experience less solar radiation than edges facing in other directions and therefore the intensity of certain abiotic edge effects may be reduced at these edges. Variation in the microclimate across edges influences vegetation structure and the distribution of resources, in turn affecting patterns in the distribution of more mobile species. Despite the confidence of Reis *et al.* (2004) at predicting the distribution of mobile species across certain types of edges, there is much less consensus on the circumstances under which certain edge effects (which ultimately drive the patterns in distribution) manifest themselves.

What do we need to know?

Murcia (1995) laments the poor design, lack of replication, inconsistent methods and inadequate descriptions of edge structure that characterize many studies on edges. Not much has changed over the 12 years since her review. Although replication and control are at the heart of scientific enquiry, they are often the hurdles landscape ecologists working at large scales find hardest to jump. However, there is little excuse not to provide clear descriptions of the edges that are being studied or, indeed, how it was decided where the edge actually occurred. The latter is not trivial and should be based on measurements of key variables (e.g. soil moisture or vegetation structure) across the landscape and using these to determine the location of the edge (rather than the often-used, arbitrary designation based on human perception).

More work needs to be done comparing the same variables across different edge types in the same landscape (e.g. Eriksson *et al.* 2001; Piper & Catterall 2004; Pauchard & Alaback 2006). Such studies provide information on how edge type influences the intensity of edge effects and associated patterns in species distribution, and contribute to identifying general trends. Edge effects are assumed to be stronger at edges with strongly contrasting adjacent patch types (Stamps *et al.* 1987). For example, edge responses at tropical forest–pasture edges should be stronger than those between native grasslands and annual crops such as wheat or barley. There is little consensus, however, on the strength of edge effects likely to occur given variation in edge contrast.

Like much of landscape ecology, edge-related studies have mostly focused on spatial variation and ignored changes over time (although this is changing;

see for example Young & Mitchell 1994; Meyer *et al.* 2001; Chalfoun *et al.* 2002a; Mazerolle & Hobson 2003). Measuring responses of species to the same edge across seasons and years (with or without concomitant changes in the adjacent landscape elements) would greatly increase our understanding of how edges influence ecological patterns and processes (e.g. Lehtinen *et al.* 2003).

Ries *et al.* (2004) argue for mapping the distribution of critical resources across edges because this is likely to strongly influence species distribution and edge responses. This requires detailed knowledge about which resources are important to which species. A logical extension of this would be to deliberately manipulate key resources to determine if predicted edge responses can be induced. For example, providing abundant food resources at edges may encourage edge-avoiding species to become edge-neutral or even edge-using species.

As discussed above, there is some evidence to suggest that edge orientation and exposure to solar radiation will influence the strength of certain edge effects. South-facing edges in the northern hemisphere may experience stronger edge effects because of increased solar radiation, with the same being true for north-facing edges in the southern hemisphere. However, studies of the latter are too rare to confirm this prediction. If this trend holds for the southern hemisphere it has important implications for how we manage edges in the landscape (see below).

In summary, much remains to be done on the variability occurring between edge types and changes over time. Current evidence suggests that these factors will influence the strength (although possibly not the direction) of edge responses and the penetration distance of associated edge effects (Reis *et al.* 2004; see also Chapter 14).

Landscape design principles

The phrase 'landscape design principle' implies a directive that assists land managers in deciding between potential designs or strategies for on-ground management. However, owing to the complexity of ecological processes occurring across landscapes and the diversity of landscape contexts, it is very difficult to identify widely applicable principles that offer clear directives for land managers. Principles that can be proffered are usually so general as to be of little practical value in a given management context, yet it is prescriptive direction that land managers desperately need. A process is required to help bridge the gap between higher-level principles and prescription. Active adaptive management offers a framework to guide this process (Walker 1998; Possingham 2001).

However, adaptive management is commonly misunderstood and misapplied (Lee 1999; Allan & Curtis 2005). This is largely because it is extremely difficult to apply true adaptive management in real-world situations. Nevertheless, it provides a transparent and defensible framework for land managers in the face of uncertainty, and allows managers to document their actions against a benchmark process to identify limitations clearly.

Active adaptive management is an ongoing, iterative procedure involving nine key steps (Fig. 16.2). I argue that land managers should also consider the context in which they are working. Management objectives are likely to include

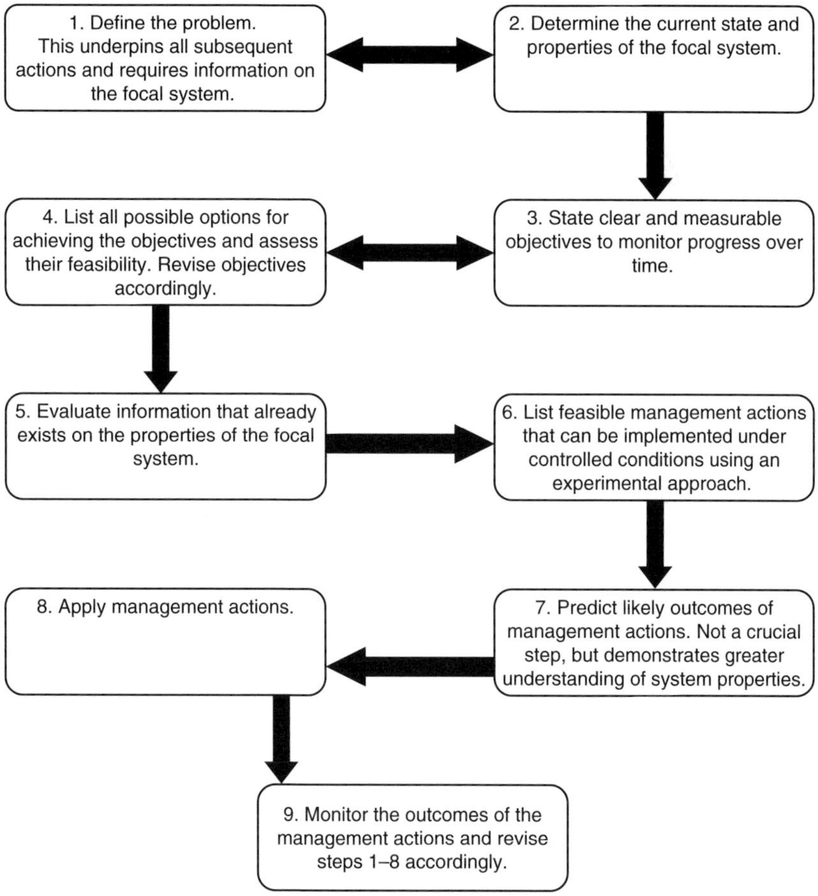

Figure 16.2 **The nine steps in the active adaptive management framework.**

desired responses of key species (e.g. population increase of a threatened plant) and actions will vary depending on the stated objectives and the species of interest (i.e. they will be context dependent). Below I provide a general example of the adaptive management framework as applied to the management of edges in highly modified landscapes. I focus on crucial steps in the process rather than elaborate on all nine steps in this example. Prescriptions for management are provided within this framework to acknowledge the contextual nature of any management actions and the uncertainty that surrounds our knowledge of species' responses.

Active adaptive management of edges

Context: Highly fragmented landscape with clear demarcation between patches of native vegetation and surrounding land uses. There are few options available to alter the predominant land use.
Problem: Edges of native vegetation patches are populated by an invasive and aggressive bird species leading to a decline in the abundance of certain desirable species ('desirable' species are those that the land manager wishes to focus attention on and may include uncommon or threatened species, or those with key ecological roles, e.g. pest control). Note here that a clear link has been established between the presence of the invasive species and the decline of the desirable species.
Objective: Increase the abundance of selected desirable species in native vegetation patches by 10% over three years.
Management options: The following options relate specifically to the management of edges and are not exhaustive (other options are available to address the problem):

1. Conduct targeted revegetation to alter the size or shape of vegetation patches to reduce the edge: area ratio (i.e. reducing the length of the patch edge while maintaining or increasing patch area). Long and thin or irregularly shaped patches have higher edge: area ratios than square patches (Patton 1975).
2. Alter the habitat structure of patch edges to make them less desirable to the invasive species.
3. Provide key resources for desirable species at patch edges (e.g. desirable species may require a dense understorey).
4. Reduce edge permeability by planting buffers around patch edges.

Apply management actions: Owing to the lack of knowledge on species' responses, it is unclear which of the above options are most appropriate. This is where the core principle of active adaptive management comes to the forefront. A land manager or group of managers/owners should implement a variety of actions under controlled conditions to determine the relative success of each action. Implementation should follow standard experimental design where actions are replicated in time and space. Experimental management is a process for learning about the properties of systems when faced with uncertainty about system or species responses.
Monitor outcomes and modify actions: Monitoring the outcomes of management actions to determine if the objectives of management are being met is a crucial step in the adaptive framework. In this example, monitoring would involve assessing the abundance of desirable species at given intervals using appropriate survey protocols. It would also involve comparing the success of the different management actions in meeting the stated objective. If all actions fail to meet the objective it would mean returning to steps one or two of the framework (Fig. 16.2) and possibly implementing new management actions. If some actions appear better than others then these should be further replicated to determine their general applicability. However, active adaptive management does not stop here – it involves continual refinement of problems, objectives and actions in an ongoing, iterative process. The problems land managers face and the actions required to address these problems are in a constant state of flux.

General considerations

The above example deals with a specific problem in a particular context, but touches on some of the options available to land managers for managing the impact of edges in their landscape. Below I list four key issues that managers should consider when assessing potential options for managing the influence of edges on landscape function. The relevance of each of these issues is context dependent (i.e. dependent on the state of the landscape, the organisms within the landscape and the desired outcomes) and addressing them should be done within an active adaptive management framework.

- **Patch size and shape:** A simple approach to reducing the amount of edge between particular landscape elements is to alter the size or shape of a given patch. This might be appropriate if desirable species are adversely affected by the presence of certain types of edges (e.g. because of changes in structure or function at the edge). It is also most relevant in particular contexts – for example, patches of remnant native vegetation in a mostly cleared landscape where the original vegetation has been replaced with a substantially different vegetation type (e.g. a rainforest patch surrounded by pasture). Carefully planned revegetation adjoining existing native vegetation patches can reduce the edge: area ratio of those patches by altering patch shape (e.g. changing a long, thin patch to a square patch).

- **Edge permeability:** The penetration of invasive plant species into vegetation patches may be influenced by, for example, edge orientation or physiognomy (Matlack 1993). Cadenasso *et al.* (2003) suggested that vegetation structure at the edge largely influences edge permeability. If invasive plants are acknowledged as a management problem, potential options for addressing this might include identifying patch edges that are most susceptible to invasion (e.g. north-facing edges) and managing these to reduce permeability (e.g. by planting buffer zones or altering adjacent land use or vegetation structure).
- **Resource access:** Desired species that require access to multiple patch types will benefit from the edges of these types being close together. Easy access to multiple patches containing complementary resources is particularly important for species that require these resources on a daily basis through their entire life cycle or for a discrete section of it (e.g. the breeding season). The issue is less critical for species that access different resources at different times in their life cycle, as long as a single patch type provides all the required resources at a given time. Reis *et al.* (2004) show that the edge responses of many species can be predicted from the distribution of the species' required resources, so managing for the resources of desired species, either within or across patches, seems crucial to ensuring their persistence.
- **Edge type and the influence of the surrounding landscape:** Altering land use adjacent to patches containing desired species might control factors adversely affecting these species. For example, certain types of land use may increase the likelihood of colonization of patches by invasive or aggressive species. Pauchard and Alaback (2006) found that forest–highway edges had greater richness and cover of alien plant species than edges created by fire. Land managers should ensure that land surrounding habitat patches is designated for uses that represent the least threat to the quality or desired characteristics of the patch, while other less suitable land uses occur away from these patches. Moreover, land adjacent to habitat patches should be managed, if possible, so that it contains resources useful to the desired species (e.g. crop types that provide nectar to invertebrates).

Key principles

1. Manage the amount, structure and/or permeability of edges to minimize the impact of undesirable species that may favour edges.
2. Ensure that the degree of edge permeability does not restrict the movement of desirable species.
3. Ensure that edges of patches containing complementary resources required by desirable species are arranged to facilitate ready access.
4. Manage land adjacent to resource patches of desirable species to minimize negative impacts on patch quality while conducting less suitable land-use activities distant from the patch.
5. Implement an adaptive management process when there is uncertainty about the outcomes of management actions.

References

Allan, C. & Curtis, A. (2005) Nipped in the bud: why regional scale adaptive management is not blooming. *Environmental Management* **36**, 414–425.

Ambuel, B. & Temple, S.A. (1983) Area-dependent changes in the bird communities and vegetation of southern Wisconsin forests. *Ecology* **64**, 1057–1068.

Baker, J., French, K. & Whelan, R.J. (2002) The edge effect and ecotonal studies: bird communities across a natural edge in southeastern Australia. *Ecology* **83**, 3048–3059.

Brittingham, M.C. & Temple, S.A. (1983) Have cowbirds caused forest songbirds to decline? *Bioscience* **33**, 31–35.

Burke, D.M. & Nol, E. (1998) Influence of food abundance, nest-site habitat, and forest fragmentation on breeding ovenbirds. *Auk* **115**, 96–104.

Cadenasso, M.L., Pickett, S.T.A., Weathers, K.C. & Jones, C.G. (2003) A framework for a theory of ecological boundaries. *Bioscience* **53**, 750–758.

Chalfoun, A.D., Ratnaswamy, M.J. & Thompson, F.R. (2002a) Songbird nest predators in forest-pasture edge and forest interior in a fragmented landscape. *Ecological Applications* **12**, 858–867.

Chalfoun, A.D., Thompson, F.R. & Ratnaswamy, M.J. (2002b) Nest predators and fragmentation: a review and meta-analysis. *Conservation Biology* **16**, 306–318.

Chasko, G.G. & Gates, J.E. (1982) Avian habitat suitability along a transmission line corridor in an oak-hickory forest region. *Wildlife Monographs* **82**, 1–41.

Clements, F.E. (1907) *Plant Physiology and Ecology*. Holt, New York.

Dasmann, R.F. (1964) *Wildlife Biology*. John Wiley & Sons, New York.

Dauber, J. & Wolters, V. (2004) Edge effects on ant community structure and species richness in an agricultural landscape. *Biodiversity and Conservation* **13**, 901–915.

Donovan, T.M., Jones, P.W., Annand, E.M. & Thompson, F.R. (1997) Variation in local-scale edge effects: mechanisms and landscape context. *Ecology* **78**, 2064–2075.

Driscoll, M.J.L. & Donovan, T.M. (2004) Landscape context moderates edge effects: nesting success of wood thrushes in central New York. *Conservation Biology* **18**, 1330–1338.

Eriksson L.M., Edenius L., Areskoug V. & Meritt D.A. (2001) Nest-predation at the edge: an experimental study contrasting two types of edges in the dry Chaco, Paraguay. *Ecography* **24**, 742–750.

Euskirchen, E.S., Chen, J. & Bi, R. (2001) Effects of edge on plant communities in a managed landscape in northern Wisconsin. *Forest Ecology and Management* **148**, 93–108.

Freemark, K.E. & Merriam, H.G. (1986) Importance of area and habitat heterogeneity to bird assemblages in temperate forest fragments. *Biological Conservation* **36**, 115–141.

Gates, J.E. & Gysel, L.W. (1978) Avian nest dispersion and fledging success in field-forest ecotones. *Ecology* **59**, 871–883.

Gates, J.E. & Mosher, J.A. (1981) A functional approach to estimating habitat edge width for birds. *American Midland Naturalist.* **105**, 189–192.

Guthery, F.S. & Bingham, R.L. (1992) On Leopold's principle of edge. *Wildlife Society Bulletin* **20**, 340–344.

Harris, L.D. (1988) Edge effects and the conservation of biotic diversity. *Conservation Biology* **2**, 2–4.

Hartley, M.J. & Hunter, M.L. (1998) A meta-analysis of forest cover, edge effects, and artificial nest predation rates. *Conservation Biology* **12**, 465–469.

Imbeau, L., Drapeau, P. & Mokkonen, M. (2003) Are forest birds categorised as "edge species" strictly associated with edges? *Ecography* **26**, 514–520.

Johnston, D.W. & Odum, E.P. (1956) Breeding bird populations in relation to plant succession on the piedmont of Georgia. *Ecology* **37**, 50–62.

Johnston, V.R. (1947) Breeding birds of the forest edge in Illinios. *Condor* **2**, 45–53.

Kapos, V. (1989) Effects of isolation on water status of forest patches in the Brazilian Amazon. *Journal of Tropical Ecology* **5**, 173–185.

Kingston, S.R. & Morris, D.W. (2000) Voles looking for an edge: habitat selection across forest ecotones. *Canadian Journal of Zoology* **78**, 2174–2183.

Kremsater, L.L. & Bunnell, F.L. (1992) Testing responses to forest edges: the example of black-tailed deer. *Canadian Journal of Zoology* **70**, 2426–2435.

Kristan, W.B., Lynam, A.J., Price, M.V. & Rotenberry, J.T. (2003) Alternative causes of edge abundance relationships in birds and small mammals of California coastal sage scrub. *Ecography* **26**, 29–44.

Kroodsma, R.L. (1982) Edge effect on breeding forest birds along a power-line corridor. *Journal of Applied Ecology* **19**, 361–370.

Lahti, D.C. (2001) The edge effect on nest predation hypothesis after twenty years. *Biological Conservation* **99**, 365–374.

Lee, K.N. (1999) Appraising adaptive management. *Conservation Ecology* **3**(2), 3 (http://www.consecol.org/vol3/iss2/art3/).

Lehtinen, R.M., Ramanamanjato, J-B. & Raveloarison, J.G. (2003) Edge effects and extinction proneness in a herpetofauna from Madagascar. *Biodiversity and Conservation* **12**, 1357–1370.

Leopold, A. (1933) *Game Management.* Charles Scribner & Sons, New York.

Lidicker, W.Z. (1999) Responses of mammals to habitat edges: an overview. *Landscape Ecology* **14**, 333–343.

Lloyd, P., Martin, T.E., Redmond, R.L., Langner, U. & Hart, M.M. (2005) Linking demographic effects of habitat fragmentation across landscapes to continental source-sink dynamics. *Ecological Applications* **15**, 1504–1514.

Luck, G.W., Possingham, H.P. & Paton, D.C. (1999) Bird responses at inherent and induced edges in the Murray Mallee, South Australia. 1. Differences in abundance and diversity. *Emu* **99**, 157–169.

Lynch, J.F. & Saunders, D.A. (1991) Responses of bird species to habitat fragmentation in the wheatbelt of Western Australia: interiors, edges and corridors. In: Saunders, D.A. & Hobbs, R.J. (eds.) *Nature Conservation 2: The Role of Corridors*, pp. 143–158. Surrey Beatty & Sons, Chipping Norton, NSW.

Lynch, J.F. & Whigham, D.F. (1984) Effects of forest fragmentation on breeding bird communities in Maryland, USA. *Biological Conservation* **28**, 287–324.

Matlack, G.R. (1993) Microenvironment variation within and among forest edge sites in the eastern United States. *Biological Conservation* **66**, 185–194.

Mazerolle, D.F. & Hobson, K.A. (2003) Do ovenbirds (*Seiurus aurocapillus*) avoid boreal forest edges? A spatiotemporal analysis in an agricultural landscape. *Auk* **120**, 152–162.

Meffe, G.K. & Carroll, C.R. (1994) *Principles of Conservation Biology*. Sinauer Associates Inc., Sunderland, MA.

Meyer, C.L., Sisk, T.D. & Covington, W.W. (2001) Microclimate changes induced by ecological restoration of ponderosa pine forests in northern Arizona. *Restoration Ecology* **9**, 443–452.

Mills, L.S. (1995) Edge effects and isolation: red-backed voles on forest remnants. *Conservation Biology* **9**, 395–402.

Murcia, C. (1995) Edge effects in fragmented forests: implications for conservation. *Trends in Ecology and Evolution* **10**, 58–62.

Odum, E.P. (1971) *Fundamentals of Ecology*. Saunders College Publishing, Philadelphia.

Orrock, J.L. & Danielson, B.J. (2005) Patch shape, connectivity, and foraging by old-field mice (*Peromyscus polionotus*). *Journal of Mammalogy* **86**, 569–575.

Paton, P.W.C. (1994) The effect of edge on avian nest success: how strong is the evidence? *Conservation Biology* **8**, 17–26.

Patton, D.R. (1975) A diversity index for quantifying habitat "edge". *Wildlife Society Bulletin* **3**, 171–173.

Pauchard, A. & Alaback, P.B. (2006) Edge type defines alien plant species invasions along *Pinus contorta* burned, highway and clearcut forest edges. *Forest Ecology and Management* **223**, 327–335.

Piper, S.D. & Catterall, C.P. (2003) A particular case and a general pattern: hyper-aggressive behavior by one species may mediate avifaunal decreases in fragmented Australian forests. *Oikos* **101**, 602–614.

Piper, S.D. & Catterall, C.P. (2004) Effects of edge type and nest height on predation of artificial nests within subtropical Australian eucalypt forests. *Forest Ecology and Management* **203**, 361–372.

Possingham, H.P. (2001) *The Business of Biodiversity: Applying Decision Theory Principles to Nature Conservation*. Australian Conservation Foundation, Fitzroy.

Rich, A.C., Dobkin, D.S. & Niles, L.J. (1994) Defining forest fragmentation by corridor width: the influence of narrow forest-dividing corridors on forest-nesting birds in southern New Jersey. *Conservation Biology* **8**, 1109–1121.

Ries, L. & Sisk, T.D. (2004) A predictive model of edge effects. *Ecology* **85**, 2917–2926.

Ries, L., Fletcher, R.J. Jr., Battin, J. & Sisk, T.D. (2004) Ecological responses to habitat edges: mechanisms, models, and variability explained. *Annual Review of Ecology, Evolution and Systematics* **35**, 491–522.

Stamps, J.A., Buechner, M. & Krishnan, V.V. (1987) The effects of edge permeability and habitat geometry on emigration from patches of habitat. *American Naturalist* **129**, 533–552.

Thomas, J.W., Maser, C. & Rodiek, J.E. (1979) Edges. In: Thomas, J.W. (ed.) *Wildlife Habitats in Managed Forests – The Blue Mountains of Oregon and Washington*, pp. 48–59. Wildlife Management Institute, US Department of Interior, Washington, DC.

Turner, M.G., Gardner, R.H. & O'Neill, R.V. (2001) *Landscape Ecology in Theory and Practice*. Springer-Verlag, New York.

Villard, M.A. (1998) On forest-interior species, edge avoidance, area sensitivity, and dogmas in avian conservation. *Auk* **115**, 801–805.

Walker, B. (1998) The art and science of wildlife management. *Wildlife Research* **25**, 1–9.

Wilder, S.M. & Meikle, D.B. (2005) Reproduction, foraging and the negative density–area relationship of a generalist rodent. *Oecologia* **144**, 391–398.

Yahner, R.H. (1988) Changes in wildlife communities near edges. *Conservation Biology* **2**, 333–339.

Young, A. & Mitchell, N. (1994) Microclimate and vegetation edge effects in a fragmented podocarp-broadleaf forest in New Zealand. *Biological Conservation* **67**, 63–72.

17

Synthesis: Edge Effects

David B. Lindenmayer and Richard J. Hobbs

Chapters 14–16 have canvassed many of the issues associated with edges and edge effects in human-modified landscapes. Like many of the topics in this book, the definition of terms and concepts is crucial to an improved shared understanding of problems and how to mitigate them. All three authors in this section have dedicated some of their essays to defining edges and edge effects. For example, Lindenmayer and Fischer (Chapter 15) describe a range of types of edges and edge effects – in particular, human/natural, primary/secondary, hard/soft and biotic/abiotic. All three sets of authors (Chapters 14–15) highlight the extraordinary range and variation in responses to these (and other) kinds of edges: between taxa, between vegetation types, between regions, etc. Sisk (Chapter 14) believes that, while edge effects are highly variable, they are not necessarily idiosyncratic and there are mechanistic approaches (based on the strength of habitat associations and resource availability) that can help better to clarify edge effects and underpin improved predictive models – at least for biotic edge effects. Luck (Chapter 16) is less optimistic and argues that there is currently no clear consensus about the circumstances in which edge effects will and will not manifest themselves. However, both Sisk and Luck agree that while in many cases the magnitude of responses to particular edge effects will be variable, often the nature of the effect (i.e. positive or negative) will not. Irrespective of such consistency and the predictive ability (or otherwise) of mechanistic models of edge effects, it is clear that it will be impossible to quantify the edge responses of every individual species in a landscape or ecosystem. Indeed, managing edges may often involve dealing with idiosyncratic problems. It is also clear that the variability in the kinds of edge effects means that simple classifications such as edge or interior species will often be an inadequate description of responses. This, in turn, has made it difficult to derive general principles – as discussed further below.

All three essays identified deficiencies in current work on edge effects. First, most workers consider edges from a human perspective, but as Luck

(Chapter 16) points out, edges and edge effects can exist at many scales – an observation that reinforces comments about the multiscaled nature of landscapes and ecological processes that are made in many places throughout this book. Similarly, how humans perceive and respond to edges will often be markedly different from other organisms. Second, the species-specific nature of edge responses has often been overlooked by researchers and landscape managers alike. Third, there are serious problems with the experimental design of many studies of edge effects, although attempts to implement controls and obtain sufficient replicates are always challenging in landscape ecology. Fourth, variation in the field methods used in different investigations also makes cross-study comparisons difficult. Moreover, the majority of studies have been cross-sectional investigations, and examinations of temporal variations in edge effects are limited (Luck, Chapter 16). Finally, although Sisk (Chapter 14) argues that edge effects have received less attention than deserved given their pervasive nature and potential for significant effects, there have been remarkably few studies assessing relationships between extinction-proneness and edge sensitivity (Lindenmayer & Fischer, Chapter 15). This knowledge gap needs addressing.

Luck (Chapter 16) believes that adaptive management experiments are needed to address the major knowledge gaps that characterize the field of edge effects. He provides some examples where specific management actions could be implemented to tackle a hypothesized problem and then tested by experimentation and monitoring to fill knowledge gaps and, in turn, evolve management practices. This suggestion has strong congruence with recommendations by Driscoll (see Chapter 11) for what he terms experimental management of biotic responses to vegetation structure and condition. However, as outlined earlier (Chapter 14), the current record on true adaptive management experiments is underwhelming to say the least. It is only very rarely done and done properly even less frequently. Distinct cultural changes towards research, experimentation and monitoring in landscape management will be needed to make the considerable hypothetical promise of adaptive management an on-ground reality.

Although the prospects for more and better management-by-experimentation-and-monitoring projects are slim at present, the proposals for such kinds of work by Driscoll (Chapter 11) and Luck (Chapter 16) emphasize the crucial importance of good problem definition, clear articulation of objectives, and clear statements of actions to achieve those objectives. This is a recurring theme throughout this book, and its relevance to studying and mitigating edge effects is no exception.

Many of the topics explored in this section are intimately linked with others from other sections. For example, how landscapes are classified and the resulting maps that are generated will influence where edges are perceived to occur in that landscape. Whether edge effects actually manifest at edges or boundaries of, for example, different vegetation types (as they are depicted on a map based on a landscape classification) is a further issue. Similarly, the magnitude of edge effects may well be strongly linked with other factors such as the amount of native vegetation remaining in the surrounding landscape (e.g. see Chapters 6 and 8) and the spatial configuration of that vegetation. Finally, we should note that the assumption that edges and edge effects are important rests partially on the perpetuation of some form of patch-based conceptualization of landscapes. In other sections, we have seen the trend towards de-emphasizing this simplistic binary model of landscapes towards a more gradient-based approach. While edges and gradients are not mutually exclusive concepts (for instance, edges can simply be viewed as discontinuities in a gradient), there is a need for some consideration of how the two can mesh together in the continued development of ideas on landscape patterns and processes.

Section 5
Total Vegetation Cover, Pattern, Patch Content

18

Emergent Properties of Land Mosaics: Implications for Land Management and Biodiversity Conservation

Andrew F. Bennett and James Q. Radford

Abstract

Over recent decades, an increased commitment to linking research with practice has resulted in many new insights for biodiversity conservation in human-dominated landscapes. However, despite concentrated research effort on 'fragmented ecosystems', there is often a mismatch between this work and the knowledge requirements of land managers. Most studies have been carried out at the site or patch level, but management questions often relate to the properties of whole mosaics. These include the total amount of habitat needed for effective conservation of a taxon, the relative value of different spatial configurations, and the optimal landscape composition. To appreciate these 'emergent' properties of land mosaics, researchers must consider the mosaic as a single entity. We discuss ways in which the emergent properties of mosaics influence biodiversity conservation, and advance seven principles for landscape-level management of biodiversity. There are many opportunities to investigate further the properties of mosaics. Key challenges include the need to investigate responses to mosaic structure for a wider range of taxa, to untangle the independent effects of different mosaic properties, to examine the influence of the spatial scale of the mosaic on responses shown by biota and to identify the form of species' responses to mosaic properties (especially non-linear responses). In most parts of the world, the future for biodiversity conservation depends on the persistence of species in human-dominated landscapes, highlighting the urgency for better understanding of how the emergent properties of mosaics influence biotic patterns and processes.

Keywords: configuration; fauna; landscape composition; landscape structure.

Introduction: Integrating science and management for sustainable landscapes

Growing awareness of the scale and magnitude of the threat to the Earth's biodiversity from human land use has stimulated a marked shift in ecological science over the last three decades. Rapid growth has occurred in disciplines such as conservation biology and landscape ecology, which make an explicit commitment to linking scientific research to improved outcomes for nature conservation and sustainable management of land (Soulé & Wilcox 1980; Hobbs 1997; Lindenmayer & Burgman 2005). The development of landscape ecology, for example, is based on the premise that conceptual advances and empirical studies of the ways in which spatial pattern affects ecological processes will deliver insights for enhanced land management (Forman 1995; Wu & Hobbs 2002).

However, sustainable land management poses great challenges, particularly in production landscapes in which humans have brought about marked changes to the pattern and composition of natural habitats, imposed an array of new land uses, introduced new species of plants and animals and altered fundamental ecological processes such as soil formation, the distribution and flow of water, and the dynamics of natural disturbances. We outline below three scenarios typical of the current issues faced by land managers in southeastern Australia. Similar kinds of issues occur in other countries.

1 **Setting long-term goals for regional land management.** In Victoria, there are 10 Catchment Management Authorities (CMAs) that each have responsibility for management of land and water resources within a defined catchment basin. In preparing a regional catchment strategy, land managers set out future directions and visions for land management, often in the form of long-term aspirational goals and interim targets. For example, the North Central CMA regional catchment strategy has an aspirational goal to increase native vegetation cover to 30% of the region and an interim target of 20% native vegetation cover by 2030 (North Central Catchment Management Authority 2003). However, such targets are rarely founded on strong empirical evidence. An urgent challenge then is to determine 'What

is an appropriate level of native vegetation cover to ensure the persistence of existing native biota and facilitate the (potential) return of species that have been locally extirpated?'

2 **Managing disturbance processes.** Fire is a major disturbance process shaping ecosystems in southern Australia, particularly in fire-prone environments such as semiarid mallee vegetation (Bradstock *et al.* 2002). Bushfires, fuel-reduction burns and fires used for ecological management all contribute to a mosaic of vegetation of different post-fire age classes. Plants and animals respond to fire in many different ways. The challenge for land managers is to ensure that all species persist within the fire-prone environment. There is widespread agreement that a mosaic of different seral stages is important, but there is little knowledge of 'which mosaic' (Bradstock *et al.* 2005). A fundamental question is 'What is an appropriate mix of post-fire age classes for this taxon, and how should they be distributed across the landscape?'
3 **Planning for landscape restoration.** Revegetation in rural landscapes is a prominent issue in southern Australia because excessive destruction of native vegetation has caused many ecological problems (Vesk & Mac Nally 2006). National programmes to encourage restoration actions are underway. Because their emphasis has been on practical actions by private land-holders, much of the revegetation effort has been in the form of linear strips or small blocks (e.g. <2 ha) on farms. Conservation agencies are now seeking ways to ensure more strategic coordination of these actions to achieve greater benefits, such as by integrated networks of habitat (e.g. Lambeck 1997; Wilson & Lowe 2003; Bennett & Mac Nally 2004). A key question therefore is 'What is the most suitable pattern of revegetated blocks or strips to achieve an effective habitat network across the landscape for target species?'

A common feature of these three scenarios is the need for landscape-level understanding of the properties of land mosaics. In the first case, the total amount of vegetation across the landscape; in the second, the composition of vegetation types and age classes in the landscape; and in the third, the primary question relates to the configuration of revegetation across the landscape. In each case these are 'emergent' properties of the land mosaic: they are a product of the numbers, types and spatial arrangement of elements or land uses in a particular land mosaic (Wiens 1995). To appreciate these emergent properties of landscapes, researchers and land managers must consider the land mosaic as a single entity, rather than focus on the individual patches or land uses that occur within it.

Properties of land mosaics

The emergent properties of land mosaics can be grouped into three main categories:

1 The **extent** of habitat in a land mosaic refers to the sum of the spatial area of all patches that potentially provide habitat for a particular taxon, regardless of the size, shape or location of patches.
2 **Composition** of a land mosaic refers to the types of different elements (land uses) present and their relative proportions. It may be measured by indices such as the richness of different elements, by diversity measures or by the use of multivariate gradients.
3 **Configuration** refers to the spatial arrangement of elements in a land mosaic. Numerous metrics have been proposed for measuring different aspects of configuration (e.g. Hargis *et al.* 1998).

How well can ecologists provide advice to land managers about the desirable properties of land mosaics, and how do such properties influence species, communities and ecological processes? We suggest that the answer, at present, is 'not very well'.

There is a vast body of research on the consequences of landscape change, much of it under the banner of 'habitat fragmentation' (McGarigal & Cushman 2002; Fahrig 2003). However, we contend that there is a significant mismatch between much of this research and the knowledge requirements of managers responsible for land mosaics. Most studies of the biota in human-dominated landscapes have been carried out at the 'site' or 'patch' level. That is, the unit of study for the response variable (e.g. species richness, presence/absence of a threatened species) is a single site or patch within the landscape. Data may be collected from multiple patches but, as noted by Fahrig (2003), the sample size for landscapes in such studies often is just one. This body of work has led to many new insights into the ways in which single species or assemblages are influenced by the types, sizes and shapes of vegetation patches (e.g. van Dorp & Opdam 1987; Bennett 1990); on how population processes and species interactions are affected by changes to landscape elements (e.g. Major *et al.* 1999; Mac Nally *et al.* 2000); and on the use of anthropogenic components of landscapes (Daily *et al.* 2001; Renjifo 2001). Yet, inference about the properties of whole mosaics is not possible because the design is not replicated at that scale.

More recently, some studies have adopted a 'focal patch' or 'patch-landscape' approach (Brennan *et al.* 2002; McGarigal & Cushman 2002), in which attributes of the surrounding landscape are measured for patches each located in an independent landscape. These studies highlight the importance of landscape context for the occurrence of particular species, or the composition of assemblages, within patches (Rosenberg *et al.* 1999; Virgós 2001; Lee *et al.* 2002). However, the response variable still is measured only for a single patch, and so the level of inference remains at the patch level.

If the land mosaic is the unit of interest – how its emergent properties influence biodiversity and how landscape structure can be manipulated to achieve conservation outcomes – then the unit of inference must reflect this. In general, this requires studies in which data are obtained for multiple independent mosaics, and that for each of these mosaics the response and predictor variables are both characterized at the mosaic level.

What do we know about the influence of land mosaics on biodiversity?

Present knowledge of the influence of land mosaics on biota comes from four main types of studies, each of which have strengths and limitations:

1 Simulation modelling approaches that use patch level data to 'build up' a model of the landscape (e.g. Fahrig 1997; With & King 2001).
2 Experimental model ecosystems, in which the structure of a particular vegetation type (such as grassland) is manipulated to test the influence of different configurations (e.g. Collinge & Forman 1998; Parker & Mac Nally 2002).
3 Analyses of data sets from Atlas projects in which landscapes are represented by grid squares (e.g. Bennett & Ford 1997; Atauri & de Lucio 2001), or from programmes involving volunteer observers who survey species across large areas (e.g. Fuller *et al.* 1997; Boulinier *et al.* 1998).
4 Empirical studies based on deliberate selection of independent land mosaics as the unit of study (Andrén 1992; McGarigal & McComb 1995; Gjerde *et al.* 2005; Radford *et al.* 2005). In these, the response variable generally represents aggregated results from a number of sampling sites within each mosaic.

Bennett *et al.* (2006) reviewed empirical studies from agricultural environments in which land mosaics were the unit of study (categories 3 and 4 above), to identify the influence of the emergent properties of mosaics on the status of animal species and assemblages. The following summary draws from that review.

Extent of habitat

The extent of suitable habitat for a taxon is frequently the primary driver of the occurrence and abundance of species, and the richness of faunal assemblages in landscapes. The occurrence of individual species of forest birds, for example, is consistently positively related to the extent of forest vegetation in the landscape (Trzcinski *et al.* 1999; Villard *et al.* 1999). A similar trend occurs for the richness of forest bird assemblages in agricultural landscapes (Bennett & Ford 1997; Boulinier *et al.* 1998; Radford *et al.* 2005). Different species have different habitats and so their response is coupled to the extent of their preferred vegetation type. For example, in a rural region near Barcelona, Spain, species richness of cropland birds was significantly correlated with extent of cropland, whereas the richness of forest birds was negatively associated with this measure but positively correlated with forest cover (Pino *et al.* 2000).

The extent of habitat influences a species primarily by its effect on population size. As the extent of its habitat decreases, overall population size will decrease. As populations become smaller they become more vulnerable to decline or local extinction from chance events. Consequently, the probability of occurrence of a species in a land mosaic will decrease as the extent of suitable habitat decreases, until eventually the population will disappear when none of its habitat remains. In landscapes with a greater extent of suitable vegetation, more species are likely to occur in populations of sufficient size for persistence, and hence species richness can be expected to be higher.

Composition of the land mosaic

Landscape composition is often an important influence, particularly when the response variable is the richness or composition of a whole taxonomic assemblage (e.g. all birds, all reptiles). Variation in soils, topography and moisture influence the distribution of natural vegetation communities, which in turn dictates habitat availability for animal species. Anthropogenic elements, such as hedges, cropland and farm dams, add further to the resources in rural landscapes. As the diversity of landscape elements increases, the beta-diversity (between-habitat diversity) of animal species generally also increases (Tews *et al.* 2004). For example, in 100-km^2 mosaics near Madrid, Spain, the richness of birds, amphibians, reptiles and butterflies was most influenced by the heterogeneity of the landscape (Atauri & de Lucio 2001). However, it is cautionary to note that species richness weights each species equally. This may not necessarily be consistent with desired conservation outcomes, especially if generalist species replace specialized or rare species in modified but heterogeneous mosaics.

Configuration

Spatial configuration of habitat is often found to be a significant correlate of the incidence of species, and of the richness and composition of assemblages (Villard *et al.* 1999; Olff & Ritchie 2002; Radford *et al.* 2005). However, measures of spatial configuration are significantly interrelated with habitat extent in most landscapes (Fahrig 2003). Identifying the independent effects of configuration has practical relevance for the efficient allocation of effort in landscape restoration. Evidence from the few studies that have partitioned the independent effects of extent and configuration demonstrate that in most cases habitat extent has greater influence than spatial configuration alone (Trzcinski *et al.* 1999; Villard *et al.* 1999; Fahrig 2003; Radford *et al.* 2005).

Animal species may be affected by spatial configuration in several ways. First, in mosaics with subdivided habitat, formerly continuous populations may be divided, disrupting population dynamics and increasing the risk of local extinction. Second, in disaggregated mosaics, individual patches may fall below the minimum size for species with large area requirements or those restricted to 'interior' habitat. Third, structural connectivity may influence the distribution and persistence of species through its fundamental relationship with functional connectivity, and hence the ability of individuals to access widespread resources and move between discrete populations. Finally, the amount of 'edge' in a mosaic is another expression of spatial configuration. Although some species thrive at edges, the processes associated with edges (e.g. changes in physical conditions, competitors and predators) have detrimental consequences for many species (Laurance & Vasconcelos 2004; see also Chapters 14–17).

Knowledge gaps and opportunities for research

The few studies for which response variables have been measured at the mosaic level, and consequently the limited understanding of the ways in which species respond to land mosaics, mean there are many opportunities for research that will offer new insights for nature conservation, especially in human-dominated landscapes.

A particular challenge is to discern the relative effects of the different properties of land mosaics. There are two main issues. First, in some studies, only one type of mosaic property has been examined, usually the mosaic composition (diversity or heterogeneity of land uses) (e.g. Böhning-Gaese 1997; Weibull *et al.* 2000). Other studies have contrasted the relative importance of two properties, notably habitat extent and configuration (Trczinski *et al.* 1999; Villard *et al.* 1999). Few studies have simultaneously considered the relative influence on biota of all three types of emergent properties – habitat extent, configuration and landscape composition – together with environmental variation among landscapes (Radford *et al.* 2005; Radford & Bennett 2007). Second, landscape attributes often are significantly intercorrelated, and it is difficult to untangle the independent effects of the different properties, particularly the confounding between habitat extent and habitat configuration (Fahrig 2003).

Most research at the landscape level has been undertaken on the occurrence of bird species and assemblages (McGarigal & Cushman 2002; Bennett *et al.* 2007). It is important to understand how other taxa respond to landscape structure, particularly less mobile species such as reptiles, small terrestrial mammals and selected invertebrate groups. It would also be informative to test the extent to which landscape-level properties influence distributional patterns of plant groups (Hobbs & Yates 2003). In addition, changes in the complement of species have flow-on effects on ecological processes that involve species interactions (e.g. Andrén 1992). There is great scope for investigating how predation, parasitism, mutualistic interactions (such as seed dispersal and pollination) and other ecological processes are affected by landscape structure, and the properties of mosaics that most influence the status of these processes.

Several studies have examined the response to landscape structure of individual species, selected from assemblages judged *a priori* to use the same vegetation type, but have found a wide range of responses (Trzcinski *et al.* 1999; Villard *et al.* 1999; Radford & Bennett 2007). Do such species, although dependent on the same vegetation type, all have idiosyncratic responses, or are there sets of species that respond to similar cues in landscape structure? Is it possible to identify groups of taxa that share similar functional responses to landscape structure, and thereby simplify the complexity of management?

The relationship between the spatial scale of land mosaics and the nature of species' responses warrants further investigation. There are two pertinent issues. First, the perception of 'landscape' varies greatly among taxa, yet most studies adopt a fixed landscape scale. Researchers therefore need to reconcile the scale of investigation with the operating scale of the taxa of interest. Second, we need to understand whether the response for a particular species changes across scales. For example, does the relative influence of habitat extent and configuration for a forest-dependent bird differ if the scale of investigation is 1 km^2, 10 km^2 or 100 km^2? Development of approaches that allow investigation of species' responses to land mosaics at different scales will be particularly useful.

The form, or shape, of the responses of species and assemblages to landscape structure also warrants further investigation, because this has profound implications for land management. For example, two species may both have a positive relationship with extent of habitat but exhibit contrasting response shapes. The abundance of one species may decrease at a constant rate with incremental loss of habitat (linear response); but the second species may follow a non-linear curve, in which abundance is unchanged as habitat decreases from 100% to 20% cover but then plummets dramatically below 20% cover (threshold response). Clearly, management strategies with regard to permitting habitat loss or setting priorities for habitat restoration will differ greatly for these two species. Given the complexity of nature, there are likely to be numerous forms of response by different species, to different mosaic properties.

Principles for design and management of land mosaics

Based on current knowledge of the ways in which biota respond to the properties of land mosaics, we offer the following principles for landscape-level management for biodiversity conservation.

1 **Identify the biodiversity goals for management of a particular landscape.** It is essential to specify conservation goals because different measures of biodiversity (e.g. species richness, occurrence of a particular species, composition of different faunal groups) are related to landscape

structure in different ways. For example, total species richness is likely to increase with greater heterogeneity of vegetation types or land uses in a landscape, but increased heterogeneity may negatively affect the richness of some subsets of species (e.g. species of conservation concern) or the status of single species associated with a particular vegetation type.

2 **Extent of habitat is generally the primary influence on the status of species in mosaics.** The most effective way to enhance the status of a species or assemblage in a landscape is to maximize the total amount of habitat for that taxon. This may require measures to increase habitat in some landscapes, and it emphasizes the importance of retaining native vegetation where it presently exists. Although it may be necessary for management purposes to set a specific goal, variation in species' responses to extent of vegetation means that there is no single 'target' or minimum value that will protect all species.

3 **The configuration of habitats in a landscape will benefit species most when the spatial pattern enhances structural and functional connectivity of the landscape.** For a fixed extent of habitat, spatial patterns that facilitate movement of individuals and continuity of otherwise isolated populations within a mosaic, will be beneficial. Species favoured by configurations that involve small dispersed patches are seldom species of conservation concern. Spatial configuration of habitat is likely to be more important in mosaics with relatively low cover of habitat (e.g. where aggregation helps surpass minimum patch-size requirements for particular species).

4 **Allow for time lags in the responses of species to landscape change.** All landscapes change through time, either from natural processes or human causes, resulting in changes to the emergent properties of the land mosaic. Time lags occur between the event of environmental change (e.g. vegetation clearance, new agricultural crop types, timber harvesting, revegetation) and the full consequences of that change being expressed in the biota of the landscape. The duration and form of time lags are poorly understood, but they are likely to be greater in duration for long-lived organisms.

5 **Non-linear responses to mosaic properties have important implications for land management.** Non-linear responses to mosaic properties, including threshold responses, mean that a minor change in landscape structure may precipitate major change in the biota. Identification of

the form of response for a range of species (and functional groups) and the level of landscape attributes at which accelerated change can be expected, will greatly assist in land-use planning.

6 **Species that occupy similar habitats may respond to different components of landscape structure.** Species are not identical in their response to landscape structure, but respond to different sets of cues in the land mosaic. While it is impractical to manage individually for every species, it is necessary to recognize that generalized guidelines will often have exceptions. Alternatively, identification of sets of species (not necessarily taxonomically related) that respond in similar ways to similar cues will assist management planning.

7 **Habitat features at both landscape- and site-level are important for effective conservation.** Landscape- and site-level attributes are both important influences on biota. Site-level attributes, such as local vegetation structure or the availability of particular resources (logs, dense shrub cover, tree hollows), influence the local occurrence of species, but are difficult to incorporate in landscape-level measures. The need for features at both levels highlights the value of a multiscale strategy for nature conservation.

Acknowledgements

Our research on land mosaics has been supported by Land and Water Australia (DUV06), the Australian Research Council (LP0560309) and the Departments of Primary Industries and Sustainability and Environment, Victoria, through the 'Our Rural Landscape' initiative. We are most grateful to each of these agencies.

References

Andrén, H. (1992) Corvid density and nest predation in relation to forest fragmentation: a landscape perspective. *Ecology* **73**, 794–804.

Atauri, J.A. & de Lucio, J.V. (2001) The role of landscape structure in species richness distribution of birds, amphibians, reptiles and lepidopterans in Mediterranean landscapes. *Landscape Ecology* **16**, 147–159.

Bennett, A.F. (1990) Land use, forest fragmentation and the mammalian fauna at Naringal, South-western Victoria. *Australian Wildlife Research* **17**, 325–347.

Bennett, A.F. & Ford, L.A. (1997) Land use, habitat change and the conservation of birds in fragmented rural environments: a landscape perspective from the Northern Plains, Victoria, Australia. *Pacific Conservation Biology* **3**, 244–261.

Bennett, A.F. & Mac Nally, R. (2004) Identifying priority areas for conservation action in agricultural landscapes. *Pacific Conservation Biology* **10**, 106–123.

Bennett, A.F., Radford, J.Q. & Haslem, A. (2006) Properties of land mosaics: implications for nature conservation in agricultural environments. *Biological Conservation* **133**, 250–264.

Böhning-Gaese, K. (1997) Determinants of avian species richness at different spatial scales. *Journal of Biogeography* **24**, 49–60.

Boulinier, T., Nichols, J.D., Hines, J.E., Sauer, J.R., Flather, C.H. & Pollock, K.H. (1998) Higher temporal variability of forest breeding bird communities in fragmented landscapes. *Proceedings of the National Academy of Sciences of the USA* **95**, 7497–7501.

Bradstock, R.A., Williams, J.E. & Gill, A.M. (eds.) (2002) *Flammable Australia: The Fire Regimes and Biodiversity of a Continent*. Cambridge University Press, Cambridge.

Bradstock, R.A., Bedward, M., Gill, A.M. & Cohn, J.S. (2005) Which mosaic? A landscape ecological approach for evaluating interactions between fire regimes, habitat and animals. *Wildlife Research* **32**, 409–423.

Brennan, J.M., Bender, D.J., Contreras, T.A. & Fahrig, L. (2002) Focal patch landscape studies for wildlife management: optimizing sampling effort across scales. In: Liu, J. & Taylor, W.W. (eds.) *Integrating Landscape Ecology into Natural Resource Management*, pp. 68–91. Cambridge University Press, Cambridge.

Collinge, S.K. & Forman, R.T.T. (1998) A conceptual model of land conversion processes: predictions and evidence from a microlandscape experiment with grassland insects. *Oikos* **82**, 66–84.

Daily, G.C., Ehrlich, P.R. & Sanchez-Azofeifa, G.A. (2001) Countryside biogeography: use of human-dominated habitats by the avifauna of southern Costa Rica. *Ecological Applications* **11**, 1–13.

Fahrig, L. (1997) Relative effects of habitat loss and fragmentation on population extinction. *Journal of Wildlife Management* **61**, 603–610.

Fahrig, L. (2003) Effects of habitat fragmentation on biodiversity. *Annual Review of Ecology and Systematics* **34**, 487–515.

Forman, R.T.T. (1995) *Land Mosaics. The Ecology of Landscapes and Regions.* Cambridge University Press, Cambridge.

Fuller, R.J., Trevelyan, R.J. & Hudson, R.W. (1997) Landscape composition models for breeding bird populations in lowland English farmland over a 20 year period. *Ecography* **20**, 295–307.

Gjerde, I., Saetersdal, M. & Nilsen, T. (2005) Abundance of two threatened woodpecker species in relation to the proportion of spruce plantations in native pine forests of western Norway. *Biodiversity and Conservation* **14**, 377–393.

Hargis, C.D., Bissonette, J.A. & David, J.L. (1998) The behavior of landscape metrics commonly used in the study of habitat fragmentation. *Landscape Ecology* **13**, 167–186.

Hobbs, R. (1997) Future landscapes and the future of landscape ecology. *Landscape and Urban Planning* **37**, 1–9.

Hobbs, R.J. & Yates, C.J. (2003) Turner Review No. 7. Impacts of ecosystem fragmentation on plant populations: generalising the idiosyncratic. *Australian Journal of Botany* **51**: 471–488.

Lambeck, R.J. (1997) Focal species: a multi-species umbrella for nature conservation. *Conservation Biology* **11**, 849–856.

Laurance, W.F. & Vasconcelos, H.F. (2004) Ecological effects of habitat fragmentation in the tropics. In: Schroth, G., da Fonseca, G.A.B., Harvey, C., Gascon, C., Vasconcelos, H.L. & Izac, A-M.N. (eds.) *Agroforestry and Biodiversity Conservation in Tropical Landscapes*, pp. 33–49. Island Press, Washington, DC.

Lee, M., Fahrig, L., Freemark, K. & Currie, D.J. (2002) Importance of patch scale vs landscape scale on selected forest birds. *Oikos* **96**, 110–118.

Lindenmayer, D.B. & Burgman, M. (2005) *Practical Conservation Biology*. CSIRO Publishing, Melbourne.

McGarigal, K. & Cushman, S.A. (2002) Comparative evaluation of experimental approaches to the study of habitat fragmentation effects. *Ecological Applications* **12**, 335–345.

McGarigal, K. & McComb, W.C. (1995) Relationships between landscape structure and breeding birds in the Oregon Coast Range. *Ecological Monographs* **65**, 235–260.

Mac Nally, R., Bennett, A.F. & Horrocks, G. (2000) Forecasting the impact of habitat fragmentation. Evaluation of species-specific predictions of the impact of habitat fragmentation on birds in the box-ironbark forests of central Victoria, Australia. *Biological Conservation* **95**, 7–29.

Major, R.E., Christie, F.J., Gowing, G. & Ivison, T.J. (1999) Age structure and density of red-capped robin populations vary with habitat size and shape. *Journal of Applied Ecology* **36**, 901–908.

North Central Catchment Management Authority (2003) *North Central Regional Catchment Strategy 2003–2007*. North Central Catchment Authority, Huntly, Victoria.

Olff, H. & Ritchie, M.E. (2002) Fragmented nature: consequences for biodiversity. *Landscape and Urban Planning* **58**, 83–92.

Parker, M. & Mac Nally, R. (2002) Habitat loss and the habitat fragmentation threshold: an experimental evaluation of impacts on richness and total abundances using grassland invertebrates. *Biological Conservation* **105**, 217–229.

Pino, J., Rodà, F., Ribas, J. & Pons, X. (2000) Landscape structure and bird species richness: implications for conservation in rural areas between natural parks. *Landscape and Urban Planning* **49**, 35–48.

Radford, J.Q., Bennett, A.F. & Cheers, G.J. (2005) Landscape-level thresholds of habitat cover for woodland-dependent birds. *Biological Conservation* **124**, 317–337.

Radford, J.Q. & Bennett, A.F. (2007) The relative importance of landscape properties for woodland birds in agricultural environments. *Journal of Applied Ecology* (in press).

Renjifo, L.M. (2001) Effect of natural and anthropogenic landscape matrices on the abundance of subandean bird species. *Ecological Applications* **11**, 14–31.

Rosenberg, K.V., Lowe, J.D. & Dhondt, A.A. (1999) Effects of forest fragmentation on breeding tanagers: a continental perspective. *Conservation Biology* **13**, 568–583.

Soulé, M.E. & Wilcox, B.A. (eds.) (1980) *Conservation Biology: An Evolutionary-Ecological Perspective*. Sinauer Associates, Sunderland, MA.

Tews, J., Brose, U., Grimm, V. *et al.* (2004) Animal species diversity driven by habitat heterogeneity/diversity: the importance of keystone structures. *Journal of Biogeography* **31**, 79–92.

Trzcinski, M.K., Fahrig, L. & Merriam, G. (1999) Independent effects of forest cover and fragmentation on the distribution of forest breeding birds. *Ecological Applications* **9**, 586–593.

van Dorp, D. & Opdam, P.F.M. (1987) Effects of patch size, isolation and regional abundance on forest bird communities. *Landscape Ecology* **1**, 59–73.

Vesk, P.A. & Mac Nally, R. (2006) The clock is ticking – Revegetation and habitat for birds and arboreal mammals in rural landscapes of southern Australia. *Agriculture, Ecosystems and Environment* **112**, 356–366.

Villard, M-A., Trzcinski, M.K. & Merriam, G. (1999) Fragmentation effects on forest birds: relative influence of woodland cover and configuration on landscape occupancy. *Conservation Biology* **13**, 774–783.

Virgós, E. (2001) Role of isolation and habitat quality in shaping species abundance: a test with badgers (*Meles meles* L.) in a gradient of forest fragmentation. *Journal of Biogeography* **28**, 381–389.

Weibull, A-C., Bengtsson, J. & Nohlgren, E. (2000) Diversity of butterflies in the agricultural landscape: the role of farming system and landscape heterogeneity. *Ecography* **23**, 743–750.

Wiens, J.A. (1995) Landscape mosaics and ecological theory. In: Hansson, L., Fahrig, L. & Merriam, G. (eds.) *Mosaic Landscapes and Ecological Processes*, pp. 1–26. Chapman & Hall, London.

Wilson, J.A. & Lowe, K.W. (2003) Planning for the restoration of native biodiversity within the Goulburn Broken Catchment, Victoria, using spatial modelling. *Ecological Management and Restoration* **4**, 212–219.

With, K.A. & King, A.W. (2001) Analysis of landscape sources and sinks: the effect of spatial pattern on avian demography. *Biological Conservation* **100**, 75–88.

Wu, J. & Hobbs, R. (2002) Key issues and research priorities in landscape ecology: an idiosyncratic synthesis. *Landscape Ecology* **17**, 355–365.

19

Assessing the Biodiversity Value of Stands and Patches in a Landscape Context

Philip Gibbons, S.V. Briggs, Andre Zerger, Danielle Ayers, Julian Seddon and Stuart Doyle

Abstract

Landscapes are often shaped by decisions taken at the scale of individual stands or patches. In this chapter we make the following points. Biodiversity assessments at the scale of individual stands or patches must be undertaken in the landscape context. Landscape measures only inform land-use decisions when they can be interpreted at the scale in which land-use decisions are taken. In a management context, landscape units must be surrogates for multiple species and ecological processes, although no single unit will suffice as a surrogate for biodiversity generally, so risk-spreading strategies such as defining landscapes at multiple scales should be employed. Vegetation cover, pattern and patch content are measures often used to assess the biodiversity value of landscapes and impacts of change in landscapes. Vegetation cover and pattern are often assessed as though vegetation or habitat is binary (present or absent) and therefore the patches are uniformly good and the matrix uniformly hostile. Patch content is rarely assessed in a landscape context. We discuss how patch content can be interpolated at landscape scales and how this information can inform the way landscapes are defined and how cover and pattern are assessed. We suggest that spatial data on habitat attributes will enable us to view and assess landscapes as continua of habitats rather than discrete patches in a hostile matrix. From this discussion, six landscape principles are derived.

Keywords: agri-environmental schemes; land clearing; rapid biodiversity assessment; scale.

Introduction

Landscapes are often shaped by decisions made at finer scales. For example, biodiversity assessments at the scale of the stand or patch often underpin the regulation of land clearing (e.g. US Fish & Wildlife Service1980; Parkes *et al.* 2003) and the distribution of financial incentives or subsidies (e.g. Ribaudo *et al.* 2001). These are major forces that shape the world's landscapes. The United Nations Food and Agricultural Organization (2005) estimated that 13 million hectares of natural forest was being cleared annually. According to Ribaudo *et al.* (2001) over US$20 billion in environmental works had been funded since 1985 in the USA's flagship agri-environmental scheme, and in 2005 approximately 25% of the total agricultural area in the European Union was covered by an agri-environmental scheme (Kleijn *et al.* 2006).

Biodiversity assessments at the stand or patch scale are more effective for biodiversity conservation if undertaken in the landscape context. Inefficient and unrepresentative conservation outcomes are more likely if a site's complementarity with other sites in the landscape is not considered when undertaking an assessment (Margules & Pressey 2000). For example, the contribution of a stand or patch to the viability of a population can only be assessed in light of the amount and configuration of surrounding habitat. Thus, landscape principles must be invoked even when assessing biodiversity at fine scales such as the stand or patch (see Briggs 2001).

In this chapter we explore some landscape principles and practical issues that are relevant for rapid biodiversity assessment at stand or patch scales. First we discuss how landscape units should be defined and then we focus on the three broad measures of landscapes that are the theme of this section of the book – cover, pattern and patch content. We define **cover** as the amount of habitat in a landscape, **pattern** as the configuration of that habitat and **patch content** as the nature of the habitat within patches. We discuss these measures in the context of their value as surrogates for biodiversity.

Case study

These issues are demonstrated using examples drawn from a case study in the state of New South Wales (NSW), Australia, in which rapid assessment techniques were developed to assess: (i) impacts on terrestrial biodiversity of proposals to clear native vegetation at the property-scale; and (ii) actions to establish native vegetation or manage existing native vegetation for biodiversity conservation at the property scale (Gibbons *et al.* 2005). Several operational requirements and constraints influenced how biodiversity could be assessed in the case study (Table 19.1). Limited time and resources, patchy data and an inability to consider the requirements of individual species are key constraints that often face land managers. We will return to these operational considerations regularly in this chapter to illustrate why some landscape measures cannot be used in biodiversity assessments and how some of these issues can be overcome.

Defining landscape units for biodiversity assessment

Landscape units should be defined using different criteria and at different scales. Surrogates such as vegetation cover, pattern and patch content should be measured at a range of scales because the biota is organized, and ecosystem processes operate, at multiple scales (e.g. Noss 1990). Rouget (2003) found that priorities for conservation vary depending on the scale and type of landscape unit used for the assessment.

In the case study we defined landscape units at local scales simply as the immediate area around the proposal, recognizing that the contribution of a stand or patch to the local area and configuration of habitat is important for the persistence of individuals and local subpopulations of many species and local ecosystem processes. For example, in southern Australia the richness of woodland birds is positively associated with the total cover of vegetation in the immediate vicinity (Radford *et al.* 2005). The scale of these landscape units should match the biota or ecological processes of interest. Because our objective was to assess the impacts of individual proposals on terrestrial biodiversity generally, we defined these landscape units as different radii around the site in recognition that different biotas and processes are influenced by

Table 19.1. **Operational requirements in the case study that are typical in many jurisdictions.**

Operational constraint	Explanation
Limited time available for each assessment	Governments are obsessed with 'cutting red tape' and therefore often require systems that lead to rapid decision-making and/or have low transaction costs (in the case study no more than a day was to be allocated to the field assessment, although this turned out to be impossible in practice)
Consistently applied across the whole jurisdiction	Assessments must often be applied consistently across the jurisdiction for equity reasons (e.g. for regulation) and to enable the measure to be tradeable across the jurisdiction in the case of market-based mechanisms (e.g. auctions, offsets)
Available data	Data (e.g. species information) were often patchy, or not available to the same standard, across the State
No influence over the timing of assessments	Assessments were required at all times of the year, so could not be confined to particular seasons or climatic events, which limits the capacity to survey individual species
Can be undertaken by individuals without specialized field skills	Comprehensive skills in flora and fauna species could not be identified among site-assessment personnel assumed
Scale of assessment	Stand- or patch-scale assessments had to be accurate at a fine scale and resolution (typically assessments were undertaken on areas <1–500 ha)
The assessment is instructive to assessors and landholders	The assessment should be useful for identifying biodiversity issues, directing management priorities that are relevant for landholders and educating assessors and land managers
Can be applied to a rangeof ecosystem types	Assessment methods must be applicable in all ecosystem or vegetation types under consideration

the area and configuration of vegetation at different scales. The largest landscape defined using this criterion (1000 ha) was the largest area in which changes in cover and pattern from an individual proposal could typically be detected given the measures of habitat we employed.

The contribution of a stand or patch to the adequate representation of biological diversity only becomes evident at broader scales. At broader scales we defined landscapes as homogeneous units of biological variation relative to the total biological variation across the region. This enabled us to assess a site's contribution to the adequate representation of biological diversity within the broader region, which cannot be done with landscapes defined using the geographical criteria applied at the local scale. Three key issues must be addressed when defining landscape units in this way: (i) the geographic extent, or total biological variation, that will be used to define the region in which individual landscapes are defined; (ii) the amount of biological variation or turnover that characterizes a discrete landscape unit; and (iii) the surrogate for biota or biological turnover.

A known or modelled distribution of an individual species is currently the finest level at which biological variation is typically assessed at broad scales (e.g. Ferrier *et al.* 2002). However, models of species distributions require species data to be collected systematically (typically information that confirms presences and absences for each species is required) across the environmental envelope represented in the study area. Data of this type for all, or even most, taxa are rarely available at these scales. In the case study, vascular plants were the only species data generally collected to a consistent standard at broad scales, although they could only be used to assess stands or patches in the context of major catchments (averaging 6 million ha) because of inconsistencies in the way these data were analysed between major catchments. In the case study the predicted pre-European distribution of vegetation communities derived from these plant species data were used as landscapes representing units of homogeneous biotic variation. Vegetation communities are useful surrogates for turnover of some fauna groups (Ferrier & Watson 1997) as well as plant communities, although there are several other methods that can be employed for analysing data for multiple species (Burgman *et al.* 2005).

Landscapes defined in the context of biological variation across scale of major catchments (c.6 million ha) were not always sufficient to assess adequately the contribution of a stand or patch to the representation of biological diversity. Some vegetation types were common in a particular catchment but under considerable threat across their range as a whole. We therefore assessed a site's contribution to the representation of biological diversity in the context of biological variation across the entire state (80 million ha). As with many large jurisdictions consistent biotic data were not available over this vast area, we drew on a classification of the state into landscapes of homogeneous biological variation based on mapped abiotic surrogates including modelled climate surfaces, terrain derived from a digital elevation model and geology (see Pressey *et al.* 2000).

Landscapes representing units of homogeneous biological variation (whether these are based on biotic or abiotic data) within large regions are typically predicted with a level of uncertainty, and mapped at a scale, that is too coarse simply to overlay with individual stands or patches that are being assessed. Uncertainty due to the scale of the mapped data was addressed by defining the landscape unit with features that could be identified by assessors on the ground. Thus, assessors did not rely on mapped data to identify what landscape a stand or patch was in. However, assessors often found it difficult to place features observed in the field (e.g. dominant plant species) into one of the predetermined landscape units (e.g. plant communities) because biotic assemblages change in response to environmental and disturbance continua whereas we define landscape units as discrete entities. Thus, in order to assess stands or patches in the landscape context, landscape units must be recognizable at the stand or patch scale.

Landscape measures for biodiversity assessments

Once landscape units are defined, they must then be assessed for their relative biodiversity conservation values and the impact of each proposal at the stand or patch scale must then be assessed in this context. Three measures that can be used to assess proposals in a landscape context – and represent the theme for this section of the book – are: vegetation cover, vegetation pattern and patch content.

Vegetation cover

The cover of native vegetation in landscape units has been used to describe: the suitability of a landscape for different biotas (e.g. Radford *et al.* 2005); ecosystem function (e.g. Ludwig *et al.* 2002); and representativeness of the conservation management network (e.g. Pressey *et al.* 2003). Generic levels of cover, around which rapid changes in species richness are predicted for different groups of species (e.g. Andrén 1994; McIntyre *et al.* 2000; Brooks *et al.* 2002; Radford *et al.* 2005), have been used widely as targets for management. These studies have found that relationships between the cover of native vegetation and species richness are not linear, so impacts on species are deemed to be greatest within certain ranges of vegetation cover change. For example, Radford *et al.* (2005) found that there was a threshold at around 10% native woody vegetation cover within landscapes in which bird species richness changed markedly, so actions that reduced or increased, the cover of woody native vegetation within landscapes below or above 10%, respectively, were predicted to have a relatively major impact on bird species richness.

Generic targets of this type have been criticized because species' responses to vegetation cover are variable (e.g. Lindenmayer *et al.* 2005). An alternative approach is to define the area of vegetation required for viable populations of individual species, functional groups of species or communities that occur in a specific area (e.g. Burgman *et al.* 2005). However, as discussed previously, suitable data (species data, population dynamics and spatial data on habitat attributes) are rarely available across large jurisdictions to enable these analyses to be undertaken on all species within the constraints outlined in Table 19.1. In the case study we therefore identified generic targets for landscape cover that represented points at which vegetation cover change was likely to have a major impact on the biota. However, assessments were also undertaken at each site for individual species of concern (threatened species) based on their individual habitat requirements in recognition that generic targets are unlikely to be sufficient for all species.

An often neglected point is that measures of vegetation cover at landscape scales are sensitive to, or limited by, the quality of the spatial data available to do the assessment. The highest-resolution vegetation cover data available over large areas within NSW at the time of our study were based on an analysis of Landsat imagery, which detected woody native vegetation to a minimum resolution of 25 × 25 m and a minimum canopy cover of approximately 20%. Non-woody native vegetation, sparsely wooded native vegetation, small patches and narrow patches of native woody vegetation – all important components of habitat in some landscapes (e.g. Fischer & Lindenmayer 2002; Gibbons & Boak 2002; van der Ree *et al.* 2003) – were not captured in these spatial data (Fig. 19.1). The limitations of these data meant that it was inappropriate to use automated (e.g. GIS-based) tools for assessing vegetation cover at local scales (up to 1000 ha) because the spatial data were often insensitive to changes brought about by individual proposals to clear or improve native vegetation. Assessors therefore had manually to digitize proposals using high-resolution aerial photography or satellite imagery as the backdrop and visually estimate vegetation cover and vegetation cover change, within the immediate landscape around the proposal. However, this meant that we could only use broad classes for assessing vegetation cover at these scales. Thus, the limitations of

Figure 19.1 **An illustration of the finest resolution native vegetation layer available across a sample of the case study (mapped at a resolution of 25 × 25 m) (shaded), overlayed with an Ikonos panchromatic satellite image (1 × 1 m grid cell resolution). Includes material ©2002 Space Imaging LLC, distributed by Raytheon Australia.**

the underlying spatial data are a crucial consideration when undertaking assessments of vegetation cover or vegetation cover change, at fine scales.

Vegetation pattern

Measures of vegetation pattern have been used as surrogates of species presence, population viability, movements of the biota, ecosystem function and threatening processes. A myriad of potential measures can be used to describe the general pattern of vegetation in a landscape. The popular software FRAGSTATS (McGarigal & Marks 1994) can calculate 59 metrics of pattern for a landscape, although the existence of a landscape metric does not confer upon it an association with ecosystem function.

There is little consensus on a robust ecological model that underpins the role of pattern on biotas generally. The patch-corridor-matrix model of Forman

(1995) has been increasingly criticized for its view of the matrix as a hostile entity and the value of corridors for the biota remains equivocal. Spatial models of populations and assemblages of species have been advanced as a potential way forward (Burgman *et al.* 2005).

In the case study we trialled a method for measuring the impacts of changing vegetation pattern on the biota using a cost-benefit algorithm grounded in metapopulation theory (Drielsma *et al.*, in press). Tools such as this can be populated with data for individual species or the requirements of multiple species. However, a lack of information on the dispersal abilities of species or assemblages, the resolution of the spatial data (Fig. 19.1) and an absence of spatial data on variation in habitat value within remnants and between remnants (or the 'matrix') meant that the tool could not be used to its full potential. Manually digitizing native vegetation and providing a qualitative assessment of vegetation condition within polygons using high-resolution imagery (e.g. aerial photography) as a backdrop may be a feasible solution to some of these issues (S. Ferrier & M. Drielsma, personal communication).

Patch content

Several measures at the scale of the stand associated with the occurrence and abundance of the biota and ecosystem functions (e.g. Lindenmayer & Franklin 2002; McElhinny *et al.* 2005) can be made within the constraints imposed by management (Table 19.1), where species-based metrics are impractical. Several methods have been developed to combine suites of habitat attributes into metrics or indices for rapid assessments at the patch-scale (Andreasen *et al.* 2001; McElhinny *et al.* 2005; Gibbons & Freudenberger 2006). However, the absence of data on patch content at landscape scales means that few attempts have been made to use measures in stands and patches in the landscape context. Assessments of patch content not undertaken in a landscape context can lead to poor decision-making. For example, Austin *et al.* (2002) found that less than 4% of remnant woodland communities in a fragmented landscape had an understorey dominated by native plants, indicating that even highly modified examples of these communities were priorities for conservation and restoration – an assessment that could not be made without data on patch content across the landscape.

Four broad approaches can be used to assess patch content at landscape scales. The first is simply sampling landscapes representatively on the ground using field-based assessment methods so the features of a stand or patch can be assessed relative to data from the representative sample. While this approach enables a site to be placed in a broader landscape context, it does not produce spatially explicit data on patch content. The second method for assessing patch content at landscape scales is a qualitative assessment based on expert knowledge or the use of surrogates such as land use. This approach can lack the rigour needed for repeatable or transparent outcomes, as is the case in a regulatory environment, and can be insensitive to change over time for monitoring purposes. The third approach is remote sensing. Inexpensive, widely available and extensively archived synoptic satellite platforms such as Landsat are suitable for assessments over large areas and for monitoring over time, but are limited in the characteristics of stands they can detect (Wallace *et al.* 2006). Higher-resolution, aircraft-mounted multispectral and hyperspectral sensors and laser scanners can detect more features of vegetation than satellite-mounted sensors, but are only practical for use over small areas because of their expense, and relatively limited coverage and archive (Gibbons & Freudenberger 2006). Multiscale remote sensing approaches, in which the strengths of different sensors are used in a complementary way, may be a solution. In the fourth approach, assessments of vegetation condition at individual sites can be spatially interpolated across landscapes at reasonable accuracies and at reasonable expense using a combination of environmental predictors (e.g. tree cover, patch dimensions, land use) and data from different remote-sensing platforms (e.g. Zerger *et al.* 2006) (Fig. 19.2).

Informing vegetation cover and pattern with patch content

Access to spatially explicit data on patch content at landscape scales will profoundly change the way we view and assess cover, pattern and even patch content. Measures of cover and pattern are typically reported as though the component vegetation can only occur in one of two states: present or absent. However, the effect on biodiversity of vegetation cover and pattern in a landscape is mediated by patch content (i.e. composition and structure of vegetation within the patch). Further, spatial data that only represent the presence/absence of vegetation locks us in the paradigm of the matrix as a uniformly hostile entity. So measures of vegetation cover and pattern based on the presence or absence of vegetation are, at best, coarse and, at worst, misleading. For example, Weinberg (2005) predicted that around 40% of the mapped vegetation

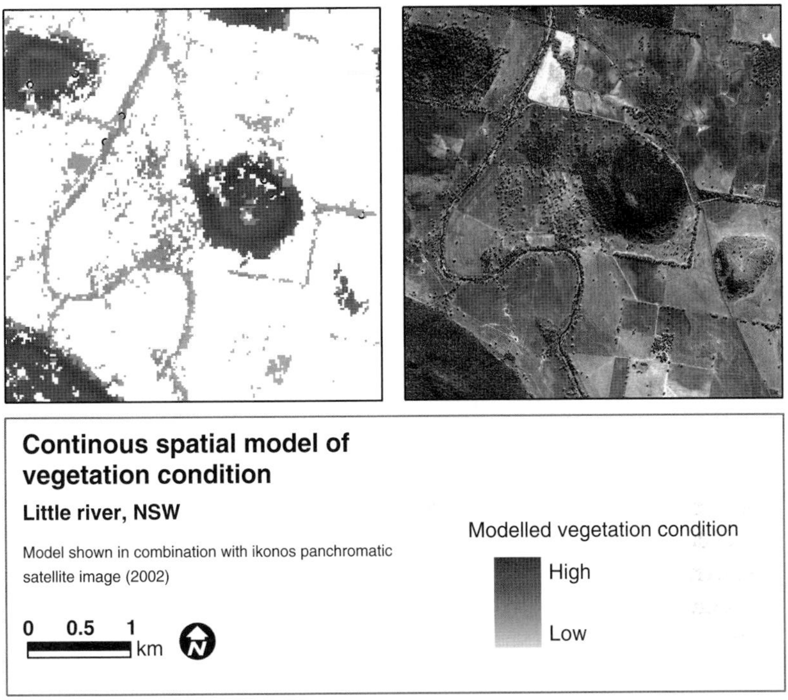

Figure 19.2 **A continuous spatial model of vegetation condition (left) based on a combination of environmental predictors and predictors derived from satellite imagery. A satellite image of the same site is provided on the right. Includes material ©2002 Space Imaging LLC, distributed by Raytheon Australia.**

cover in an agricultural region in Australia was not regenerating, indicating that substantially less vegetation cover than the amount mapped was likely to contribute to the viability of the biota in the long term.

Spatially explicit data on changes to habitat at fine scales can advance the way we view and assess biodiversity in many ways. Spatial representations of habitat enable patch content to be expressed not only in terms of the quantity of different habitat attributes, but also as the spatial arrangement of these habitat attributes at different scales. The latter is likely to represent more closely the habitat requirements of species. The uncertainty associated with species, assemblage and population modelling (see Burgman *et al.* 2005) will be therefore be enhanced with such information. Authors with access to continuous spatial habitat data have demonstrated that the matrix can be crucial to the functioning of landscapes as a whole (e.g. Ludwig *et al.* 2002). Continuous spatial representations of habitat allow alternative approaches for assessing fragmentation impacts (e.g. Drielsma *et al.* 2007) to be used to their potential. Continuous spatial representations of habitat at landscape scales may ultimately render measures of cover and pattern redundant, instead of allowing us to view landscapes simply as continua of different habitats (see Manning *et al.* 2004) rather than collections of discrete patches in a hostile matrix.

Key principles

1. Assessments of biodiversity at the scale of individual stands or patches must be made in the landscape context.
2. Landscape measures inform land-use decisions only if they can be interpreted at the scale that land-use decisions are taken.
3. Landscapes should be defined at multiple scales. No single landscape unit suffices for assessing biodiversity.
4. Some biodiversity issues can be addressed only if landscape units are defined as homogeneous units of biological variation.
5. Landscape measures such as cover and pattern should be informed by patch content mapped at landscape scales.
6. Landscapes are continua of habitats, not collections of discrete patches in a uniformly hostile matrix.

References

Andreasen, J.K., O'Neill, R.V., Noss, R. & Slosser, N.C. (2001) Considerations for the development of a terrestrial index of ecological integrity. *Ecological Indicators* **1**, 21–35.

Andrén, H. (1994) Effects of habitat fragmentation on birds and mammals in landscapes with different proportions of suitable habitat: a review. *Oikos* **71**, 355–366.

Austin, M.F, Cawsey, E.M., Baker, B.L., Yialeloglou, M.M., Grice, D.J. & Briggs, S.V. (2000) *Predicted Vegetation Cover in the Central Lachlan Region*. CSIRO Sustainable Ecosystems, Canberra.

Briggs, S.V. (2001) Linking ecological scales and institutional frameworks for landscape rehabilitation. *Ecological Management and Restoration* **2**, 28–35.

Brooks, T.M., Mittermeier, R.A., Mittermeier, C.G. *et al.* (2002) Habitat loss and extinction in the hotspots of biodiversity. *Conservation Biology* **16**, 909–923.

Burgman, M.A., Lindenmayer, D.B. & Elith, J. (2005) Managing landscapes for conservation under uncertainty. *Ecology* **86**, 2007–2017.

Drielsma, M., Manion, G. & Ferrier, S. (2007) The spatial links tool: Automated mapping of habitat linkages in variegated landscape. *Ecological Modelling*, **200**, 403–411.

Ferrier, S. & Watson, G. (1997) *An Evaluation of the Effectiveness of Environmental Surrogates and Modelling Techniques in Predicting the Distribution of Biological Diversity*. Environment Australia, Canberra.

Ferrier, S., Watson, G., Pearce, J. & Drielsma, M. (2002) Extended statistical approaches to modelling spatial pattern in biodiversity in northeast New South Wales. I. Species-level modelling. *Biodiversity and Conservation* **11**, 2275–2307.

Fischer, J. & Lindenmayer, D.B. (2002) Small patches can be valuable for biodiversity conservation: two case studies on birds in southeastern Australia. *Biological Conservation* **106**, 129–136.

Forman, R.T. (1995) *Land Mosaics. The Ecology of Landscapes and Regions*. Cambridge University Press, New York.

Gibbons, P. & Boak, M. (2002) The value of paddock trees for regional conservation in an agricultural landscape. *Ecological Management and Restoration* **3**, 205–210.

Gibbons, P. & Freudenberger, D. (2006) An overview of methods used to assess vegetation condition at the scale of the site. *Ecological Management and Restoration* **7**, S10–S17.

Gibbons, P., Ayers, D., Seddon, J., Doyle, S. & Briggs, S. (2005) *BioMetric Version 1.8. A Terrestrial Biodiversity Assessment Tool for the NSW Property Vegetation Plan Developer. Operational Manual*. Department of Environment and Conservation (NSW), Canberra (www.nationalparks.nsw.gov.au/npws.nsf/content/ biometric_tool).

Gibbons, P., Zerger, A., Jones, S. & Ryan, P. (2006) Mapping vegetation condition in the context of biodiversity conservation. *Ecological Management and Restoration* **7**, S1–S2.

Kleijn, D., Baquero, R.A., Clough, Y. *et al.* (2006) Mixed biodiversity benefits of agri-environment schemes in five European countries. *Ecology Letters* **9**, 243–254.

Lindenmayer, D.B. & Franklin, J. F. (2002) *Conserving Forest Diversity: A Comprehensive Multiscaled Approach*. Island Press, Washington, DC.

Lindenmayer, D.B., Fischer, J. & Cunningham, R.B. (2005) Native vegetation cover thresholds associated with species responses. *Biological Conservation* **124**, 311–316.

Ludwig, J.A., Eager, G.N., Bastin, V.H., Chewings, V.H. & Liedloff, A.C. (2002) A leakiness index for assessing landscape function using remote sensing. *Landscape Ecology* **17**, 157–172.

McElhinny, C., Gibbons, P., Brack, C. & Bauhus, J. (2005) Forest and woodland stand structural complexity: Its definition and measurement. *Forest Ecology and Management* **218**, 1–24.

McGarigal, K. & Marks, B.J. (1994) *FRAGSTATS. Spatial Pattern Analysis Program for Quantifying Landscape Structure.* Version 2.0. Forest Science Department, Oregon State University, Corvallis, OR.

McIntyre, S., McIvor, J.G. & MacLeod, N.D. (2000) Principles for sustainable grazing in eucalypt woodlands: landscape-scale indicators and the search for thresholds. In: Hale, P., Petrie, A., Moloney, D. & Sattler, P. (eds.) *Management for Sustainable Ecosystems*, pp. 92–100. University of Queensland, Brisbane.

Manning, A.D., Lindenmayer, D.B. & Nix, H.A. (2004) Continua and Umwelt: novel perspectives on viewing landscapes. *Oikos* **104**, 621–628.

Margules, C.R. & Pressey, R.L. (2000) Systematic conservation planning. *Nature* **415**, 243–253.

Noss, R.F. (1990) Indicators for monitoring biodiversity: a hierarchical approach. *Conservation Biology* **4**, 355–364.

Parkes, D., Newell, G. & Cheal, D. (2003) Assessing the quality of native vegetation: the 'habitat hectares' approach. *Ecological Management and Restoration* **4**, S29–S38.

Pressey, R.L., Hagar, T.C., Ryan, K.M. *et al.* (2000) Using abiotic data for conservation assessments over extensive regions: quantitative methods applied across New South Wales, Australia. *Biological Conservation* **96**, 55–82.

Pressey, R.L., Cowling, R.M. & Rouget, M. (2003) Formulating conservation targets for biodiversity pattern and process in the Cape Floristic Region, South Africa. *Biological Conservation* **112**, 99–127.

Radford, J.Q., Bennett, A.F. & Cheers, G.J. (2005) Landscape-level thresholds of habitat cover for woodland-dependent birds. *Biological Conservation* **124**, 317–337.

van der Ree, R., Bennett, A.F. & Gilmore, D.C. (2003) Gap-crossing by gliding marsupials: Thresholds for use of isolated woodland patches in an agricultural landscape. *Biological Conservation* **115**, 241–249.

Ribaudo, M.O., Hoag, D.L., Smith, M.E. & Heimlich, R. (2001) Environmental indices and the politics of the Conservation Reserve Program. *Ecological Indicators* **1**, 11–20.

Rouget, M. (2003) Measuring conservation value at fine and broad scales: implications for a diverse and fragmented region, the Agulhas Plain. *Biological Conservation* **112**, 217–232.

United Nations Food and Agriculture Organization (2005) *Global Forest Resources Assessment.* UNFAO, Rome.

US Fish and Wildlife Service (1980) *Habitat Evaluation Procedures.* Department of the Interior, Washington, DC.

Wallace, J., Behn, G. & Furby, S. (2006) Vegetation condition assessment and monitoring from sequences of satellite imagery. *Ecological Management and Restoration* **7**, S31–S36.

Weinberg, A.Z. (2005) Eucalypt regeneration in a fragmented agricultural landscape. Honours Thesis, University of New South Wales, Sydney.

Zerger, A., Gibbons, P., Jones, S. *et al.* (2006) Spatially modelling native vegetation condition. *Ecological Management and Restoration* **7**, S37–S44.

20

Avoiding Irreversible Change: Considerations for Vegetation Cover, Vegetation Structure and Species Composition

Joern Fischer and David B. Lindenmayer

Abstract

An important goal in landscape design is to avoid potentially irreversible ecosystem changes. Such changes have been discussed in the context of thresholds, regime shifts and extinction cascades. Thresholds occur where small changes in one variable result in a large change in another variable. Regime shifts occur when a system 'flips' from one state to another. Extinction cascades occur where the extinction of one species triggers the loss of one or more other species, which in turn leads to further extinctions. Potentially irreversible changes may occur as a result of changes in many variables. Three variables are discussed here: (i) the amount of native vegetation cover; (ii) the structure of native vegetation; and (iii) species composition. Species extinctions may occur more rapidly at particularly low levels of native vegetation cover. However, negative effects may be partly mitigated in heterogeneous landscapes and where the matrix resembles natural vegetation structure. The structure of native vegetation is often related to disturbance regimes. Extinction cascades are more likely to occur following the loss of structural attributes that many species depend upon, such as features typical of old-growth forest or other ecosystem-specific keystone structures. Changes to species composition per se also may result in extinction cascades. This risk is particularly high when entire functional groups or keystone species are lost. Landscape design should attempt to maintain: (i) high levels of natural vegetation cover embedded within a heterogeneous matrix; (ii) structurally characteristic native vegetation, including keystone structures; and (iii) a diversity of species within and across functional groups, including keystone species.

Keywords: extinction cascades; keystone species; keystone structures; regime shifts; thresholds.

Introduction

The design of economically productive, sustainable and resilient landscapes is a major challenge for scientists, policy-makers and land managers (Mattison & Norris 2005). Many landscapes around the world are currently managed unsustainably. Several recent reviews have expressed concerns about losses of native species due to agricultural intensification in Europe (Benton *et al.* 2003), Australia (Vesk & Mac Nally 2006) and at the global scale (Robertson & Swinton 2005). Widespread land-use intensification poses a severe threat to the ongoing supply of vital ecosystem services that are needed to sustain human and other life on Earth (Foley *et al.* 2005). Concerns are also increasing that some apparently sustainable landscapes may lack the resilience to retain their function in the face of global change, and particularly climate change (Bengtsson *et al.* 2003; Schröter *et al.* 2005). For example, an increase in the incidence of drought has been predicted to alter fundamentally North American pinon pine ecosystems due to increased risks from insect infestation

(Breshears *et al.* 2005). A range of studies has emphasized the need to pay particular attention to possible irreversible ecosystem changes before they occur (Milton *et al.* 1994; Brown *et al.* 1999; Millennium Ecosystem Assessment 2003; Walker *et al.* 2004). However, addressing this need is hampered by poor knowledge about which management actions are likely to lead to irreversible changes (Balmford & Bond 2005). This essay attempts to address this knowledge gap with respect to key landscape design principles. Three themes are discussed, for which sound landscape design is particularly important to avoid irreversible ecosystem changes: (i) the amount and configuration of native vegetation cover; (ii) vegetation structure and associated disturbance regimes; and (iii) species composition. Following a brief section defining extinction cascades, thresholds and regime shifts, key considerations and their underlying logic will be discussed for each theme. A summary of emerging principles for landscape design concludes this chapter.

Thresholds, regime shifts and extinction cascades

The notions of thresholds, regime shifts, and extinction cascades are related. However, because the links between them are not always clear, each notion is briefly introduced here (see Fig. 1 for a schematic representation). "**Thresholds**" are widely discussed in the ecological literature (Walker & Meyers 2004; Huggett 2005). However, not all authors use the term in the same way (Lindenmayer & Luck 2005). Strictly speaking, a threshold is "the magnitude or intensity that must be exceeded for a certain reaction or phenomenon to occur" (Oxford English Dictionary 1989). Some authors have relaxed this requirement for a sudden change point, and instead refer to thresholds as "transition ranges across which small changes in ... [one variable] ... produce abrupt shifts in ecological responses" (With & Crist 1995). This relaxed definition is more likely to have practical value, since most ecological phenomena are characterized by complex interactions where it is difficult to define a single specific point of change (as opposed to a narrow range) (Lindenmayer *et al.* 2005). Thresholds have been discussed for a range of ecological phenomena (Walker & Meyers 2004), including for the effect of land cover pattern on species distribution (With & Crist 1995), the maximum gliding distance of arboreal marsupials in relation to habitat connectivity (van der Ree *et al.* 2003), the travel costs for birds through modified landscapes (Graham 2001), or the effects of inbreeding on the probability of population extinction (Frankham 1995). Some authors have also used the term with specific reference

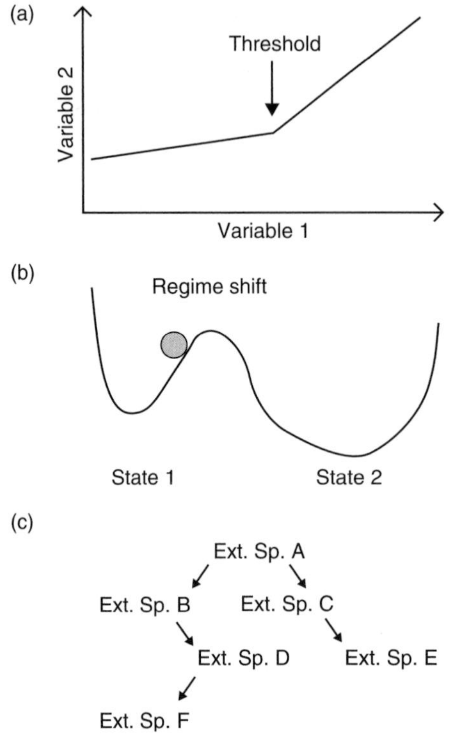

Figure 20.1 **Diagrammatic representation of three interrelated concepts that relate to potentially irreversible change. (a) A threshold occurs when the relationship between two variables includes a sudden change point. (b) A regime shift occurs when a system changes from one stable state to an alternative stable state. (c) An extinction cascade occurs when the extinction of one species triggers the loss of one or more other species, which in turn leads to further extinctions.**

to land management, where a "threshold" is a specific recommended value (e.g. 30% of native vegetation cover) (McIntyre *et al.* 2000).

Regime shifts have been considered to be broadly synonymous with threshold effects by some authors (Folke *et al.* 2004; Mayer & Rietkerk 2004; Walker & Meyers 2004). However, the term 'regime shift' is usually reserved to ecosystem-level change – for example, the maximum distance an arboreal marsupial can glide between trees may be considered a 'threshold' (van der Ree *et al.* 2003), but it is unlikely that the inaccessibility of trees beyond this distance would be considered a 'regime shift'. Regime shifts occur when a small change in one variable causes a fundamental change in an ecosystem property that cannot be readily reversed by returning the first variable to its original level (Folke *et al.* 2004). Examples of regime shifts and associated alternative stable states have been documented for lakes (clear vs. turbid), coral reefs (corals vs. macroalgae), woodlands (herbaceous vs. woody vegetation) and deserts (perennial vegetation vs. bare soil with ephemeral plants) (Scheffer *et al.* 2001; Folke *et al.* 2004). The link between thresholds, regime shifts and extinction cascades has received limited attention to date. **Extinction cascades** arise when the extinction of one species causes the extinction of another species, which in turn threatens other species, and so on. The most widely discussed case of extinction cascades is that of 'trophic cascades', where the extinction of apex predators changes species composition at lower levels of the food chain (Soulé *et al.* 2005). However, cascading effects on other species also have been discussed in different contexts. For example, Vitt *et al.* (1998) suggested that the increase of heliothermic lizards in selectively logged Amazonian forest may have negative cascading effects on other forest lizards. In this essay, extinction cascades are discussed alongside thresholds and regime shifts because they often characterize or result from a regime shift, and because certain extinction events (e.g. of keystone species) may, in themselves, be seen as small changes in one variable that have drastic effects on overall ecosystem identify, function and feedbacks.

Amount and configuration of native vegetation cover

Effects for single species

The amount of habitat cover as well as its spatial arrangement in a given landscape are considered to affect the persistence of a given individual species. This logic arises because of the well-known positive association between habitat patch size and population persistence (Bender *et al.* 1998), and because populations are more likely to persist if an exchange of individuals between patches is possible (Hanski 1998). Many modelling studies have assessed the combined effect of the total amount of habitat cover and its spatial arrangement on the populations of individual species (Bascompte & Sole 1996; Fahrig 1997; With *et al.* 1997). Modelling studies have shown that the spatial arrangement of habitat becomes increasingly important as the total amount of habitat decreases, because of a compounding of the negative effects of habitat loss and habitat subdivision (Fahrig 2003). Other parameters likely to influence the persistence of a given population are the reproductive output of the species (With & King 1999), as well as emigration rate and probability of survival in the matrix between core habitat patches (Fahrig 2003). Several empirical studies also support the notion that the spatial arrangement of habitat patches can be important. An example is the long-tailed tit (*Aegithalos caudatus*) in boreal forest patches in central Sweden (Jansson & Angelstam 1999). The distance between biophysically suitable patches was a key determinant of them being occupied, especially in landscapes where the total amount of forest cover was low (5%). Patches within 100 m of another occupied patch had an 80% chance of being occupied, whereas patches within 500 m of another occupied patch had only a 10% chance of being occupied. Similarly pronounced changes with a relatively small increase in isolation distance were identified for the white-browed treecreeper (*Climacteris affinis*) in northwestern Victoria, Australia (Radford & Bennett 2004).

Effects for multiple species

Importantly, what constitutes suitable habitat cover will vary between species, and different levels of isolation will negatively affect different species with respect to different vegetation communities (Andrén *et al.* 1997; Villard *et al.* 1999; Homan *et al.* 2004). A major review of both individual species' populations and species richness concluded that the effects of habitat subdivision were likely to be particularly pronounced when native vegetation cover in a given landscape decreased below a proposed threshold of 30% (Andrén 1994). Although this review has been very influential, it has also attracted some criticism because it did not carefully describe what constituted suitable habitat for different species, and because it did not sufficiently emphasize the important role of matrix conditions (Mönkkönnen & Reunanen 1999). Other workers also have assessed species richness in relation to the amount and spatial arrangement of vegetation. For example, Parker and Mac Nally (2002) assessed invertebrate richness in an experimental model system but found no evidence of threshold effects (see also Lindenmayer *et al.* 2005). In contrast, Radford *et al.* (2005) noted a pronounced decline in the species richness of woodland birds in fragmented woodland landscapes in northern Victoria, Australia, when woodland cover declined below a proposed threshold of 10%.

Implications for landscape design

With respect to landscape design principles, a key insight is that compounding negative effects of the loss of native vegetation and its subdivision are particularly likely when levels of native vegetation cover are low. However, it appears unlikely that there is a universal rule about how much native vegetation is required in any given landscape for it to be sustainable. First, species are likely to be lost even above a possible threshold where the effects of vegetation loss and subdivision compound one another (Radford *et al.* 2005). Second, different species are likely to respond to landscape modification in different ways (Lindenmayer *et al.* 2005), and some species are more important for overall ecosystem function than others (see below). Third, conditions outside areas of native vegetation are likely to play a crucial role in determining species' persistence. The maintenance of landscape heterogeneity and matrix management therefore are particularly important design aspects in landscapes with little remaining native vegetation cover (Fahrig 2003; Fischer *et al.* 2006). Conversely, a binary view of landscapes that values native vegetation but assigns a zero habitat value to modified areas is likely to be overly pessimistic and may inadvertently escalate the conflict between production and conservation, thereby preventing their successful integration (Polasky *et al.* 2005). This is problematic because the integration of production and conservation is widely accepted as a key challenge for the sustainable management of modified landscapes in the future (Foley *et al.* 2005; Robertson & Swinton 2005). Finally, the practical utility of simple vegetation cover thresholds to manage complex ecosystems may be limited by the high variability that characterizes many ecological datasets. Such variability, in turn, can make it difficult or impossible to identify correctly possible threshold values, even if they occur (Lindenmayer *et al.* 2005).

Vegetation structure and associated disturbance regimes

In addition to the size and spatial arrangement of patches of native vegetation, vegetation structure is a key component affecting species distribution patterns (Gilmore 1985; Antvogel & Bonn 2001). Changes to vegetation structure in modified landscapes often result from altered disturbance regimes. For example, anthropogenic disturbance regimes like logging (Johns 1988; Vitt *et al.* 1998;

Williams *et al.* 2001) or grazing (Bromham *et al.* 1999; Ludwig *et al.* 2000; McIntyre *et al.* 2003) can fundamentally alter vegetation structure and species composition. In addition, human activities can alter natural disturbance regimes, such as fire regimes (Bowman *et al.* 2004) or hydrological flows (Hazell *et al.* 2003). From the perspective of landscape design, it is often useful to attempt to imitate historical disturbance regimes as closely as possible, because local organisms are most likely to respond favourably to disturbances that are similar to typical historical disturbances in spatial extent, frequency and intensity, and create a similar vegetation structure (Lindenmayer & Franklin 2002).

The consideration of thresholds, regime shifts and possible extinction cascades is relevant to managing disturbance regimes and vegetation structure. In particular, key structural attributes may be lost under inappropriate disturbance regimes. For example, clearfelling with short rotation cycles prevents the development of structural features that are characteristic of old-growth forest, such as large quantities of coarse woody debris or tree hollows (Lindenmayer *et al.* 1991; Williams *et al.* 2001). Similarly, high-intensity grazing pressure can alter the vegetation structure and species composition of grasslands (McIntyre *et al.* 2003) or may prevent the natural regeneration of trees in grazed woodlands (Spooner *et al.* 2002; Saunders *et al.* 2003; Dorrough & Moxham 2005). The loss of key structural attributes can lead to extinction cascades, thereby fundamentally altering local species composition in a way that can be difficult or impossible to reverse, especially if certain species have become rare across an entire landscape or region.

Tews *et al.* (2004) defined structural attributes that were particularly important to maintain local species diversity as 'keystone structures'. Examples include tree hollows in many Australian forests, without which many vertebrate species could not survive (Gibbons & Lindenmayer 2002), but also a wide range of other, less well-recognized structural attributes. For example, after heavy spring or winter precipitation, temporary wetlands occur in conventionally ploughed depressions in agricultural fields in northeastern Germany. These temporary wetlands provide habitat for a diverse community of plants and carabid beetles, and local species diversity would be substantially reduced without them (Tews *et al.* 2004). Scattered trees in many grazing landscapes also fulfil disproportionately important ecological roles (Manning *et al.* 2006). For example, in African savannas, scattered trees are known to alter their local abiotic environment through water and nutrient redistribution. These abiotic changes, in turn, lead to increased local plant diversity (Vetaas 1992; Belsky 1994). The structural complexity thus created by scattered trees and their diverse understorey provides habitat for a range of animal species, including raptors that build nests in the canopy, herbivores like the springbok (*Antidorcas marsupialis*), which shelter under scattered trees, and arboreal mammals like the tree rat (*Thallomys paedulcus*), which use hollows in old trees (Dean *et al.* 1999).

In general, a diverse range of structural attributes is likely to lead to increased species diversity (MacArthur & MacArthur 1961), and historical disturbance regimes provide a useful guide with respect to what may constitute sustainable human disturbance regimes (Lindenmayer & Franklin 2002). To avoid undesirable extinction cascades, landscape design particularly needs to consider keystone structures that are of disproportionate importance to many species.

Species composition

In addition to the structure and spatial pattern of vegetation, the species composition of ecosystems needs to be considered in landscape design. Species composition is important because undesirable extinction cascades or regime shifts may occur if certain species are lost, even if native vegetation cover is intact.

A useful framework to conceptualize the importance of species composition is to consider the different functional roles that species play in an ecosystem. Functional roles may vary widely, and include nitrogen fixation, the provision of food for herbivores, the excavation of tree hollows and the predation of large herbivores. Functional diversity refers to the diversity of different groups that fulfil broadly similar functions (Elmqvist *et al.* 2003; Srivastava & Vellend 2005). If an entire functional group is lost, this may lead to undesirable and potentially irreversible changes in ecosystem function (Walker 1992, 1995).

Two related situations can be identified where the risk of extinction cascades is particularly high: (i) if 'response diversity' within a given functional group is low (Elmqvist *et al.* 2003), or (ii) if a given functional group comprises a single 'keystone species' (Power *et al.* 1996). Response diversity refers to the range of different responses to a particular external change exhibited by species within a given functional group (Elmqvist *et al.* 2003). For example, Walker *et al.* (1999) assigned grasses in an Australian rangeland to different functional groups (see also Chapter 34). The system exhibited high response diversity to grazing pressure. While some species within a given functional group declined with grazing pressure, others increased in abundance and thus were able to compensate for the loss of their functional equivalents (Walker *et al.* 1999). Without this compensation, that is, if all species had declined in response to grazing pressure, response diversity would have been low, and some important functions would have been lost from the ecosystem.

Keystone species represent a particular case of low response diversity. A keystone species, or strongly interacting species (Soulé *et al.* 2005), is 'one whose impact on its community or ecosystem is large, and disproportionately large relative to its abundance' (Power *et al.* 1996). Usually, the large impact of keystone species arises because they have no functional equivalents, and their loss therefore would amount to the loss of an entire functional group. Extinction cascades resulting from the loss of keystone species are well documented (Power *et al.* 1996). Examples of keystone species include: the carnivorous starfish *Pisaster ochraceus* in rocky intertidal areas of the west coast of North America through its effect on marine species composition (Paine 1969); the beaver (*Castor* spp.) through its effects on habitat structure for a wide range of other species (Soulé *et al.* 2005); the bison (*Bos bison*) through its effects on the species composition of native grasslands (Knapp *et al.* 1999); and the red-naped sapsucker (*Sphyrapicus nuchalis*) through its cavity-excavating role (Daily *et al.* 1993). In each of these cases, the loss of the keystone species would result in fundamental (and often undesirable) ecosystem change.

An appreciation of the role of species composition in maintaining ecosystem function has important implications for landscape design principles. Special attention in landscape design is required to maintain representatives of species in all important functional groups. Maintaining a diversity of species within a given functional group increases the likelihood of at least some species within the group persisting in the event of an untried management practice or an unforeseen environmental change – thereby providing an 'insurance' or 'resilience' value to the ecosystem (Elmqvist *et al.* 2003). Keystone species have no functional equivalents and therefore require particular attention in landscape design.

Implications for landscape design

Many considerations are important for the design of sustainable landscapes. A key consideration is the avoidance of undesirable, and potentially irreversible, changes to ecosystem function that may arise as a result of 'thresholds' being crossed and manifest themselves as extinction cascades and regime shifts.

Principles

1 **High levels of native vegetation cover should be maintained. Where this is not possible, the matrix should be heterogeneous and should incorporate elements that are structurally similar to native vegetation.** (*Rationale:* Theoretical and empirical studies on single species have found increased probabilities of extinction at low levels of habitat cover. Native vegetation is suitable habitat for many species. However, many native species also will be able to persist in modified environments, especially if their structure resembles native vegetation.)

2 **Structurally characteristic native vegetation should be maintained by using appropriate disturbance regimes that attempt to mirror historical disturbance regimes. Particular attention should be paid to 'keystone structures' that are important to many species.** (*Rationale:* Vegetation structure is a key determinant of habitat suitability for many species. Structural complexity tends to enhance species richness, especially in forest ecosystems. Some structural features, such as those characteristic of old-growth forests, are particularly useful habitat features for many species. Hence, their maintenance is disproportionately important.)

3 **A diversity of species should be maintained within and across functional groups. Particular care should be taken not to lose keystone species or entire functional groups.** (*Rationale:* Different species fulfil different ecosystem functions. If an entire functional group is lost, this is particularly likely to have negative consequences for ecosystem functioning. Because keystone species have no functional equivalents, their maintenance is disproportionately important.)

4 **A modified landscape is most likely to be sustainable if it is characterized by patterns and processes that resemble those typical of the landscape before modification.** Patterns that result in low levels of native vegetation cover and low landscape connectivity are more likely to trigger extinction cascades than patterns with extensive native vegetation; reduced structural complexity is more likely to be linked to extinction cascades than high structural complexity; and species-rich systems with many species fulfilling a given ecosystem function are less likely to suffer from cascading effects than ecosystems where functional groups are represented by only a few species. For all three considerations – vegetation patches, vegetation structure and species composition – some elements are more important than others. Such 'keystone features' of ecosystems should be a high priority in landscape design.

Acknowledgements

Comments and discussions with various workshop participants have greatly helped to improve the quality of this essay.

References

Andrén, H. (1994) Effects of habitat fragmentation on birds and mammals in landscapes with different proportions of suitable habitat – a review. *Oikos* **71**, 355–366.

Andrén, H., Delin, A. & Seiler, A. (1997) Population response to landscape changes depends on specialization to different landscape elements. *Oikos* **80**, 193–196.

Antvogel, H. & Bonn, A. (2001) Environmental parameters and microspatial distribution of insects: a case study of carabids in an alluvial forest. *Ecography* **24**, 470–482.

Balmford, A. & Bond, W. (2005) Trends in the state of nature and their implications for human well-being. *Ecology Letters* **8**, 1218–1234.

Bascompte, J. & Sole, R.V. (1996) Habitat fragmentation and extinction thresholds in spatially explicit models. *Journal of Animal Ecology* **65**, 465–473.

Belsky, A.J. (1994) Influences of trees on savanna productivity: tests of shade, nutrients, and tree-grass competition. *Ecology* **75**, 922–932.

Bender, D.J., Contreras, T.A. & Fahrig, L. (1998) Habitat loss and population decline: a meta-analysis of the patch size effect. *Ecology* **79**, 517–529.

Bengtsson, J., Angelstam, P., Elmqvist, T. *et al.* (2003) Reserves, resilience and dynamic landscapes. *Ambio* **32**, 389–396.

Benton, T.G., Vickery, J.A. & Wilson, J.D. (2003) Farmland biodiversity: is habitat heterogeneity the key? *Trends in Ecology and Evolution* **18**, 182–188.

Bowman, D., Walsh, A. & Prior, L.D. (2004) Landscape analysis of Aboriginal fire management in Central Arnhem Land, north Australia. *Journal of Biogeography* **31**, 207–223.

Breshears, D.D., Cobb, N.S., Rich, P.M. *et al.* (2005) Regional vegetation die-off in response to global-change-type drought. *Proceedings of the National Academy of Sciences of the USA* **102**, 15144–15148.

Bromham, L., Cardillo, M., Bennett, A.F. & Elgar, M.A. (1999) Effects of stock grazing on the ground invertebrate fauna of woodland remnants. *Australian Journal of Ecology* **24**, 199–207.

Brown, J.R., Herrick, J. & Price, D. (1999) Managing low-output agroecosystems sustainably: the importance of ecological thresholds. *Canadian Journal of Forest Research/Revue Canadienne De Recherche Forestière* **29**, 1112–1119.

Daily, G.C., Ehrlich, P.R. & Haddad, N.M. (1993) Double keystone bird in a keystone species complex. *Proceedings of the National Academy of Sciences of the USA* **90**, 592–594.

Dean, W.R.J., Milton, S.J. & Jeltsch, F. (1999) Large trees, fertile islands, and birds in arid savanna. *Journal of Arid Environments* **41**, 61–78.

Dorrough, J. & Moxham, C. (2005) Eucalypt establishment in agricultural landscapes and implications for landscape-scale restoration. *Biological Conservation* **123**, 55–66.

Elmqvist, T., Folke, C., Nystrom, M. *et al.* (2003) Response diversity, ecosystem change, and resilience. *Frontiers in Ecology and the Environment* **1**, 488–494.

Fahrig, L. (1997) Relative effects of habitat loss and fragmentation on population extinction. *Journal of Wildlife Management* **61**, 603–610.

Fahrig, L. (2003) Effects of habitat fragmentation on biodiversity. *Annual Review of Ecology, Evolution and Systematics* **34**, 487–515.

Fischer, J., Lindenmayer, D.B. & Manning, A.D. (2006) Biodiversity, ecosystem function and resilience: ten guiding principles for off-reserve conservation. *Frontiers in Ecology and the Environment* **4**, 80–86.

Foley, J.A., DeFries, R., Asner, G.P. *et al.* (2005) Global consequences of land use. *Science* **309**, 570–574.

Folke, C., Carpenter, S., Walker, B. *et al.* (2004) Regime shifts, resilience, and biodiversity in ecosystem management. *Annual Review of Ecology, Evolution and Systematics* **35**, 557–581.

Frankham, R. (1995) Inbreeding and extinction – a threshold effect. *Conservation Biology* **9**, 792–799.

Gibbons, P. & Lindenmayer, D.B. (2002) *Tree Hollows and Wildlife Conservation in Australia*. CSIRO Publishing, Collingwood.

Gilmore, A.M. (1985) The influence of vegetation structure on the density of insectivorous birds. In: Keast, A., Recher, H.F., Ford, H. & Saunders, D. (eds.) *Birds of Eucalypt Forests and Woodlands: Ecology, Conservation and Management*, pp. 21–31. Royal Australasian Ornithologists Union and Surrey Beatty and Sons, Chipping Norton, NSW.

Graham, C.H. (2001) Factors influencing movement patterns of keel-billed toucans in a fragmented tropical landscape in southern Mexico. *Conservation Biology* **15**, 1789–1798.

Hanski, I. (1998) Metapopulation dynamics. *Nature* **396**, 41–49.

Hazell, D., Osborne, W. & Lindenmayer, D. (2003) Impact of post-European stream change on frog habitat: southeastern Australia. *Biodiversity and Conservation* **12**, 301–320.

Homan, R.N., Windmiller, B.S. & Reed, J.M. (2004) Critical thresholds associated with habitat loss for two vernal pool-breeding amphibians. *Ecological Applications* **14**, 1547–1553.

Jansson, G. & Angelstam, P. (1999) Threshold levels of habitat composition for the presence of the long-tailed tit (*Aegithalos caudatus*) in a boreal landscape. *Landscape Ecology* **14**, 283–290.

Johns, A.D. (1988) Effects of 'selective' timber extraction on rain forest structure and composition and some consequences for frugivores and folivores. *Biotropica* **20**, 31–37.

Knapp, A.K., Blair, J.M., Briggs, J.M. *et al.* (1999) The keystone role of bison in north American tallgrass prairie – Bison increase habitat heterogeneity and alter a broad array of plant, community, and ecosystem processes. *Bioscience* **49**, 39–50.

Lindenmayer, D.B. & Franklin, J. (2002) *Conserving Forest Biodiversity.* Island Press, Washington, DC.

Lindenmayer, D.B., Cunningham, R.B., Tanton, M.T., Smith, A.P. & Nix, H.A. (1991) Characteristics of hollow-bearing trees occupied by arboreal marsupials in the montane ash forests of the Central Highlands of Victoria, south-east Australia. *Forest Ecology and Management* **40**, 289–308.

Lindenmayer, D.B., Cunningham, R.B. & Fischer, J. (2005) Vegetation cover thresholds and species responses. *Biological Conservation* **124**, 311–316.

Ludwig, J.A., Eager, R.W., Liedloff, A.C. *et al.* (2000) Clearing and grazing impacts on vegetation patch structures and fauna counts in eucalypt woodland, Central Queensland. *Pacific Conservation Biology* **6**, 254–272.

MacArthur, R.H. & MacArthur, J.W. (1961) On bird species diversity. *Ecology* **42**, 594–598.

McIntyre, S., Heard, K.M. & Martin, T.G. (2003) The relative importance of cattle grazing in subtropical grasslands: does it reduce or enhance plant biodiversity? *Journal of Applied Ecology* **40**, 445–457.

Manning, A.D., Fischer, J. & Lindenmayer, D.B. (2006) Scattered trees are keystone structures – implications for conservation. *Biological Conservation* **206**, 311–321.

Mattison, E.H.A. & Norris, K. (2005) Bridging the gaps between agricultural policy, land-use and biodiversity. *Trends in Ecology and Evolution* **20**, 610–616.

Mayer, A.L. & Rietkerk, M. (2004) The dynamic regime concept for ecosystem management and restoration. *Bioscience* **54**, 1013–1020.

Millennium Ecosystem Assessment (2003) *Ecosystems and Human Well-Being.* Island Press, Washington, DC.

Milton, S.J., Dean, W.R.J., Duplessis, M.A. & Siegfried, W.R. (1994) A conceptual model of arid rangeland degradation – the escalating cost of declining productivity. *Bioscience* **44**, 70–76.

Mönkkönen, M. & Reunanen, P. (1999) On critical thresholds in landscape connectivity: a management perspective. *Oikos* **84**, 302–305.

Paine, R.T. (1969) A note on trophic complexity and community stability. *American Naturalist* **103**, 91–93.

Parker, M. & Mac Nally, R. (2002) Habitat loss and the habitat fragmentation threshold: an experimental evaluation of impacts on richness and total abundances using grassland invertebrates. *Biological Conservation* **105**, 217–229.

Polasky, S., Nelson, E., Lonsdorf, E., Fackler, P. & Starfield, A. (2005) Conserving species in a working landscape: Land use with biological and economic objectives. *Ecological Applications* **15**, 1387–1401.

Power, M.E., Tilman, D., Estes, J.A. *et al.* (1996) Challenges in the quest for keystones. *Bioscience* **46**, 609–620.

Radford, J.Q. & Bennett, A.F. (2004) Thresholds in landscape parameters: occurrence of the white-browed treecreeper *Climacteris affinis* in Victoria, Australia. *Biological Conservation* **117**, 375–391.

Radford, J.Q., Bennett, A.F. & Cheers, G.J. (2005) Landscape-level thresholds of habitat cover for woodland-dependent birds. *Biological Conservation* **124**, 317–337.

van der Ree, R., Bennett, A.F. & Gilmore, D.C. (2003) Gap-crossing by gliding marsupials: thresholds for use of isolated woodland patches in an agricultural landscape. *Biological Conservation* **115**, 241–249.

Robertson, G.P. & Swinton, S.M. (2005) Reconciling agricultural productivity and environmental integrity: a grand challenge for agriculture. *Frontiers in Ecology and the Environment* **3**, 38–46.

Saunders, D.A., Smith, G.T., Ingram, J.A. & Forrester, R.I. (2003) Changes in a remnant of salmon gum *Eucalyptus salmonophloia* and York gum *E. loxophleba* woodland, 1978 to 1997. Implications for woodland conservation in the wheat-sheep regions of Australia. *Biological Conservation* **110**, 245–256.

Scheffer, M., Carpenter, S., Foley, J.A., Folke, C. & Walker, B. (2001) Catastrophic shifts in ecosystems. *Nature* **413**, 591–596.

Schröter, D., Cramer, W., Leemans, R. *et al.* (2005) Ecosystem service supply and vulnerability to global change in Europe. *Science* **310**, 1333–1337.

Soulé, M.E., Estes, J.A., Miller, B. & Honnold, D.L. (2005) Strongly interacting species: conservation policy, management, and ethics. *Bioscience* **55**, 168–176.

Spooner, P., Lunt, I. & Robinson, W. (2002) Is fencing enough? The short-term effects of stock exclusion in remnant grassy woodlands in southern NSW. *Ecological Management and Restoration* **3**, 117–126.

Srivastava, D.S. & Vellend, M. (2005) Biodiversity–ecosystem function research: is it relevant to conservation? *Annual Review of Ecology, Evolution and Systematics* **36**, 267–294.

Tews, J., Brose, U., Grimm, V. *et al.* (2004) Animal species diversity driven by habitat heterogeneity/diversity: the importance of keystone structures. *Journal of Biogeography* **31**, 79–92.

Vesk, P.A. & Mac Nally, R. (2006) The clock is ticking – revegetation and habitat for birds and arboreal marsupials in rural landscapes of southern Australia. *Agriculture, Ecosystems and Environment* **112**, 356–366.

Vetaas, O.R. (1992) Micro-site effects of trees and shrubs in dry savannas. *Journal of Vegetation Science* **3**, 337–344.

Villard, M-A., Trzcinski, M.K. & Merriam, G. (1999) Fragmentation effects on forest birds: relative influence of woodland cover and configuration on landscape occupancy. *Conservation Biology* **13**, 774–783.

Vitt, L.J., Avila-Pires, T.C.S., Caldwell, J.P. & Oliveira, V.R.L. (1998) The impact of individual tree harvesting on thermal environments of lizards in Amazonian rain forest. *Conservation Biology* **12**, 654–664.

Walker, B.H. (1992) Biodiversity and ecological redundancy. *Conservation Biology* **6**, 18–23.

Walker, B. (1995) Conserving biological diversity through ecosystem resilience. *Conservation Biology* **9**, 747–752.

Walker, B. & Meyers, J.A. (2004) Thresholds in ecological and social-ecological systems: a developing database. *Ecology and Society* **9**, 3 [online]. URL:http://www.ecologyandsociety.org/vol9/iss2/art3/.

Walker, B., Kinzig, A. & Langridge, J. (1999) Plant attribute diversity, resilience, and ecosystem function: The nature and significance of dominant and minor species. *Ecosystems* **2**, 95–113.

Walker, B., Hollin, C.S., Carpenter, S.R. & Kinzig, A. (2004) Resilience, adaptability and transformability in social–ecological systems. *Ecology and Society* **9**.

Williams, M.R., Abbott, I., Liddelow, G.L., Vellios, C., Wheeler, I.B. & Mellican, A.E. (2001) Recovery of bird populations after clearfelling of tall open eucalypt forest in Western Australia. *Journal of Applied Ecology* **38**, 910–920.

With, K.A. & Crist, T.O. (1995) Critical thresholds in species responses to landscape structure. *Ecology* **76**, 2446–2459.

With, K.A. & King, A.W. (1999) Extinction thresholds for species in fractal landscapes. *Conservation Biology* **13**, 314–326.

With, K.A., Gardner, R.H. & Turner, M.G. (1997) Landscape connectivity and population distributions in heterogeneous environments. *Oikos* **78**, 151–169.

21

Synthesis: Total Vegetation Cover, Pattern and Patch Content

David B. Lindenmayer and Richard J. Hobbs

Chapters 18–20 focus on a range of issues linked to the total amount of native vegetation cover, the spatial pattern of that cover, and the content of the vegetation patches that comprise that cover. Like the other sections in this book, the three essays contrast strongly in their content and style, but also share some areas of commonality.

A key theme of the very challenging essay written by Bennett and Radford (Chapter 18) is that much of the research in landscape ecology and conservation biology has overlooked the importance of mosaics of patches for biodiversity. They contend that the vast majority of studies have focused on individual patches or sites within patches, and that the role and importance of ensembles of patches (mosaics) and landscapes remain poorly understood. Management of landscape to provide the 'right' kinds of mosaics is therefore presently without a strong scientific underpinning. Given such uncertainty, it seems appropriate that an aim of management practices might be to create not one but many different kinds of mosaics. Ensuring that 'the same thing is not done everywhere' is a risk-spreading strategy that attempts to limit the chances of making the same mistake in all places.

Bennett and Radford (Chapter 18) outline some of the ways that biotic responses to mosaics might be quantified, and this is valuable information for guiding new work to address this important knowledge gap. However, there are strengths and limitations of the different approaches canvassed by Bennett and Radford, and some key issues remain unresolved. How to design robust field counting protocols that generate high-quality data on species inhabiting mosaics is one of these (see also the comments on landscape-level sampling in Chapter 9). Another is what kind of statistics is most appropriate for analyses of compositional data derived from work on biotic responses to mosaics.

Collaborative efforts between landscape ecologists and those with true expertise in statistical methodology and analytical methods will be required to tackle some of these issues.

Two important additional points made by Bennett and Radford are that: (i) most of the work on landscapes is focused at a single spatial scale (typically one relevant for birds), but different organisms respond to factors at different scales and multiple scales; and (ii) issues of mosaics and composition often cannot be easily divorced from those of the total amount of native vegetation cover. The total native vegetation cover is usually the most important factor – because vegetation loss has been the biggest driver of species loss (see also Mac Nally, Chapter 6) – but configuration is also not unimportant.

The essay by Gibbons *et al.* – (Chapter 19) takes a very different approach from the other chapters in this section. Unlike the primarily academic perspectives of the majority of contributors to this book, the work by Gibbons *et al.* is written from the perspective of landscape managers. Their case study from New South Wales addresses practical, on-ground problems related to the assessment of native vegetation cover and, in turn, using those assessments to make decisions on vegetation clearing versus vegetation retention. Like most authors in this book, Gibbons *et al.* are cognizant of scale issues and highlight a need for multiscaled assessments. Indeed, they tackle all three themes of this section in their case study – quantifying total vegetation cover, quantifying spatial vegetation cover and appraising vegetation condition as a measure of patch content. However, they add the important caveat that many of the practical measures that are used must be relevant to the spatial scale at which land-use decisions are made. Some of the outcomes of the case study resonate closely with issues raised in the first section of this book on landscape models and landscape classification (see Chapters 2–5). Two examples are given below:

1. The problems of defining landscapes as being composed of discrete patches when in fact many attributes are continuous and sharp boundaries often do not occur.
2. The reality that any single landscape model and associated landscape classification cannot address all of the complexities and biotic responses in a given landscape (see also Franklin & Swanson, Chapter 12).

In many respects, the case study presented by Gibbons *et al.* (Chapter 19) is a sobering one that demonstrates not only the true complexity of real-world landscapes but also the immense challenges associated with converting our increasingly complex understanding of landscapes into effective on-ground landscape management.

For Fischer and Lindenmayer (Chapter 20) the focus is primarily on vegetation cover, vegetation structure and species composition and their interrelationships with thresholds, cascading impacts, regime shifts and resilience. This chapter shares many common elements with that by Walker (Chapter 34), although the focus is less tightly linked with disturbance. Fischer and Lindenmayer (Chapter 20) examine these issues from the broader perspective of attempting to ensure that management practices do not create irreversible changes in ecosystems, landscapes, ecological processes and species. Unfortunately, as outlined elsewhere (see Chapter 36), it is extremely difficult to identify the critical change points for thresholds and to anticipate (and prevent) regime shifts and extinction cascades before they occur. Data to guide identification are limited and robust statistical methods to explore these problems are in urgent need of further development. Fischer and Lindenmayer offer some 'rules of thumb' given current data and methodological limitations. They argue that thresholds are more likely to be crossed, extinction cascades more likely to be triggered and regime shifts more likely to occur when there are low levels of native vegetation cover in a landscape. This conclusion appears to have support from the empirical literature for both terrestrial ecosystems (Bennett and Radford, Chapter 18) as well as aquatic ecosystems (Cullen, Chapter 39). Therefore, a crucial aspect of landscape management must be to avoid low levels of native vegetation cover.

Fischer and Lindenmayer also suggest that human disturbance regimes should have some congruence with natural disturbance regimes. This is important not only because of the need to maintain 'structurally characteristic' vegetation but also to maintain spatial vegetation patterns and ecological processes typical of areas less subject to human modification. This proposal is explored in detail by Hunter (Chapter 35) and has advantages, but also limitations. Of course, all of the 'rules of thumb' should be treated as working hypotheses and vigorously tested as part of the integration of research and management (see Driscoll, Chapter 11; Luck, Chapter 16).

Section 6
Connectivity, Corridors, Stepping Stones

22

Corridors, Connectivity and Biological Conservation

F.K.A. Schmiegelow

Abstract

Substantial investments in corridor research over the past two decades have yielded an impressive body of theoretical and empirical work, but generalizations remain elusive. In some instances, there is compelling evidence to suggest that corridors are effective in enhancing movement between patches; a few studies have further demonstrated that increased movement rates contribute to a greater probability of population persistence. Sorely lacking, however, are evaluations of the relative effectiveness of corridors in comparison with other conservation measures that address connectivity. As a result, corridors continue to be promoted as a widespread conservation tool, when their utility may in fact be limited to specific landscape states and conditions. Given the relative ease with which corridors can be incorporated into conservation planning exercises there is a danger that more effective measures to maintain connectivity are not being considered or implemented, resulting in the potential for reduced conservation success in the long term. Emerging approaches to landscape ecology and biological conservation that address the heterogeneity inherent in landscapes, and the opportunities afforded to address functional connectivity through means other than the construction of physical corridors, hold considerable promise. There is an urgent need for research that investigates the influence of variable landscape con-ditions, particularly that of the matrix, on connectivity and other fragmentation-related concerns.

Keywords: connectivity; landscape corridors; matrix management.

Introduction

Few topics in conservation biology and landscape ecology have generated as much controversy as that of conservation corridors. The primary rationale for establishment and maintenance of conservation corridors is the promotion of connectivity within and among populations of animals distributed in heterogeneous environments – to facilitate movement and prevent isolation (Merriam 1984). The ultimate objective is to enhance or maintain the viability of populations (Beier & Noss 1998). Formulation of the corridor concept as a tool in conservation derives from interpretations of the equilibrium theory of island biogeography (MacArthur & Wilson 1967) relative to the design of nature reserves. Based on theoretical considerations, Wilson and Willis (1975) asserted that 'extinction will be lower when fragments can be connected by corridors of natural habitat, no matter how thin the corridors'. Similarly, Diamond (1975) stated that 'corridors between reserves may dramatically increase dispersal rates over what would otherwise be negligible values.' The intuitive appeal of physically connecting patches of habitat, and links to theoretical pillars of conservation biology and landscape ecology, resonated with both scientists and practitioners. As a result, Diamond's (1975) schematic of 'better' (connected) and 'worse' (unconnected) reserve scenarios found a receptive audience, and the graphic and associated principles can still be found in most introductory textbooks on conservation science.

Assumptions regarding the ubiquitous benefits of corridors went largely unchallenged in the primary literature for over a decade. From 1987 to 2000, however, a lively debate of the issues transpired. Simberloff and Cox (1987) provided one of the first reviews of potential benefits and costs of corridors, challenging the doctrine that corridors confer only advantages. Proponents had identified benefits of corridors to include increased immigration rates (Brown & Kodric-Brown 1977), reduced threats of inbreeding depression (Harris 1984, 1985) and demographic stochasticity (Fahrig & Merriam 1985), increased accessibility of resources for far-ranging species (Harris 1984, 1985) and provision of additional habitat. Proposed disadvantages included transmission of disease, fire and other catastrophes (Simberloff & Abele 1982), increased predation risk (Soulé & Simberloff 1986), provision of conduits for invasive species

(Noss & Harris 1986) and increased road kill (O'Neil *et al.* 1983; Harris 1985) along road rights-of-way. In addition to these concerns, Simberloff and Cox (1987) added consideration of the economic costs and trade-offs associated with corridor establishment and maintenance relative to other conservation strategies and the lack of empirical evaluation and substantiation as further reasons for vigilance in broad-scale application of the corridor concept.

Respondents to these criticisms acknowledged the paucity of data to support generalizations, but advocated a precautionary approach to corridor management based on the connectedness inherent in natural landscapes (Noss 1987), asserted that the potential advantages outweighed the disadvantages (Noss 1987; Soulé *et al.* 1988) and argued that the uncertainties favour the null hypothesis that corridors have value (Hobbs 1992). Further exchanges emphasized the complexity of considerations in corridor design and evaluation, including species- and site-specific variation in response (e.g. Beier & Loe 1992; Lindenmayer & Nix 1993). Later reviews of accumulating empirical studies reached similar conclusions regarding the importance of species life history and highlighted the additional challenges in interpreting results from studies with confounding covariates, such as habitat area, corridor quality and matrix condition (Rosenberg *et al.* 1997; Beier & Noss 1998). Nevertheless, there was general agreement that limitations of observational data for inferring corridor effectiveness continued to impede rigorous evaluation. Disagreements persisted, however, about the potential for experimental studies, particularly those of experimental model systems, to inform land-management strategies at relevant scales (Haddad *et al.* 2000; Noss & Beier 2000).

Lack of strong evidence for corridor effectiveness continues to be advanced as a general criticism of the concept (e.g. Orrock & Damschen 2005; Rantalainen *et al.* 2005), without careful discrimination of the specific conditions associated with individual studies. Results of recent empirical work (>100 studies in the primary literature from 1999 to 2005; T. Gartner & F. Schmiegelow, unpublished review) tend to support the intermediate position of Haddad *et al.* (2000), who concluded that corridors are valuable conservation tools for some species and landscapes, but that trade-offs exist with other landscape management strategies that promote connectivity. Few ecologists would advocate establishment of corridors as a sole conservation measure, consistent with increasing recognition that reserve-based approaches in isolation will be insufficient to stem widespread biodiversity loss. While we seek empirical generalizations to guide conservation approaches, oversimplification leads to apparently divergent positions when different conclusions may originate from underlying variation in the systems studied. It is also not simply a matter of considering trade-offs in strategies; rather, there is a need to develop a framework to guide determination of the most appropriate suite of strategies to maintain or enhance connectivity of wildlife populations at multiple spatial and temporal scales, given landscape condition, system dynamics and species life history. Incorporation of corridors into landscape planning is not a prerequisite for maintaining connectivity, but rather a prescriptive measure to treat specific symptoms of fragmentation.

Habitat fragmentation and loss

Habitat fragmentation has been the subject of extensive study for nearly four decades and has often been cited as a primary driver of population declines and species loss from degraded systems. Empirical generalizations from fragmentation studies have also proven elusive (Debinski & Holt 2000), for many of the same reasons that no universal truths exist concerning corridors: species- and system-specific variation in response is great, and the logistical challenges of implementing experiments over appropriate spatial and temporal scales are considerable. Furthermore, landscape degradation involves the parallel processes of habitat loss and habitat fragmentation (Fahrig 2003). When habitat is lost, individuals are also lost, resulting in population declines. From a landscape perspective, loss of one habitat type also implies increases in other landscape elements, which can strongly influence dynamics in remaining habitat patches (e.g. Lindenmayer & Fischer 2006). Habitat fragmentation can compound habitat loss through negative area and edge effects, and as a result of habitat isolation.

Organisms become isolated within habitat patches because the intervening landscape – the matrix – is both unsuitable and impermeable. Species can persist in isolated patches if resources are sufficient (i.e. if the patch is large enough or of sufficiently quality) to support viable populations. However, isolation can compromise persistence when patches are too small to support self-sustaining local populations and movement of individuals between patches is impeded. Maintaining or re-establishing physical corridors to connect patches is one potential solution to this problem. However, improvements to either matrix quality or permeability can also lessen isolation and enhance connectivity. Much as the fragmentation debate detracted attention from the larger issue of stemming overall habitat loss (e.g. Fahrig 1997), the controversy surrounding corridors has diverted attention from the crucial need for conservation strategies that focus on improved matrix management.

Matrix- and corridor-based approaches

The matrix is broadly defined as the dominant land type for a given landscape. Because conservation and landscape ecology research has often focused on human-dominated systems, where much of the native vegetation cover has been permanently converted for other uses, the matrix has been characterized as a largely hostile environment, contributing little or negatively to conservation objectives. Further, the matrix has generally been treated as a static entity, often binary in nature (e.g. suitable or unsuitable habitat), without appropriate consideration of landscape context. Context includes system-specific factors such as natural disturbance regimes, land-use history and current land uses. Such limited treatment is unnecessarily constraining from the standpoint of landscape and conservation planning, and far too simplistic from an ecological perspective (e.g. Schmiegelow & Mönkönnen 2002). Lindenmayer and Franklin (2002) provide an excellent overview of interactions between the matrix and major themes in conservation biology and landscape ecology. Here, I focus on the relative potential for matrix- and corridor-based strategies to address issues of connectivity, as well as other factors associated with habitat loss and fragmentation, as part of a more comprehensive approach to conservation planning. I assume that the overarching objective is to maintain connectivity across multiple spatial and temporal scales. In this context, structural connectivity refers to the physical connectedness of landscape elements, whereas functional connectivity refers to ecological flows across landscapes (*sensu* Taylor *et al.* 1993). Such these flows include biological processes such as dispersal, as well as physical processes such as water movement.

Cover and connectivity

McIntyre and Hobbs' (1999) classification of landscapes into broad states relative to conversion of original cover is summarized in Table 22.1. Within each class, landscape condition can vary as a function of modification of remaining original cover and management of new cover types. I use the term **cover** rather than habitat as a neutral measure of landscape state (e.g. amount of forest cover). **Habitat** will herein refer to landscape conditions relative to their suitability for certain species or suites of species (after Hall *et al.* 1997). This distinction is important because a given landscape state can represent a range of habitat conditions depending on the species of interest. For example, a landscape with a high proportion of original forest cover (Fig. 22.1a)

Table 22.1 **General landscape states relative to conversion of original cover (after McIntyre & Hobbs 1999).**

Landscape state	Intact	Variegated	Fragmented	Relictual
Original cover	>90%	60–90%	10–60%	<10%

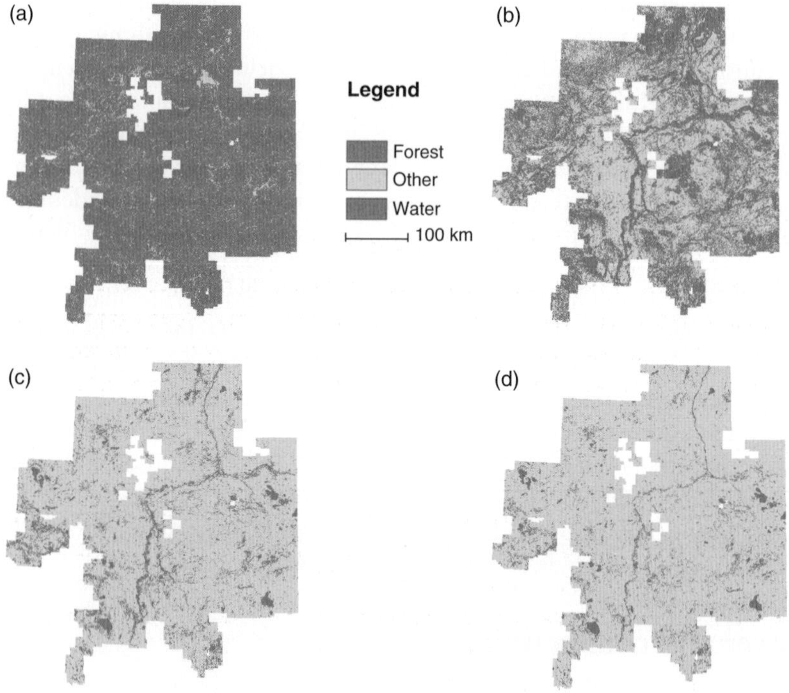

Figure 22.1 **Illustration of the distribution of (a) all forest cover, (b) mesic forest, (c) old mesic forest and (d) old deciduous forest in landscapes in northeastern Alberta, Canada. For species that require older forest, the natural distribution of habitat is highly fragmented, even in intact landscapes.**

may represent nearly continuous habitat for forest generalist species, whereas it presents diminishing conditions of habitat availability and connectivity for species with increasingly specialized requirements for particular forest types (Fig. 22.1b–d). Hence, the structural connectivity of habitat for certain species may be naturally low, even in intact landscapes. Landscape state also has implications for attributes other than habitat of target species, such as hydrological connectivity, thus the term 'cover' has broader application in conservation planning and landscape management.

While it is possible to quantify structural connectivity relative to landscape state, as a general function of amount of remaining cover (e.g. through the use of neutral landscape models and application of percolation theory; Gardner *et al.* 1987; With 1997), these approaches are sensitive to parameters such as landscape grain. Evaluating functional connectivity for organisms relative to landscape state requires further specification of habitat requirements, dispersal ability, movement patterns and other attributes that influence landscape flows. As earlier identified, such data are rarely available. Nevertheless, the broad landscape states described by McIntyre and Hobbs (1999) generally correspond with ranges in thresholds in structural and functional connectivity based on percolation theory (e.g. With & King 1999) and fragmentation thresholds from empirical and simulation studies of habitat loss (e.g. Andrén 1994; Fahrig 1998).

In both intact and variegated landscape states, the matrix is dominated by original cover, which supports maintenance of natural levels of landscape connectivity. Where landscape condition has been altered due to human uses, or there has been selective loss of certain cover types that represent specialized habitat, both physical and functional connectivity may be compromised. However, appropriate management of the matrix can address connectivity and may confer greater overall conservation gains than corridor-based strategies by directly influencing the quantity, quality and spatial juxtaposition of all landscape elements over time, including novel cover types. Similarly, enhancing the quality of the converted matrix in fragmented landscape states can not only improve connectivity, but also increase the amount of habitat available to sensitive species and decrease suitability for competitors and predators. Even in the most degraded landscapes, where only relictual patches of original cover remain, the potential to enhance connectivity through matrix management exists. While it is often assumed that movement habitats for organisms are similar or identical to habitats where they permanently reside, empirical studies have demonstrated that animals may, in fact, select different attributes for these functions (e.g. Palomares 2001; Saher & Schmiegelow 2005). Further, the efficacy of corridors can be strongly influenced by matrix conditions (e.g. Perault & Lomolino 2000; Baum *et al.* 2004). Thus maintaining or improving connectivity requires approaches beyond simply mapping and preserving apparently suitable habitat for species of concern. Failure to consider these approaches results in reduced conservation opportunities.

The promotion of matrix-based approaches to managing structural and functional connectivity is not intended to dismiss the value of physical corridors designed specifically to maintain or enhance movement of target organisms between otherwise isolated habitat patches. However, it is intended to highlight the narrower set of conditions under which such strategies should be the focus of conservation efforts. Certainly, there are relatively immutable landscape features, such as roads, urban infrastructures and intensive industrial developments, that disrupt connectivity and may not be effectively managed by other approaches. Measures of human influence on global ecosystems nevertheless suggest that these extreme conditions are not as common as the academic attention conferred on corridors would suggest. While less than 20% by area of global terrestrial biomes remain fully intact, those areas most severely impacted by humans also occupy only a relatively small proportion of the Earth's surface (Sanderson *et al.* 2002), suggesting there remain significant opportunities to maintain or enhance connectivity and other conservation values through matrix management, in the majority of the world's biomes.

Strategies for connectivity

Determining the most effective strategies to address connectivity involves careful consideration of the life-history characteristics of target species, the state, condition and dynamics of the landscapes they occupy, and the spatial and temporal scales of concern. Permeability can be a transitional attribute of landscapes, dependent on the condition of vegetation cover, and reduced movement for some period of time may not translate into population isolation. For example, in forest landscapes managed for timber production, recently harvested forest may constrain movement of species associated with older forest stages. Maintenance of forested corridors can increase movement between patches of remnant forest in the short term (e.g. Machtans *et al.* 1996), but as conditions of the matrix change with forest regeneration, permeability may increase (e.g. Robichaud *et al.* 2002), and enhancing matrix conditions or leaving larger forest patches (e.g. Hannon & Schmiegelow 2002; Selonen & Hanski 2003) may be more effective in the long term in not only addressing connectivity, but also in mitigating other effects of habitat loss and fragmentation. However, corridor use alone cannot be assumed as evidence of the need for corridors, nor can use of the matrix be assumed to contribute to conservation without demonstration of its contribution to population viability. Evaluating the actual, relative and potentially synergistic contributions of different landscape components and management strategies to connectivity requires integrated approaches incorporating carefully designed field studies coupled with empirical and theoretical simulation modelling. Greater understanding of patterns of connectivity in naturally complex landscapes can inform design of management strategies for the continuum of landscape states and conditions created by human activities, although some human alterations may have no natural analogues. I nevertheless concur with Soulé and Gilpin (1991) that 'Corridors are bandages for wounded landscapes, and at best can only partly compensate for the denaturing activities of humans.' We must aggressively pursue the development and implementation of pre-emptive strategies to maintain natural connectivity where such opportunities exist, and similarly pursue restoration strategies that recognize the potential contributions of all landscape elements.

Key principles

1. Connectivity is an emergent property of landscapes that is both scale-dependent and species-specific.
2. A given landscape state can represent varying levels of connectivity dependent on land-use history and natural dynamics, as reflected by condition of the landscape mosaic.
3. Loss of connectivity can compromise population viability through demographic and genetic isolation, when movements within and among populations are reduced.
4. Establishing physical corridors is one approach to facilitating movement and reducing isolation; improvements to landscape permeability can also enhance connectivity.
5. Insufficient attention has been paid to the potential for improved matrix management to support a wide range of conservation values, including connectivity.
6. Generalizations regarding corridor design and effectiveness are not possible; design must reflect the life-history and demographic characteristics of target species, and effectiveness will vary with landscape state and condition.
7. Pre-emptive strategies that address maintenance of connectivity through matrix management can reduce the need for corridor prescriptions;

matrix management can also improve the effectiveness of corridors and address other symptoms of habitat loss and fragmentation.

8 Management approaches should recognize the opportunities afforded by landscape heterogeneity to address connectivity at multiple spatial and temporal scales using a variety of strategies.

9 Research approaches should integrate field studies with simulation modelling to explore the potential efficacy of and trade-offs between different management strategies, given varying species and landscape attributes.

References

Andrén, H. (1994) Effects of habitat fragmentation on birds and mammals in landscapes with different proportions of suitable habitat: a review. *Oikos* **71**, 355–366.

Baum, K.A., Haynes, K.J., Dillemuth, F.P., & Cronin, J.T. (2004) The matrix enhances the effectiveness of corridors and stepping stones. *Ecology* **85**, 2671–2676.

Beier, P. & Loe, S. (1992) A checklist for evaluating impacts to wildlife movement corridors. *Wildlife Society Bulletin* **20**, 434–440.

Beier, P. & Noss, R.F. (1998) Do habitat corridors provide connectivity? *Conservation Biology* **12**, 241–252.

Brown, J.H. & Kodric-Brown, A. (1977) Turnover rates in insular biogeography: effect of immigration on extinction. *Ecology* **58**, 445–449.

Debinski, D.M. & Holt, R.D. (2000) A survey and overview of habitat fragmentation studies. *Conservation Biology* **14**, 342–355.

Diamond, J.M. (1975) The island dilemma: lessons of modern biogeographic studies for the design of natural reserves. *Biological Conservation* **7**, 129–146.

Fahrig, L. (1997) Relative effects of habitat loss and fragmentation on population extinction. *Journal of Wildlife Management* **61**, 603–610.

Fahrig, L. (1998) When does fragmentation of breeding habitat affect population survival? *Ecological Modeling* **105**, 273–292.

Fahrig, L. (2003) Effects of habitat fragmentation on biodiversity. *Annual Reviews of Ecology and Systematics* **34**, 487–515.

Fahrig, L. & Merriam, G. (1985) Habitat patch connectivity and population survival. *Ecology* **66**, 1762–1768.

Gardner, R.H., Milne, B.T., Turner, M.G. & O'Neill (1987) Neutral models for the analysis of broad-scale landscape pattern. *Landscape Ecology* **1**, 19–28.

Haddad, N.M., Rosenberg, D.K & Noon, B.R. (2000) On experimentation and the study of corridors: response to Beier and Noss. *Conservation Biology* **14**, 1543–1545.

Hall, L.S., Krausman, P.R. & Morrison, M. (1997) The habitat concept and a plea for standard terminology. *Wildlife Society Bulletin* **25**, 173–182.

Hannon, S.J. & Schmiegelow, F.K.A. (2002) Corridors may not improve the conservation of small reserves for most boreal birds. *Ecological Applications* **12**, 1457–1468.

Harris, L.D. (1984) *The Fragmented Forest: Island Biogeographic Theory and the Preservation of Biotic Diversity*. University of Chicago Press, Chicago, IL.

Harris, L.D. (1985) *Conservation Corridors: a Highway System for Wildlife*. Report 85–5. Environmental Information Center of Florida Conservation Foundation, Inc., Winter Park, FL.

Hobbs, R.J. (1992) The role of corridors in conservation – solution or bandwagon. *Tree* **7**, 389–392.

Lindenmayer, D.B. & Fischer, J. (2006) *Habitat Fragmentation and Landscape Change: an Ecological and Conservation Synthesis*. Island Press, Washington, DC.

Lindenmayer, D.B. & Franklin, J.F. (2002) *Conserving Forest Biodiversity: a Comprehensive Multi-Scaled Approach*. Island Press, Washington, DC.

Lindenmayer, D.B. & Nix, H.A. (1993) Ecological principles for the design of wildlife corridors. *Conservation Biology* **7**, 627–630.

MacArthur, R.H. & Wilson, E.O. (1967) *The Theory of Island Biogeography*. Princeton University Press, NJ.

Machtans, C.S., Villard, M-A. & Hannon, S.J. (1996) Use of riparian strips as movement corridors by forest birds. *Conservation Biology* **10**, 1366–1379.

McIntyre, S. & Hobbs, R. (1999) A framework for conceptualizing human effects on landscapes and its relevance to management and research models. *Conservation Biology* **13**, 1282–1292.

Merriam, G. (1984) Connectivity: a fundamental ecological characteristic of landscape pattern. In Brandt, J. & Agger, P. (eds) *Proceedings of the First International Seminar on the Methodology in Landscape Ecological Resources and Planning*, pp. 5–15. International Association for Landscape Ecology, Roskilde.

Noss, R.F. (1987) Corridors in real landscapes: a reply to Simberloff and Cox. *Conservation Biology* **1**, 159–164.

Noss, R.F. & Beier, P. (2000) Arguing over little things: response to Haddad et al. *Conservation Biology* **14**, 1546–1548.

Noss, R.F. &. Harris, L.D. (1986) Nodes, networks, and MUMs: preserving diversity at all scales. *Environmental Management* **10**, 299–309.

O'Neil, D.H., Robel, R.J. & Dayton, A.D. (1983) Lead contamination near Kansas highways: implications for wildlife enhancement programs. *Wildlife Society Bulletin* **11**,152–160.

Orrock, J.L. & Damschen, E.I. (2005) Corridors cause differential seed predation. *Ecological Applications* **15**, 793–798.

Palomares, F. (2001) Vegetation structure and prey abundance requirements of the Iberian lynx: implications for the design of reserves and corridors. *Journal of Applied Ecology* **38**, 9–18.

Perault, D. & Lomolino, M.V. (2000) Corridors and mammal community structure across a fragmented, old-growth forest landscape. *Ecological Monographs* **70**, 401–422.

Rantalainen, M.L., Fritze, H., Haimi, J., Pennanen, T. & Setala, H. (2005) Species richness and food web structure of soil decomposer community as affected by the size of habitat fragment and habitat corridors. *Global Change Biology* **11**, 1614–1627.

Robichaud, I., Villard, M-A. & Machtans, C.S. (2002) Effects of forest regeneration on songbird movements in a managed forest landscape of Alberta, Canada. *Landscape Ecology* **17**, 247–262.

Rosenberg, D.K., Noon, B.R. & Meslow, E.C. (1997) Biological corridors: form, function, and efficacy. *BioScience* **47**, 677–687.

Saher, J.D. & Schmiegelow, F.K.A. (2005) Movement pathways and habitat selection by woodland caribou during spring migration. *Rangifer*, Special Issue No. 16, 143–154.

Sanderson, E.W., Jaiteh, M., Levy, M.A., Redford, K.H., Wannebo, A.V. & Woolmer, G. (2002) The human footprint and the last of the wild. *BioScience* **52**, 891–904.

Schmiegelow, F.K.A. & Mönkönnen, M. (2002) Habitat loss and fragmentation in dynamic landscapes: avian perspectives from the boreal forest. *Ecological Applications* **12**, 375–389.

Selonen, V. &. Hanski, I.K. (2003) Young flying squirrels (*Pteromys volans*) dispersing in fragmented forests. *Behavioral Ecology* **15**, 564–571.

Simberloff, D. & Abele, L.G. (1982) Refuge design and island biogeographic theory: effects of fragmentation. *American Naturalist* **120**, 41–50.

Simberloff, D. &. Cox, J. (1987) Consequences and costs of conservation corridors. *Conservation Biology* **1**, 63–71.

Soulé, M.E. & Gilpin, M.E. (1991) The theory of wildlife corridor capability. In Saunders, D.A. & Hobbs, R.J. (eds.) *Nature Conservation 2: The Role of Corridors*, pp. 3–8. Surrey Beatty and Sons, Chipping Norton, Australia.

Soulé, M.E. & Simberloff, D. (1986) What do genetics and ecology tell us about the design of nature reserves? *Biological Conservation* **35**, 19–40.

Soulé, M.E., Bolger, D.T., Roberts, A.C., Wright, J., Sorice, M. & Hill, S. (1988) Reconstructed dynamics of rapid extinctions of chaparral-requiring birds in urban habitat. *Conservation Biology* **2**, 75–92.

Taylor, P.D., Fahrig, L., Henein, K. & Merriam, G. (1993) Connectivity is a vital element of landscape structure. *Oikos* **68**, 571–573.

Wilson, E.O. & Willis, E.O. (1975) Applied biogeography. In: Cody, M.L. & Diamond, J.M. (eds.) *Ecology and Evolution of Communities*, pp. 522–534. Belknap Press, Cambridge.

With, K. (1997) The application of neutral landscape models in conservation biology. *Conservation Biology* **11**, 1069–1080.

With, K. & King, A.W. (1999). Extinction thresholds for species in fractal landscapes. *Conservation Biology* **13**, 314–326.

23

Focal Species for Determining Connectivity Requirements in Conservation Planning

Reed F. Noss

Abstract

Connectivity is a key conservation planning principle because a well-connected network of reserves might support viable populations of species that could not be supported within single isolated reserves. Functional connectivity is determined by the intersection of an organism's life history and the structure of the landscape. Although probably no two species have identical connectivity requirements, fragmentation-sensitive focal species identified at multiple spatial scales and representing a variety of habitat types may capture the movement needs of many other species. I review some focal species that are potentially useful for determining connectivity requirements at several scales of planning in Florida, USA. Empirical data and models suggest species-specific sensitivity to the diverse impacts of habitat fragmentation. Assembling an efficient set of complementary focal species remains challenging. A promising approach is to use existing knowledge to build *a priori* models and develop alternative connectivity designs, coupled with field measurements of responses of organisms to various corridors, landscape configurations and road-crossing structures. This approach will be difficult to apply in regions where the biota is poorly known and where funding for research is scarce. In the broader sphere of conservation planning a focal-species approach must be complemented by other approaches.

Keywords: connectivity; conservation planning; corridors; focal species.

Introduction

Early attention to habitat corridors centred on their potential role as movement routes for game animals in human-altered landscapes (Sumner 1936; Baumgartner 1943), then more theoretically as avenues for colonization of habitat patches (Preston 1962; MacArthur & Wilson 1967). The explosion of interest in the application of island biogeographic theory to the design of nature reserves in the 1970s (e.g. Diamond 1975; Wilson & Willis 1975; Simberloff & Abele 1976) was followed by consideration of corridors as fundamental structural elements of landscapes (Forman & Godron 1981). Most early studies of the role of corridors in promoting population connectivity and viability were inconclusive, however, often due to poor experimental design (Simberloff *et al.* 1992; Beier & Noss 1998).

Today we understand that connectivity can be provided by means other than corridors. Metapopulation theory (Levins 1969; Gilpin & Hanski 1991; Hanski & Simberloff 1997) strengthened interest in the broader concept of functional connectivity, where suitable conditions in the landscape matrix – in addition to or instead of corridors – may promote colonization or recolonization of habitat patches and rescue effects for small populations. Perhaps a connected system of reserves might be a whole greater than the sum of its parts (Noss & Harris 1986). That is, although no single reserve is large enough to maintain a viable population, a well-connected network of reserves might maintain a viable metapopulation. Observations of harmful effects of habitat fragmentation on the persistence of populations of area-dependent and other vulnerable species (Saunders *et al.* 1991; Andrén 1994; Woodroffe & Ginsberg 1998) helped promote connectivity as a fundamental design principle in conservation planning.

Connectivity research remains focused largely on corridors. Well-designed empirical studies began to be published in the 1990s, with compelling evidence of the effects of corridors appearing over the last few years (e.g. Haddad *et al.* 2003; Tewksbury *et al.* 2002; Levey *et al.* 2005; Rantalainen *et al.* 2005). Most documented effects have been positive, in the sense of promoting natural movements, dispersal and population persistence, confirming the conclusion of Beier and Noss (1998). Importantly, however, for logistical reasons most empirical studies are of organisms that (i) have small body size, (ii) are of little or no conservation concern and (iii) are moving through experimental landscapes that show little resemblance to the real landscapes where fragmentation problems have been observed or are likely to occur. The value of 'ecological model systems' (in this case, experiments based on small-bodied organisms) for testing predictions from theory and extrapolating results to larger organisms in real landscapes is unclear (Noss & Beier 2000). Although realistic studies on species vulnerable to fragmentation on a range of spatial scales – especially broad scales – will be more difficult to replicate and control for confounding factors, the idiosyncratic species-specific and landscape-specific nature of the problem demands them.

In this essay I summarize an empirical focal-species approach to connectivity planning. My examples are mainly from Florida, USA, where extensive habitat destruction creates barriers to wildlife movement at spatial scales varying from local to regional (Fig. 23.1). Within conservation biology and landscape

Figure 23.1 **In Florida, riparian areas are often the least developed portions of the landscape (because they flood regularly) and thus serve as natural corridors that offer paths of least resistance for wildlife movement through the landscape. Many upland species, however, may not use these lowland corridors. Photo of Myakka River by Reed Noss.**

ecology generally, the familiar 'patch-matrix-corridor' model of landscape structure is not as broadly applicable as once assumed. Most real landscapes are too complex in structure to be described by this simple model. Nevertheless, the patch-matrix-corridor model is by no means obsolete. Indeed, this model is becoming more applicable in parts of the world where intensive development creates a striking habitat/non-habitat dichotomy for many species. Although agricultural or managed forest landscapes can be made more permeable through retention and restoration of key structural features (Lindenmayer & Franklin 2002), in Florida such lands are being rapidly converted to mile after mile of dense housing subdivisions and strip malls with little or no vegetation. Patches and corridors of protected or yet-undeveloped lands are embedded in this hostile matrix. Hence, Florida offers a case study of how connectivity must be addressed in regions with runaway human population growth and development.

Functional connectivity for focal species

Functional connectivity is a condition of landscapes that: (i) provides for daily and seasonal movements of animals; (ii) facilitates dispersal, gene flow and rescue effects of animals and animal-dispersed or animal-pollinated plants; (iii) allows for range shifts or other distributional responses to environmental change; and (iv) maintains flows of ecological processes such as fire and hydrological processes. Historically, wildlife biologists have focused on the first of these functions, and theoretical ecologists and conservation biologists on the second. The third and fourth functions have received only modest attention. I concentrate here on species-specific connectivity in the relative near term, recognizing that ultimately this concern must be expanded to consider connectivity of ecological processes and to address long-term environmental (e.g. climate) change.

Functional connectivity is determined by: (i) the mobility and dispersal behaviour of a species; (ii) other autecological characteristics, such as the need for cover or food when travelling through a landscape; (iii) the interaction of behaviour and autecology with the structural characteristics and spatial pattern of the landscape; (iv) the distance between patches of suitable habitat, as related to dispersal capacity; (v) the presence of features, such as highways, that serve as barriers to movement; and (vi) potential interference from humans or natural predators (Noss & Cooperrider 1994).

Conservation planners recognize corridors in the field or from topographic maps, aerial photographs, satellite images and other sources, and then draw those corridors into reserve network designs. Such 'seat-of-the-pants' approaches make assumptions about functional connectivity that may never be tested (Noss & Daly 2006). Corridor designs based on subjective best-guesses or expert opinion are not necessarily flawed, however, and may sometimes approximate the results of sophisticated analyses. For some species, for example puma (*Puma concolor*) in southern California, radiotelemetry studies and other data have confirmed that travel routes constitute corridors that are clearly recognizable by humans (e.g. Beier 1995). On the other hand, whereas humans focus on the visible connections among habitats, other animals might follow scent trails or rely on other sensory cues during dispersal.

Because protection of land for corridors and mitigation of road impacts through engineering are expensive, rigorous methods for determining functional connectivity are preferable to intuitive or opportunistic approaches based on landscape structure (Noss & Daly 2006). I see no defensible substitute for planning that takes into account the mobility, dispersal characteristics and other autecological features of individual species. Probably no two species have identical requirements for connectivity. The dilemma is that landscapes contain many more species than can be considered individually; in addition, little is known of the movement requirements and dispersal capacity of most of them.

As proposed by Lambeck (1997; but see earlier usage, e.g. Wright & Hubbell 1983), **focal species** can be defined as the most sensitive species within four main vulnerability groups: area-limited, dispersal-limited, resource-limited and process-limited species. The most sensitive species in each group would presumably serve as an umbrella species for others within that group. This is a useful refinement of the otherwise vague concept of umbrella species (see Simberloff 1998). Lindenmayer *et al.* (2002) criticized Lambeck's approach, citing such problems as the lack of data for use in selecting focal species, and argued that a mix of strategies should be adopted to spread the risk of failure of any single approach. A focal-species approach alone is unlikely to meet all conservation objectives; for this reason most modern conservation plans apply multiple selection and design criteria (e.g. Kintsch & Urban 2002; Noss *et al.* 2002; Cowling *et al.* 2003; Pressey *et al.* 2003). Nevertheless, focal species are a useful item in the conservation planning toolbox.

For identifying focal species that are potentially useful in connectivity planning, detailed consideration of life-history and behavioural traits (Table 23.1) is necessary (Noss *et al.* 2005). Much regional and taxonomic variation exists in the association of life-history traits with sensitivity to fragmentation (Hansen & Rotella 2000). Loose generalizations should be avoided in favour of region-specific assessments. Ideally, the sensitivity of species to landscape-scale fragmentation should be tested experimentally (e.g. Terborgh *et al.* 2001;

Table 23.1 **Life-history or behavioural traits associated with sensitivity to fragmentation and with positive responses to enhanced connectivity (albeit correlation does not necessarily mean causation). Examples are of species possessing such traits that might be useful at some scale of connectivity planning.**

Life-history or behavioural trait	**Examples**
Wide-ranging (area-limited)	Large carnivores (Mladenoff *et al.* 1995; Woodroffe & Ginsberg 1998; Carroll *et al.* 2001) Migratory ungulates (Berger 2004; Thirgood *et al.* 2004) Nearshore marine mammals, e.g. manatees (Flamm *et al.* 2005) Seasonal migratory tropical birds or long-distance migratory birds with distinct stopover areas (Meyer 1995; Powell & Bjork 1995)
Nonvagile (dispersal-limited)	Large-seeded plants (Ehrlen & Eriksson 2000) Carrion, dung and other ground beetles (Mader 1984; Klein 1989) Small vertebrates (birds, mammals, salamanders) hesitant to cross gaps (e.g. roads) or enter a dissimilar landscape matrix (Oxley *et al.* 1974; van Dorp & Opdam 1987; Woolfenden & Fitzpatrick 1996; DeMaynadier & Hunter 2000; Develey & Stouffer 2001;Viveiros de Castro & Fernandez 2004)
Specialized habitat requirements	Butterflies restricted to particular grassland types or host plants and with limited patch perception distances (Schultz & Crone 2005)
Highly vulnerable to road mortality	Snakes (Bernardino & Dalrymple 1992; Rosen & Lowe 1994; Rudolph *et al.* 1999; Smith & Dodd 2003; Breininger *et al.* 2004) Turtles (Gibbs & Shriver 2002; Smith & Dodd 2003; Steen & Gibbs 2004: Aresco 2005) Amphibians (Smith & Dodd 2003; Mazerolle 2004) Large mammals (Foster & Humphrey 1995; Ruediger 1996)

Laurance *et al.* 2002). Unfortunately, opportunities for rigorous empirical tests on a broad scale are limited, and time is the essence; thus, candidate focal species must usually be selected through expert-based approaches. Biologists should not sell themselves short. A great deal of natural historic knowledge exists to form reasonable hypotheses about the sensitivity of species to landscape change. Especially in well-studied regions, selection of focal species arguably can be quite robust.

I recommend an approach that combines expert knowledge, habitat modelling and empirical tests to identify focal species at several spatial scales and for each major habitat type. Focal species should include both habitat specialists and generalists. Matching species to the spatial scale of planning can be achieved largely by considering the spatial scale of population processes, in turn a function of trophic level, diet, body size and mobility. Details of autecology, such as dispersal behaviour and sensitivity to particular barriers, must be considered as thoroughly as available information allows.

Focal species for connectivity planning in Florida

I consider three spatial scales of conservation planning: (i) a broad regional or statewide scale of millions of hectares; (ii) a 'meso' scale of tens to hundreds of thousands of hectares (i.e. county to multi-county); and (iii) a fine scale of thousands of hectares down to very small projects (e.g. individual wildlife-crossing structures).

Broad scale

For planning conservation networks at regional and inter-regional scales, large mammals are commonly applied as focal species (Noss *et al.* 1996; Carroll *et al.* 2001). Several large carnivores in North America appear to display metapopulation structure (Beier 1993; Sweanor *et al.* 2000; Carroll *et al.* 2003). Local or project-by-project mitigation will not conserve these species, which often require either immense and intact core areas with minimal edge effects (Woodroffe & Ginsberg 1998) or, alternatively, networks of reasonably large areas connected by substantial dispersal zones (Noss & Harris 1986; Dixon *et al.* 2006). Long-distance migrations by some ungulates occur over vast linear scales (i.e. hundreds of kilometres); more attention to these species in broad-scale connectivity planning is warranted (Berger 2004; Thirgood *et al.* 2004).

For wide-ranging terrestrial animals in general, connectivity is mainly an issue of circumventing barriers to movement (e.g. highways, developed areas) and minimizing human-caused mortality. The dependence of the Florida panther (*Puma concolor coryi*) on habitat connectivity to allow safe movement within and between home ranges is well recognized (Noss & Harris 1986; Maehr *et al.* 2002), as for other subspecies of puma (Beier 1993, 1995; Sweanor *et al.* 2000; Dickson *et al.* 2005). Roadkills are the largest known source of mortality for panthers (Buergelt *et al.* 2002) and have increased recently as traffic volume has exploded. True biological recovery (i.e. long-term population viability) of the panther may depend on restored connectivity across a large portion of its original range, with connections ranging from individual wildlife crossings (Foster & Humprey 1995) to inter-regional corridors (Noss & Cooperrider 1994). The Florida black bear (*Ursus americanus floridanus*) is similar to the panther in habitat requirements and sensitivity to roads. Significant variables explaining its distribution include forest cover, patch size, road density, land-use intensity and distance from intensive land uses (Hoctor 2003). Both habitat models (Hoctor 2003; Larkin *et al.* 2004) and empirical studies including genetic analysis (Dixon *et al.* 2006) have demonstrated the value of regional-scale habitat corridors for this species. Ideally, such corridors would be wide enough to maintain reproducing populations within them.

Aquatic animals may also serve as focal species for connectivity planning. Coastal development, fragmentation of waterways and collisions with watercraft are key threats to the persistence of the Florida population of the manatee (*Trichechus manatus latirostris*), which moves regularly between freshwater, estuarine and marine systems. Manatee movement corridors are defined as routes selected by many manatees to move among aggregation sites; although influenced by water depth and intensity of human use, they are more flexible than terrestrial corridors (Flamm *et al.* 2005).

Meso scale

Among the species vulnerable to fragmentation at a meso scale are snakes, especially large-bodied species (Rosen & Lowe 1994; Enge & Wood 2002). Large snakes are highly mobile and require relatively large areas to maintain viable populations. Their slow locomotion, tendency to thermoregulate on road surfaces, and unpopularity with humans places them at high risk of mortality on and adjacent to roads and developed areas (Rudolph *et al.* 1999). A key focal snake species in Florida is the eastern indigo snake (*Drymarchon corais couperi*), which occupies large home ranges and is vulnerable to road mortality and edge effects (Breininger *et al.* 2004). Turtles, by virtue of their relatively slow locomotion, low levels of recruitment and slow rates of population growth, are also highly vulnerable to fragmentation, especially by roads (Gibbs & Shriver 2002). Because female aquatic turtles move from wetland sites to uplands to deposit eggs, they are disproportionately vulnerable to roadkill, resulting in male-biased sex ratios (Steen & Gibbs 2004; Aresco 2005). In Florida, the gopher tortoise (*Gopherus polyphemus*) – a keystone species with nearly 400 species of invertebrates found in its burrows, many of them obligates (Jackson & Milstrey 1989) – is an ideal focal species for upland habitat connectivity.

Some birds are sensitive to fragmentation of their breeding habitat. The familiar area and edge effects (e.g. Robbins *et al.* 1989; Robinson *et al.* 1995) are not of direct concern with respect to connectivity. Rather, the issue is that some bird species are oddly unwilling to cross gaps – sometimes quite narrow ones – of unsuitable habitat (van Dorp & Opdam 1987; Develey & Stouffer 2001). Both natal dispersal and adult dispersal may be affected by fragmentation, and cooperative-breeding species are particularly vulnerable because of their restricted movements (Walters 1998). In Florida, the Florida scrub-jay (*Aphelocoma coerulescens*) and red-cockaded woodpecker (*Picoides borealis*) are imperilled cooperative-breeding species with extreme habitat specificity and poor dispersal capacity (Stith *et al.* 1996; Schiegg *et al.* 2002). Fragmentation of scrub-jay populations often follows an interesting chain of events: roads and other artificial firebreaks disrupt the spread of fire, which leads to increased forest cover (which the jays generally will not fly over) across the landscape, in turn reducing the connectivity of scrub habitat and isolating populations of jays (D. Breininger, personal communication). Moreover, Florida scrub-jays are vulnerable to roadkill, such that territories bordering roads can be demographic sinks (Mumme *et al.* 2000). Because of their legal protected status, both the Florida scrub-jay and the red-cockaded woodpecker are intensively monitored, which enhances their utility as focal species.

Fine scale

Habitat fragmentation on a fine scale, such as by roads and other linear features, potentially affects a wide variety of species, from large-seeded plants with limited dispersal capacities, to small invertebrates and vertebrates sensitive to narrow linear gaps, to the broad range of animals subject to high rates of road mortality (Table 23.1). Mitigation of road impacts, for example by wildlife-crossing structures, is perhaps the most crucial element of a connectivity strategy in a region with such high traffic volume as Florida. As of September 2005 there were at least 460 terrestrial and 300 aquatic wildlife crossings in North America (Cramer & Bissonette 2006).

Wildlife crossings at carefully selected locations may allow animals to cross highways successfully and maintain connectivity and gene flow within and among populations (Forman *et al.* 2003). Geographical information systems (GIS)-based habitat models and least-cost path analysis, data on roadkill locations, radiotelemetry, remote camera photography and animal signs such as tracks, can identify useful sites for crossing structures (Foster & Humphrey 1995; Clevenger *et al.* 2002: Lyren & Crooks 2002). Culverts and other structures not designed for wildlife movement may nevertheless be useful, especially for small-bodied species, when suitable habitat exists on either side of the highway (Ng *et al.* 2004). However, poorly designed crossings, such as small or flooded culverts, may not be used (Beier 1993; Tigas *et al.* 2002).

Despite some promising studies, knowledge of the effectiveness of various designs for wildlife-crossing structures is limited (Transportation Research Board 2002). Species differ in their requirements and behavioural preferences for crossings, such that a given crossing will be permeable to some species but not to others, potentially causing changes in predator–prey relationships and other community- or ecosystem-level properties (Clevenger & Waltho 2000, 2005). Therefore, monitoring of crossing structures, roadkills, and successful crossings of highways must encompass multiple species. Moreover, because crossings constructed for wildlife will inevitably affect hydrological processes (e.g. stream flow and sheet flow), especially in wetland-rich landscapes, crossings should be multifunctional.

Conclusions

Assembling a minimum set of complementary focal species for addressing connectivity in landscape design remains challenging. Whether selected focal species function as true umbrella species is a question that can only be answered through additional research. Nevertheless, attention to species known or suspected to be at high risk of decline due to fragmentation is necessary regardless of whether the umbrella assumption holds true. The ultimate objective is to restore a landscape that approaches natural levels of connectivity or permeability for all native organisms, but we must start with those fragmentation-sensitive species that we know reasonably well.

The best argument against habitat corridors is that they are expensive, often unproven, and potentially detract from other conservation investments in a world of limited funds (Simberoff *et al.* 1992). I have argued here that selection of multiple, fragmentation-sensitive focal species at several spatial scales and representing multiple habitat types is a way to make connectivity planning more cost-effective. Increased use of quantitative habitat and population modelling, combined with extensive empirical tests, will increase the reliability of connectivity planning. A promising approach is to use existing expert knowledge to build models and develop alternative connectivity designs, which serve as multiple working (*a priori*) hypotheses. Then, the efficacy of these designs can be evaluated through field tests of the responses of organisms to different corridors, landscape configurations and wildlife-crossing structures.

The major limitation to the approach I have outlined in this chapter is knowledge. In some regions the biota is too poorly known for candidate focal species to be reliably identified, evaluated and applied to conservation planning. Money and time for research may be extremely scarce. In such cases analysing structural connectivity of landscape elements remotely (e.g. through satellite imagery) and trying to maintain it by habitat protection and appropriate management may be the best option. In the broader sphere of conservation planning a focal-species approach must be complemented by other approaches, including identification and protection of special elements (e.g. rare species occurrences, unusual habitat types, spatial surrogates of ecological processes) and representation of an adequate area of all land-cover classes (Noss *et al.* 2002).

Landscape design principles for connectivity

1 A well-designed network of reserves or habitat areas can be a whole greater than the sum of its parts.

2 **Create a connectivity design based on the landscape of the present and the future.** This design must be particular to each region, but should be informed by case studies and analogies from other regions.

3 **Some regions allow a primary focus on landscape matrix permeability to provide functional connectivity, whereas in other, more dichotomous landscapes, the patch-matrix-corridor model applies well, and discrete corridors should be identified.** Many regions allow some combination of these approaches.

4 **Identify a minimum set of focal species that are sensitive to the connectivity-disrupting effects of fragmentation at several spatial scales and for multiple habitat types.** A combination of expert judgment, quantitative habitat and population modelling, and empirical studies will be necessary to refine lists of candidate focal species.
5 The strategy is to maintain natural connections in the landscape or, where possible, to restore connections that have been severed by human activity.

References

Andren, H. (1994) Effects of habitat fragmentation on birds and mammals in landscapes with different proportions of suitable habitat: a review. *Oikos* **71**, 355–366.

Aresco, M.J. (2005) The effect of sex-specific terrestrial movements and roads on the sex ratio of freshwater turtles. *Biological Conservation* **123**, 37–44.

Baumgartner, L. (1943) Fox squirrels in Ohio. *Journal of Wildlife Management* **7**, 193–202.

Beier, P. (1993) Determining minimum habitat areas and habitat corridors for cougars. *Conservation Biology* **7**, 94–108.

Beier, P. (1995) Dispersal of cougars in fragmented habitat. *Journal of Wildlife Management* **59**, 228–237.

Beier, P. & Noss, R.F. (1998) Do habitat corridors provide connectivity? *Conservation Biology* **12**, 1241–1252.

Berger, J. (2004) The last mile: how to sustain long-distance migration in mammals. *Conservation Biology* **18**, 320–331.

Bernardino, F.S., Jr. & Dalrymple, G.H. (1992) Seasonal activity and road mortality of the snakes of the Pa-hay-okee wetlands of Everglades National Park, USA. *Biological Conservation* **62**, 71–75.

Breininger, D.R., Legare, M.L. & Smith, R.B. (2004) Eastern indigo snake (*Drymarchon corais couperi*) in Florida. In: Akcakaya, H.R., Burgman, M.A., Kindvall, O. *et al.* (eds.) *Species Conservation and Management: Case Studies*, pp. 299–311. Oxford University Press, Oxford.

Buergelt, C.D., Homer, B.L. & and Spalding, M.G. (2002) Causes of mortality in the Florida panther (*Felis concolor coryi*). *Annals of the New York Academy of Sciences* **969**, 350–353.

Carroll, C., Noss, R.F. & and Paquet, P.C. (2001) Carnivores as focal species for conservation planning in the Rocky Mountain region. *Ecological Applications* **11**, 961–980.

Carroll, C., Noss, R.F., Paquet, P.C. & Schumaker, N.H. (2003) Integrating population viability analysis and reserve selection algorithms into regional conservation plans. *Ecological Applications* **13**, 1773–1789.

Clevenger, A.P. & Waltho, N. (2000) Factors influencing the effectiveness of wildlife underpasses in Banff National Park, Alberta, Canada. *Conservation Biology* **14**, 47–56.

Clevenger, A.P. & Waltho, N. (2005) Performance indices to identify attributes of highway crossing structures facilitating movement of large mammals. *Biological Conservation* **121**, 453–464.

Clevenger, A.P., Wierzchowski, J., Chruszcz, B. & Gunson, K. (2002) GIS-generated, expert-based models for identifying wildlife habitat linkages and planning mitigation passages. *Conservation Biology* **16**, 503–514.

Cowling, R.M., Pressey, R.L., Rouget, M. & Lombard, A.T. (2003) A conservation plan for a global biodiversity hotspot – the Cape Floristic Region, South Africa. *Biological Conservation* **112**, 191–216.

Cramer, P.C. & Bissonette, J.A. (2006) Wildlife crossings in North America: the state of the science and practice. In: Irwin, C.L., Garrett, P. & McDermott, K.P. (eds.) *Proceedings of the 2005 International Conference on Ecology and Transportation*, pp. 442–447. Center for Transportation and the Environment, North Carolina, State University, Raleigh, NC.

DeMaynadier, P.G. & Hunter, M.L., Jr. (2000) Road effects on amphibian movements in a forested landscape. *Natural Areas Journal* **20**, 56–65.

Develey, P.F. & Stouffer, P.C. (2001) Effects of roads on movements of understory birds in mixed-species flocks in Central Amazonian Brazil. *Conservation Biology* **15**, 1416–1422.

Diamond, J.M. (1975) The island dilemma: Lessons of modern biogeographic studies for the design of natural preserves. *Biological Conservation* **7**, 129–146.

Dickson, B.G., Jenness, J.S. & Beier, P. (2005) Influence of vegetation, topography, and roads on cougar movement in southern California. *Journal of Wildlife Management* **69**, 264–276.

Dixon, J.D., Oli, M.K., Wooten, M.C., Eason, T.H., McCown, J.W. & Paetkau, D. (2006) Effectiveness of a regional corridor in connecting two Florida black bear populations. *Conservation Biology* 20, 155–162.

van Dorp, D. & Opdam, P.F.M. (1987) Effects of patch size, isolation and regional abundance on forest bird communities. *Landscape Ecology* **1**, 59–73.

Ehrlen, J. & Eriksson, O. (2000) Dispersal limitations and patch occupancy in forest herbs. *Ecology* **81**, 1667–1674.

Enge, K.M. & Wood, K.N. (2002) A pedestrian road survey of an upland snake community in Florida. *Southeastern Naturalist* **1(4)**, 365–380.

Flamm, R.O., Weigle, B.L., Wright, I.E., Ross, M. & Agliettia, S. (2005) Estimation of manatee (*Trichechus manatus latirostris*) places and movement corridors using telemetry data. *Ecological Applications* **15**, 1415–1426.

Forman, R.T.T. & Godron, M. (1981) Patches and structural components for a landscape ecology. *BioScience* **31**, 733–740.

Forman, R.T.T., Sperling, D., Bissonette, J. *et al.* (2003) *Road Ecology: Science and Solutions*. Island Press, Washington, DC.

Foster, M.L. & Humphrey, S.R. (1995) Use of highway underpasses by Florida panthers and other wildlife. *Wildlife Society Bulletin* **23**, 95–100.

Gibbs, J.P. & Shriver, W.G. (2002) Estimating the effects of road mortality on turtle populations. *Conservation Biology* **16**, 1647–1652.

Gilpin, M. & Hanski, I. (1991) *Metapopulation Dynamics: Empirical and Theoretical Investigations*. Academic Press, London.

Haddad, N.M., Bowne, D.R., Cunningham, A. *et al.* (2003) Corridor use by diverse taxa. *Ecology* **84**, 609–615.

Hansen, A.J. & Rotella, J.J. (2000) Bird responses to forest fragmentation. In: Knight, R.L., Smith, F.W., Buskirk, S.W., Romme, W.H. & Baker, W.L. (eds.) *Forest Fragmentation in the Southern Rocky Mountains*, pp. 201–219. University Press of Colorado, Boulder, CO.

Hanski, I. & Simberloff, D. (1997) The metapopulation approach: its history, conceptual domain, and application to conservation. In Hanski, I.A. & Gilpin, M.E. (eds.) *Metapopulation Biology*, pp. 5–25. Academic Press, San Diego, CA.

Harris, L.D. (1984) *The Fragmented Forest: Island Biogeography Theory and the Preservation of Biotic Diversity*. University of Chicago Press, Chicago, IL.

Hoctor, T.S. (2003) Regional landscape analysis and reserve design to conserve Florida's biodiversity. PhD Dissertation, University of Florida, Gainesville, FL.

Jackson, D.R. & Milstrey, E.G. (1989) The fauna of gopher tortoise burrows. In: Diemer, J.E., Jackson, D.R., Landers, J.L., Layne, J.N. & Wood, D.A. (eds.) *Gopher Tortoise Relocation Symposium Proceedings*, pp. 86–98. Nongame Wildlife Program Technical Report No. 5. Florida Game and Fresh Water Fish Commission, Tallahassee, FL.

Kintsch, J.A. & Urban, D.L. (2002) Focal species, community representation, and physical proxies as conservation strategies: a case study in the Amphibolite Mountains, North Carolina, U.S.A. *Conservation Biology* **16**, 936–947.

Klein, B.C. (1989) Effects of forest fragmentation on dung and carrion beetle communities in central Amazonia. *Ecology* **70**, 1715–1725.

Lambeck, R.J. (1997) Focal species: a multi-species umbrella for nature conservation. Conservation Biology 11, 849–856.

Larkin, J.L., Maehr, D.S., Hoctor, T.S., Orlando, M.A. & Whitney, K. (2004) Landscape linkages and conservation planning for the black bear in west-central Florida. *Animal Conservation* **7**, 23–34.

Laurance, W.F., Lovejoy, T.E., Vasconcelos, H.L. *et al.* (2002) Ecosystem decay of Amazonian forest fragments: a 22-year investigation. *Conservation Biology* **16**, 605–618.

Levey, D.J., Bolker, B.M., Tewksbury, J.J., Sargent, S. & Haddad, N.M. (2005) Effects of landscape corridors on seed dispersal by birds. *Science* **309**, 146–148.

Levins, R. (1969) Some demographic and genetic consequences of environmental heterogeneity for biological control. *Bulletin of the Entomological Society of America* **15**, 237–240.

Lindenmayer, D.B. & Franklin, J.F. (2002) *Conserving Forest Biodiversity: A Comprehensive Multiscaled Approach*. Island Press, Washington, DC.

Lindenmayer, D.B., Manning, A.D., Smith, P.L. *et al.* (2002) The focal-species approach and landscape restoration: a critique. *Conservation Biology* **16**, 338–345.

Lyren, L.M. & Crooks, K.R. (2002) Factors influencing the movement, spatial patterns and wildlife underpass use of coyotes and bobcats along State Route 71 in Southern California. *Proceedings of the International Conference on Ecology and Transportation*. Center for Transportation and the Environment, Raleigh, NC.

MacArthur, R.H. & Wilson, E.O. (1967) *The Theory of Island Biogeography*. Princeton University Press, Princeton, NJ.

Mader, H.J. (1984) Animal habitat isolation by roads and agricultural fields. *Biological Conservation* **29**, 81–96.

Maehr, D.S., Land, E.D., Shindle, D.B., Bass, O.L. & Hoctor, T.S. (2002) Florida panther dispersal and conservation. *Biological Conservation* **106**, 187–197.

Mazerolle, M.J. (2004) Amphibian road mortality in response to nightly variations in traffic intensity. *Herpetologica* **60**, 45–53.

Meyer, K.D. (1995) Swallow-tailed kite (*Elanoides forficatus*). In: Poole, A. & Gill, F. (eds.) *The Birds of North America*, No. 138. The Academy of Natural Sciences, Philadelphia, and The American Ornithologists' Union, Washington, DC.

Mladenoff, D.J., Sickley, T.A., Haight, R.G. & Wydeven, A.P. (1995) A regional landscape analysis and prediction of favorable gray wolf habitat in the northern Great Lakes region. *Conservation Biology* **9**, 279–294.

Mumme, R.L., Schoech, S.J., Woolfenden, G.E. & Fitzpatrick, J.W. (2000) Life and death in the fast lane: demographic consequences of road mortality in the Florida scrub-jay. *Conservation Biology* **14**, 501–512.

Ng, S.J., Dole, J.W., Sauvajot, R.M., Riley, S.P.D. & Valone, T.J. (2004) Use of highway undercrossings by wildlife in southern California. *Biological Conservation* **115**, 499–507.

Noss, R.F. & Beier, P.B. (2000) Arguing over little things: A reply to Haddad *et al. Conservation Biology* **14**, 1546–1548.

Noss, R.F. & Cooperrider, A. (1994) *Saving Nature's Legacy: Protecting and Restoring Biodiversity*. Island Press, Washington, DC.

Noss, R.F. & Daly, K. (2006) Incorporating connectivity into broad-scale conservation planning. In: Crooks, K. & Sanjayan, M. (eds.) *Connectivity Conservation: Maintaining Connections for Nature*, pp. 587–619. Cambridge University Press, Cambridge, UK.

Noss, R.F. & Harris, L.D. (1986) Nodes, networks, and MUMS: preserving diversity at all scales. *Environmental Management* **10**, 299–309.

Noss, R.F., Quigley, H.B., Hornocker, M.G., Merrill, T. & Paquet, P.C. (1996) Conservation biology and carnivore conservation. *Conservation Biology* **10**, 949–963.

Noss, R.F., O'Connell, M.A. & Murphy, D.D. (1997) *The Science of Conservation Planning*. Island Press, Washington, DC.

Noss, R.F., Carroll, C. Vance-Borland, K. & Wuerthner, G. (2002) A multicriteria assessment of the irreplaceability and vulnerability of sites in the Greater Yellowstone Ecosystem. *Conservation Biology* **16**, 895–908.

Noss, R.F., Csuti, B. & Groom. M.J. (2005) Habitat fragmentation. In: Groom, M.J., Meffe, G.K. & Carroll, R.C. (eds.) *Principles of Conservation Biology*, 3rd edn., pp. 213–251. Sinauer Associates, Sunderland, MA.

Oxley, D.J., Fenton, M.B. & Carmody, G.R. (1974) The effects of roads on populations of small animals. *Journal of Applied Ecology* **11**, 51–59.

Powell, G.V.N. & Bjork, R. (1995) Implications of intratropical migration on reserve design: A case study using *Pharomachrus mocinno*. *Conservation Biology* **9**, 354–362.

Pressey, R.L., Cowling, R.M. & Rouget, M. (2003) Formulating conservation targets for biodiversity pattern and process in the Cape Floristic Region, South Africa. *Biological Conservation* **112**, 99–127.

Preston, F.W. (1962) The canonical distribution of commonness and rarity: Part II. *Ecology* **43**, 410–432.

Rantalainen, M.-L., Fritze, H., Haimi, J., Pennanen, T. & Setälä. H. (2005) Species richness and food web structure of soil decomposer community as affected by the size of habitat fragment and habitat corridors. *Global Change Biology* **11**, 1614–1627.

Robbins, C.S., Dawson, D.K. & Dowell, B.A. (1989) Habitat area requirements of breeding forest birds of the Middle Atlantic states. *Wildlife Monographs* **103**, 1–34.

Robinson, S.K., Thompson, F.R. III, Donovan, T.M., Whitehead, D.R. & Faaborg, J. (1995) Regional forest fragmentation and the nesting success of migratory birds. *Science* **267**, 1987–1990.

Rosen, P.C. & Lowe, C.H. (1994) Highway mortality of snakes in the Sonoran Desert of southern Arizona. *Biological Conservation* **68**, 143–148.

Rudolph, C., Burgdorf, S., Conner, R. & Schaefer, R. (1999) Preliminary evaluation of the impact of roads and associated vehicular traffic on snake populations in eastern Texas. In: Evink, G.L., Garrett, P. & Ziegler, D. (eds.) *Proceedings of the Third International Conference on Wildlife Ecology and Transportation*, pp. 129–136. FL-ER-73–99. Florida Department of Transportation, Tallahassee, FL.

Ruediger, B. (1996) The relationship between rare carnivores and highways. In: Evink, G.L., Garrett, P., Ziegler, D. & Berry, J. (eds.) *Trends in Addressing Transportation Related Wildlife Mortality*, pp. 1–7. Florida Department of Transportation, Tallahassee, FL.

Saunders, D.A., Hobbs, R.J & Margules, C.R. (1991) Biological consequences of ecosystem fragmentation: a review. *Conservation Biology* **5**, 18–32.

Schiegg, K., Walters, J.R. & Priddy, J.A. (2002) The consequences of disrupted dispersal in fragmented red-cockaded woodpecker *Picoides borealis* populations. *Journal of Animal Ecology* **71**, 710–721.

Schultz, C.B. & Crone, E.E. (2005) Patch size and connectivity thresholds for butterfly habitat restoration. *Conservation Biology* **19**, 887–896.

Simberloff, D.A. (1998) Flagships, umbrellas, and keystones: Is single-species management passé in the landscape era? *Biological Conservation* **83**, 247–257.

Simberloff, D. & Abele, L.G. (1976) Island biogeography theory and conservation practice. *Science* **191**, 285–286.

Simberloff, D., Farr, J.A., Cox, J. & Mehlman, D.W. (1992) Movement corridors: conservation bargains or poor investments? *Conservation Biology* **6**, 493–504.

Smith, L.L. & Dodd, C.K. (2003) Wildlife mortality on U.S. Highway 441 across Paynes Prairie, Alachua County, Florida. *Florida Scientist* **66**, 128–140.

Steen, D.A. & Gibbs, J.P. (2004) Effects of roads on the structure of freshwater turtle populations. *Conservation Biology* **18**, 143–148.

Stith, B.M., Fitzpatrick, J.W., Woolfenden, G.E. & Pranty, B. (1996) Classification and conservation of metapopulations: a case study of the Florida scrub-jay (*Aphelocoma coerulescens*). In: McCullough, D.R. (ed.) *Metapopulations and Wildlife Conservation*, pp. 187–216. Island Press, Washington, DC.

Sumner, E. (1936) *A Life History of the California Quail, with Recommendations for Conservation and Management*. California State Printing Office, Sacramento, CA.

Sweanor, L.L., Logan, K.A. & Hornocker, M.G. (2000) Cougar dispersal patterns, metapopulation dynamics, and conservation. *Conservation Biology* **14**, 798–808.

Terborgh, J., Lopez, L., Nuñez, P. *et al.* (2001) Ecological meltdown in predator-free forest fragments. *Science* **294**, 1923–1926.

Tewksbury, J.J., Levey, D.J., Haddad, N.M. *et al.* (2002) Corridors affect plants, animals, and their interactions in fragmented landscapes. *Proceedings of the National Academy of Sciences* **99**, 12923–12926.

Thirgood, S., Mosser, A., Tham, S. *et al.* (2004) Can parks protect migratory ungulates? The case of the Serengeti wildebeest. *Animal Conservation* **7**, 113–120.

Tigas, L.A., Van Vuren, D.H. & Sauvajot, R.M. (2002) Behavioral responses of bobcats and coyotes to habitat fragmentation and corridors in an urban environment. *Biological Conservation* **108**, 299–306.

Transportation Research Board (2002) *Surface Transportation Environmental Research: a Long-Term Strategy*. Special Report 268. National Academy Press, Washington, DC.

Viveiros de Castro, E.B. & Fernandez, F.A.S. (2004) Determinants of differential extinction vulnerabilities of small mammals in Atlantic forest fragments in Brazil. *Biological Conservation* **119**, 73–80.

Walters, J.R. (1998) The ecological basis of avian sensitivity to habitat fragmentation. In: Marzluff, J.M. & Sallabanks, R. (eds.) *Avian Conservation: Research and Management*, pp. 181–192. Island Press, Washington, DC.

Wilson, E.O. & Willis, E.O. (1975) Applied biogeography. In: Cody, M.L. & Diamond, J.M. (eds.) *Ecology and Evolution of Communities*, pp. 522–534. Belknap Press/Harvard University Press, Cambridge, MA.

Woodroffe, R. & Ginsberg, J.R. (1998) Edge effects and the extinction of populations inside protected areas. *Science* **280**, 2126–2128.

Woolfenden, G.E. & Fitzpatrick, J.W. (1996) Florida scrub-jay (*Aphelocoma coerulescens*). In: Poole, A. & Gill, F. (eds.) *The Birds of North America*, No. 228. The Academy of Natural Sciences, Philadelphia, and The American Ornithologists' Union, Washington, DC.

Wright, S.J. & Hubbell, S.P. (1983) Stochastic extinction and reserve size: a focal species approach. *Oikos* **41**, 466–476.

24

Connectivity, Corridors and Stepping Stones

Denis A. Saunders

Abstract

There are many publications on wildlife corridors in the scientific and general literature. While there are fewer publications on landscape connectivity and stepping stones for wildlife, all of these terms are used in relation to landscape linkages. The issue of landscape connectivity is one of great scientific and conservation concern; however, there is a lack of understanding about the purposes of such connectivity and how it should be constituted. Landscape design, including landscape linkages, is viewed very much from a cultural perspective. A landscape designed by someone from one continent would look markedly different from one designed by someone from another continent. In this chapter, 11 general principles for landscape design are presented as a basis for developing a 'checklist' to guide thinking in developing landscape connectivity. These principles include the need to align boundaries of human influence with ecological boundaries, and the need to develop networks with those in the corridors of power and influence.

Keywords: landscape connectivity; wildlife corridors; wildlife stepping stones.

Introduction

The issue of landscape connectivity has been extremely well aired in the scientific and the general literature. This is illustrated clearly by searching Google for 'wildlife corridors'. I did this on 13 January 2006 and there were more than 209,000 sites to be explored. I certainly did not explore them all but a quick scan showed they range from general leaflets on conservation benefits of corridors to highly specific scientific papers relating to connectivity for single species. The term 'landscape connectivity' is not as widely used, but nevertheless there were more than 26,700 'hits'. 'Stepping stones for wildlife' was not in common use, with only 8361 hits.

The issue of fragmentation of landscapes and the need for connectivity is of considerable scientific and conservation management concern and has been for several decades. During that period there have been some excellent research papers, reviews and composite publications on corridors and landscape connectivity. For examples, see Fahrig and Merriam (1985), Simberloff and Cox (1987), Bennett (1990, 1998), Saunders and Hobbs (1991), Simberloff *et al.* (1992), Forman (1995), Lindenmayer (1998), and Jongman and Pungetti (2004). There is little point in going over this well-worked ground yet again.

The organizers of the workshop on which this book is based have asked essayists to provide key insights into the assigned topic, draw out what remains unknown or unresolved, show where key ideas might work best and where they would fail and why, and conclude with 5–10 key landscape design principles that relate to the topic area. Essayists were charged to do all this in 3000–4000 words. This is a tall order and one that I intend to follow only partially. What I have done is examine the issue of landscape connectivity from a practical point of view and set out a series of guidelines to help put connectivity in a practical landscape perspective.

Lack of understanding of corridor function

Although it is widely accepted that there is the need to maintain linkages in landscapes fragmented by removal of the original native vegetation, there is considerable ignorance about what the purposes of such linkages are, and how they should be constituted. I shall resort to personal anecdote to illustrate this point. Although this anecdote may seem to be history, it still illustrates the woolly thinking on the subject.

In September 1995, I was one of the invited speakers at a Landcare field day in the central wheatbelt of Western Australia. I have written at length about this particular field day in Saunders (1996) because my experiences that day had a major impact on my perceptions of the understanding among those involved in the management of landscape functions and reconstruction. The organizers of the field day were a Landcare Group consisting of farmers interested in combining nature conservation with their agricultural production. There was still a resident population of the endangered megapode, the malleefowl (*Leipoa ocellata*), in the district and the members of the group undertook active control of foxes (*Vulpes vulpes*) and cats (*Felis catus*), the two main predators of the malleefowl. The owner of the property on which the main group of birds resided and bred undertook intermittent surveys of the bird, but unfortunately he did not keep records. Nevertheless, he was convinced that there were less predators and more malleefowl in the district than there were when predator control was begun. By 1995, it was not unusual to see a mob of 14 malleefowl crossing agricultural fields to drink at a water storage dam surrounded by cleared land.

The stated aim of the field day was to use the expertise of the speakers to plan a corridor to link the major patches of native vegetation on the high ground in the district via a road reserve to a lake in a nature reserve at the lowest point of the district. The road reserve was a gazetted public road that had never been cleared of native vegetation and so was not used for the purpose it had been set aside. The native vegetation along the road reserve was badly degraded by domestic livestock and the nature reserve in which the lake occurred was mainly salmon gum (*Eucalyptus salmonophloia*) woodland. Most of the woodland trees showed distinct signs of 'staghorning' (Saunders *et al.* 2003), a good indication that they were under considerable stress, almost certainly from a rising hypersaline water table, a major environmental problem in southwestern Australia (George *et al.* 1995). When the issues of the design and development of the corridor were discussed, the major aim stated was to provide the malleefowl with a linear strip of native vegetation linking the patches on the hill with the salmon gum woodland at the bottom (Fig. 24.1).

The questions I posed to the Landcare Group were 'why does the malleefowl need the linkages to move?' and 'why do you want malleefowl going to the woodland at the bottom?' The first question was asked because the Landcare Group members stated the birds commonly crossed open agricultural fields (provided foxes and cats were kept under control). From that it seemed that the birds did not need linear strips of native vegetation to foster

Figure 24.1 **Linear strips of native vegetation provide landscape linkages in the central wheatbelt of Western Australia. Photo by Denis Saunders.**

their movements. The second was asked because the birds' preferred habitat is mallee, which occurred in the patches on the high ground, not the more open salmon gum woodland found in the nature reserve at the end of the road reserve. I suggested to the organizers that the conservation resources may be better placed in strategic revegetation to address the rising hypersaline water table to protect the woodland at the bottom, maintaining control of foxes and cats, protecting malleefowl breeding mounds, monitoring the malleefowl population and revegetating to provide more breeding habitat around the existing patches. I do not know how much of this advice was followed in subsequent management.

This story illustrates a number of problems associated with practical landscape design. The first problem is deciding whether there is a need for landscape connectivity and if so, deciding which species the landscape is being designed for and understanding the life-history characteristics of those species. A close analysis of landscape design would show that much is designed for *Homo sapiens*, although we often claim to be doing it for other elements of the biota. The species chosen as guiding landscape design are those that particular human communities hold dear, such as the malleefowl. These species are often charismatic elements of the biota, but in many cases they have no clear ecological functional significance.

Landscape linkages, like beauty, are in the eyes of the beholder. What the beholder sees is very much related to his or her cultural heritage and understanding (or lack of it) of landscape ecology. Consequently, the concepts of landscape design have strong cultural roots. The result is that a landscape designed by the Dutch would look markedly different from that designed by Australians.

The issues of spatial and temporal scale

Besides a lack of understanding about corridor function, a second problem relates to the issue of spatial scale, and this is dictated by our culture and our understanding of the organisms of interest. A third problem relates to the issue of time. I have not sufficient experience of countries other than Australia to enable me to speak on corridor or landscape linkage design of any other continent than Australia. It is obvious in Australia that neither spatial scale nor time are taken into account in most landscape reconstructions because the size of the connections will not stand the test of time. They are too narrow to be anything more than cosmetic and will not be able to resist the processes that lead to clearing by ecological actions (Saunders *et al.* 1991). Space and time would clearly be considered if those involved in landscape design were guided by the suggestions of Harris and Scheck (1991, pp. 203–204), who set out the following three design principles in relation to corridor width:

1 When the movement of individual animals is being considered, when much is known about their behaviour, and when the corridor is expected to function in terms of weeks or months, then the appropriate corridor width can be measured in metres (c.1–10 m).
2 When the movement of a species is being considered, when much is known about its biology, and when the corridor is expected to function in terms of years, then the corridor width should be measured in hundreds of metres (c.100–1000 m).
3 When the movement of entire assemblages of species is being considered, and/or when little is known about the biology of the species involved, and/or if the faunal dispersal corridor is expected to function over decades then the appropriate width must be measured in kilometres.

I can think of no area of revegetation in Australia where the designers have followed those guiding principles.

Eleven general principles for landscape design

There is a need for a set of guiding principles for practical landscape design, recognizing that connectivity is but only one element of successful design. I have presented a draft set of guidelines. In doing so, I have drawn heavily on those proposed by Saunders and Briggs (2002). They can be used as a basis for developing a 'checklist' to guide thinking in developing landscape connectivity.

1 **Shared visions of landscapes of the future:** There needs to be a shared vision of the landscapes of the future and how they should function ecologically, socially and economically in order to be sustainable. There is little point considering only the ecological aspects because the social and economic aspects must be addressed for successful landscape reconstruction. It is important that the vision is developed with the local community because that vision (or visions) must encompass the needs and aspirations of local people and ecological communities and their functions. Too much landscape design is carried out without regard to those who must live and work in those landscapes and who will bear the brunt of the work associated with landscape reconstruction and subsequent management.
2 **What are the impediments to successful landscape reconstruction?** It is necessary to define and understand the environmental, social, institutional and economic problems that must be addressed to achieve the sustainable future implied in the landscapes envisioned in the point above. Much of the failure in landscape reconstruction in Australia is a direct result of the failure to examine the social, institutional and economic problems that stand in the way of much environmental action. Lack of environmental understanding is but one of a string of issues that need to be addressed for successful landscape design.
3 **History must be taken into account:** It has often been said that it is essential to understand the past in order to plan for the future; if only to learn from past mistakes. The importance of history cannot be overemphasized in landscape design. In order to plan for future landscapes it is essential to

determine the functional ecological elements that were present in the landscape before development changed the landscape and what they are now. Are they functioning as they were or has their function changed? In most cases, the changes in ecological functioning are such that the forces causing these changes must be addressed. It is still relatively easy in Australia to find patches of well-intentioned revegetation aimed to address rising water tables that are positioned such that the very changes they are supposed to be addressing will destroy them.

4 **Is there a skeleton on which landscapes can be built?** This point has a strong cultural element. For example, some British landscapes have been so extensively modified that there are very few of the original pre-human elements remaining in the landscapes. In these landscapes, often the aim is to retain a cultural landscape that has important conservation benefits. Ponds in agricultural landscapes are a good case in point (Hull & Boothby 1995). In Australia, usually some elements of the original landscape are present and they can be very useful to determine what is needed to build the future landscape upon.

5 **No further loss of the native biota:** Given that the major cause of loss of biodiversity is the removal of native vegetation it is essential to retain, protect and manage all remnant vegetation to prevent further loss of dependent biota.

6 **Alignment of ecological and management boundaries:** Ecological boundaries and boundaries relating to areas of human management almost never coincide. In any landscape there will be a range of boundaries of human influence, all of which have important impacts on the environment. Any reconstruction plan needs to take cognizance of this mismatch of boundaries of influence. Accordingly, reconstruction plans need to be based on ecological zonings and functional human and ecological communities and organizations, with consideration of how account may be taken of these mismatches.

7 **Develop realistic objectives:** It is essential to establish goals, structures (or frameworks) and timelines for developing the landscapes of the future. The goals need to be realistic and include the resources necessary for their achievement. In Australia, much landscape revegetation to address conservation goals is insufficient to reach thresholds for success and so is essentially a waste of valuable conservation resources. In developing the plan for landscape design it is essential to link human and ecological scales and use the best local knowledge, science and experience available.

8 **Implement the plan:** The world is full of conservation plans that have been developed with the best of intentions, but they have not been implemented.

It is a statement of the obvious that having developed a plan for landscape design, it must be implemented.

9 **Measure progress towards the agreed goals:** Much landscape design fails because of the lack of monitoring progress and adjusting management. In order to follow the principles of adaptive management it is necessary to design and implement a realistic monitoring and evaluation programme that allows those involved in landscape reconstruction to assess progress, record results and adapt their management accordingly.

10 **Lead by example and communicate widely:** Throughout the world there have been imaginative and innovative approaches to landscape design and implementation. Some have been successful and some have not. It is equally important to know what has not been successful as it is to know what has succeeded, so that action is also informed by failures. This means showing others what has been done and communicating the results so others can learn from those results.

11 **Follow the corridors of power and influence:** In the final analysis in landscape design, the most important corridors to develop are the linkages

Checklist for landscape design

1 Develop shared visions of the landscapes of the future in conjunction with those who must live and work in those landscapes and who will bear the brunt of the work associated with landscape reconstruction and subsequent management.

2 Establish the impediments to successful landscape design and implementation.

3 Take ecological history into account in landscape design.

4 Establish the skeleton on which landscape design and reconstruction can be based.

5 Prevent any further loss of the native biota.

6 Align boundaries of human influence with ecological boundaries.

7 Develop realistic objectives for implementation of landscape design.

8 Implement the landscape design.

9 Design and implement a realistic monitoring and evaluation programme to measure progress in implementation of landscape design.

10 Lead by example and communicate successes and failures widely.

11 Network effectively with those in the corridors of power and influence.

with the corridors of power and influence in order to link theoretical landscape design with practical landscape management and to obtain the will and the resources to have the design implemented.

Acknowledgements

I am grateful to David Lindenmayer and Richard Hobbs for the invitation to attend the workshop on 'Vegetation Management and Landscape Design: from Principles to Practice' and submit my small contribution to a meeting of influential international conservation biologists and landscape ecologists.

References

Bennett, A.F. (1990) *Habitat Corridors: the Role in Wildlife Management and Conservation.* Department of Conservation and Environment, Victoria.

Bennett, A.F. (1998) *Linkages in the Landscape. The Role of Corridors and Connectivity in Wildlife Conservation.* IUCN, Gland.

Fahrig, L., Merriam, G. (1985) Habitat patch connectivity and population survival. *Ecology* **66**, 1762–1768.

Forman, R.T.T. (1995) *Land Mosaics. The Ecology of Landscapes and Regions.* Cambridge University Press, Cambridge.

George, R.J., McFarlane, D.J. & Speed, R.J. (1995) The consequences of a changing hydrologic environment for native vegetation in southwestern Australia. In: Saunders, D.A., Craig, J.L. & Mattiske, E.M. (eds.) *Nature Conservation 4: the Role of Networks*, pp. 9–22. Surrey Beatty & Sons, Chipping Norton, NSW.

Harris, L.D. & Scheck, J. (1991) From implications to applications: the dispersal corridor principle applied to the conservation of biological diversity. In: Saunders, D.A. & Hobbs, R.J. (eds.) *Nature Conservation 2: the Role of Corridors*, pp. 189–220. Surrey Beatty & Sons, Chipping Norton, NSW.

Hull, A.P. & Boothby, J. (1995) Networking, partnership and community conservation in north-west England. In: Saunders D.A., Craig, J.L. & Mattiske, E.M. (eds.) *Nature Conservation 4: the Role of Networks*, pp. 373–384. Surrey Beatty & Sons, Chipping Norton, NSW.

Jongman, R. & Pungetti, G. (eds.) (2004) *Ecological Networks and Greenways: Concept, Design, Implementation.* Cambridge University Press, Cambridge.

Lindenmayer, D.B. (1998) *The Design of Wildlife Corridors in Wood Production Forests.* Forest Issues 4. New South Wales National Parks and Wildlife Service, Hurstville, NSW.

Saunders, D.A. (1996) Does our lack of vision threaten the viability of the reconstruction of disturbed ecosystems? *Pacific Conservation Biology* **2**, 321–326.
Saunders, D.A. & Briggs, S.V. (2002) Nature grows in straight lines – or does she? What are the consequences of the mismatch between human-imposed linear boundaries and ecosystem boundaries? An Australian example. *Landscape and Urban Planning* **61**, 71–82.
Saunders, D.A. & Hobbs, R.J. (eds.) (1991) *Nature Conservation 2. The Role of Corridors*. Surrey Beatty & Sons, Chipping Norton, NSW.
Saunders, D.A., Hobbs, R.J. & Margules, C.R. (1991) Biological consequences of ecosystem fragmentation: a review. *Conservation Biology* **5**, 18–32.
Saunders, D.A., Smith, G.T., Ingram, J.A. & Forrester, R.I. (2003) Changes in a remnant of salmon gum *Eucalyptus salmonophloia* and York gum *E. loxophleba* woodland, 1978 to 1997. Implications for woodland conservation in the wheat-sheep regions of Australia. *Biological Conservation* **110**, 245–256.
Simberloff, D.S. & Cox, J. (1987) Consequences and costs of conservation corridors. *Conservation Biology* **1**, 63–71.
Simberloff, D.S., Farr, J.A., Cox, J. & Mehlman, D.W. (1992) Movement corridors: conservation bargains or poor investments? *Conservation Biology* **6**, 493–504.

25

Synthesis: Corridors, Connectivity and Stepping Stones

David B. Lindenmayer and Richard J. Hobbs

The topics of corridors, connectivity and stepping stones form part of a massive literature in the fields of conservation biology and landscape ecology. Saunders (Chapter 24) underscores this with the extraordinary number of websites he found that cited these topics. The three essays in this theme provide markedly different but nevertheless highly complementary perspectives on issues associated with connectivity in three quite different parts of the world. As in many themes throughout this book, the key topics addressed spill over into other themes. There are clear connections between terrestrial ecosystems and aquatic ones. Connectivity and, conversely, the loss of connectivity, is as important in streamscapes as it is in landscapes (see Chapters 37–40). Similarly, issues of connectivity cannot be readily divorced from those associated with the amount of native vegetation cover remaining in the landscape (see Chapters 6–9).

The chapter by Saunders (Chapter 24) is a fascinating narrative about the key questions that should always be posed (but unfortunately only rarely are) before embarking on establishing networks of wildlife corridors. Does the species of concern actually need corridors to move? Why is it important for that species to move from place X to place Y? Are corridors the best option to achieve the desired objectives? In the example of the malleefowl outlined by Saunders, it was clear that other approaches were superior to establishing corridors. In particular, patterns of habitat use, tackling the key threatening process (predation by feral animals) and the use by the species of the cleared agricultural areas (the 'matrix') outside designated linear strips of native vegetation, indicated where the best returns for conservation efforts were to be gained. In this case, the matrix clearly contributed significantly to connectivity for the malleefowl – a theme developed further by Schmiegelow (Chapter 22).

She contends that although human presence is widespread throughout almost all of the world's terrestrial ecosystems, the matrix in many of these systems nevertheless still both provides habitat for many species and contributes significantly to connectivity. This reinforces many of the key points made in the first section on landscape classification (see Chapters 2–5) about the inherent problems associated with overly simplistic binomial perspectives of landscapes that equate vegetation cover and semicleared or cleared areas with 'habitat' and 'non-habitat' respectively. It also reinforces the fact that the way landscapes are classified can influence management decisions and that the way humans perceive landscapes can be markedly different from the way other organisms perceive the same landscape.

Schmiegelow (Chapter 22) further argues that given the habitat and connectivity roles of the matrix in many ecosystems, appropriate management of the matrix, rather than physically established corridors, will be the most important factor influencing connectivity for most species. She contends that the ease with which corridors can be implemented in theoretical planning exercises like reserve design can fail to consider appropriately the connectivity function of the matrix, with less than optimal conservation outcomes in the longer term. Noss (Chapter 23) extends this debate and, while acknowledging that connectivity involves much more than physical corridors, suggests that the rate and extent of human landscape modification in some regions (e.g. Florida, USA) means that physical corridors will become essential for the movement and persistence of dispersal-limited species. He argues that some individual taxa may act as focal or umbrella species to guide effective corridor design, and that these species should be selected on at least three spatial scales – the regional scale, the meso-scale and the local (fine) scale. Expert knowledge built around particular focal taxa could be used to develop plausible alternative designs for corridor networks in any given region or landscape. However, as Simberloff (Chapter 26) argues, the potential for surrogacy delivered via this approach remains a hypothesis that requires significant further testing.

The chapters by Schmiegelow, Noss and Saunders in many ways represent a continuum of landscape cover conditions in which connectivity and physical corridors might be considered. In the boreal forest systems where Schmiegelow has worked, much of the cover remains, and matrix management will be fundamental to the maintenance of connectivity for the vast majority of species. In the ecosystems that are rapidly being modified in Florida, Noss is making a plea for the retention of areas of remaining vegetation to act as physical corridors for some particular species of conservation concern.

Saunders describes work in an already heavily disturbed region where native vegetation cover is now minimal. He contends that for species that may respond positively to corridor establishment, then the residual 'skeletons' of the original landscape cover could form the basis of corridor reconstruction efforts. He also argues that the attributes of those corridors (e.g. width) need to take longer-term biotic use and ecological functions into account.

Some sobering conclusions arise from all three chapters (and the subsequent group discussions) around the theme of 'Corridors, connectivity and stepping stones'. The first is that although an increasing number of studies are showing that physical corridors can contribute to connectivity for some taxa, these taxa tend to be small species of limited conservation concern (Noss, Chapter 23). Second, the loss of connectivity is probably **not** a serious issue for the vast majority of species. Nevertheless, a subset of taxa remains for which it is important. Third, there are few generalizations that can be applied to corridor design. What constitutes an effective corridor will be a function of many things, particularly life-history and demographic attributes of the individual species of concern. This means that where physical corridors are considered to be the best approach to establishing or maintaining connectivity, their design will need to be guided by specific considerations – the particular goals of corridor establishment, the target species for which corridors are intended, and a wide range of other factors. One of these key factors is the amount of native vegetation remaining in a landscape: this significantly influences issues of habitat subdivision, habitat isolation and connectivity. They are likely to be less important where the amount of remaining native vegetation cover is high and more important when vegetation cover is low (see Chapters 6–9 on 'Habitat, habitat loss and patch sizes'). The extreme situation is where there is insufficient vegetation cover remaining to connect up again in a sensible way.

Finally, the absence of generic recipes for corridor design that can be applied uncritically anywhere emphasizes the importance of setting the goals for corridor establishment, defining problems and clearly articulating objectives. Perhaps not surprisingly, this is a recurrent theme across almost all of the topics tackled in this book, but one that is often still ignored in practice.

Section 7
Individual Species Management – Threatened Taxa and Invasive Species

26

Individual Species Management: Threatened Taxa and Invasive Species

Daniel Simberloff

Abstract

Any management programme should be tailored to a specific landscape and have unambiguous objectives, and these objectives will largely determine whether management should target single species, entire ecosystems or both. Most successful existing management programmes for both threatened and invasive species target single species, despite much recent publicity about possible benefits of managing entire ecosystems. There is no reason why a landscape could not be managed for overall ecosystem objectives but include elements aimed at particular species; just such a programme has evolved to deal with a dwindling habitat type and a woodpecker that inhabits it in the southeastern USA. Whether management tools deal with single species or entire ecosystems (e.g. by regulation of ecosystem processes), the presence in many landscapes of several species of particular concern, each concern seemingly requiring a different treatment, will present great challenges. However, a growing catalogue of both single-species and ecosystem management projects should aid in meeting those challenges.

Keywords: adaptive management; biological invasions; ecosystem management; single-species management; threatened species.

Introduction

Any resource management, including management of a landscape, must have clear objectives. What those objectives are cannot be scientifically determined but is a societal issue. All scientists can do is to address whether particular objectives can be achieved, what are the most effective means of achieving them, and what are the various consequences of achieving them, such as whether achievement of one objective conflicts with achievement of others. A corollary to this principle is that resource management must be tailored to a specific system. No generic management prescription is possible for all landscapes.

Whereas single-species management generally has a clearly defined objective – maintenance, increase or decrease of a particular species – ecosystem management has at times been associated with vague, undefined goals, such as 'ecosystem health' (Simberloff 1998). If the chief goal for a certain landscape is simply monetary value of sustainable agricultural production, it may be that no particular species will be associated with this goal. Although some of the literature on ecosystem management regards ecosystem processes in their own right as the goal, those processes are usually a means to an end – the maintenance of one or more particular species. Sometimes large populations of particular species are the overwhelmingly dominant goal, as when lakes are continually stocked with large numbers of some exotic fish. In other words, it is sometimes the case that, ultimately, 'ducks are everything' (see Chapter 37). If this is so, then the appropriate management tool may still be ecosystem management, but this is not automatic. Single-species management might be the most effective or even the only useful approach in particular cases.

Nowadays it is fashionable to attempt to achieve economies of scale by not engaging in single-species management. Instead, managers are encouraged to manage entire regions and ecosystems so as to simulate the array of natural processes, such as disturbance regimes, as closely as possible. This is the underlying principle of ecosystem management, which has become the official policy of many resource agencies (Christensen *et al.* 1996; Meffe *et al.* 2006), but it is nonetheless problematic. It is often associated with the concept of adaptive management – project as experiment (Walters 1986; Walters & Holling 1990) – which is also problematic. Ecosystem management is new, so there is no extensive catalogue of attempts and results with which to judge its success. Also, its main operational precept – maintain a semblance of natural processes – sounds straightforward but is often technically difficult, especially given the fact that ecosystem management is usually employed in situations where a system is explicitly designated for some degree of multiple use, not just conservation (Meffe *et al.* 2006). Adaptive management is so loosely and variously defined (cf. Lee 2001; James 2004) that, although it has become a resource management buzzword, it is difficult to find cases where it has been rigorously practised (Stankey *et al.* 2003), much less where it has been successful.

Therefore, although single-species management is often cast as passé in the rush to manage ecosystems (Simberloff 1994, 1998) it is important to continue to manage at least some individual species and not to jettison management approaches that work, simply because they seem old-fashioned.

Threatened species

One fruitful focus of single-species management is species consisting of one or a few very small populations. As noted by Caughley (1994), conservation approaches to such species tend to fall into two paradigms, the small population paradigm and the declining population paradigm. The former sees problems associated with such species as generic and inherent in any small population – demographic and genetic stochasticity and the like. The declining population paradigm, by contrast, seeks the idiosyncratic reasons that led what were presumably once widely distributed species to become sparse and restricted, then attempts to remedy or compensate for them. A few species probably never had more than one or a few small populations. One example is the Devil's Hole pupfish (*Cyprinodon diabolis*), endemic to a single pool in Nevada, USA, with a surface area of c.200 m^2; its population hovers between c.200 and 600 individuals (United States Fish and Wildlife Service 1980). The population fell in the late 1960s because irrigation pumping lowered the water level. Even with a puzzling recent population decline, there is no evidence that generic small population threats (e.g. demographic and genetic stochasticity) are at play, but a variety of management procedures have aimed specifically at the pupfish, which would surely be extinct without them – maintenance of water level, establishment of a captive 'insurance' population, security fencing, etc. If there were other components of the ecosystem to worry about, some of these measures would aid all of them, but others (e.g. establishment of the insurance population) can only benefit the pupfish.

I can think of no rescue effort that clearly saved a tiny population of an endangered species from an unambiguous threat from one of the generic factors, but many species have been successfully saved (at least for now) by management that targets specific, idiosyncratic causes of decline; Caughley and Gunn (1996) provide many examples. The classic case is the Chatham Island black robin (*Petroica traversi*), once reduced to five individuals and saved by a yeoman effort including moving individuals to new islands and foster-rearing by another species (Butler & Merton 1992); there are now c.250 individuals. None of these activities could have aided other species.

Another impressive example of single-species management from New Zealand is the recovery of a declining passerine, the North Island kokako, *Callaeas cinerea wilsoni* (Innes *et al.* 1999). This case is particularly interesting because it entailed a good faith effort to employ adaptive management. Predation of eggs and/or chicks by introduced predators and competition with introduced mammalian herbivores were suspected causes of the decline. Three forests (1000–3000 ha) were assigned intensive pest-management treatments for eight years. There was no complete control and little replication, but partial control was achieved by applying pest management to one of the forests only for the first four years, and to a second forest only for the second four years. These results implicated predation as the key threat, and extensive predator management has now become standard. More complete experimental designs were considered but rejected for want of more study areas and because of expense.

The key tool used in conservation biology to predict extinction risks and compare management options for threatened species is population viability analysis (PVA), which is fundamentally a single-species approach. It has performed quite well in some settings (Brook *et al.* 2000) and poorly in others (Lindenmayer *et al.* 2003), although, as currently developed and parameterized, it is unlikely to accommodate adequately aspects of movement and connectivity of particular populations in particular landscapes (Chapter 7). Nevertheless, PVA will surely continue to be used in many landscape management programmes, and improvements in prediction accuracy will still be in the context of single-species management.

Often a longstanding, essentially single-species management approach has been transmogrified, under the aegis of the general trend toward ecosystem management, into an effort that includes procedures to mimic aspects of natural ecosystem cycles. The long battle to save the red-cockaded woodpecker (*Picoides borealis*) in the southeastern USA is an example (Simberloff 2004). One of the first species listed under the US Endangered Species Act, it has long generated conflict between the forest industry and conservationists. It declined from being a common, widespread bird in the Southeast in the 19th century to a population size of c.15,000 by 1970, spread among several 'islands' of suitable habitat that are all remnants of what had been formerly a largely continuous mass of c.25 million hectares of mature longleaf pine (*Pinus palustris*) forests. Suitable habitat now consists of a few hundred hectares in scattered small tracts, plus roughly 4 million hectares of variously acceptable second-growth forest. The woodpecker nests in laboriously constructed cavities in old, dying longleaf pine trees, which became increasingly rare as longleaf forest was converted to farmland, production forests of faster growing trees and other uses (Tebo 1985). Specific factors implicated in the decline include hardwood midstorey encroachment causing abandonment of cavities, loss of old trees, forest fragmentation and difficulties in finding mates in isolated populations (James *et al.* 1997, Conner *et al.* 2001).

The United States Fish and Wildlife Service (2003) has elaborated a management plan touted as a mixture of single-species management and ecosystem management, but it emphasizes management of the woodpecker as an umbrella species – one with habitat requirements so demanding that managing this species by managing its habitat will save all other species (Simberloff 1998). Though the species-rich longleaf community includes several species of special concern, only the woodpecker is targeted for particular activities, such as moving individuals and installing artificial cavities. The other key part of the programme – more frequent burning and retaining more old trees – is very much in the ecosystem management tradition of mimicking natural processes. An adaptive management approach (Provencher *et al.* 2001) has suggested effective routes toward the goal of reducing hardwood encroachment by various means (prescribed fire, herbicides, mechanical removal), but these have yet to be tested further.

However salubrious the burning regime is for the longleaf community as a whole (the suitability of the woodpecker as an umbrella species remains to be established), it will probably not suffice to sustain the woodpecker. James *et al.* (2003) argue that, even with the suggested prescribed fire programme, recruitment of saplings will not lead to an adequate supply of large, old trees. So individual species management will have to be pursued in perpetuity for the woodpecker.

Noss (Chapter 23) suggests the possibility of using a focal species approach to deal with maintaining adequate connectivity while managing an entire landscape; in essence, this is an analogue of the umbrella species approach. As with umbrella species, managing such a connectivity focal species with highly targeted interventions – such as moving individuals from site to site, or constructing pathways that are unlikely to be suitable for most other species – would defeat the purpose of designating such a species in the first place. Also, as with umbrella species (cf. Simberloff 1998), substantial population research would be necessary to demonstrate the utility of a particular focal species in understanding connectivity requirements for much of a community. However, one could imagine some melding of single-species and ecosystem management with respect to connectivity in much the same way that the US Fish and Wildlife Service is attempting to unite conservation of the red-cockaded woodpecker and conservation of the longleaf pine community, as described above.

Invasive introduced species

Although there is a tendency among the public and some resource managers to view management of invasive introduced species as hopeless, there have been many successes, almost all of them due to managing single species. Of course preventing the introduction of non-indigenous species in the first place by shutting down pathways through which they arrive is a key goal nowadays (Ruiz & Carlton 2003), and this strategy targets all species using various pathways, not just single species. There is also occasionally discussion of managing entire ecosystems so that the habitat will be inimical to invaders. The ancient agricultural practice of cultural control was based on the idea that management should minimize invasion by all weeds (Stiling 1985), and it is still a viable option. For example, managing North American pastures properly, and especially eliminating overgrazing, favours native grasses and limits the growth and spread of invaders such as musk thistle (*Carduus nutans*) (Louda 2000). In the longleaf pine forests discussed above, approximating a natural fire regime may be responsible for the relatively small impact of non-indigenous plants on the groundcover flora (Simberloff 2001). However, a recent, rapidly spreading, fire-adapted invader, cogon grass (*Imperata cylindrica*), may break down this ecosystem defence. Overall, there has been little research on ecosystem management of introduced species.

Eradication

Once a non-indigenous species has established a population, it is often possible to eradicate it before it spreads. Although eradication has a bad name in some management circles (e.g. Dahlsten 1986), there have been many successful eradications (Simberloff 2002a) as well as some highly publicized failures, such as attempts in the USA to eliminate the red imported fire ant *Solenopsis invicta* (Buhs 2004) and the gypsy moth *Lymantria dispar* (Spear 2005). There are several key features that separate successful eradications from failures (Myers *et al.* 2000).

1 There must be adequate resources to complete the eradication. Because populations of the target organism fall to near invisibility in the latter stages of an eradication campaign, costs may increase, but the job must be completed.
2 Some entity must have the authority to undertake the project and to enforce cooperation. Eradication is an all-or-none approach, and individual stakeholders must not be able to opt out of the programme.
3 There must be adequate knowledge of the target organism to suggest a vulnerable point in its life cycle and to allow a reasonable probability of success. However, it is important to recognize that often very basic natural history (e.g. can the species self-fertilize?) is adequate, and sometimes a rapid scorched-earth approach using blunt tools, such as axes, guns and chemicals, can succeed (Simberloff 2003). It is usually not a good excuse to fail to act quickly on the grounds that we need better scientific knowledge, because any eradication campaign becomes vastly more difficult when species are widespread.
4 Reinvasion by an eradicated target must be either an acceptable outcome or of such low probability as to be discounted (this is often the case on islands, though some species, such as rats, can be surprisingly adept at reinvading; Russell *et al.* 2005).
5 There must not be a high probability that the eradication of one pest will enable a worse one to flourish. For example, on Santa Cruz Island, California, resurgence of highly flammable exotic fennel (*Foeniculum vulgare*) followed removal of large introduced grazers (Dash & Gliessman 1994).

When these conditions have been met, many successes have ensued, such as the eradication of the giant African snail (*Achatina fulica*) from Florida (Mead 1979) and Queensland (Colman 1978), rats from many islands (e.g. Taylor *et al.* 2000; Towns & Broome 2003), the melon fly (*Bactrocera cucurbitae*) from the Ryukyu Archipelago (Kuba *et al.* 1996), karoo thorn (*Acacia karoo*) from Western Australia and Victoria (Weiss 1999) and Taurian thistle (*Onopordum tauricum*) from Victoria (Weiss 1999). The Island Conservation group, based in Santa Cruz, California, has eradicated various combinations of rats, cats, mice, goats, rabbits and burros from over 30 Mexican islands (Tershy *et al.* 2002).

Two of the most impressive recent eradications have been in Australia – the elimination of the Caribbean black-striped mussel (*Mytilopsis sallei*) from Darwin Harbor (Wittenberg & Cock 2001) and the extirpation of kochia (*Bassia scoparia*) from over 3000 ha spread out over 900 km in Western Australia (Randall 2001).

Maintenance management

Several individual-species-management procedures have succeeded in keeping established introduced species at acceptably low levels in particular cases. These fall into three categories (Simberloff 2002b): mechanical or physical control, chemical control and biological control.

Mechanical or physical management includes hand-picking of plants and animals, trapping and shooting animals, cutting and burning invasive or infested plants, and, increasingly, using machinery specifically designed to manage invasive plants. Mechanical or physical management can be highly effective in the long term, but it is often labour intensive. In wealthy nations, volunteers have frequently been mobilized for this purpose – in the USA the Nature Conservancy has many such volunteer programmes (Randall *et al.* 1997), and the Grand Canyon Wildlands Council in the Southwest has cleared many areas of tamarisk (*Tamarix ramosissima*) (Grand Canyon Wildlands Council, Inc. 2005). Also in the USA many state agencies use convict labour. For example, in Kentucky, the State Nature Preserves Commission manages Eurasian musk thistle (*Carduus nutans*) using volunteers convicted of driving under the influence of alcohol (J. Bender, personal communication), while Florida inmates are a major part of a varied management programme that has reduced infestations of Australian paperbark (*Melaleuca quinquenervia*) by c.40% (Campbell & Carter 1999; Silvers 2004).

Paid labour is also often used for maintenance management that has major mechanical and physical components. For instance, the Florida management programme for *Melaleuca* employs many state and regional agency personnel (Silvers 2004). The Alberta Rat Patrol, employed by the province of Alberta in Canada, has kept Norway rat populations at a minimal level for several decades throughout the province (Bourne 2000). Perhaps the most striking use of paid labour in long-term physical control of invasive species is the South African Working for Water Programme, a public works project employing thousands of individuals to fight exotic plant invasions (McQueen *et al.* 2000; van Wilgen *et al.* 2000).

Chemical control, using herbicides, pesticides and poison baits, occasionally targets groups of invaders – e.g. with use of the microbial pesticide Btk (*Bacillus thuringiensis* var. *kurstaki*) – although it generally aims at individual introduced species and aims to avoid non-target impacts. Occasionally, it engenders opposition on the grounds that vertebrates suffer a painful death; for example, control of the brushtail possum (*Trichosurus vulpecula*) in New Zealand with the toxicant 1080 (sodium monofluoroacetate) aroused public concern (Eason *et al.* 2000, Parliamentary Commissioner for the Environment 2000). Chemical control is often controversial because of substantial non-target impacts, including health effects and biological magnification. Rachel Carson's *Silent Spring* (1962) highlighted such problems, and the failed fire ant eradication campaign in the USA, using chlorinated hydrocarbons, is an example (Buhs 2004). Although many modern chemical controls have far fewer non-target impacts, there is substantial chemophobia, particularly among conservationists (Williams 1997). In addition, the expense of chemically treating large areas is often prohibitive, particularly where the goal is conservation rather than agriculture, and target species eventually evolve resistance to chemicals (Simberloff 2002b).

However, in many instances, chemical control has succeeded in the long term in maintaining an invasive pest at tolerably low densities. For instance, in Florida, water hyacinth (*Eichhornia crassipes*) is adequately managed with 2,4-D, and the amounts of chemical used have fallen drastically as the plant has receded (Schardt 1997), while the state of Oregon has eradicated many infestations of the gypsy moth *Lymantria dispar* for over 25 years by spraying Btk, even though continual reintroduction requires ongoing treatment (Oregon Department of Agriculture 2006).

Biological control is often seen as a green alternative to chemical control (e.g. McFadyen 1998), and it almost always targets single species. In the past, generalized predators have occasionally been used to attempt to control all pestiferous members of a particular class (e.g. all aphids), but professional biocontrol practitioners now usually eschew using generalized predators because these have produced some legendary fiascos – e.g. introduction of the small Indian mongoose (*Herpestes javanicus*) and rosy wolf snail (*Euglandina rosea*) (Simberloff 2002b), which have between them caused the extinction of at least 60 species. Similarly, generalized herbivores are usually frowned upon – one, the cactus moth (*Cactoblastis cactorum*), is currently of great concern as it marches towards the US Southwest and Mexico threatening 79 prickly pear (*Opuntia*) species (Zimmermann *et al.* 2001).

However, biological control has had many successes, primarily with natural enemies of much more restricted host range (Thomas & Willis 1998). For example, in Africa, a devastating South American mealybug pest of cassava (*Phenacoccus manihoti*) was brought under control by an imported South American wasp parasitoid, *Epidinocarsis lopezi* (Bellotti *et al.* 1999), while on the island of St Helena, the endemic gumwood tree, *Commidendrum robustum*, was threatened with extinction by the South American scale *Orthezia insignis* and saved by introduction of a lady beetle, *Hyperaspis pantherina* (Wittenberg & Cock 2001). When it works, biocontrol has the advantage over chemical and mechanical control in that it will operate in perpetuity without continued treatment. Also, the control agent often spreads on its own far from the release site, to wherever the target pest is established.

One key disadvantage of biological control is that it usually does not work; most releases of biocontrol agents do not yield substantial control of the target pest (Williamson 1996). Another disadvantage is the risk of non-target impacts, such as the many extinctions caused by the mongoose and rosy wolf snail on islands worldwide. The Eurasian weevil *Rhinocyllus conicus*, introduced to North America to control musk thistle, threatens native thistles, including a federally endangered species (Simberloff 2002b), while the spread of the cactus moth towards Mexico, noted above, threatens several native cactus species (Stiling & Simberloff 2000). Finally, unlike chemical control, biological control is essentially irreversible. Even if an unintended non-target impact arises, eradication of the control agent is unlikely.

Discussion

Despite the appeal of adaptive ecosystem management, management of individual species, including adaptive management, still plays a key role in both conservation and in control of invasive species. In conservation, certainly when a species has extremely low numbers and a narrow range, and society has determined that the species should be saved, it would be stubbornly ideological to insist on ecosystem management as the *modus operandi*. In such instances, there are often no other species of concern, and the goal of saving the threatened species mandates tailoring a management approach to fit the particulars of the case. As argued by Caughley (1994), the great majority of such species are in decline, and the key remedy is to stop whatever is causing the decline. Although such an activity may affect other species in the system (e.g. the prescribed fire in the longleaf-woodpecker system), often the benefit will accrue almost exclusively to the targeted species (for instance, predator management for the kokako). It is possible that, as the science of ecosystem management matures, procedures and consequences will be sufficiently understood that some species currently requiring individual management can be managed as part of more comprehensive schemes. But the field is not nearly at that point now.

For established introduced invasive species, almost all management is currently of single species, and it is often quite effective. Eradication sometimes succeeds, and it is being attempted at even larger scales. For maintenance management, there is no single panacea, but successes abound for mechanical/physical, chemical and biological control. There are, of course, many failures for all three approaches, but there is every reason to believe that increasing experience, and especially increasing publication and knowledge-sharing (Simberloff 2002b), will increase the success rates. It is likely that certain ecosystems can be managed so as to favour native species over non-indigenous ones, but it is doubtful that such an approach will ever be proof against all invaders – there will always be a need for single-species management.

Even in instances in which the technology of managing individual species is well developed, managing all species that require such management in a system may be challenging. It is possible that a method that would save one threatened species would worsen the plight of another. For example, habitat management to aid the Devil's Hole pupfish, described above, has harmed populations of an insect on the federal Endangered Species list, the Ash Meadows naucorid *Ambrysus amargosus* (Polhemus 1993). Or it may be difficult to manage a threatened species so as to maintain or increase its population while decreasing that of an invader (see Chapter 27). For instance, Schiffman (1994) found in California valley grasslands that burrowing by the endangered giant kangaroo rat (*Dipodomys ingens*) facilitates invasion by exotic ruderal Mediterranean annuals, and the rodent caches the exotic seeds. She suggests this mutualism greatly complicates restoration of grasslands in the range of the kangaroo rat. As with the possibility of improved technologies for ecosystem management of introduced invasives, it is likely that continued experiment and an ever-increasing catalogue of attempted simultaneous single-species management of more than one pecies will lead to advances in this area.

General principles

The following principles are clear from the above examples and discussion:

1 The specific objectives of a management programme must be clear and unambiguous.
2 Any management programme must be tailored to a specific landscape and set of objectives.
3 Depending on the landscape and objectives, management techniques may target single species, groups of species, or whole ecosystems. A programme may employ a mix of techniques.
4 Two or more objectives may be incompatible with presently available techniques. Such a situation necessitates ranking objectives and suggests a search for improved techniques.

References

Bellotti, A.C., Smith, L. & Lapointe, S.L. (1999) Recent advances in cassava pest management. *Annual Review of Entomology* **44**, 343–370.

Bourne, J. (2000) *A History of Rat Control in Alberta*. Alberta Agriculture, Food and Rural Development, Edmonton.

Brook, B.W., O'Grady, J.J., Chapman, A.P., Burgman, M.A., Akçakaya, H.R. & Frankham, R. (2000) Predictive accuracy of population viability analysis in conservation biology. *Nature* **404**, 385–387.

Buhs, J.B. (2004) *The Fire Ant Wars*. University of Chicago Press, Chicago, IL.

Butler, D. & Merton, D. (1992) *The Black Robin: Saving the World's Most Endangered Bird*. Oxford University Press, Auckland.

Campbell, C. & Carter, F.D. (1999) The Florida Department of Corrections involvement in exotic pest plant control. In: Jones, D.T. & Gamble, B.W. (eds.) *Florida's Garden of Good and Evil*, pp. 147–149. Florida Exotic Pest Plant Council, West Palm Beach, FL.

Carson, R. (1962) *Silent Spring*. Houghton Mifflin, Boston.

Caughley, G. (1994) Directions in conservation biology. *Journal of Animal Ecology* **63**, 215–244.

Caughley, G. & Gunn, A. (1996) *Conservation Biology in Theory and Practice*. Blackwell Science, Cambridge, MA.

Christensen, N.L., Bartuska, A.M., Brown, J.H. *et al.* (1996) The report of the Ecological Society of America on the scientific basis for ecosystem management. *Ecological Applications* **6**, 665–691.

Colman, P.H. (1978) An invading giant. *Wildlife in Australia* **15**, 46–47.

Conner, R.N., Rudolph, D.C. & Walters, J.R. (2001) *The Red-Cockaded Woodpecker: Surviving in a Fire-Maintained Ecosystem.* University of Texas Press, Austin, TX.

Dahlsten, D.L. (1986) Control of invaders. In: Mooney, H.A. & Drake, J.A. (eds.) *Ecology of Biological Invasions of North America and Hawaii*, pp. 275–302. Springer-Verlag, New York.

Dash, B.A. & Gliessman, S.R. (1994) Nonnative species eradication and native species enhancement: Fennel on Santa Cruz Island. In: Halvorson, W.L. & Maender, G.J. (eds.) *The Fourth California Islands Symposium: Update on the Status of Resources, Santa Barbara, California*, pp. 505–512. Santa Barbara Museum of Natural History, Santa Barbara, CA.

Eason, C., Warburton, B. & Henderson, R. (2000) Toxicants used for possum control. In: Montague, T.L. (ed.) *The Brushtail Possum. Biology, Impact and Management of an Introduced Marsupial*, pp. 164–174. Manaaki Whenua Press, Lincoln, New Zealand.

Grand Canyon Wildlands Council, Inc. (2005) Glen and Grand Canyon riparian restoration project: Final report of Arizona Water Protection Fund (http://www.grandcanyonwildlands.org/GIC%20and20%GC%20Riparian%20Restoration%20Project.pdf).

Innes, J., Hay, R., Flux, I., Bradfield, P., Speed, H. & Jansen, P. (1999) Successful recovery of North Island kokako *Callaeas cinerea wilsoni* populations, by adaptive management. *Biological Conservation* **87**, 201–214.

James, F.C. (2004) Adaptive management in the Nature Conservancy. *Science Chronicles* **2**, 2–5.

James, F.C., Hess, C.A. & Kufrin, D. (1997) Species-centered environmental analysis: Indirect effects of fire history on red-cockaded woodpeckers. *Ecological Applications* **7**, 118–129.

James, F.C., Richards, P.M., Hess, C.A., McCluney, K.E., Walters, E.L. & Schrader, M.S. (2003) Sustainable forestry for the red-cockaded woodpecker ecosystem. In: Costa, R. & Daniels, S.J. (eds.) *Red-cockaded Woodpecker: Road to Recovery*, pp. 60–69. Hancock House, Blaine, WA.

Kuba, H., Kohama, T., Kakinohana, H. *et al.* (1996) The successful eradication programs of the melon fly in Okinawa. In: McPheron, B.A. & Steck, G.J. (eds.) *Fruit Fly Pests: A World Assessment of Their Biology and Management*, pp. 534–550. St. Lucie Press, Delray Beach, FL.

Lee, K.N. (2001) Appraising adaptive management. In: Buck, L.E., Geisler, C.C., Schelhas, J. & Wollenberg, E. (eds.) *Biological Diversity: Balancing Interests through Adaptive Collaborative Management*, pp. 3–26. CRC Press, Boca Raton, FL.

Lindenmayer, D.B., Possingham, H.P., Lacy, R.C., McCarthy, M.A. & Pope, M.L. (2003) How accurate are population models? Lessons from landscape-scale tests in a fragmented system. *Ecology Letters* **6**, 41–47.

Louda, S.M. (2000) Negative ecological effects of the musk thistle biological control agent, *Rhinocyllus conicus*. In: Follett, P.A. & Duan, J.J. (eds.) *Nontarget Effects of Biological Control*, pp. 215–243. Kluwer, Boston, MA.

McFadyen, R.E.C.(1998) Biological control of weeds. *Annual Review of Entomology* **43**, 369–393.

McQueen, C., Noemdoe, S. & Jezile, N. (2000) The Working for Water Programme. In: Preston, G., Brown, G. & van Wyk, E. (eds.) *Best Management Practices for Preventing and Controlling Invasive Alien Species. Symposium Proceedings*, pp. 51–54. The Working for Water Programme, Cape Town.

Mead, A.R. (1979). Ecological malacology: with particular reference to *Achatina fulica*. In: Fretter, V., Fretter, J. & Peake, J. (eds.) *Pulmonates*, Vol. 2b, pp. i–x and 1–150 Academic Press, London.

Meffe, G.K., Groom, M.J. & Carroll, C.R. (2006) Ecosystem approaches to conservation: Responses to a complex world. In: Groom, M.J., Meffe, G.K. & Carroll, C.R. (eds.) *Principles of Conservation Biology*, 3rd edn., pp. 467–507. Sinauer, Sunderland, MA.

Myers, J.H., Simberloff, D., Kuris, A.M. & Carey, J.R. (2000) Eradication revisited – dealing with exotic species. *Trends in Ecology and Evolution* **15**, 316–320.

Oregon Department of Agriculture (2006) Gypsy moth eradication program (http://egov.oregon.gov/ODA/PLANT/ippm_control_gm.shtml).

Parliamentary Commissioner for the Environment (2000) *Caught in the Headlights: New Zealanders' Reflections on Possums, Control Options and Genetic Engineering*. Parliamentary Commissioner for the Environment, Wellington, New Zealand.

Polhemus, D.A. (1993) Conservation of aquatic insects. Worldwide crisis of localized threats. *American Zoologist* **33**, 588–598.

Provencher, L., Herring, B.J., Gordon, D.R. *et al.* (2001) Longleaf pine and oak responses to hardwood reduction techniques in fire-suppressed sandhills in northwest Florida. *Forest Ecology and Management* **148**, 63–77.

Randall, J.M., Lewis, R.R. III & Jensen, D.B. (1997) Ecological restoration. In: Simberloff, D., Schmitz, D.C. & Brown, T.C. (eds.) *Strangers in Paradise. Impact and Management of Nonindigenous Species in Florida*, pp. 205–219. Island Press, Washington, DC.

Randall, R. (2001) Eradication of a deliberately introduced plant found to be invasive. In: Wittenberg, R. & Cock, M.J.W. (eds.) *Invasive Alien Species: A Toolkit of Best Prevention and Management Practices*, p. 174. CAB International, Wallingford.

Ruiz, G.M. & Carlton, J.T. (eds.) (2003) *Invasive Species. Vectors and Management Practices*. Island Press, Washington, DC.

Russell, J.C., Towns, D.R., Anderson, S.H. & Clout, M.N. (2005) Intercepting the first rat ashore. *Nature* **437**, 1107.

Schardt, J.D. (1997) Maintenance control. In: Simberloff, D., Schmitz, D.C. & Brown, T.C. (eds.) *Strangers in Paradise. Impact and Management of Nonindigenous Species in Florida*, pp. 229–243. Island Press, Washington, DC.

Schiffman, P.M. (1994) Promotion of exotic weed establishment by endangered kangaroo rats (*Dipodomys ingens*) in a California grassland. *Biodiversity and Conservation* **3**, 524–537.

Silvers, C.S. (2004) Status and impacts of the melaleuca biological control program. *Wildland Weeds* **7**, 8–10.

Simberloff, D. (1994) How forest fragmentation hurts species and what to do about it. In: Covington, W.W. & DeBano, L. (eds.) *Sustainable Ecological Systems: Implementing an Ecological Approach to Land Management*. pp. 85–90. Rocky Mountain Forest and Range Experiment Station, US Dept. of Agriculture, Flagstaff, AZ, and Fort Collins, CO.

Simberloff, D. (1998) Flagships, umbrellas, and keystones: is single-species management passé in the landscape era? *Biological Conservation* **83**, 247–257.

Simberloff, D. (2001) Biological invasions – How are they affecting us, and what can we do about them? *Western North American Naturalist* **61**, 308–315.

Simberloff, D. (2002a) Today Tiritiri Matangi, tomorrow the world! – Are we aiming too low in invasives control? In Veitch, C.R. & Clout, M.N. (eds.) *Turning the Tide: The Eradication of Invasive Species*, pp. 4–12. IUCN Species Survival Commission, Gland, Switzerland.

Simberloff, D. (2002b) Managing existing populations of alien species. In: Claudi, R., Nantel, P. & Muckle-Jeffs, E. (eds.) *Alien Invaders in Canada's Waters, Wetlands, and Forests*, pp. 269–278. Natural Resources Canada, Canadian Forest Service, Ottawa.

Simberloff, D. (2003) How much population biology is needed to manage introduced species? *Conservation Biology* **17**, 83–92.

Simberloff, D. (2004) Community ecology: Is it time to move on? *American Naturalist* **163**, 787–799.

Spear, R.J. (2005) *The Great Gypsy Moth War*. University of Massachusetts Press, Amherst.

Stankey, G.H., Bormann, B.T., Ryan, C. *et al.* (2003) Adaptive management and the Northwest Forest Plan: Rhetoric and reality. *Journal of Forestry* **101**, 40–46.

Stiling, P.D. (1985) *An Introduction to Insect Pests and their Control*. Macmillan, London.

Stiling, P.D. & Simberloff, D. (2000) The frequency and strength of nontarget effects of invertebrate biological control agents of plant pests and weeds. In: Follett, P.A. & Duan, J.J. (eds.) *Nontarget Effects of Biological Control*, pp. 31–43. Kluwer, Boston, MA.

Taylor, R.H., Kaiser, G.W. & Drever, M.C. (2000) Eradication of Norway rats for recovery of seabird habitat on Langara Island, British Columbia. *Restoration Ecology* **8**, 151–160.

Tebo, M.E. (1985) The Southeastern Piney Woods: describers, destroyers, survivors. MA Thesis, Florida State University, Tallahassee, FL.

Tershy, B.R., Donlan, C.J., Keitt, B.S. *et al.* (2002) Island conservation in north-west Mexico: a conservation model integrating research, education, and exotic mammal eradication. In Veitch, C.R. & Clout, M.N. (eds.) *Turning the Tide: The Eradication of Invasive Species*, pp. 293–300. IUCN Species Survival Commission, Gland, Switzerland.

Thomas, M.B. & Willis, A.J. (1998) Biological control – risky but necessary? *Trends in Ecology and Evolution* **13**, 325–329.

Towns, D.R. & Broome, K.G. (2003) From small Maria to massive Campbell: forty years of rat eradications from New Zealand islands. *New Zealand Journal of Zoology* **30**, 377–398.

United States Fish and Wildlife Service (1980) *Devil's Hole Pupfish Recovery Plan.* Prepared in cooperation with the Devil's Hole pupfish recovery team. US Fish and Wildlife Service, Washington, DC.

United States Fish and Wildlife Service (2003) *Recovery Plan for the Red-Cockaded Woodpecker* (Picoides borealis), 2nd revision. US Fish and Wildlife Service, Southeast Region, Atlanta, GA.

Walters, C.J. (1986) *Adaptive Management of Natural Resources.* McGraw-Hill, New York.

Walters, C.J. & Holling, C.S. (1990) Large-scale management experiments and learning by doing. *Ecology* **71**, 2060–2068.

Weiss, J. (1999) Contingency planning for new and emerging weeds in Victoria. *Plant Protection Quarterly* **14**, 112–114.

van Wilgen, B., Richardson, D. & Higgins, S. (2000) Integrated control of invasive alien plants in terrestrial ecosystems. In: Preston, G., Brown, G. & van Wyk, E. (eds.) *Best Management Practices for Preventing and Controlling Invasive Alien Species. Symposium Proceedings*, pp. 118–128. The Working for Water Programme, Cape Town.

Williams, T. (1997) Killer weeds. *Audubon* **99**, 24–31.

Williamson, M. (1996) *Biological Invasions.* Chapman & Hall, London.

Wittenberg, R. & Cock, M.J.W. (eds.) (2001) *Invasive Alien Species: A Toolkit of Best Prevention and Management Practices.* CAB International, Wallingford.

Zimmermann, H.G., Moran, V.C. & Hoffmann, J.H. (2001) The renowned cactus moth, *Cactoblastis cactorum* (Lepidoptera: Pyralidae): its natural history and threat to native *Opuntia* floras in Mexico and the United States of America. *Florida Entomologist* **84**, 543–551.

27

Managing Landscapes for Vulnerable, Invasive and Disease Species

Erika Zavaleta and Jae Ryan Pasari

Abstract

As the focus of conservation moves from protected areas to include working landscapes, management must accommodate multiple goals. In this context, management focused on individual species must be well justified. Conservation practices that target carefully selected individual species can complement pattern-based criteria needed to maintain species richness and ecosystem function by providing mechanistic rationale and quantifiable targets for conservation actions. However, most treatments of landscape design do not consider the compatibility of management for individual disease-causing species, invasive species and threatened species simultaneously. We summarize recommendations from landscape epidemiology and invasion management and evaluate whether they are congruent with general principles for vulnerable species' protection. Many, but not all, broad strategies for controlling invasive and disease species appear compatible with strategies for protecting vulnerable species. Local circumstances, scale considerations and the relative importance of landscape and other interventions should guide management of trade-offs among conflicting goals.

Keywords: conservation; invasive species; landscape epidemiology; trade-offs; vulnerable species.

Introduction

Landscape design more than ever must target multiple goals as the focus of conservation moves from protected areas to include working landscapes (Daily 1997) and as fewer landscapes remain free of human activities (Vitousek *et al.* 1997). In this context, landscape management focused on individual species needs to be well justified. Yet individual species-based approaches remain important in conservation, invasive species management and disease control. We examine the rationale for landscape design targeted at individual species. We summarize landscape design recommendations from the fields of landscape epidemiology and invasive species management. Although invasive species include those that harm human health (Invasive Species Council 2001), the invasion literature deals relatively little with human diseases and their vectors. From an ecosystem services perspective, landscape management that effectively reduces human disease threats as well as other impacts of invasive species would be highly desirable (Kremen & Ostfeld 2006). By including landscape approaches to disease management from the medical and public health literature, we sought useful design principles across two literatures concerned with harmful species.

Management to control invasive and disease organisms usually occurs in landscapes that also harbour native and threatened taxa. To address whether simultaneous landscape management for invasive, disease and vulnerable taxa is even feasible, we compare common interventions aimed at suppressing undesired species with landscape management goals typical of vulnerable species conservation. Landscape management both for and against particular classes of species focuses on common key themes including habitat protection; biodiversity maintenance; and management of disturbance, connectivity and fragmentation. In most – but not all – of these areas, broadly defined management strategies for managing invasions and diseases appear compatible with strategies for protecting and enhancing native species.

Why base management on individual species?

Individual species approaches are widely used to both control undesired species and enhance desired species. Vectors that spread life-threatening diseases like malaria and sleeping sickness can often be more successfully controlled with landscape-management than other techniques in many phases of an epidemic, especially over long time scales (Bos & Mills 1987; Utzinger *et al.* 2001; Campbell-Lendrum *et al.* 2005). Control and eradication of invasive species are also usually approached species by species, with emphasis on species that cause significant harm such as introduced mammals on islands (Atkinson 1989) and agricultural weeds (Bossard *et al.* 2000). Eradications of introduced species have protected scores of threatened taxa (Veitch & Clout 2002). Single-species management efforts are not problem-free, however, because focus on removal of individual species can sometimes lead to unexpected effects or neglect of broader ecosystem restoration goals (Zavaleta *et al.* 2001). In complex landscapes where ecosystems are under frequent threat of new invasion, management for ecosystem resistance as well as problem species removal could improve outcomes.

Single-species approaches are commonly pursued for vulnerable or threatened taxa, especially those unique to a particular landscape. Single or small groups of species are also a focus of landscape conservation when management for the species in question is expected to meet a range of other needs or defined goals (Noss *et al.* 1996; Kotliar *et al.* 1999; Freudenberger & Brooker 2004). The utility of such approaches, variously termed focal or indicator species approaches among others, has been debated for decades (e.g. Landres *et al.* 1988; Lindenmayer *et al.* 2002). Focal species approaches have been criticized for ad hoc species selection (Landres *et al.* 1988) and lack of evidence that conservation priorities based on focal taxa adequately protect other species and values (Simberloff 1998; Andelman & Fagan 2000; Lindenmayer *et al.* 2002; Smith & Zollner 2005). Methodologies range widely – 'focal species' may or may not need to possess umbrella, keystone, flagship and indicator traits or combinations thereof (Simberloff 1998), so the term only broadly refers to individual species whose needs are highlighted in conservation planning.

Limited findings on the performance of focal species approaches suggest that individual species targets can provide mechanistic rationales for landscape design to complement the pattern-based criteria on which species richness and ecosystem function targets are generally based (Sanderson *et al.* 2002; Coppolillo *et al.* 2004; Taylor *et al.* 2005; but see also Andelman & Fagan 2000). Landscape planning that targets both individual species and broader conservation goals might therefore provide the best and most measurable outcomes (Poiani *et al.* 1998; Dan Doak, personal communication). This area, however, deserves much more study.

Managing both for and against species

Broadly speaking, vulnerable species are thought to benefit from landscape features like corridors, large reserves, protection from anthropogenic disruption, and maintenance of natural disturbance regimes. Are these needs fundamentally different from those of invasive and disease organisms? While invasive species tend to share traits like generalist habits (Patz & Wolfe 2002, Marvier *et al.* 2004), early age of first reproduction, and small seed or offspring size (Rejmanek & Richardson 1996) that differ from traits typical of vulnerable species (specialist habits, restricted distributions, late age of first reproduction, few offspring, large adult size), exotic and disease species can sometimes move through habitat corridors intended to enhance habitat value for native species (Simberloff & Cox 1987; Loney & Hobbs 1991; Hess 1996). Some exotic species can invade relatively undisturbed forest understoreys or montane parks (Pauchard *et al.* 2003; Chornesky *et al.* 2005). Many species threatened in their home ranges are abundant or invasive where introduced, highlighting the limits to generalizing about biological differences between invasive and threatened taxa. However, some commonly suggested landscape management approaches for vulnerable taxa and against harmful species resemble one another – such as maintaining high biodiversity (Shea & Chesson 2002) and minimizing novel disturbances (Hobbs & Huenneke 1992).

Disease species management

Reviews of landscape epidemiology identify 34 diseases with known or suspected ties to landscape change (Patz *et al.* 2004) and innumerable environmental disturbances that trigger their emergence (McMichael 2001; Patz *et al.* 2005). For example, Lyme disease illustrates disease emergence driven by habitat fragmentation and biodiversity loss, while malaria exhibits complex responses to disturbance. Lyme disease, the most common vector-borne disease in the USA (Centers for Disease Control and Prevention 2000), has a complex transmission cycle involving many vertebrate hosts of varying competence (transmission ability) and environmental sensitivity. The abandonment of agriculture in the northeastern USA during the twentieth century allowed widespread reforestation and a subsequent explosion of the adult vectors' most favoured food resource, white-tailed deer (Fish 1993). Today, urban sprawl increases disease risk by fragmenting these forests and thereby increasing the population of a highly competent host, the mouse *Peromyscus leucopus,* while simultaneously reducing the abundance of less competent hosts (Ostfeld *et al.* 2002). This landscape-driven loss of biodiversity produces a risk 'amplication effect' (Keesing *et al.* 2006) that may be applicable to many other disease systems including leishmaniasis, Chagas disease, babesiosis and plague (Ostfeld & Keesing 2000). Malaria has a much simpler transmission cycle involving only mosquitoes and people. Difficulty in predicting the effects of environmental change on malaria arise because each of the 30–40 malaria-carrying mosquito species has different competence levels, habitat preferences and adaptability to environmental changes (World Health Organization 1982; Patz & Wolfe 2002). Landscape changes that decrease the abundance of one malaria-carrying mosquito can increase the abundance of another (Lines 1993). Varying responses of malaria-carrying mosquitoes to urbanization, agriculture, deforestation and other landscape changes around the world (Lines 1993) suggest that the landscape epidemiology of malaria is site and species-specific. This phenomenon is even more apparent when multiple diseases are concerned, as in West Africa, where environmental changes that reduced malaria have also increased the spread of tick-borne relapsing fever (Trape *et al.* 1996).

However, landscape management can successfully control disease organisms. Environmental management strategies have successfully reduced malaria in India (Rajagopalan *et al.* 1990), Honduras (Reid 1997) and northern Europe (McMichael 2001). Various strategies, including draining swamps in Indonesia and Zambia, removing algae in Oaxaca, managing river flow in Zambia, clearing forest in Africa (Molyneux 2002), altering irrigation in Asia (Campbell-Lendrum 2005), and a suite of landscape management techniques used by mosquito abatement districts in the USA, Singapore and Cuba have successfully reduced many diseases (Bos & Mills 1987). Still, the idiosyncratic environmental responses of many diseases challenge land managers and policy-makers who want to incorporate landscape epidemiology into management. A few region-, disease- or development-specific health assessment guides exist (World Health Organization 1982; Jobin 1999) but are limited in their applicability. We synthesized known mechanisms linking environmental change and disease emergence (summarized in Patz *et al.* 2005) to suggest a focused management and assessment guide (Box 27.1).

Box 27.1 Summary guide to landscape disease management and assessment (based on Patz *et al.* 2005)

When designing a landscape, one should consider the effects of anthropogenic drivers (A) on the mechanisms of ecological change (B) that can affect the dynamics of environmentally sensitive and dangerous diseases (C).

A. Anthropogenic drivers that affect disease risk

1 Wildlife habitat destruction, conversion or encroachment, particularly through deforestation and reforestation.
2 Changes in the distribution and availability of surface waters, such as through dam construction, irrigation and stream diversion.
3 Agricultural land-use changes, including proliferation of both livestock and crops; deposition of chemical pollutants, including nutrients, fertilizers and pesticides.
4 Uncontrolled urbanization, including urban sprawl, climate variability and change, migration and international travel and trade.
5 Accidental or intentional human introduction of pathogens.

B. Mechanisms of effects of ecological change on infectious diseases

1 Altered habitats or breeding sites for disease vectors or reservoirs.
2 Niche invasions or transfer of interspecies hosts.
3 Biodiversity change (including loss of predator species and changes in host population density).
4 Human-induced genetic changes in disease vectors or pathogens (such as pesticide and antibiotic resistance).
5 Environmental contamination by infectious disease agents.

C. Diseases sensitive to ecological change or with high potential disease burden

1 Malaria across most ecological systems.
2 Schistosomiasis, lymphatic filariasis and Japanese encephalitis in cultivated and inland water systems in the tropics.
3 Dengue fever in tropical urban centres.

4 Leishmaniasis and Chagas disease in forest and dryland systems.
5 Meningitis in the Sahel.
6 Cholera in coastal, freshwater and urban systems.
7 West Nile virus and Lyme disease in urban and suburban systems of Europe and North America.
8 Cryptosporidiosis in agricultural systems.

Although Box 27.1 provides a synthesis of major historical drivers and threats, it cannot predict the emergence of unknown pathogens or suggest testable hypotheses in the search for ecological principles underlying disease invasion. The following generalities move in those directions, and have each been tested, observed or hypothesized more than once, with those higher in the list appearing more widely accepted and generalizable.

1 Disruption of natural systems can increase the emergence of disease agents whose ecosystem disservices offset the services gained by the disruption, and vice versa. Human actions can both enhance and curb the density-dependent mechanisms that maintain equilibrium in disease transmission cycles (Campbell-Lendrum *et al.* 2005).
2 Limiting contact between humans, hosts and vectors can reduce disease. This can be accomplished by reducing host and vector habitat in human settlements and siting human settlements away from these habitats (World Health Organization 1982; Epstein 2002; Patz & Wolfe 2002).
3 Climate change could increase the global infectious disease burden by altering landscapes and vector distribution and behaviour (McMichael *et al.* 2004).
4 Culture and behaviour that change the environment to a state other than that in which we evolved increase disease risk (the evodeviationary hypothesis, McMichael 2001). Alteration of natural landscapes should be minimized because human populations often lack resistance to new vectors that invade the perturbed area.
5 Greater abundances and diversity of vector and host predators (Ostfeld *et al.* 2002) and less competent hosts (Keesing *et al.* 2006), and reduced abundances of generalist vectors and hosts (Molyneux 2003), reduce disease risk.
6 Vectors tend to travel along roads, rivers, valleys and other linear landscape features (Timischl 1984; Molyneux 2003).

7 Minimizing edge between different habitat types (including human communities) reduces vector and host habitat and contact with humans (Epstein *et al.* 1994; McMichael 2001).

Managing landscapes for invasion resistance

Species invasions are so numerous that a single-species approach is no longer feasible in many landscapes. In ecosystems that harbour many exotics or frequently receive new ones, single-species management cannot keep pace and may simply allow one exotic to replace another (Zavaleta *et al.* 2001). For example, managers of tamarisk (*Tamarix* spp.) invasion in the western USA increasingly recognize that they must also address exotic Russian olive (*Eleagnus angustifolia*) and Siberian elm (*Ulmus pumila*), which are both replacing tamarisk over growing areas (NISC Tamarisk Analysis Team, unpublished). Yet most invasive species work still focuses on individual species (Marvier *et al.* 2004). Landscape interventions for individual species, as opposed to straightforward control and removal approaches, most frequently involve manipulating disturbance – including herbivory by wild or domestic animals, fire, flooding and tillage. Interventions more broadly targeted at building landscape resistance to invasion could, but do not yet, include focal species approaches similar to those for conservation.

The limited literature on landscape design for controlling invasions suggests several coarse-scale principles similar to those for disease management. Several authors emphasize the importance of protecting habitat, minimizing fragmentation and minimizing novel disturbances (Hobbs 2000; Marvier *et al.* 2004; With 2004). However, a habitat conservation approach is limited by inevitable conversion, fragmentation and disturbance even in protected landscapes. Global-change-mediated habitat disturbance will extend the reach of invaders (see Chapter 31), for example into closed-canopy forest ecosystems (Corlett 1992), which will probably experience greater mortality and gap formation with climate change (Graham *et al.* 1990). Positive invader interactions can also drive cycles of habitat change and invasion – as when gap formation by exotic insect pests creates opportunities for plant invasions in forests.

Beyond general habitat conservation, particular spatial patterns of disturbance might help limit the spread of invasions (With 2004). With's models suggest that poor dispersers spread more quickly in landscapes where disturbances are clumped or concentrated, but good dispersers spread more quickly through landscapes with dispersed disturbances that serve as stepping stones for movement. The implications for landscape design thus depend on the dispersal abilities of invaders relative to desired native species. With's models also suggest minimizing edge to reduce invader movement from matrix habitats into less-disturbed fragments. However, these recommendations need to be empirically tested before they are widely applied.

Restoring or maintaining site-specific historical disturbance regimes, including flooding, fire and herbivory, could also help bolster invasion resistance (Hobbs & Huenneke 1992; Stromberg & Chew 2002) (Fig. 27.1). This approach is complicated, though, by the ability of some invaders to capitalize on any disturbance. Fire appears to promote many more invasions than it reduces, even in fire-adapted systems like South African fynbos (D'Antonio *et al.* 1998; D'Antonio 2000; see also Chapter 11). In some cases, maintaining natural disturbance regimes could limit invader establishment or spread at low abundances but might only accelerate the spread of widespread or abundant invaders. For example, pulse flooding is considered crucial to maintaining and restoring riparian communities in the western USA (Stromberg & Chew 2002) but also contributes to rapid spread of invasive tamarisk (T. Carlson, personal communication). In some cases, reintroducing disturbance could boost long-term control only if the invader's ability to spread is also impaired

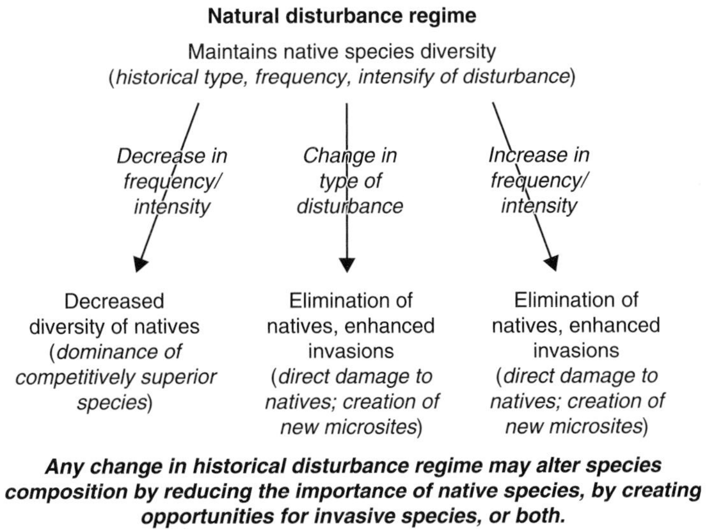

Figure 27.1 **Altered disturbance regimes can change communities via multiple pathways. Adapted from Hobbs and Huenneke 1992.**

through traditional control methods such as spraying, biocontrol or mechanical removal. Where disturbance management targets individual invasive species, one must also monitor the effects on other exotics and natives. Controlled burns are used to suppress yellow star thistle in several parts of California, but at one site ecologists found that these fires were also suppressing native forbs and increasing overall exotic cover (C. Christian, unpublished). The effectiveness of and trade-offs involved in manipulating disturbance ultimately depend on local circumstances.

Finally, maintaining diversity and composition of native communities appears to enhance resistance to at least some invasions. Invasibility generally increases as a function of extrinsic site factors that support a more diverse native biota and species similar to potential invaders, but declines within a given site if more species are present to compete directly or indirectly with the invader (Shea & Chesson 2002). The mechanisms linking biodiversity maintenance to increased biotic resistance include an enhanced competitive environment, greater resource use that makes fewer resources available to exotics, and more rapid or complete ecosystem recovery from disturbances that could permit invasions (Alpert *et al.* 2000). Ecologists have successfully explored seeding native competitors into California grasslands, for instance, to suppress individual invaders like yellow star thistle (K. Hulvey, unpublished) or to boost resistance generally (Seabloom *et al.* 2003), but managers have not yet applied these findings at large scales.

A holistic approach targeted at reducing invasions should include prevention measures as well as containment, control and resistance features on the landscape. New introductions can probably best be prevented through interventions in trade, transport and early detection at points of entry into a region. Consistent detection efforts are also crucial to prevent nascent foci of exotics from spreading (Chornesky & Randall 2003). Exotics that initially fail to establish or spread because of modest incompatibilities with local environmental conditions, susceptibility to local diseases and predators, or inability to compete can take advantage of temporary windows of opportunity or evolve the capacity to overcome these barriers (Cody & Overton 1996; Thompson 1998). Established exotics that fail to spread should not, therefore, be regarded as benign.

Managing for and against species: General principles

Coarse-scale landscape management suggestions emerging from the disease and biological invasions literature are very similar to each other (Table 27.1).

Table 27.1 **Summary of coarse-scale landscape design suggestions for three species-based goals.**

Disease control	Invasive species control	Vulnerable species protection
Minimize anthropogenic disruption in general, but some disturbances can improve disease control	**Minimize anthropogenic disruption** in general, but natural disturbance regimes can sometimes improve invader control	**Minimize anthropogenic disruption** in general, but natural disturbance regimes can sometimes benefit vulnerable taxa
Anticipate climate change effects on landscape and species dynamics	**Anticipate climate change effects** on landscape and species dynamics	**Anticipate climate change effects** on landscape and species dynamics
Maintain biodiversity to dilute and slow disease transmission	**Maintain biodiversity** for direct and indirect competitive resistance	**Maintain biodiversity** to maintain species' interactions and resilience
Minimize edge to reduce edgehabitat for vectors/hosts and reduce human contact	**Minimize edge** between disturbed and undisturbed patches to reduce spread into new habitats	**Minimize edge**: area ratio between disturbed and undisturbed habitat to maximize habitat value for species
Anticipate and break up movement along linear landscape features, especially if they involve disturbance (roads, rivers) but even if they consist of relatively undisturbed habitat	**Anticipate and break up movement** along linear landscape features, especially if they involve disturbance (roads, rivers) but even if they consist of relatively undisturbed habitat	**Anticipate need for movement** through habitat corridors and stepping stones, and need to break up linear disturbance features such as roads
	Clump disturbances and conversion to limit spread of good dispersers, **and/or disperse disturbances** to limit spread of poor dispersers	
Limit contact between humans, hosts and vectors by keeping their respective habitats separate		

This should not be surprising; landscape epidemiology explicitly treats disease organisms and vectors as invaders, and epidemiological frameworks are used in the ecological invasions literature, particularly to describe and model spread (Hastings 1996, Higgins *et al.* 2000). Management suggestions emerging from the disease and invasions literature are also similar, at the conceptual level, to conservation practices typically targeted at protection of vulnerable and threatened taxa. A potential conflict area between desired and undesired species is around corridors and connectivity – which are to be maximized for the protection of vulnerable species but broken up to prevent disease and invasive spread (Hobbs 1992). Relatively little empirical work has been done on the functioning of corridors, and generalizations across species, scales and ecosystem types are difficult (see Chapters 22–25). However, differences between the ways in which vulnerable and invasive species use corridors might be exploited to minimize conflict. With the notable exception of wildlife diseases borne by large mammals such as brucellosis (Bienen 2002), undesired species often make use of linear disturbed features like roads, while vulnerable taxa often require corridors or stepping stones made up of less-disturbed habitat (Beier & Noss 1998). Minimizing connectivity via roads, power lines and other anthropogenic features to reduce disease and exotic spread can be compatible with maximizing connectivity of relatively undisturbed habitat. Connectivity at different scales might also be managed differently to limit disease and exotic species spread while maintaining desired taxa. For instance, to protect an ecosystem type from invasions or disease, it can be useful to protect examples of that ecosystem in more than one area. Across these discrete areas, connectivity may be minimized or intentionally limited to prevent contagion of harmful organisms (Bienen 2002). Within each protected area, however, connectivity may be enhanced to maintain functional communities and viable populations of large, wide-ranging and top predator species.

Landscape principles

Is general concordance between management principles for disease, invasion and vulnerable species wishful thinking?

1 **There are inevitably trade-offs among the goals of controlling disease, limiting invasive species, and protecting or enhancing vulnerable taxa and communities** – for example, when forest clearance abates infectious

disease or when enhancement of riparian function enables invasive species spread.

2 **Sometimes the most critical intervention to protect one or more species will not be landscape design. Managers should tackle multiple goals, addressing landscape interventions for those goals that are most responsive or sensitive to landscape configuration relative to other drivers, and managing goals most affected by other factors through other means.** The importance of landscape design versus other interventions must be evaluated when trade-offs arise in managing for and against different values and species. For instance, biological invasions can be managed at the stages of establishment and spread through landscape design as well as at the introduction stage through other measures. Similarly, in some cases it is most effective to use pharmaceuticals, bed nets, quarantines and insecticides rather than landscape design to control disease. Finally, vulnerable taxa may in some cases be conserved best through mitigating the direct impacts of harvesting or invasive species, pollution and/or disease as well as or instead of habitat change (Dirzo & Raven 2003).

3 **Consider whether interventions at more than one scale can reduce conflicts and trade-offs among goals.** Landscapes experience processes at multiple, nested scales and should be managed at more than one scale (Lindenmeyer & Franklin 2002). Hence . . .

4 **General principles cannot substitute for an understanding of local particulars** – rather, they can guide which local particulars need to be addressed.

Finally, we return to the overarching question of when and whether individual species considerations should be part of landscape design and stewardship. We suggest the following general guidelines:

1 **Carefully selected species should be part – but not all – of landscape targets to provide a mechanistic (process) basis for management.** Methodologies for including individual species targets vary, and many have not been evaluated for effectiveness. We argue that including both species-level and ecosystem-level or higher-order targets in landscape design allows management to capture complementary process and pattern-based values.

2 **Individual species should be selected with complementarity in mind – vulnerable taxa with contrasting needs, as well as species to be managed against.** No single species, or any small set of species, will act as a complete umbrella for other species and desired functions on a landscape. Management that targets different kinds of species, including invasive or disease species as well as threatened or vulnerable species, will provide more complete coverage than management based on one or a few flagship species.

Acknowledgements

We thank Kris Hulvey, Joanna Nelson and the editors for helpful suggestions and comments on earlier drafts of this chapter.

References

Alpert, P., Bone, E. & Holzapfel, C. (2000) Invasiveness, invasibility and the role of environmental stress in the spread of non-native plants. *Perspectives in Plant Ecology, Evolution & Systematics* **3**, 52–66.

Andelman, S.J. & Fagan, W.F. (2000) Umbrellas and flagships: efficient conservation surrogates or expensive mistakes? *Proceedings of the National Academy of Sciences of the U. S. A.* **97**, 5954–5959.

Atkinson, I. (1989) Introduced animals and extinctions. In: Western, D. & Pearl, M. (eds.) *Conservation for the Twenty-First Century*, pp. 54–69. Oxford University Press, New York.

Beier, P. & Noss, R.F. (1998) Do habitat corridors provide connectivity? *Conservation Biology* **12**, 1241–1252.

Bienen, L. (2002) Informed decisions: conservation corridors and the spread of infectious disease. *Conservation in Practice* **3**, 10–19.

Bos, R. & Mills, A. (1987) Financial and economic aspects of environmental-management for vector control. *Parasitology Today* **3**, 160–163.

Bossard, C.C., Randall, J.M. & Hoshovsky, M.C. (2000) *Invasive Plants of California's Wildlands.* University of California Press, Berkeley, CA.

Campbell-Lendrum, D., Molyneux, D.H., Amerasinghe, F. *et al.* (2005) Ecosystems and vector-borne disease control. In: Chopra, K., Leemans, R., Kumar, P. *et al.* (eds.) *Ecosystems and Human Well-Being: Policy Responses, Findings of the*

Responses Working Group of the Millennium Ecosystem Assessment, pp. 353–372. Island Press, Washington, DC.

Centers for Disease Control and Prevention (2000) Surveillance for Lyme disease – United States, 1992–1998. *Morbidity and Mortality Weekly Report* **49**, 1–11.

Chornesky, E.A. & Randall, J.M. (2003) The threat of invasive alien species to biological diversity: setting a future course. *Annals of the Missouri Botanical Garden* **90**, 67–76.

Chornesky, E.A., Bartuska, A.M., Aplet, G. H. *et al.* (2005) Science priorities for reducing the threat of invasive species to sustainable forestry. *Bioscience* **55**, 335–348.

Cody, M.L. & Overton, J.M. (1996) Short-term evolution of reduced dispersal in island plant populations. *J. Ecology* **84**, 53–61.

Coppolillo, P., Gomez, H., Maisels, F. *et al.* (2004) Selection criteria for suites of landscape species as a basis for site-based conservation. *Biological Conservation* **115**, 419–430.

Corlett, R. (1992) The ecological transformation of Singapore, 1819–1900. *Journal of Biogeography* **19**, 411–420.

Daily, G.C. (1997) Countryside biogeography and the provision of ecosystem services. In: Raven, P. (ed.) *Nature and Human Society: The Quest for a Sustainable World*, pp. 104–113 National Research Council, National Academy Press, Washington, DC.

D'Antonio, C.M. (2000) Fire, plant invasions, and global changes. In: Mooney, H.A. & Hobbs, R.J. (eds.) *Invasive Species in a Changing World*, pp. 65–93. Island Press, Washington, DC.

D'Antonio, C.M. & Thomsen, M. (2004) Ecological resistance in theory and practice. *Weed Technology* **18**, 1572–1577.

D'Antonio, C.M., Dudley, T. & Mack, M. (1998) Disturbance and biological invasion: direct effects and feedbacks. In: Walker, L. (ed.) *Ecosystems of Disturbed Ground*, pp. 429–468. Elsevier, Amsterdam.

Dirzo, R. & Raven, P. (2003) Global state of biodiversity and loss. *Annual Review of Environment and Resources* **28**, 137–167.

Epstein, P.R. (1997) Climate, ecology, and human health. Consequences **3**(2), 2–19.

Epstein, P.R. (2002) Biodiversity, climate change, and emerging infectious diseases. In: Aguirre, A.A., Ostfeld, R.S., Tabor, G.M. *et al.* (eds.) *Conservation Medicine: Ecological Health in Practice*, pp. 27–39. Oxford University Press, Oxford.

Epstein, P.R., Ford, T.E., Puccia, C. *et al.* (1994) Marine ecosystem health: implications for public health. In: Wilson, M.E., Levins, R. & Spielman, A. (eds.) *Disease in Evolution: Global Changes and Emergence of Infectious Diseases*, pp. 13–23. The New York Academy of Sciences, New York.

Fish, D. (1993) Population ecology of *Ixodes dammini*. In: Ginsberg, H.S. (ed.) *Ecology and Environmental Management of Lyme Disease*, pp. 25–42. Rutgers University Press, New Brunswick, NJ.

Freudenberger, D. & Brooker, L. (2004) Development of the focal species approach for biodiversity conservation in the temperate agricultural zones of Australia. *Biodiversity and Conservation* **13**, 253–274.

Graham, R.L., Turner, M.G. & Dale, V.H. (1990) How increasing CO_2 and climate change affect forests. *BioScience* **40**, 575–587.

Hastings, A.M. (1996) Models of spatial spread: is the theory complete? *Ecology* **77**, 1675–1679.

Hess, G. (1996) Disease in metapopulation models: Implications for conservation. *Ecology* **77**, 1617–1632.

Higgins, S.I., Richardson, D.M. & Cowling, R.M. (2000) Using a dynamic landscape model for planning the management of alien plant invasions. *Ecological Applications* **10**, 1833–1848.

Hobbs, R.J. (1992) The role of corridors in conservation: solution or bandwagon? *Trends in Ecology and Evolution* **7**, 389–392.

Hobbs, R.J. (2000) Land use changes and invasion. In: Mooney, H.A. & Hobbs, R.J. (eds.) *Invasive Species in a Changing World*, pp. 55–64. Island Press, Washington, DC.

Hobbs, R.J. & Huenneke, L.F. (1992) Disturbance, diversity, and invasion: implications for conservation. *Conservation Biology* **6**, 324–337.

Invasive Species Council (2001) *Meeting the Invasive Species Challenge: National Invasive Species Management Plan*. US Government Printing Office, Washington, DC.

Jobin, W.R. (1999) *Dams and Disease: Ecological Design and Health Impacts of Large Dams, Canals and Irrigation Systems*. E & FN Spon, London and New York.

Keesing, F., Holt, R.D., Ostfeld, R.S. *et al.* (2006) Effects of species diversity on disease risk. *Ecology Letters* **9**, 485–498.

Kessler, C.C. (2002) Eradication of feral goats and pigs and consequences for other biota on Sarigan Island, Commonwealth of the Northern Mariana Islands. In: Veitch, C.R. & Clout, M.N. (eds.) *Turning the Tide: The Eradication of Invasive Species*, pp. 132–140. IUCN Species Survival Commission, Gland, Switzerland, and Cambridge.

Kotliar, N.B., Baker, B.W., Whicker, A.D. *et al.* (1999) A critical review of assumptions about the prairie dog as a keystone species. *Environmental Management* **24**, 177–192.

Kremen, C. & Ostfeld, R.S. (2006) A call to ecologists: measuring, analyzing, and managing ecosystem services. *Frontiers in Ecology and the Environment* **3**, 540–548.

Landres, P., Verher, J. & Thomas, J. (1988) Ecological uses of vertebrate indicator species: a critique. *Conservation Biology* **2**, 316–328.

Lindenmayer, D.B. & Franklin, J.F. (2002) *Conserving Forest Biodiversity: A Comprehensive Multiscaled Approach*. Island Press, Washington, DC.

Lindenmayer, D.B., Manning, A.D., Smith, P.L. *et al.* (2002) The focal-species approach and landscape restoration: a critique. *Conservation Biology* **16**, 338–345.

Lines, J. (1993). The effect of climatic and land-use changes on the insect vectors of human disease. In: Harrington, R. & Stork, N.E. (eds.) *Insects in a Changing Environment*, pp. 158–173. Academic Press, London.

Loney, B. & Hobbs, R.J. (1991) Management of vegetation corridors: maintenance, rehabilitation, and establishment. In: Saunders, D.A. & Hobbs, R.J. (eds.) *Nature Conservation 2: The Role of Corridors*, pp. 299–311. Surrey Beatty & Sons, Chipping Norton, NSW.

McMicheal, A.J. (2001) *Human Frontiers, Environments, and Disease: Past Patterns, Uncertain Futures*. Cambridge University Press, Cambridge.

McMichael, A., Campbell-Lendrum, D., Kovats, S. *et al.* (2004). *Global Climate Change. Comparative Quantification of Health Risks: Global and Regional Burden of Disease Attributable to Selected Major Risk Factors*. World Health Organization, Geneva.

Marvier, M., Kareiva, P. & Neubert. M.G. (2004) Habitat destruction, fragmentation, and disturbance promote invasion by habitat generalists in a multispecies metapopulation. *Risk Analysis* **24**, 869–878.

Molyneux, D.H. (2002) Vector-borne infections and health related to landscape changes. In: Aguirre, A.A., Ostfeld, R.S., Tabor, G.M. *et al.* (eds.) *Conservation Medicine: Ecological Health in Practice*, pp. 194–206. Oxford University Press, Oxford.

Molyneux, D.H. (2003) Common themes in changing vector-borne disease scenarios. *Transactions of the Royal Society of Tropical Medicine and Hygiene* **97**, 129–132.

Mooney, H.A. & Hobbs, R.J. (eds.) (2000) *Invasive Species in a Changing World*. Island Press, Washington, DC.

Noss, R.F., Quigley, H.B., Hornocker, M.G. *et al.* (1996) Conservation biology and carnivore conservation in the Rocky Mountains. *Conservation Biology* **10**, 949–963.

Ostfeld, R.S. & Keesing, F. (2000) Biodiversity and disease risk: The case of lyme disease. *Conservation Biology* **14**, 722–728.

Ostfeld, R.S., Keesing, F., Schauber, E.M. & Schmidt K.A. (2002) Ecological context of lyme disease. In: Aguirre, A.A., Ostfeld, R.S., Tabor, G.M. *et al.* (eds.) *Conservation Medicine: Ecological Health in Practice*, pp. 207–219. Oxford University Press, Oxford.

Patz, J.A. & Wolfe, N.D. (2002) Global ecological change and human health. In: Aguirre, A.A., Ostfeld, R.S., Tabor, G.M. *et al.* (eds.) *Conservation Medicine: Ecological Health in Practice*, pp. 167–181. Oxford University Press, Oxford.

Patz, J.A., Daszak, P., Tabor, G.M. *et al.* (2004) Unhealthy landscapes: policy recommendations on land use change and infectious disease emergence. *Environmental Health Perspectives* **112**, 1092–1098.

Patz, J.A., Confalonieri, U.E.C., Amerasinghe, F. *et al.* (2005) Human health: ecosystem regulation of infectious diseases. In: Hassan, R., Scholes, R. & Ash, N. (eds.) *Ecosystems and Human Well-being: Current State and Trends: Findings of the Condition and Trends Working Group*, pp. 393–411. Island Press, Washington, DC.

Pauchard, A., Alaback, P.B. & Edlund, E.G. (2003) Plant invasions in protected areas at multiple scales: *Linaria vulgaris* (Scrophulariaceae) in the west Yellowstone area. *Western North American Naturalist* **63**, 416–428.

Poiani, K.A., Baumgartner, J.V., Buttrick, S.C. *et al.* (1998) A scale-independent, site conservation planning framework in The Nature Conservancy. *Landscape and Urban Planning* **43**, 143–156.

Rajagopalan, P.K., Das, P.K., Panicker, K.N. *et al.* (1990) Environmental and water management for mosquito control. In: Curtis, C.F. (ed.) *Appropriate Technology in Vector Control*, pp. 122–137. CRC Press, Boca Raton, FL.

Reid, W.V. (1997) Opportunities for collaboration between the biomedical and conservation communities. In: Grifo, F. & Rosenthal, J. (eds.) *Biodiversity and Human Health*, pp. 334–352. Island Press, Washington, DC.

Rejmanek, M. & Richardson, D.M. (1996) What attributes make some plant species more invasive? *Ecology* **77**, 1655–1661.

Sanderson, E.W., Redford, K.H., Vedder, A. *et al.* (2002) A conceptual model for conservation planning based on landscape species requirements. *Landscape and Urban Planning* **58**, 41–56.

Seabloom, E.W., Borer, E.T., Boucher, V.L. *et al.* (2003) Competition, seed limitation, disturbance, and reestablishment of California native annual forbs. *Ecological Applications* **13**, 575–592.

Shea, K. & Chesson, P. (2002) Community ecology theory as a framework for biological invasions. *Trends in Ecology and Evolution* **17**, 170–176.

Simberloff, D. (1998) Flagships, umbrellas, and keystones: Is single-species management passé in the landscape era? *Biological Conservation* **83**, 247–257.

Simberloff, D. & Cox, J. (1987). Consequences and costs of conservation corridors. *Conservation Biology* **6**, 63–71.

Smith, W.P. & Zollner, P.A. (2005) Sustainable management of wildlife habitat and risk of extinction. *Biological Conservation* **125**, 287–295.

Stromberg, J.C. & Chew, M.K. (2002) Flood pulses and restoration of riparian vegetation in the American Southwest. In: Middleton, B.A. (ed.) *Flood Pulsing in Wetlands: Restoring the Natural Hydrological Balance*, pp. 11–49. John Wiley & Sons, New York.

Taylor, M.F.J., Suckling, K.F. & Rachlinski, J.J. (2005) The effectiveness of the Endangered Species Act: a quantitative analysis. *Bioscience* **55**, 360–367.

Thompson, J.N. (1998) Rapid evolution as an ecological process. *Trends in Ecology and Evolution* **13**, 329–332.

Timischl, W. (1984) Influence of landscape on the spread of an infection. *Bulletin of Mathematical Biology* **46**, 869.

Trape, J., Godeluck, B., Diatta, G. *et al.* (1996). The spread of tick-borne borreliosis in West Africa and its relationship to sub-Saharan drought. *American Journal of Tropical Medicine and Hygiene* **54**, 289–293.

Utzinger, J., Tozan, Y. & Singer, B.H. (2001) Efficacy and cost-effectiveness of environmental management for malaria control. *Tropical Medicine and International Health* **6**, 677–687.

Veitch, C.R. & Clout, M.N. (eds.) (2002) *Turning the Tide: The Eradication of Invasive Species*. IUCN Species Survival Commission, Gland, Switzerland, and Cambridge.

Vitousek, P.M., Mooney, H.A., Lubchenco, J. *et al.* (1997) Human domination of Earth's ecosystems. *Science* **277**, 494–499.

With, K.A. (2004) Assessing the risk of invasive spread in fragmented landscapes. *Risk Analysis* **24**, 803–815.

World Health Organization (1982) *Manual on Environmental Management for Mosquito Control, with Special Emphasis on Malaria Vectors.* WHO, Geneva.

Zavaleta, E.S., Hobbs, R.H. & Mooney, H.A. (2001) Viewing invasive species removal in a whole-ecosystem context. *Trends in Ecology and Evolution* **16**, 454–459.

28

Tools for Conserving Managing Individual Plant Species in Dynamic Landscapes

Mark Burgman, Jane Elith, Emma Gorrod and Bonnie Wintle

Abstract

Many conservation management strategies focus on individual species. Emerging landscape-scale processes that affect large numbers of species make spatially explicit strategies essential. Spatial models of species habitat are important elements of planning. There are numerous modelling options to suit a variety of data constraints. No modelling option alone accounts easily for the dynamic nature of habitat change. Coupled with dynamic habitat and population models, they provide a relatively complete framework for testing ideas and exploring conservation options.

Keywords: habitat models; population models; single species conservation; dispersal.

Introduction

Conservation strategies in most jurisdictions focus on individual species, a consequence of social mandates, despite evidence of landscape-scale threats and the need to conserve ecosystems and their functions. Single-species conservation management is concerned with a species' dynamics and habitat within the envelope of its distribution. The purpose of this chapter is to outline the imperatives for and pitfalls of single-species conservation management. It will focus on evaluating the strengths and weaknesses of different approaches to defining a species' habitat, and of modelling species in dynamic landscapes, discussing general principles that emerge from these considerations.

Land clearance for agriculture (grazing and cropping) and urbanization have been primary causes of range contractions and habitat loss globally in the past (Hunter 2002). The number of species in small, remnant populations in Australia (the 'few populations' category in Fig. 28.1) is mostly a legacy of land clearing in the past. Most emerging threats operate at landscape scales and affect many species simultaneously, including changed disturbance regimes, salinization, weeds, fragmentation, flooding, pollinator disruption, limited sexual reproduction, lack of habitat support and extreme environmental conditions.

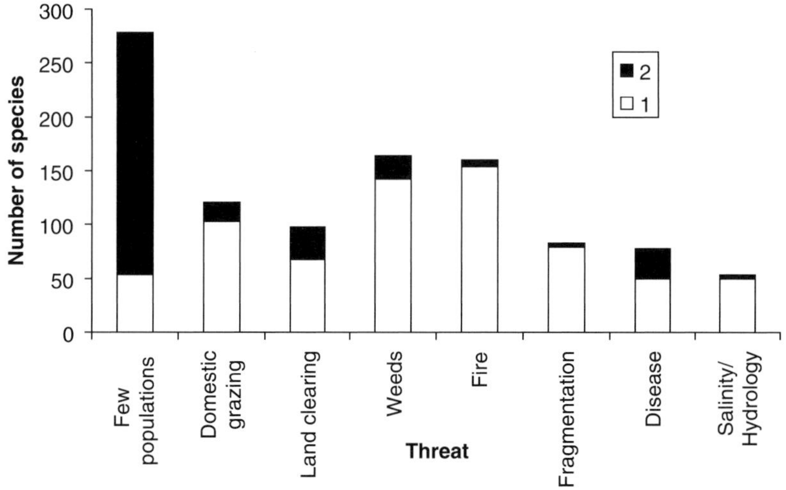

Figure 28.1 **The number of endangered and critically endangered Australian vascular plant species subject to severe (2) and substantial (1) existing or future threats from landscape factors including weeds, fire, fragmentation, disease, and changing salinity and hydrology (after Burgman *et al.* 2007). In the lefthand bar, numbers of species threatened by land clearing, grazing and demographic processes in few populations are also shown. Other emerging landscape-scale processes (not shown here) include disrupted pollination, lack of disturbance and extreme environmental conditions; they affect smaller but important numbers of species.**

Managing these factors will require new strategies, including social and legal initiatives.

Decisions about the management of individual species depend on maps of their habitat, usually represented as ranges of potential or realised distributions, sometimes with a representation of variation in habitat quality. Within these ranges, managers evaluate options for reserve design, rehabilitation, vegetation offsets, harvesting schedules or development. The following sections outline habitat and population models for single species, sketching the objectives that can and cannot be achieved with a single-species focus.

Modelling habitat for single species

During the last two decades, statistical modelling has been used increasingly to relate the abundance or occurrence of a species to environmental and/or geographic predictors (e.g. Goolsby 2004; Mac Nally & Fleishman 2004; Thuiller *et al.* 2005). The simplest methods for predicting species' distributions use presence-only (occurrence) records, assessing new sites by their environmental similarity to sites of known species presence. The environmental distributions of species are described using subjective models (habitat suitability indices), environmental envelopes (BIOCLIM, Nix & Switzer 1991) or distance measures (DOMAIN, Carpenter *et al.* 1993). This information can be combined with mapped environmental data to identify sites that are suitable for the species of interest (Elith & Burgman 2002). These simple models focus on presence-only data and do not identify reliably environments that are unsuitable for a species (Elith *et al.* 2006).

In ecology, spatial models based on presence/absence data usually include generalized linear regression models (GLMs) and their non-parametric equivalent, generalized additive models (GAMs) (Hastie *et al.* 2001). These techniques are reliable and reasonably flexible (Barry & Elith, 2006). In the last decade, genetic algorithms have been used to model data from natural history collections (e.g. GARP, Peterson 2003). Neural network applications are most common in remote sensing and fisheries research (e.g. Olden 2003). Other techniques have emerged from the machine learning literature including classification and regression trees (CART), multivariate adaptive regression splines (MARS – a hybrid of CART and more conventional regression methods) and boosted regression trees (BRT, Hastie *et al.* 2001).

Some techniques focus on modelling more than one species. A multi-response ('community') version of multivariate adaptive regression splines simultaneously relates variation in the occurrence of all species to the environmental predictors, identifying the most parsimonious set of predictor variables and then estimating model coefficients for each species individually (Leathwick *et al.* 2005). Generalized dissimilarity models (GDM, Ferrier 2002) model spatial turnover in community composition (or 'compositional dissimilarity') between pairs of sites as a function of environmental differences between these sites.

Burgman *et al.* (2005) emphasized that predictions often are different for different modelling methods (Fig. 28.2). Yet, the choice of method is usually guided by what is familiar and available, rather than by model performance. In part, this reflects the time required to become expert enough with a method to use it competently. Choice of method is also limited by data. Presence-only data are generally modelled with one of a few techniques specifically developed for that purpose (e.g. BIOCLIM, DOMAIN, GARP), even though several other methods are suitable and generally more successful for prediction (Elith *et al.* 2006). Regression methods can deal with a range of data types, and specialized applications address problems typical in ecological data

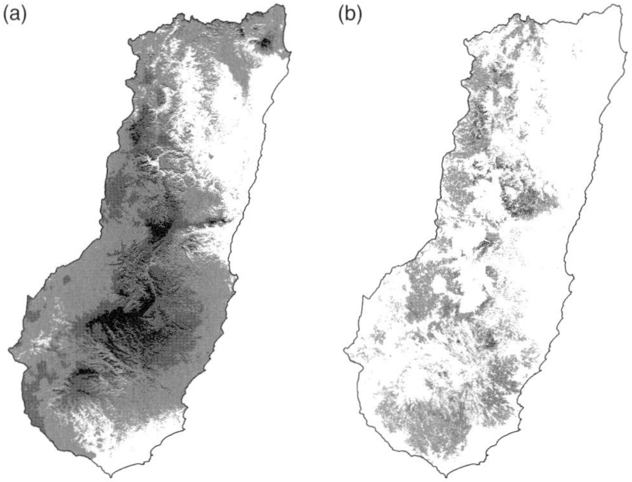

Figure 28.2 **Habitat models for *Poa sieberiana* in the coastal region of New South Wales, Australia. (a) A Generalized Additive Model prediction for the habitat of *Poa sieberiana* and (b) a habitat model based on BIOCLIM, a bioclimatic envelope model (after Elith *et al.* 2006). Darker grey areas represent areas of higher predicted probability of occurrence.**

such as zero-inflation and overdispersion (e.g. Martin *et al.* 2005). Work is ongoing to develop techniques suitable for the kinds of data generally available in conservation (Elith *et al.* 2006; Gelfand *et al.* 2006; Pearce & Boyce, 2006).

For species management, there is a more fundamental problem with habitat models. All methods take a static view of the landscape. While samples taken over time may integrate long-term change, typically, models match a snapshot of observations of species with a snapshot of observations of the environment. There may be temporal mismatches between the two. Samples may not capture the full successional dynamics of a species, resulting in a biased picture of its distribution. For instance, many large data sets include observations made over years or decades (Fig. 28.3) but analyses ignore the age of the samples. The reliability of such observations depends on the dynamics of succession, the ecological role and longevity of the species, and the resilience of the ecosystem.

Modellers interested in predicting species' responses to dynamic change sometimes use a strategy of changing relevant independent variables to values that represent the anticipated change. The ecological consequences of change are then represented in a new habitat map (e.g. Araújo *et al.* 2004). However, both versions of the species habitat (the map of the status quo and expected future condition) assume implicitly that the species reaches equilibrium with

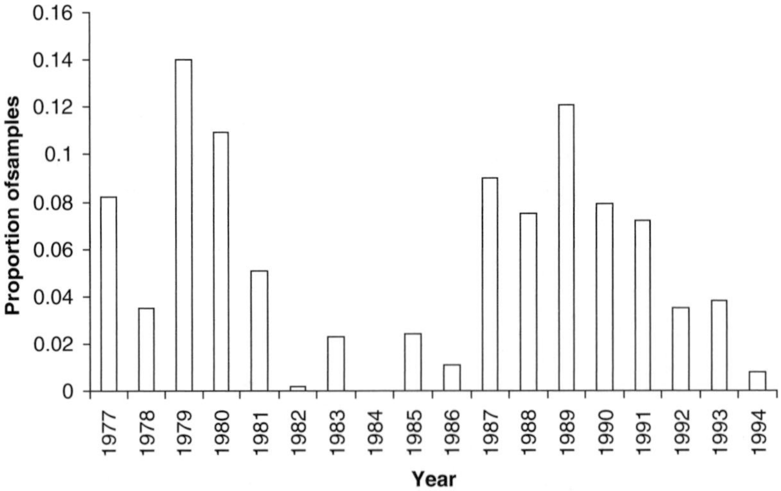

Figure 28.3 **Observation year of floral data surveys for the period 1997–1994, from 3522 floral survey sites from the Central Highlands of southeastern Australia (DSE, in Elith 2002).**

its environment, ignoring the dynamic processes that link the two states, and the time and spatial changes needed to achieve them. All such habitat models ignore non-equilibrium theories of landscape ecology (Connell & Slayter 1977; Westoby *et al.* 1989), which suggest multiple successional pathways and alternative metastable states (Yates & Hobbs 1997).

Modelling species in dynamic habitats

There are many options for creating models of population dynamics (Burgman *et al.* 2005). They range from incidence functions, diffusion approximations and structured population models to cellular automata and individual-based models. Irrespective of the modelling frame, management questions with a spatial context usually can only be answered with a spatially explicit model. Inevitably, they rest on habitat models created using one of the techniques outlined above.

Most metapopulation models rely on a static representation of habitat, or vary habitat by manipulating carrying capacity. A few implementations of these models (ALEX, RAMAS/Landscape; see Lindenmayer *et al.* 1995) provide options for modellers to specify the details of the way in which habitat fluctuates through time (deterministically and stochastically). These models reflect the relationship between habitat quality and stochastic environmental and biological events (fire, extreme weather, flood events, timber harvesting, vegetation dynamics). Recovery is usually modelled as a response function of expected habitat suitability against time since disturbance (e.g. Regan *et al.* 2001; Fig. 28.4). Decision theory provides a platform for structuring alternatives and exploring preferences in such models (Maguire 2004). For example, stochastic habitat models may be linked to population models to generate optimal decisions for landscape management (e.g. Richards *et al.* 1999).

Limitations of single-species models

Conventional assumptions can become embedded in the thinking of modellers. That is, it can be hard to know what assumptions modellers have made. Such assumptions can lead to biases and generalizations that are contrary to empirical observation, and yet are difficult to change in modelling practice.

For example, dispersal is an important part of the ecology and management of threatened and invasive species (With 2004; Matlack 2005; Trakhtenbrot

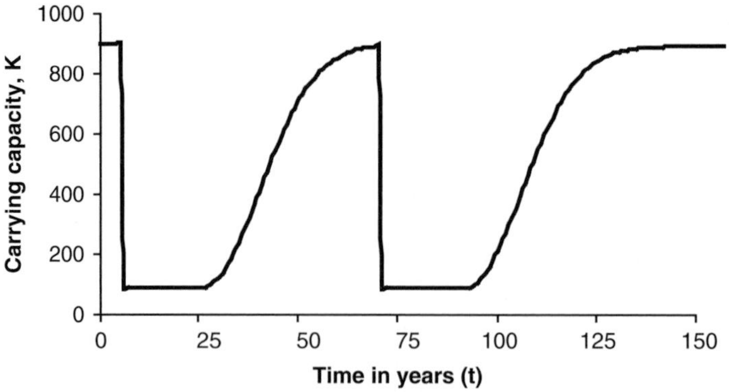

Figure 28.4 **Example of deterministic changes in carrying capacity (*K*) versus time (*t*). In this example of the predatory snail, *Tasmaphena lamproides*, harvest occurs at year 7 and year 72 with 10% of the habitat retained (after Regan *et al.* 2001).**

et al. 2005). Dispersal models for single species rely typically on exponential or inverse power functions (e.g. Hill *et al.* 1996; Hanski *et al.* 2000; Parvinen *et al.* 2003) to estimate the proportion of individuals from one patch that disperse to another. Several studies have compared exponential and inverse power functions. Other dispersal models have been used from time to time, including the gamma, Weibull and normal distributions (e.g. Collingham & Huntley 2000; Yamamura 2002) and are sometimes based on explicit physical models (e.g. Schurr *et al.* 2005).

Despite the importance of modelling dispersal dynamics correctly, most conventional choices to represent dispersal have characteristically 'short' tails (Higgins *et al.* 1999). That is, they tend to predict that long-distance dispersal events are rare. One of the reasons for this convention is that parameters for 'exponentially bounded' distributions can be estimated from data using standard statistical techniques. Researchers tend to overlook the problems with these distributions because the standard procedures for measuring fit do not necessarily provide a reliable guide to the tails of the distribution.

Palaeoecological evidence about the dispersal of plants in Europe and North America is at odds with the qualitative predictions of exponentially bounded ('short-tailed') dispersal models (e.g. Lavorel *et al.* 1995; Collingham & Huntley 2000; Russo 2005; Schurr *et al.* 2005). Alternative, 'fat-tailed' distributions predict plant dispersal much closer to observed patterns. These models do not approach constant rates of spread. Instead, spread accelerates as rare, long-distance dispersal produces outlying populations. The population front is noisy and hard to characterize as outliers establish and coalesce (Clark 1998; Higgins *et al.* 1999).

This example of the limitations of distributional assumptions for dispersal is just one of many aspects of single-species models that reflect conventionally acceptable but poorly tested assumptions that often are not apparent to those who rely on the models for guidance. The solution is to evaluate the full spectrum of uncertainties in a model by thorough sensitivity analysis. The importance of assumptions should then be evaluated by how sensitively decisions might be affected by changes in the assumptions. Applications of decision theory (Ben-Haim 2006) in this context have particular promise (e.g. Regan *et al.* 2005).

Even if complete assessments of uncertainty in single-species habitat and population models were routine, building detailed single-species models is limited ultimately by time, technical skills and data. It may take up to 3–6 months for a person with excellent skills to build reliable, detailed habitat and metapopulation models, link them to disturbance and management scenarios, and generate complete sensitivity analyses. Such relatively long time-scales and large resource investments limit most applications to culturally or economically important species, those motivated by social conflict over competing demands for scarce environmental resources.

While single-species management satisfies social objectives, in such contexts it is too slow, too focused on narrow ecological and taxonomic subsets and too susceptible to the capriciousness of public opinion to keep pace with the ecological effects of urban development, invasive species, pathogens and other landscape-scale environmental change.

However, this view overlooks the potential benefits of rapid prototyping and exploratory modelling to help structure ideas, explore data and assumptions, and guide thinking about potential solutions (Starfield 1997). Thus, even though distributional data and ecological knowledge for most species are lacking (e.g. data on invertebrates, non-vascular and most vascular plants are very scarce), the social imperatives for single-species conservation will continue for the foreseeable future. Simple, rapidly developed models for habitat and population dynamics provide a viable, efficient platform for testing ideas, establishing which crucial pieces of information may be missing, and evaluating alternative conservation decisions.

Alternatives to single-species conservation include landscape-planning principles that aim to conserve species by managing landscape patterns and surrogates ('indicator' or 'focal' species) to represent a broader suite of species (Simberloff 1998; Burgman *et al.* 2005). Unfortunately, landscape-planning alternatives proposed to date have had limited ecological support (Simberloff 1998).

Models for biological diversity have the potential to link data from all sampled species to environmental data (Ferrier 2002). Vegetation condition assessments (Taft *et al.* 1997; Parkes *et al.* 2003; Gibbons *et al.* 2005) and indices of 'biotic integrity' (Karr 1991) aim to measure the characteristics and range of variability of environmental attributes that provide habitat based on features important for some of the better known groups of species and those thought to reflect habitat degradation (Keith & Gorrod, 2006). These approaches have had variable success (see Lindenmayer *et al.* 2000, 2002; Ferrier 2002; Lobo *et al.* 2004; Mac Nally *et al.* 2004; Ferrier & Guisan, 2006).

Single-species conservation is likely to remain important for the foreseeable future because of social imperatives to conserve species, and because ecological theory and predictive capabilities are better developed for single species than for any other level of ecological organization. Models for single species will be most useful if simple, rapid approaches to building dynamic habitat models and population models are readily available, and the skills to use them and to understand their limitations and assumptions are widely disseminated.

Principles for landscape ecology

1 Detailed, spatially explicit population and dynamic habitat models will have a role in landscape ecology for the foreseeable future to integrate data and explore management options for high-profile, culturally or economically important species and contentious management decisions.
2 Detailed, explicit spatial habitat models and population models are too cumbersome for routine application for all species of conservation concern.
3 Instead, single habitat and population models should be used strategically in conjunction with structured decision-making to integrate information, identify crucial data and provide a coherent platform for exploring the consequences of management alternatives through sensitivity analyses.
4 Statistical habitat models offer substantially improved predictive capabilities over polygons, multivariate models, climate models and

habitat suitability indices and should be applied in landscape ecology practice wherever possible.

5 Dynamic, stochastic habitat models encompass uncertainities that static models ignore.

6 Decision theory provides a platform to integrate stochastic population and habitat models with management planning decisions.

7 Community or biodiversity modelling and measures of ecological condition may provide new opportunities for conserving species. However, they have not been validated broadly tested with independent ecological data and their utility for planning for the full spectrum of biodiversity is unknown.

Acknowledgements

We thank David Lindenmayer and two anonymous reviewers for comments that substantially improved the manuscript. We thank NCEAS and Simon Ferrier for the data that support Fig. 28.2, and Rebecca Montague-Drake for her editorial and logistic support. This is contribution from the Australian Centre of Excellence for Risk Analysis and was supported in part by the Australian Research Council.

References

Araújo, M., Cabeza, M., Thuiller, W., Hannah L. & Williams, P.H. (2004) Would climate change drive species out of reserves? An assessment of existing reserve-selection methods. *Global Change Biology* **10**, 1618–1626.

Barry, S.C. & Elith, J. (2006) Error and uncertainty in habitat models. *Journal of Applied Ecology* **43**, 413–423.

Ben-Haim, Y. (2006) *Info-Gap Decision Theory*. Academic Press, Amsterdam, London.

Burgman, M.A., Lindenmayer, D.B. & Elith, J. (2005) Managing landscapes for conservation under uncertainty. *Ecology* **86**, 2007–2017.

Burgman, M.A., Keith, D.A., Hopper, S.D., Widyatmoko, D. & Drill, C. 2007. Threat syndromes and conservation of the Australian flora. *Biological Conservation* **134**, 73-82.

Carpenter, G., Gillison, A.N. & Winter, J. (1993) DOMAIN: a flexible modelling procedure for mapping potential distribution of plants and animals. *Biodiversity and Conservation* **2**, 667–680.

Clark, J.S. (1998) Why trees migrate so fast: confronting theory with dispersal biology and the paleorecord. *American Naturalist* **152**, 204–224.

Collingham, Y.C. & Huntley, B. (2000) Impacts of habitat fragmentation and patch size upon migration rates. *Ecological Applications* **10**, 131–144.

Connell J.H. & Slayter R.O. (1977) Mechanisms of succession in natural communities and their role in community stability and organisation. American Naturalist 111, 1119–1144.

Elith, J. (2002) Predicting the distribution of plants. PhD Thesis, University of Melbourne, Melbourne.

Elith, J. & Burgman, M.A. (2002) Predictions and their validation: rare plants in the Central Highlands, Victoria, Australia. In: Scott, J.M., Heglund, P.J., Morrison, M.L., Raphael, M.G., Wall W.A. & Samson, F.B. (eds.) *Predicting Species Occurrences: Issues of Accuracy and Scale*, pp. 303–314. Island Press, Covelo, CA.

Elith, J., Graham, C.H., Anderson, R.P. *et al.* (2006) Novel methods improve prediction of species' distributions from occurrence data. *Ecography* **29**, 129–151.

Ferrier, S. (2002) Mapping spatial pattern in biodiversity for regional conservation planning: where to from here? *Systematic Biology* **51**, 331–363.

Ferrier, S. & Guisan, A. (2006) Spatial modelling of biodiversity at the community level. *Journal of Applied Ecology* **43**, 393–404.

Gelfand, A.E., Silander, J.A.J., Wuz, S. *et al.* (2006) Explaining species distribution patterns through hierarchical modeling. *Bayesian Analysis* **1**, 41–92.

Gibbons, P., Ayers, D., Seddon, J., Doyle, S. & Briggs, S. (2005) *BioMetric, A Terrestrial Biodiversity Assessment Tool for the NSW Property Vegetation Plan Developer, Operational Manual.* Version 1.8. NSW Department of Environment and Conservation, CSIRO Sustainable Ecosystems, Canberra.

Goolsby, J.A. (2004) Potential distribution of the invasive old world climbing fern, *Lygodium microphyllum* in north and south America. *Natural Areas Journal* **24**, 351–353.

Hanski, I., Alho, J. & Moilanen, A. (2000) Estimating the parameters of survival and migration of individuals in metapopulations. *Ecology* **81**, 239–251.

Hastie, T., Tibshirani, R. & Friedman, J.H. (2001) *The Elements of Statistical Learning: Data Mining, Inference, and Prediction*. Springer-Verlag, New York.

Higgins, S.I., Richardson, D.M., Cowling, R.M. & Trinder-Smith, T.H. (1999) Predicting the landscape-scale distribution of alien plants and their threat to plant diversity. *Conservation and Biology* **13**, 303–313.

Hill, J.K., Thomas, C.D. & Lewis, O.T. (1996) Effects of habitat patch size and isolation on dispersal by *Hesperia* comma butterflies: Implications for metapopulation structure *Journal of Animal Ecology* **65**, 725–735.

Hunter, M.L. (2002). *Fundamentals of Conservation Biology*. Blackwell Science, Melbourne.

Karr, J.R. (1991) Biological integrity: a long-neglected aspect of water resource management. *Ecological Applications* **1**, 66–84.

Keith, D. & Gorrod, E. (2006) The meanings of vegetation condition. *Ecological Management and Restoration* **7**(S1), S7–S9.

Lavorel, S., Gardner, R.H. & O'Neill, R.V. (1995) Dispersal of annual plants in hierarchically structured landscapes. *Landscape Ecology* **10**, 277–289.

Leathwick, J.R., Rowe, D., Richardson, J., Elith, J. & Hastie, T. (2005) Using multivariate adaptive regression splines to predict the distributions of New Zealand's freshwater diadromous fish. *Freshwater Biology* **50**, 2034–2052.

Lindenmayer, D.B., Burgman, M.A., Akcakaya, H.R., Lacy, R.C. & Possingham, H.P. (1995) A review of the generic computer programs ALEX, RAMAS/space and VORTEX for modelling the viability of metapopulations. *Ecological Modelling* **82**, 161–174.

Lindenmayer, D.B., Margules, C.R. & Botkin, D. (2000) Indicators of forest sustainability biodiversity: the selection of forest indicator species. *Conservation Biology* **14**, 941–950.

Lindenmayer, D.B., Cunningham, R.B., Donnelly, C.F. & Lesslie, R. (2002) On the use of landscape indices as ecological indicators in fragmented forests. *Forest Ecology and Management* **159**, 203–216.

Lobo, J.M., Jay-Robert, P. & Lumaret, J.P. (2004) Modelling the species richness distribution for French Aphodiidae (Coleoptera, Scarabaeoidea). *Ecography* **27**, 145–156.

Mac Nally, R. & Fleishman, E. (2004) A successful predictive model of species richness based on indicator species. *Conservation Biology*, **18**, 646–654.

Mac Nally, R., Fleishman, E., Bulluck, L.P. & Betrus, C.J. (2004) Comparative influence of spatial scale on beta diversity within regional assemblages of birds and butterflies. *Journal of Biogeography* **31**, 917–929.

Maguire, L.A. (2004) What can decision analysis do for invasive species management? *Risk Analysis* **24**, 859–868.

Martin, T.G., Wintle, B.A., Rhodes, J.R. *et al.* (2005) Zero tolerance ecology: improving ecological inference by modelling the source of zero observations. *Ecology Letters* **8**, 1235–1246.

Matlack, G. (2005) Slow plants in a fast forest: local dispersal as a predictor of species frequencies in a dynamic landscape. *Journal of Ecology* **93**, 50–59.

Nix, H.A. & Switzer, M.A. (1991) Rainforest animals. Atlas of vertebrates endemic to Australia's wet tropics. *Kowari* **1**, 1–112.

Olden, J.D. (2003) A species-specific approach to modeling biological communities and its potential for conservation. *Conservation Biology* **17**, 854–863.

Parkes, D., Newell, G. & Cheal, D. (2003) Assessing the quality of native vegetation: The 'habitat hectares' approach. *Ecological Management and Restoration* **4** (suppl.), S29–S38.

Parvinen, K., Dieckmann, U., Gyllenberg, M. & Metz, J.A.J. (2003) Evolution of dispersal in metapopulations with local density dependence and demographic stochasticity. *Journal of Evolutionary Biology* **16**, 143–153.

Pearce, J.L. & Boyce, M.S. (2006) Modelling distribution and abundance with presence-only data. *Journal of Applied Ecology* **43**(3), 405–412.

Peterson, A.T. (2003) Predicting the geography of species' invasions via ecological niche modeling. *Quarterly Review of Biology* **78**, 419–433.

Regan, T.J., Regan, H.M., Bonham, K., Taylor, R.J. & Burgman, M.A. (2001) Modelling the impact of timber harvesting on a rare carnivorous land snail (*Tasmaphena lamproides*) in northwest Tasmania, Australia. *Ecological Modelling* **139**, 253–264.

Regan, H.M., Ben-Haim, Y., Langford, B. *et al.* (2005) Robust decision making under severe uncertainty for conservation management. *Ecological Applications* **15**, 1471–1477.

Richards, S.A., Possingham, H.P. & Tizard, J. (1999) Optimal fire management for maintaining community diversity. *Ecological Applications* **9**, 880–892.

Russo, S.E. (2005) Linking seed fate to dispersal patterns: identifying factors affecting predation and scatter-hoarding of seeds of *Virola calophylla* in Peru. *Journal of Tropical Ecology* **21**, 243–253.

Schurr, F.M., Bond, W.J., Midgley, G.F. & Higgins, S.I. (2005) A mechanistic model for secondary seed dispersal by wind and its experimental validation. *Journal of Ecology* **93**, 1017–1028.

Simberloff, D. (1998) Flagships, umbrellas and keystones: is single-species management passé in the landscape era? *Biological Conservation* **83**, 247–257.

Starfield, A.M. (1997) A pragmatic approach to modeling for wildlife management. *Journal of Wildlife Management* **61**, 261–270.

Taft, J.B., Wilhelm, G.S., Ladd, D.M. & Masters, L.A. (1997) Floristic quality assessment for vegetation in Illinois, a method for assessing vegetation integrity. *Erigenia* **15**, 3–95.

Thuiller, W., Lavorel, S., Araújo, M.B., Sykes, M.T. & Prentice, I.C. (2005) Climate change threats to plant diversity in Europe. *Proceedings of the National Academy of Sciences of the USA* **102**, 8245–8250.

Trakhtenbrot, A., Nathan, R., Perry, G. & Richardson, D.M. (2005) The importance of long-distance dispersal in biodiversity conservation. *Diversity and Distributions* **11**, 173–181.

Westoby, M., Walker, B. & Noy Meir, I. (1989) Opportunistic management for rangelands not at equilibrium. *Journal of Range Management* **42**, 266–274.

With, K.A. (2004) Assessing the risk of invasive spread in fragmented landscapes. *Risk Analysis* **24**, 803–815.

Yamamura, K. (2002) Dispersal distance of heterogeneous populations. *Population Ecology* **44**, 93–101.

Yates C.J. & Hobbs R.J. (1997) Woodland restoration in the Western Australian Wheatbelt: A conceptual framework using a state and transition model. *Restoration Ecology* **5**, 28–35.

29

Synthesis: Individual Species Management – Threatened Taxa and Invasive Species

David B. Lindenmayer and Richard J. Hobbs

Ecosystem management (Chapters 26–28) may achieve many of the key goals and objectives of managing landscapes and ecosystems. However, sometimes there will be a need for more focused efforts on particular species – individual threatened taxa, individual invasive species or individual species will exert considerable influence over the rest of the landscape or ecosystem (e.g. keystone species). The three chapters of this section have explored some of the issues relating to the management of individual species, including those that are threatened and those that are invasive.

Zavaleta and Pasari (Chapter 27) take quite a novel approach in exploring the intersection of concepts and approaches to manage threatened species, invasive species and disease agents in landscapes. Many of the strategies used in an attempt better to conserve threatened species (like countering the effects of 'fragmentation') will be the antithesis of approaches to limit the spread of invasive species or disease-causing species. Hence, there is considerable value in appraising similarities and differences among theoretical underpinning and practical approaches. For example, enhanced connectivity may be important to promote the persistence of a threatened species (e.g. see Noss, Chapter 23), but at the same time it may promote the spread of invasive species and diseases. The contrasting influences of disturbance on threatened species and invasive species is another example of this issue explored by Zavaleta and Pasari (Chapter 27).

Burgman *et al.* (Chapter 28) note that despite imperatives to manage ecosystems, landscapes and multiple species, much of current conservation focus remains on managing individual species. They note that informed management of any individual species requires knowledge of its distribution and abundance and also knowledge of its habitat requirements. However, there are many ways of defining habitat and many ways of attempting to quantify both habitat requirements and the spatial distribution of a species. Burgman *et al.* provide a short overview of the vast array of methods for tackling these key tasks and rightly acknowledge that there are important strengths, but also key assumptions and limitations associated with each approach. Moreover, different approaches can produce markedly different outcomes for the same species, even when approximately the same data are used! Modelling species in dynamic habitats is the second key theme of the essay by Burgman *et al.* They discuss some of the issues associated with models for population viability analysis (see also Fahrig, Chapter 7) as well as modelling dispersal, particularly for plants. Most appropriately, they acknowledge the high degree of uncertainty associated with all the methods they discuss.

Burgman *et al.* (Chapter 28) recognize the sheer impossibility of defining the habitat requirements and distribution patterns of all species, a task made even more difficult when habitats are dynamic (e.g. in response to disturbance; see also Hunter, Chapter 35; Cramer Chapter 32). They suggest the use of metrics and benchmarks for vegetation assessments as a possible alternative to single-species approaches, but here too there are some significant hurdles to be cleared (see also Woinarski, Chapter 10). Their value in guiding the management of the full array of biota is unknown but likely to be problematic for a range of reasons.

Simberloff (Chapter 26) is perhaps less optimistic about the prospects of success of multispecies and ecosystem management approaches than Burgman *et al.* (Chapter 28). He notes that whereas single-species management has often (but certainly not always) been associated with clear and well-defined goals, ecosystem management has not. Moreover, he believes that managing particular ecological processes is especially difficult, in part because defining what such processes entail is often not straightforward (e.g. ecosystem health) and partly because developing robust metrics to assess congruence or departure from some benchmark for a process can be problematic (see Woinarski, Chapter 10).

Simberloff (Chapter 26) further argues that although single-species management is not fashionable, there are many examples illustrating the success of targeted management to recover threatened species and to control and/or eradicate invasive species. Whether the successes of these single-species programmes have benefited other taxa is not well established, but Simberloff argues that there are cases where they probably have not.

All authors in Chapters 26–28 acknowledge the difficulties of managing every individual species in a landscape or ecosystem, but also highlight the potential shortcomings of multispecies and ecosystem approaches. From a practical perspective, it is clear that a range of management strategies will nearly always be required; some will be focused on individual species, others on suites of species and yet others on entire landscapes or ecosystems (Simberloff, Chapter 26). This re-emphasizes the comments of a number of other authors in this book about the need to conduct research, develop a greater understanding, and implement management at a range of spatial scales (e.g. Walker, Chapter 34).

Like many other authors in this book, Simberloff (Chapter 26) and Zavaleta and Pasari (Chapter 27) argue strongly that irrespective of what is being managed (species, assemblages, landscapes), the goals and objectives must be carefully formulated, priorities well articulated and rigorous mechanisms implemented to assess progress (see also Luck, Chapter 16). Unfortunately, as also highlighted by many contributors, this is often not the case. Such arguments for well-articulated objectives and priorities are further reinforced by Zavaleta and Pasari (Chapter 27), who highlight, in particular, the tensions (and hence the trade-offs) that arise with management strategies (such as promoting connectivity or implementing disturbance regimes) that benefit threatened species but also may benefit invasive species and diseases.

Section 8
Ecosystems and Ecosystem Processes

30

Ecosystems, Ecosystem Processes and Global Change: Implications for Landscape Design

Adrian D. Manning

Abstract

Ecosystems are dynamic in time and space and are complex and adaptive. Human-induced global change has caused, and is causing, major changes to ecosystem structure, function and species composition. Ecosystem processes include energy flows, nutrient and hydrological cycles, and biotic interactions and they are structured by ecosystem components such as internal and external disturbances, trophic cascades and interactions, functional groups and keystone species. Humans depend on vital ecosystem services derived from ecosystem processes and functions. Global change can reduce resilience and cause regime shifts in ecosystems and is having a profound effect on organisms, processes and ecosystems. To meet the challenges of global change and its effects on ecosystems and ecosystem processes, eight principles are suggested for landscape design: (i) consider the human – Earth relationship; (ii) protect what is already there; (iii) build and enhance resilience; (iv) restore ecosystems; (v) enhance 'landscape fluidity'; (vi) use stretch-goals to achieve ambitious conservation targets; (vii) implement long-term research and monitoring; and (viii) use predictive modelling to anticipate species' responses and plan conservation and ecological restoration.

Keywords: ecosystem management; ecosystem processes; ecosystems; ecosystem services; global change.

> In reality, it is species that are constant elements and ecosystems that are transitory (Lawton 1997, p. 5).

Introduction

The Earth is undergoing unprecedented human-induced changes to biotic diversity and the structure and function of ecosystems (Kareiva *et al.* 1993; Vitousek *et al.* 1997; Sala *et al.* 2000). While ecosystems are dynamic in both time and space (Levin 1992), the predicted magnitude and rapidity of human-induced global environmental change demands a radical re-evaluation of how humans manage, design and restore ecosystems and landscapes. In this chapter, global change is briefly explained. An explanation of ecosystems, ecosystem processes and services, and ecosystem management is outlined and, within this context, the implications for ecosystem management and landscape design are discussed. Recommendations for landscape-design principles are then outlined.

What is an ecosystem?

An ecosystem is 'a spatially explicit unit of the Earth that includes all of the organisms, along with all of the components of the abiotic environment within its boundaries' (Likens 1992, p. 7). Tansley is attributed with coining the term 'ecosystem' in 1935 (Christensen *et al.* 1996). He believed that an ecosystem was the basic unit of nature. However, it is now recognized that ecosystems vary in time and space and that the identification of ecosystem units and boundaries is an essentially arbitrary decision (Likens 1992). In the early part of the 20th century, many ecologists believed that the biological component of ecosystems (i.e. communities) progressed steadily along a successional pathway towards a self-sustaining 'climax' state – the 'normal' condition. However, in the 1970s, the role of natural disturbance in ecosystems was increasingly recognized, and it was realized that it was unrealistic to assume that it was 'normal' for most ecosystems to be in a 'climax' state (Sprugel 1991). It is now widely understood that ecosystems are not static but are dynamic in both time and space (Levin 1992; Chapin *et al.* 1996; Christensen *et al.* 1996). Further, ecosystems are adaptive systems that are highly complex as a result of factors such as non-linear biotic interactions, threshold effects, evolutionary history, assembly history, past environmental disturbances, hysteresis and high levels of stochastic events (Levin 1999 cited by Folke *et al.* 2004; Sinclair & Byrom 2006).

What are ecosystem processes?

There are three broad categories of ecological processes:

1 Energy flows – the flow of energy through an ecosystem. Most energy is originally derived from the sun through photosynthesis by plants.
2 Nutrient cycles – the movement of elements, such as nitrogen, carbon and phosphorus, through the biotic and abiotic components of the system.
3 Hydrological cycles – the movement of water from the ocean to the land and back (Aber & Melillo 1991; Noss & Cooperrider 1994).

These processes are structured or facilitated by ecosystem components including internal and external disturbance regimes, trophic cascades (such as top-down predator effects) and interactions and functional groups (such as decomposers and pollinators) and 'keystone species' (Sprugel 1991; Aber & Melillo 1991; Power *et al.* 1996; Pace *et al.* 1999; Folke *et al.* 2004). Ecosystems also have numerous feedback and equilibrium processes that affect the overall structure and function of the ecosystem (see discussion of resilience and regime shifts below).

Organisms play an important role in the structuring and function of ecosystem processes, and the composition of the biota influences these processes (Chapin *et al.* 1996). The result is that disturbances that directly affect the biota can also affect ecosystem processes and functions (Sinclair & Byrom 2006). Biological diversity is thought to be very important in maintaining ecosystem resilience (see below) under changing environmental conditions (Peterson *et al.* 1998). Species do not have equal functional roles in communities (Power *et al.* 1996; Sinclair & Byrom 2006). Those species that have a disproportionate functional role in an ecosystem are called 'keystone species' (Paine 1966, 1980; Power *et al.* 1996). The loss or addition of keystone species can have dramatic effects on an ecosystem (Chapin *et al.* 1996). For example, the beaver (*Castor* spp.) is known to have a major influence on riparian and surrounding terrestrial ecosystems through dam building and associated coppicing of trees. Beavers affect ecosystem function through creation of wetland habitats, modification of stream hydrology, geomorphology, sediment and water chemistry, retention of sediment and organic material by dams, changes to forest structure and dynamics (Naiman *et al.* 1988, 1994; Smith *et al.* 1991).

Ecosystems also contain suites of different functional groups of species, such as those that pollinate, graze, consume prey, fix nitrogen, decompose, generate soil and modify water flows (Folke *et al.* 2004). Therefore, any phenomenon that alters species composition is likely to alter ecosystem processes through changes in functional traits of the biota (Chapin *et al.* 1996). The variability in the responses of species within functional groups is called 'response diversity', which Elmqvist *et al.* (2003, p. 488) define as 'the diversity of responses to environmental change among species that contribute to the same ecosystem function' (see also Chapter 34). The likelihood of renewal and reorganization following a disturbance is thought to increase if an ecosystem has high response diversity (Chapin *et al.* 1996; Elmqvist *et al.* 2003).

Ecosystem resilience and regime shifts

Historically, humans have reduced the capacity of ecosystems to handle change through a combination of impacts on top-down and bottom-up processes and alteration of disturbance regimes, including climate change (Folke *et al.* 2004). 'Resilience' is the 'the capacity of a system to absorb disturbance and reorganize while undergoing change so as to retain essentially the same function, structure, identity and feedbacks' (Folke *et al.* 2004, p. 558). Ecosystems can have multiple alternative states and, under the same environmental conditions, an ecosystem can exist with different combinations of species abundance (Scheffer *et al.* 2001; Sinclair & Byrom 2006). 'Regime shifts' are the movements between states, and occur when an ecosystem crosses a 'threshold' (Folke *et al.* 2004; Mayer & Rietkerk 2004). The resulting changes to the ecosystem can be gradual or sudden, catastrophic and even irreversible (Scheffer *et al.* 2001; Folke *et al.* 2004). Sharp regime shifts may occur more readily if resilience is reduced by human actions such as:

- removal of functional groups and response diversity (e.g. whole trophic levels);
- impact on ecosystems through waste and pollutant emissions, and climate change;
- alteration of the magnitude, frequency and duration of disturbance regimes to which organisms are adapted (Folke *et al.* 2004).

A major concern is that ecosystems may show no major response to gradual changes in drivers for some time (i.e. no early warning signal), before experiencing sudden and catastrophic regime shifts (Scheffer *et al.* 2001). Failure to recognize the possibility of regime shifts could have major costs to society (Scheffer *et al.* 2001), specifically through the loss of 'ecosystem services' (*sensu* Daily 1997).

What are ecosystem services?

Daily (1997, p. 3) defines ecosystem services as 'the conditions and processes through which natural ecosystems, and the species that make them up, sustain and fulfill human life.' Ecosystem services are those that humans derive, directly or indirectly, from ecosystem functions such as habitat, biological and system properties or ecological processes (Costanza *et al.* 1997). Costanza *et al.* (1997) calculated the economic value of 17 ecosystem services across 16 biomes. They estimated that the value was between US$16 and 54 trillion per annum, with an average of US$33 trillion per annum. This is as much as 1.8 times the Gross World Product (Costanza *et al.* 1997). An important approach for maintaining ecosystems, ecosystem processes and services is 'ecosystem management'.

What is ecosystem management?

The concept of 'ecosystem management' recognizes that approaches that focus on species, subspecies and population are insufficient alone to preserve biological diversity (Franklin 1993). Single-species research and conservation remains vitally important (Franklin 1993; Sinclair & Byrom 2006), but for the vast majority of species, habitat preservation and/or ecosystem conservation is the only viable conservation approach. Also, 90% of the Earths's total species are invertebrates and the majority of these taxa are unknown (Franklin 1993). Invertebrates, fungi and bacteria provide crucial ecosystem functions, including nitrogen fixation and decomposition (Franklin 1993). Conservation of organisms in poorly known habitats and ecological subsystems (e.g. below ground) can only be achieved through an ecosystem approach (Franklin 1993).

In 1996, the Committee on the Scientific Basis for Ecosystem Management of the Ecological Society of America defined ecosystem management as: 'Ecosystem management is management driven by explicit goals, executed by policies, protocols, and practices, and made adaptable by monitoring and research based on our best understanding of ecological interactions and processes necessary to sustain ecosystem composition, structure, and function' (Christensen *et al.* 1996).

Some recurrent goals of ecosystem management are to:

1 Maintain viable populations of all native species.
2 Protect representative examples of all native ecosystem types across their natural range of variation.
3 Maintain evolutionary and ecological processes.
4 Manage landscapes and species to be responsive to both short- and long-term environmental change (Grumbine 1994; Wilcove & Blair 1995).

What is global change?

Global change is much broader than the more commonly understood concept of climate change (La Riviere 1994; Steffen *et al.* 2004). Global change can also include socioeconomic changes (e.g. globalization, urbanization) (Steffen *et al.* 2004); however, the main focus of this chapter is biophysical changes. Vitousek (1994) lists three key components of human-induced global change: (i) increasing concentrations of atmospheric carbon dioxide (CO_2); (ii) alterations in the biogeochemistry of the global nitrogen cycle; and (iii) ongoing land use/land cover change. These, and other global changes, are leading to major alterations in the functioning of the Earth's systems, and in particular they are driving global climatic change and loss of biological diversity (Kareiva *et al.* 1993; Vitousek 1994; Pimm *et al.* 1995; Vitousek *et al.* 1997; Sala *et al.* 2000; Foley *et al.* 2005). The magnitude of biodiversity change is now so large (Pimm *et al.* 1995) that it is considered a global change in its own right (Walker & Steffen 1996, cited by Sala *et al.* 2000).

Ecosystems and global change

Global change is expected to have profound effects on ecosystems and ecosystem processes (Vitousek 1994; Pimm *et al.* 1995; Vitousek *et al.* 1997; Sala *et al.* 2000; Foley *et al.* 2005). The profound interconnected and interacting effects that human-induced global change is having on the Earth's ecosystems mean that past distinctions between pristine ecosystems and human-altered areas no longer exist (Vitousek 1994). Within terrestrial ecosystems, by 2100, it is predicted that land-use change will have the largest effect on biodiversity, followed by climate change, nitrogen deposition, biotic exchange (introduction of plants and animals to an ecosystem) and elevated CO_2 concentrations (Sala *et al.* 2000). However, the dominant drivers could change (Millennium Ecosystem Assessment 2005), and climate change may become the dominant threat to biodiversity in many regions (Thomas *et al.* 2004). At all scales of ecological organization (genes, populations/species and biomes), climate change and invasions of invasive alien species are two drivers of changes that it is thought will be most difficult to reverse (Millennium Ecosystem Assessment 2005).

Climate change and land-use change are having a major effect on the distribution and abundance of global biodiversity (Warren *et al.* 2001). Further, the interaction of the two is expected to have a far more detrimental effect on biodiversity than either factor in isolation (Peters 1990; Erasmus *et al.* 2002). In response to climate change, ecological communities are expected to disassemble, and organisms will respond individualistically, with differing rates of movement (Peters 1990; Erasmus *et al.* 2002; Thuiller 2004). There is already growing evidence that climate change is having major effects on species' range shifts (Parmesan & Yohe 2003; Root *et al.* 2003; Hickling *et al.* 2005), as well as other key biological functions such as commencement of breeding (Crick *et al.* 1997). As climate change proceeds, these effects are expected to intensify. For example, it has been predicted that if the warming trend in the North Atlantic Oscillation (a meridional displacement of atmospheric mass between low- and high-pressure cells centred, respectively, close to Iceland and the Azores) continues, over the next 50 years the start of the growing season in Europe could occur as much as 13 days earlier than it currently does (Cook *et al.* 2005).

Climate changes can have major effects on organisms and their interactions, which in turn affect the structure of the ecosystem. It has been argued that some predicted effects of climate change on ecosystems may be too conservative, because they do not consider the interplay and feedback from higher trophic levels (Schmitz *et al.* 2003). In some cases, the composition of key species in some ecosystems may remain the same, but species abundance, interaction strength and ecosystem function will be altered (Schmitz *et al.* 2003). During certain phases, the North Atlantic Oscillation has climatic effects that are similar to the predicted effects of rising CO_2 (Schmitz *et al.* 2003) and can hint at some of the effects such global change scenarios might have on interactions within ecosystems. For example, Post *et al.* (1999) found that increased snow depth on Isle Royale in Lake Superior (USA), related to the North Atlantic Oscillation, resulted in wolves (*Canis lupus*) hunting in larger packs. This resulted in triple the number of moose (*Alces alces*) being killed. This reduction in moose numbers resulted in a reduction in browsing of balsam fir (*Abies balsamea*) and an increase in understorey growth. Hence, the climate fluctuations led to a cascading behavioural response linking climate change to changes in ecosystem structure and/or function. The result of further global change on these sorts of interactions is unknown.

Global change is also expected to have major effects on ecosystem physiology – the interacting physiological processes determining exchanges of carbon, energy, water and nutrients in an ecosystem (Mooney *et al.* 1999). Some key factors that are expected to change include: atmospheric CO_2 concentrations, temperature, water availability, nitrogen deposition, ultraviolet B radiation (UV-B) and tropospheric ozone (Mooney *et al.* 1999). Increased atmospheric CO_2 concentrations can have a number of effects: (i) increased photosynthesis and primary productivity, (ii) reduced plant transpiration and improved plant water status; (iii) increases to average global temperatures; and (iv) changed patterns of precipitation (Mooney *et al.* 1999; Weltzin *et al.* 2003; Morgan *et al.* 2004; Emmerson *et al.* 2005). In some ecosystems, increased photosynthesis and primary productivity leads to greater above-ground biomass accumulation (Mooney *et al.* 1999). Increased CO_2 concentrations may also have effects on foliar chemistry, which can affect abundances of folivorous animals (e.g. Kanowski 2001).

In many terrestrial ecosystems, nitrogen is naturally the most limiting nutrient (Scholes *et al.* 1999). Alteration of the nitrogen cycle has major effects on ecosystem physiology, particularly those systems that are nitrogen limited. Effects include increased productivity and carbon storage, increased losses of nitrogen and cations from soils, decreases in biological diversity and eutrophication of aquatic systems (Vitousek *et al.* 1997). Higher temperatures may lead to greater release CO_2 from decomposition and respiration rates of soil organic matter at a rate surpassing the raised levels of photosynthesis and net primary productivity, possibly leading to positive feedback (Mooney *et al.* 1999). Changes in seasonality and variability of precipitation regimes, as a result of increased CO_2 concentrations, are expected to have major effects on soil moisture and in turn on species and their interactions (Weltzin *et al.* 2003).

Drivers of change in ecosystem physiology also can have significant effects on plants and patterns of rates of consumption of them by animals (e.g. Peters *et al.* 2006).

Implications for ecosystems and landscape management

The speed and magnitude of global change and potential regime shifts will present acute problems for landscape managers. Although the spatial and temporal dynamics of ecosystems are often explicitly or implicitly acknowledged (see above), the response on the ground is still often spatially and temporally fixed. This landscape-management approach is often underpinned by spatially and temporally fixed landscape models, which fail to account for changes in landscape processes and trajectories through time (Manning *et al.* 2004). The advent of ecosystem management has had the advantage that it has forced land managers to consider ecologically significant time frames (Wilcove & Blair 1995). However, 'it will be exceedingly difficult to get land managers and politicians to think in terms of centuries' (Wilcove & Blair 1995, p. 345).

In the context of global change, as outlined above, it may not be feasible to sustain current ecosystem composition, structure and function. Accordingly, the adaptability of ecosystems may need greater emphasis. This, in turn, raises questions about the types of goals, policies, protocol and practices that are formulated. In light of this, it is important to reassess landscape design principles in response to global change.

Landscape design principles

1. Consider the human–Earth relationship

'A culture that destroys its environment is suicidal' (Noss & Cooperrider 1994, p. 15). It is important to recognize the nature of the human–Earth relationship. A major challenge for humanity is to manage trade-offs between short-term human needs, and the capacity of the biosphere to provide ecosystem services in the long-term (Foley *et al.* 2005). Although humans are a major component of Earth systems, they are not vital to the survival of life on Earth *per se*. That is, in the absence of humans, the Earth's ecosystems would continue. However, in the absence of the many specific ecosystem services provided by the Earth, the continuation of human life would be impossible. Humans have a vested interest in successful ecosystem management, and are vital to achieving its success. However, this should not be confused with the notion that humans are vital to the planet. Hence, if humanity wishes to survive, the use and management of the Earth's ecosystems must be based on a fundamental understanding of humanity's place in this relationship.

2. Protect what is already there

Existing ecosystems, even if they are highly modified, are vitally important as habitat for many organisms (e.g. scattered trees, Manning *et al.* 2006a). Existing ecosystems provide biological legacies (*sensu* Franklin *et al.* 2000) and ecological memory (*sensu* Bengtsson *et al.* 2003). Existing ecosystems, at all states of modification, provide the building blocks for future ecosystem restoration.

3. Build and enhance ecosystem resilience

Much ecosystem management concentrates on minimizing perturbations; however, disturbance is a natural part of ecosystems (Scheffer *et al.* 2001). It is actually the gradual loss of resilience that makes ecosystems vulnerable to regime shifts, not stochastic disturbances *per se*. Furthermore, while stochastic events are difficult to predict and control, building and maintaining ecosystem resilience is practically achievable (Scheffer *et al.* 2001). However, the key challenge is to determine how actually to build and enhance resilience in ecosystems.

4. Restore ecosystems

Ecosystem restoration will be a vital tool in the maintenance and restoration of resilience. An important part of this will be the maintenance or restoration of functional groups and response diversity and keystone species. Ecosystem repair and restoration is now essential to avoid ecological collapse (Dobson *et al.* 1997; Hobbs & Harris 2001; Millennium Ecosystem Assessment 2005). This is particularly important as climate change progresses in the future, and there are shifts in the ranges of biota (Harris *et al.* 2006; Sinclair & Byrom 2006). While the loss of keystone species can precipitate the 'unravelling' of ecosystem structure and function, the restoration of keystone species and associated positive ecosystem effects will be an important part of ecosystem reconstitution or 'ravelling up' (Sinclair & Byrom 2006).

5. Enhance 'landscape fluidity'

'Landscape fluidity is a term used to describe "*the ability of organisms to ebb and flow in response to environmental change through space and time*" (Manning et al. in prep). In response to the challenges that global change poses for organisms, a goal for landscape managers might be to enhance landscape fluidity. Enhancing landscape fluidity would involve the maintenance and restoration of habitat, landscape and ecological connectivity through time and space in anticipation of ongoing global change. This could involve some of the following management actions:'

- **Ecosystem restoration.** The use of conventional ecological restoration to arrest declines of ecosystems and organisms, with particular efforts to restore less well-represented ecosystems.
- **Anticipatory restoration.** This is restoration that occurs in advance of a future reintroduction of an organism, particularly a keystone species (Manning *et al.*, 2006b). It could also be used to anticipate future range shifts and changes in ecosystem processes.
- **Static and dynamic reserves.** The use of traditional spatially static reserves in conjunction with reserves that are dynamic in time and space (*sensu* Bengtsson *et al.* 2003).
- **Integration of conservation and production.** To produce more continuous habitats and facilitate range shifts in human-dominated landscapes, more integrated landscape management is needed.
- **Facilitated range shifts and inoculations.** The juxtaposition of ecosystems to provide sources or 'inoculations' of organisms ('ecological memory' *sensu* Bengtsson *et al.* 2003) and facilitate movements through time. For those species or genotypes that are unable to move quickly enough, or not at all, in response to global change, their movements may need to be facilitated by translocations (Peters 1990).

6. Use stretch-goals to achieve ambitious conservation targets

Management of landscapes in response to global change and enhancing landscape fluidity are very ambitious goals (Manning *et al.* 2006b), particularly on the necessary timescales. One way to achieve seemingly impossible goals is to use stretch-goals. Stretch-goals are 'ambitious, long-term goals used to inspire creativity and innovation to achieve outcomes that currently seem impossible' (Manning *et al.* 2006b). Stretch-goals can be used to achieve shorter-term milestones (Manning *et al.* 2006b).

7. Implement long-term research and monitoring

'Conservation action without good science to underpin it is like alchemy, or faith healing. Both sometimes produce desirable results, but you have no idea why, and mostly they do not' (Lawton 1997, p. 3). Long-term management of landscapes and longer-term processes, such as disturbance, in a changing global environment must be informed by long-term ecological research (Hobbie *et al.* 2003; Kratz *et al.* 2003; Turner *et al.* 2003) and monitoring. Long-term ecological research allows the tracking of long-term trends, and this would facilitate adaptive management approaches. Consequently, long-term research projects should be built into landscape management projects.

8. Use predictive modelling to anticipate species' responses and plan conservation and ecological restoration

Landscape management in the future will require sound predictive modelling to determine which species will need assistance to establish and the identification and purchase of key areas for conservation (Sinclair & Byrom 2006). This will be especially important when planning anticipatory restoration, for keystone species, and facilitated range shifts (see above).

Landscape design principles for ecosystems and ecosystem processes in the context of global change

1 Consider the human–Earth relationship.
2 Protect what is already there.
3 Build and enhance resilience.
4 Restore ecosystems.
5 Enhance 'landscape fluidity'. Key elements of this include: ecosystem restoration, anticipatory restoration, static and dynamic reserves, integration of conservation and production, and facilitated range shifts and inoculations.
6 Use stretch-goals to achieve ambitious conservation targets.
7 Implement long-term research and monitoring.
8 Use predictive modelling to anticipate species' responses and plan conservation and ecological restoration.

Acknowledgements

Many thanks to Pat Werner and Joern Fischer for helpful discussions and/or supplying literature and information. Thanks to Richard Hobbs, David Lindenmayer, Henry Nix, Will Steffen and Pat Werner for helpful comments on earlier drafts of this chapter.

References

Aber, J.D. & Melillo, J.M. (1991) *Terrestrial Ecosytems*, Saunders College Publishing, Orlando, FL.

Bengtsson, J., Angelstam, P., Elmqvist, T. *et al.* (2003) Reserves, resilience and dynamic landscapes. *Ambio* **32**, 389–396.

Chapin F.S. III, Torn, M.S. & Tateno, M. (1996) Principles of ecosystem sustainability. *American Naturalist* **148**, 1016–1037.

Christensen, N.L., Bartuska, A.M., Brown, J.H. *et al.* (1996) The report of the Ecological Society of America committee on the scientific basis for ecosystem management. *Ecological Applications* **6**, 665–691.

Cook, B.I., Smith, T.M. & Mann, M.E. (2005) The North Atlantic Oscillation and regional phenology prediction over Europe. *Global Change Biology* **11**, 919–926.

Costanza, R., d' Arge, R., de Groot, R. *et al.* (1997) The value of the world's ecosystem services and natural capital. *Nature* **387**, 253–260.

Crick, H.Q.P., Dudley, C., Glue, D.E. & Thompson, D.L. (1997) UK birds are laying eggs earlier. *Nature* **388**, 526.

Daily, G.C. (ed.) (1997) *Nature's Services: Societal Dependence on Natural Ecosystems.* Island Press, Washington, DC.

Dobson, A.P., Bradshaw, A.D. & Baker, A.J.M. (1997) Hopes for the future: restoration ecology and conservation biology. *Science* **277**, 515–522.

Elmqvist, T., Folke, C., Nyström, M. *et al.* (2003) Response diversity and ecosystem resilience. *Frontiers in Ecology and the Environment* **1**, 488–494.

Emmerson, M., Bezemer, M., Hunter, M.D. & Jones, T.H. (2005) Global change alters the stability of food webs. *Global Change Biology* **11**, 490–501.

Erasmus, B.F.N., Van Jaarsveld, A.S., Chown, S.L., Kshatriya, M. & Wessels, K.J. (2002) Vulnerability of South African animal taxa to climate change. *Global Change Biology* **8**, 679–693.

Foley, J.A., DeFries, R., Asner, G.P. *et al.* (2005) Global consequences of land use. *Science* **309**, 570–574.

Folke, C., Carpenter, S., Walker, B. *et al.* (2004) Regime shifts, resilience, and biodiversity in ecosystem management. *Annual Review of Ecology, Evolution and Systematics* **35**, 557–581.

Franklin, J.F. (1993) Preserving biodiversity: species, ecosystems, or landscapes? *Ecological Applications* **3**, 202–205.

Franklin, J.F., Lindenmayer, D.B., MacMahon, J.A. *et al.* (2000) Threads of continuity: ecosystem disturbances, biological legacies and ecosystem recovery. *Conservation Biology in Practice* **1**, 8–16.

Grumbine, R.E. (1994) What is ecosystem management? *Conservation Biology* **8**, 27–38.

Harris, J.A., Hobbs, R.J., Higgs, E. & Aronson, J. (2006) Ecological restoration and global climate change. *Restoration Ecology* **14**, 170–176.

Hickling, R., Roy, D.B., Hill, J.K. & Thomas, C.D. (2005) A northward shift of range margins in British Odonata. *Global Change Biology* **11**, 502–506.

Hobbie, J.E., Carpenter, S.R., Grimm, N.B., Gosz, J.R. & Seastedt, T.R. (2003) The US long term ecological research program. *Bioscience* **53**, 21–32.

Hobbs, R.J. & Harris, J.A. (2001) Restoration ecology: repairing the Earth's ecosystems in the new Millennium. *Restoration Ecology* **9**, 239–246.

Kanowski, J. (2001) Effects of elevated CO_2 on the foliar chemistry of seedlings of two rainforest trees from north-east Australia: implications for folivorous marsupials. *Austral Ecology* **26**, 165–172.

Kareiva, P.M., Kingsolver, J.G. & Huey, R.B. (eds.) (1993) *Biotic Interactions and Global Change*. Sinauer Associates Inc, Sunderland, MA.

Kratz, T.K., Deegan, L.A., Harmon, M.E. & Lauenroth, W.K. (2003) Ecological variability in space and time: insights gained from the US LTER program. *BioScience* **53**, 57–67.

La Riviere, J.W.M. (1994) The role of the International Council of Scientific Unions in Biodiversity and Global change. In: Solbrig, O.T., van Emden, H.M. & van Oordt, P.G.W.J. (eds.) *Biodiversity and Global Change*, pp. 13–20. CAB International, Wallingford.

Lawton, J.H. (1997) The science and non-science of conservation biology. *Oikos* **79**, 3–5.

Levin, S.A. (1992) The problem of pattern and scale in ecology. *Ecology* **73**, 1943–1967.

Likens, G.E. (ed.) (1992) *The Ecosystem Approach: its Use and Abuse*. Ecology Institute, Oldendorf/Luhe, Germany.

Manning, A.D., Lindenmayer, D.B. & Nix, H.A. (2004) Continua and *Umwelt*: novel perspectives on viewing landscapes. *Oikos* **104**, 621–628.

Manning, A.D., Fischer, J. & Lindenmayer, D.B. (2006b) Scattered trees are keystone structures – implications for conservation. *Biological Conservation* **206**, 311–321.

Manning, A.D., Lindenmayer, D.B. & Fischer, J. (2006b) Stretch-goals and backcasting: approaches for overcoming barriers to large-scale ecological restoration. *Restoration Ecology* **14**, 487–492.

Mayer, A.L. & Rietkerk, M. (2004) The dynamic regime concept for ecosystem management and restoration. *BioScience* **54**, 1013–1020.

Millennium Ecosystem Assessment (2005) *Ecosystems and Human Well-Being: Synthesis Report*. Island Press, Washington, DC.

Mooney, H.A., Canadell, J., Chapin F.S. III *et al.* (1999) Ecosystem physiology responses to global change. In: Walker, B.H. & Steffen, W.L. (eds.) *The Terrestrial Biosphere and Global Change. Implications for Natural and Managed Ecosystems*, 141–149. Cambridge University Press, Cambridge.

Morgan, J.A., Pataki, D.E., Körner, C. *et al.* (2004) Water relations in grassland and desert ecosystems exposed to elevated atmospheric CO_2. *Oecologia* **140**, 11–25.

Naiman, R.J., Johnston, C.A. & Kelley, J.C. (1988) Alteration of North American streams by beaver. *BioScience* **38**, 753–762.

Naiman, R.J., Pinay, G., Johnston, C.A. & Pastor, J. (1994) Beaver influences on the long-term biogeochemical characteristics of boreal forest drainage networks. *Ecology* **75**, 905–921.

Noss, R.F. & Cooperrider, A.Y. (1994) *Saving Nature's Legacy: Protecting and Restoring Biodiversity*. Island Press, Washington, DC.

Pace, M.L., Cole, J.J., Carpenter, S.R. & Kitchell, J.F. (1999) Trophic cascades revealed in diverse ecosystems. *Trends in Ecology and Evolution*, **14**(12), 483–488.

Paine, R.T. (1966) Food web complexity and species diversity. *American Naturalist* **100**, 65–75.

Paine, R.T. (1980) Food webs: linkage, interaction strength and community infrastructure. *Journal of Animal Ecology* **49**, 667–685.

Parmesan, C. & Yohe, G. (2003) A globally coherent fingerprint of climate change impacts across natural systems. *Nature* **421**, 37–42.

Peters, H.A., Cleland, E.E. & Field, C.B. (2006) Herbivore control of annual grassland composition in current and future environment. *Ecology Letters* **9**, 86–98.

Peters, R.L. (1990) Effects of global warming on forests. *Forest Ecology and Management* **35**, 13–33.

Peterson, G., Allen, C.R. & Holling, C.S. (1998) Ecological resilience, biodiversity and scale. *Ecosystems* **1**, 6–18.

Pimm, S.L., Russell, G.J., Gittleman, J.L. & Brooks, T.M. (1995) The future of biodiversity. *Science* **269**, 347–350.

Post, E., Peterson, R.O., Stenseth, N.C. & McLaren, B.E. (1999) Ecosystem consequences of wolf behavioural response to climate. *Nature* **401**, 905–907.

Power, M.E., Tilman, D., Estes, J.A. *et al.* (1996) Challenges in the quest for keystones. *BioScience* **46**, 609–620.

Root, T.L., Price, J.T., Hall, K.R., Schneider, S.H., Rosenzweig, C. & Pounds, J.A. (2003) Fingerprints of global warming on wild animals and plants. *Nature* **421**, 57–60.

Sala, O.E., Chapin, F.S., III, Armesto, J.J. *et al.* (2000) Global biodiversity scenarios for the year 2010. *Science* **287**, 1770–1774.

Scheffer, M., Carpenter, S., Foley, J.A., Folke, C. & Walker, B. (2001) Catastrophic shifts in ecosystems. *Nature* **413**, 591–596.

Schmitz, O.J., Post, E., Burns, C.E. & Johnston, K.M. (2003) Ecosystem responses to global climate change: moving beyond color mapping. *BioScience* **53**, 1199–1205.

Scholes, R.J., Schulze, E-D., Pitelka, L.F. & Hall, D.O. (1999) Biogeochemistry of terrestrial ecosystems. In: Walker, B.H. & Steffen, W.L. (eds.) *The Terrestrial Biosphere and Global Change. Implications for Natural and Managed Ecosystems*, pp. 271–303. Cambridge University Press, Cambridge.

Sinclair, A.R.E. & Byrom, A.E. (2006) Understanding ecosystem dynamics for conservation of biota. *Journal of Animal Ecology* **75**, 64–79.

Smith, M.E., Driscoll, C.T., Wyskowski, B.J., Brooks, C.M. & Cosentini, C.C. (1991) Modification of stream ecosystem structure and function by beaver (*Castor canadensis*) in the Adirondack Mountains, New York. *Canadian Journal of Zoology* **69**, 55–61.

Sprugel, D.G. (1991) Disturbance, equilibrium, and environmental variability: What is 'natural' vegetation in a changing environment? *Biological Conservation* **58**, 1–18.

Steffen, W., Sanderson, A., Tyson, P. *et al.* (2004) *Global Change and the Earth System.* Springer-Verlag, Berlin.

Thomas, C.D., Cameron, A., Green, R.E. *et al.* (2004) Extinction risk from climate change. *Nature* **427**, 145–148.

Thuiller, W. (2004) Patterns and uncertainties of species' range shifts under climate change. *Global Change Biology* **10**, 2020–2027.

Turner, M.G., Collins, S.L., Lugo, A.E., Magnuson, J.J., Rupp, T.S. & Swanson, F.J. (2003) Disturbance dynamics and ecological response: the contribution of long-term ecological research. *BioScience* **53**, 46–56.

Vitousek, P.M. (1994) Beyond global warming: ecology and global change. *Ecology* **75**, 1861–1876.

Vitousek, P.M., Mooney, H.A., Lubchenco, J. & Melillo, J.M. (1997) Human domination of Earth's ecosystems. *Science* **277**, 494–499.

Warren, M.S., Hill, J.K., Thomas, J.A. *et al.* (2001) Rapid responses of British butterflies to opposing forces of climate and habitat change. *Nature* **414**, 65–69.

Weltzin, J.F., Loik, M.E., Schwinning, S. *et al.* (2003) Assessing the response of terrestrial ecosystems to potential changes in precipitation. *BioScience* **53**, 941–952.

Wilcove, D.S. & Blair, R.B. (1995) The ecosystem management bandwagon. *Trends in Ecology and Evolution* **10**(8), 345.

31

The Costs of Losing and of Restoring Ecosystem Services

H.A. Mooney

Abstract

In the practice of conservation restoration or rehabilitation greater attention to valuation of ecosystem services can be important in guiding operational strategies. Further, in considering management options the rapidly changing nature of the physical and biological realms has to be considered.

Keywords: conservation; ecosystem services; restoration; valuation.

Introduction

The biotic world is undergoing enormous change. The direct effects of humans on organisms on the land and in the oceans have been massive and increasingly well documented. Human populations continue to increase. Before they stabilize there will be demands for the utilization of even more natural resources to support them, including food and water. This will result in the further disruption of landscapes and of the 'plumbing' of nations and regions with accompanying biotic degradation. These are the direct effects. Adding to these impacts are the growing indirect effects of greenhouse gas emissions on climate change and soon the rising of oceanic water levels. Agricultural activities are resulting in major disruptions of the nitrogen and phosphorus cycles with increasing off-site effects. Acceleration in global trade is fuelling the movement of biological material around the world with accompanying invasive species. All of these changes are resulting in a new biotic world. This is difficult to manage because of the rapid pace of change. Putting a fence around a piece of land will no longer by itself be an adequate approach to conserving the status quo. Restoring a piece of landscape to a given 'target' will become increasingly difficult because the target will be a moving one due to the continuously changing nature of the forces structuring the ecosystems.

Putting it back together piece by piece: The difficulties of restoration

This volume considers not only the rules for best conserving biodiversity from a landscape perspective but also how to conserve or maintain ecosystem functioning of landscapes. There are, of course, important and significant efforts at restoring systems that have been degraded. The science of restoration ecology is gaining considerable momentum, fuelled in part by the requirements of legislation regarding mitigation of environmental impacts of development. Through time the concepts of targets for restoration – that is, to bring back what was previously extant in a given location – have been developed. Although the field is very active there has been a disappointing post-project analysis of what really works and how effective the efforts have been in attaining the original goals. Bernhardt *et al.* (2005) note that only 10% of the 3700 projects they reviewed actually evaluated their effectiveness in achieving pre-project goals.

The difficulty of achieving restoration goals has been well demonstrated by the careful and long-term studies of Joy Zedler and her students in marsh restoration. In a summary of their work, Zedler and Callaway (1999) noted that 'compensation' sites may never fully replace natural wetland functions and that post-project monitoring projects are generally too short to determine what the final outcome of any particular restoration efforts will be. Their studies (Zedler *et al.* 2001) also show the complexity of striving to attain maximum species diversity in their marsh restoration projects and how a single species with unique properties – unusual sequestration of nitrogen – can control the plant diversity of the whole system. These results support those of others in trying to understand the biodiversity – ecosystem functioning relationship, showing that not all species have an equally dramatic impact on system functioning. The relevance of their results to restoration attempts is to show that a great deal of basic biology is needed in order to achieve particular target structure–function relationships. More importantly, these studies, as well as many others, demonstrate that keeping crucial service-delivery systems intact to begin with is a lot more cost effective than trying to put them, or some reasonable facsimile of them, back together again. The lesson we should draw from these efforts is 'protect what works – it is difficult to replace'.

Related to this lesson is the fact that there can be considerable economic loss as the result of the destruction of natural systems. The following example illustrates this in a powerful way.

The economic costs of losing ecological services

Tsunamis and mangroves

The disastrous tsunami of 26 December 2004 killed 200,000 people, made 2 million homeless and caused economic losses of 6 billion dollars (Kathiresan & Rajendran 2005). Was the extensive decimation of mangrove forests in the previous decades responsible for some of these losses? A number of studies indicate that this was indeed the case, although the analyses are difficult because of variability in topography, wave height and the condition of the protecting vegetation. However, models do show a dramatic effect of tree vegetation in reducing the energy of waves (Danielsen *et al.* 2005).

Mangroves provide a host of ecosystem services, as enumerated by Chong (2005) and illustrated in Box 31.1. Why is it then that we have been losing mangrove forests at such an extraordinary rate? Agardy *et al.* (2005) calculate that presently mangrove forests cover 166–181 $\times$ 10^3 km^2, and are primarily found in Asian coastal systems. Of those regions where good data are available on multiyear fluctuations in coverage (just over half of the total area), 35% of the mangroves have been lost in the past two decades, representing a rate of 2834 km^2/year. Most of this loss (52%) is due to aquaculture activities; with shrimp farming representing 38% of the total mangrove loss. In some countries more than 80% of the mangrove forests have been lost. Why?

A particularly revealing case of the interplay between mangrove development and ecosystem service loss has been documented by Barbier and coworkers (Barbier & Cox 2002), who analysed the factors driving shrimp farming in Thailand. Between 1975 and 1993 Thailand lost nearly half of its mangrove forests, mainly as a consequence of the rapid increase in shrimp

Box 31.1 Mangrove ecosystem services (from Chong 2005)

- Shoreline stabilization
- Storm protection
- Water quality
- Microclimate regulation
- Groundwater recharge and discharge
- Flood and flow control
- Sediment and nutrient retention
- Habitat protection and biodiversity
- Biomass, productivity and resilience
- Gene bank
- Recreation, tourism and culture
- Hunting and fishing
- Forestry products
- Water transport

farms. The total export market for shrimp in the late 1990s was 1–2 billion US dollars. This resulted from the easy access to lucrative global markets from the port of Bangkok, the high economic return on shrimp due to low labour costs, inexpensive feed, availability of government-controlled land and outside capital for this market. However, the lack of government incentives did not facilitate sustainable practices.

Sathirathai and Barbier (2001) calculate that the value of mangroves to the local community ranges from 27,264 to 35,921 US dollars per hectare over a 20-year period per hectare, considering a 75 metre coastine, for replacement costs of mangroves or coastal protection. This includes the value of timber and non-timber products of $88/ha, $20–69/ha for offshore fishery linkages and the very large sum of $3680 for coastline protection. This analysis does not take into account other values, such as tourism, carbon fixation and non-use values. And yet when the economic benefits from shrimp farming are calculated the returns are only $194–209/ha when such factors as shrimp farm abandonment is included. Clearly, mining the mangrove habitat for shrimp farming is not economical when both public and private goods are taken into account.

These events represent a case of buying ecosystem services at an undervalued rate (private investment) in relation to the true value that the public could enjoy.

An addendum lesson that we can draw from this second example is 'protect what works since not only is it difficult to replace but also considerable economic losses can be incurred with its destruction.'

The economic costs of restoring services

The above example shows what economic losses may be incurred by the loss of a system. What about the costs of attempting to replace the functioning of an ecosystem that has been lost? To illustrate this point I use some very large-scale and recent examples of rehabilitation attempts.

Mississippi watershed

The Mississippi river watershed can be considered an ecological disaster area. It has lost its capacity for flood control and nutrient retention, and at its mouth the loss of wetlands of the delta has reduced the capacity for storm protection. Since the watershed covers such a vast area these losses in services represent considerable financial losses. The basin drains 41% of the contiguous part of the USA, as well as parts of Canada, and covers more than 3,225,000 km^2. It is the third largest river basin in the world.

With the environmental costs of these ecosystem service losses, plans are being put forward to rehabilitate some of the services lost. One of the problems is that the source of sediment that feeds the delta, and which kept up with subsidence, has been cut off by a series of dams. Since the 1700s the sediment load has been halved, with the greatest decrease occurring after 1950 with the construction of large reservoirs upstream (Sparks 2006). Now there is a controversy about whether it is sensible to try to restore the levees and save New Orleans and whether the city should be moved to a more secure position. The political pressures, however, are such that the levees will be strengthened, and at a cost that is not trivial. The US government has agreed to spend 3.1 billion dollars in strengthening the levees. This is putting a 'band-aid' on the problem. The whole Mississippi river basin needs replumbing. There has been a delta wetland restoration project on the books for years and a two billion dollar effort is underway. It is estimated that it would take 14 billion dollars in total to complete the job, which includes reconnecting the river to its historic delta, removing some canals (more than 90% of the lower Mississippi river has been leveed and drained; Sparks 2006) and putting dredging spoils in localities that would let the storm-protective wetlands and barrier islands re-emerge from the

Gulf waters. There has been concern about the siltation of the flood control reservoirs upstream. Apparently, plans for sediment bypass structures around the dams were abandoned as they were considered to be not cost effective in terms of the savings in water-storage capacity. However, the disastrous effects of the loss of sediments to the delta were not taken into account (Sparks 2006).

Then there is the problem of the nitrate runoff from the agricultural fields of the upper reaches of the Mississippi that is the source of the dead zone problem in the Gulf of Mexico. In 2002 (Rabalais *et al.* 2002) the hypoxic zone was about 21,000 km^2. Seventy-four percent of the nitrate delivered to the Gulf is from agricultural nonpoint sources.

A government task force has been studying the problem for some time and issued their first action plan in 2001 (http://www.epa.gov/msbasin/taskforce/pdf/actionplan.pdf). This plan called for voluntary measures and depends on existing programmes to accomplish its ambitious goals of reducing the dead zone by two-thirds by the year 2015 while at the same time improving economic conditions throughout the Basin.

A new assessment is now beginning to determine how the action plan has made a difference in the reduction in the extent of the dead zone. Considerable research has been underway for ecological engineering solutions to the problem. Mitsch *et al.* (2001), in a comprehensive analysis of the sources of nitrogen entering the Mississippi river basin, calculate that 31% comes from fertilizer. Total inputs to the basin amount to $20,979 \times 10^3$ t/year. Thus, solutions to the dead zone lie in part with more efficient agricultural practices that reduce losses to the watershed. They calculate that improved agricultual practices could reduce nitrate input by between 1900 and 2400×10^3 t/year. Creating and restoring wetlands and riparian buffers could result in another $600–1600 \times 10^3$ t/year nitrogen reduction. However, the latter would require the designation of between 100,000 and 253,000 km^2 for wetlands and riparian buffers.

Florida Everglades

Another example of attempts to bring back the ecological services that have been lost through development is the Florida Everglades. In a recent book Grunwald (2006) notes that 'half of the original Everglades has disappeared, the remainder is slowly dying and the pressures of population and development in Florida continue unabated. Good intentions, and lots of government money, may not be enough'.

This is a pretty bleak prognosis. At the same time it is remarkable that the government is dedicating some 8.3 billion dollars to a restoration programme of unprecedented scope for what remains of the system (Sklar *et al.* 2005). Most of this money will be spent on land acquisition and replumbing, with much less on other ecological issues. The challenges of making this programme work are very large because of the size and nature of the problems that need to be solved.

What happened? Well, the 'swamp' began to be drained for agriculture over a century ago with increasing efforts in recent times. Canals were built and waters diverted away from the Everglades. Thus the plumbing of this unique system was massively altered. The lands newly released for agriculture were fertilized leading to off-site impacts of nutrient enrichment, particularly phosphorus. Subsidence of soils of over 2 m in the agricultural areas exacerbated the alteration of water flow. The goals of the restoration programme are to restore the region's hydrological system and reduce nutrient enrichment. If this weren't a big enough challenge, there is the issue of dealing with new invasive species, including native plants (*Typha*), as well as exotics such as *Melaleuca quinquenervia*, which have altered biogeochemistry and food webs.

Sklar *et al.* (2005) note the great challenges for the restoration project. Each of the proposed restoration projects must be justified in terms of both ecological and economic benefits before they can start. Dealing with the uncertainties of outcomes becomes 'magnified by conflicting interpretations, non-linear feedback mechanisms, slow response times and lack of data.' So, we broke it, but fixing it is a massive challenge for our knowledge base of what approaches will work best. In these cases we call for adaptive management. But in these systems the response times are slow and the result of a given alteration takes a long time to perceive. On top of this, the public attention span is limited. Nonetheless, there is a non-partisan public recognition of the problem, and the dedicated resources to work on the problem. Now the science of restoration ecology and engineering has to meet the challenge. But the challenge is not only fixing it, but doing so with imperfect knowledge as well as the reality of increasing changes in the drivers of ecosystem processes as noted below.

The Mississippi and Everglades examples give us an additional addendum to the rule, which now reads 'protect what works because not only is it difficult to replace but also considerable economic losses can be incurred from its destruction as well as in attempts at restoration.'

Beyond conservation

Unfortunately, we do not usually have the option of saving intact systems that have had a long evolutionary history. The massive impacts of humans on the landscape have produced large regions of novel ecosystems. The rate of direct human-driven disruption of natural systems is increasing (Millennium Ecosystem Assessment 2005) with an increased presence of short-lifespan species. The indirect effects of human activities on ecosystems are also very large and are increasing. These include the now easily discernible impacts of climate change on Arctic ecosystems with strong positive feedbacks to the climate system (Chapin *et al.* 2005). We are seeing a biotic world where the local players are changing rapidly, not only due to impacts of disturbance, but also due to the global biotic homogenization that is occurring (Mooney & Hobbs 2000). Landscapes are being impacted by toxic chemicals as well as by increased nutrient deposition as our food production systems increase with the increasing population. We are seeing more cases of ecological regime shifts where increasingly stressed ecosystems cross resilience thresholds (Millennium Assessment 2005; see also Chapter 34). The hydrology of landscapes is continually being altered as the search for fresh water intensifies, and coastal areas are facing the threat of sea level rise; all these changes are having major impacts on ecosystem structure and functioning.

Management of landscapes will entail not only preservation, or restoration or even adaptive management, but also perhaps a move towards more design and engineering of landscapes. If so, what sorts of design principles will we utilize? These of course will depend on specific objectives, but some general principles might also be considered. They will, of necessity, depend on our understanding of how ecosystems operate because we need to work within a system context. But, we will have to go one step further. We need to understand how ecosystem processes deliver ecosystem services. This is because, increasingly, efforts to conserve or rebuild ecosystem processes will have to be done in the context of the benefits delivered to society.

Rebuilding ecosystems and their services

Ecologists have a large task ahead in quantifying the links between ecosystem functioning and services on the one hand and between ecosystem services and well-being on the other. Part of the task of the former is unpacking ecosystems into components and showing their relationships to services. The Millennium Assessment provides a beginning for this task by relating functioning to diversity (Fig. 31.1) (Diaz *et al.* 2005). As ecologists have long noted, not all organisms within an ecosystem have equal impact in processing materials. Dominant plant species have a controlling impact on carbon fixation and water transfer,

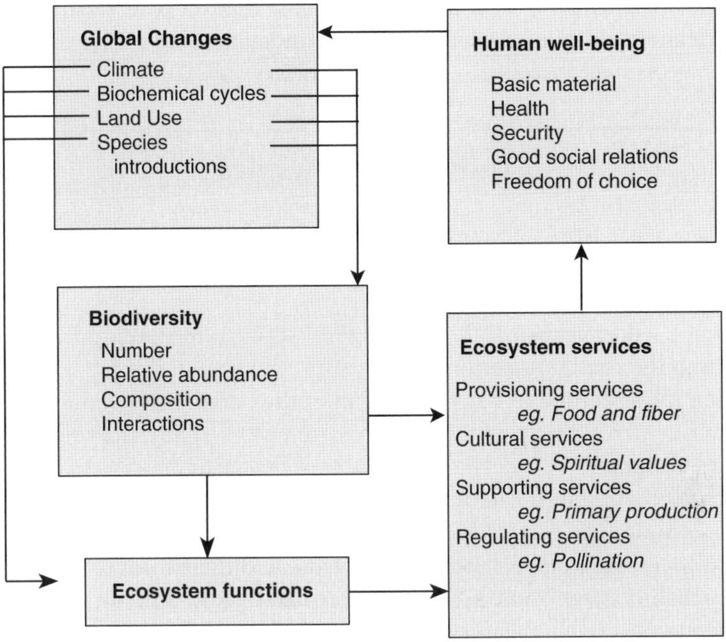

Figure 31.1 **The interrelationships between biodiversity, ecosystem functioning and ecosystem services. Also indicated are the driving forces of change and the benefits that society derives from ecosystem services. (Modified from Diaz *et al.* 2005.)**

for example. Yet, minor species can also have large impacts on ecosystem structure if they serve a keystone role. So number of species alone is not necessarily a good indicator of the capacity of the resident organisms to process the available resources, although species richness of similar functional types is related to resource acquisition and processing capacity. Thus, in Fig. 31.1 the number of species is noted as being important in providing services, but the abundance of the various species, their functional composition and their interactions are also important.

Biodiversity represents more than species richness, as the Convention on Biological Diversity acknowledges. It also represents genes, populations, ecosystems and landscapes, where all of the diversity metrics apply, such as numbers of kinds, abundance, composition and interactions. Biodiversity represents an analytical challenge of relating this diversity to functioning and hence to services and especially to efforts at restoration and rehabilitation. This also represents a challenge to land managers. Simple metrics of diversity, such as species richness, do not necessarily yield an understanding of the capacity of any given ecosystem to deliver services that society values.

> **Landscape principles**
>
> 1 Protect what works because not only is it difficult to replace but also considerable economic losses can be incurred from its destruction as well as in attempts at restoration.
> 2 Do not sell low when trading ecosystem services in development, e.g. shrimp for coastal protection.
> 3 Quantify, as completely as possible, all of the benefits derived from the services provided by any given ecosystem for which alterations are being proposed.
> 4 When rehabilitating, adopt an ecosystem approach in design that incorporates the ecosystem service paradigm and considers the range of services that can be restored, their value and their comparative costs for rehabilitation. Some services are very costly and difficult to replace whereas other valuable services may be less expensive and more likely to become self-sustaining.
> 5 In either conservation or restoration or rehabilitation, prepare for a very uncertain future:
>
> - a world of rapid change in climate and vegetation;
> - a world of increasingly extreme events;
> - a world of weeds and diseases;
> - a world of regime shifts;
> - a world of rising seas and an acidifying ocean;
> - a world that has been diced and replumbed;
> - a world of increasing nitrogen and phosphorus redistribution.

References

Agardy, T., Alder, J., Dayton, P. et al. (2005) Coastal systems. In: Hassan, R., Scholes, R. & Ash, N. (eds.) *Ecosystems and Human Well-Being. Current State and Trends*, vol. 1 pp. 515–549. Island Press, Washington, DC.

Barbier, E. & Cox, M. (2002) Economic and demographic factors affecting mangrove loss in the coastal provinces of Thailand, 1979–1996. *Ambio* **31**, 351–357.

Bernhardt, E.S., Palmer, M.A., Allan, J.D. *et al.* (2005) Ecology: synthesizing US river restoration efforts. *Science* **308**, 636–657.

Chapin, F.S., Sturm, M., Serreze, M.C. *et al.* (2005) Role of land-surface changes in Arctic summer warming. *Science* **310**, 657–660.

Chong, J. (2005) *Protective Values of Mangrove and Coral Ecosystems: A Review of Methods and Evidence.* IUCN, Gland, Switzerland.

Danielsen, F., Sorensen, M.K., Olwig, M.F. *et al.* (2005) The Asian tsunami: A protective role for coastal vegetation. *Science* **310**, 643.

Diaz, S., Tilman, D., Fargione, J. *et al.* (2005) Biodiversity regulation of ecosystem services. In: Hassan, R., Scholes, R. & Ash, N. (eds.) *Millennium Ecosystem Assessment. Ecosystems and Human Well-being. Current State and Trends*, pp. 297–329. Island Press, Washington, DC.

Grunwald, M. (2006) *The Swamp: The Everglades, Florida, and the Politics of Paradise.* Simon & Schuster, New York.

Kathiresan, K. & Rajendran, N. (2005) Coastal mangrove forests mitigated tsunami. *Estuarine, Coastal and Shelf Science* **65**, 601–606.

Millennium Ecosystem Assessment (2005) *Ecosystems and Human Well-Being. Synthesis.* Island Press, Washington, DC.

Mitsch, W.J., Day, J.W. Gilliam, J.W. *et al.* (2001) Reducing nitrogen loading to the Gulf of Mexico from the Mississippi River Basin: Strategies to counter a persistent ecological problem. *Bioscience* **51**, 373–388.

Mooney, H.A. & Hobbs, R.J. (2000) *Invasive Species in a Changing World.* Island Press, Washington, DC.

Rabalais, N.N., Turner, R.E. & Wiseman, W.J. (2002) Gulf of Mexico hypoxia, A.K.A. 'the dead zone'. *Annual Review of Ecology and Systematics* **33**, 235–263.

Sathirathai, S. & Barbier, E.B. (2001) Valuing mangrove conservation in Southern Thailand. *Contemporary Economic Policy* **19**, 109–122.

Sklar, F.H., Chimney, M.J., Newman, S.P. *et al.* (2005) The ecological-societal underpinnings of Everglades restoration. *Frontiers in Ecology and the Environment* **3**, 161–169.

Sparks, M.E. (2006) Rethinking, then rebuilding New Orleans. *Issues in Science and Technology* **22**, 33–39.

Zedler, J.B. & Callaway, J.C. (1999) Tracking wetland restoration: Do mitigation sites follow desired trajectories? *Restoration Ecology* **7**, 79–84.

Zedler, J.B., Callaway, J.C. & Sullivan, G. (2001) Declining biodiversity: Why species matter and how their functions might be restored in Californian tidal marshes. *Bioscience* **51**, 1005–1017.

32

Managing Disturbance Across Scales: An Essential Consideration for Landscape Management and Design

Viki A. Cramer

Abstract

Landscape management has historically aimed to maintain ecosystems in a largely static, 'desirable' state through minimizing moderate levels of change. Yet ecosystems are now recognized as complex and non-linear systems, and our mangement of them should aim to retain crucial types and ranges of natural variation in the system. One of the most important sources of variation and change in ecosystems is disturbance. Disturbance occurs across all scales, from small and frequent to large and rare events, and the complementarity or redundancy of species' responses to these events indicates stability in the ecosystem. Yet the management of disturbance in the landscape is rarely a purely ecological problem. Ecological disturbances, particularly those that are large and infrequent, have broader social and economic repercussions that also must be taken into consideration in landscape design. Increased policy sophistication and institutional integration, fostered through greater engagement between research, policy and institutional settings, is essential to the successful management, both ecologically and socially, of disturbance in the landscape.

Keywords: complex systems; ecological connectivity; ecosystem processes; wicked problems.

Introduction

The essential constant of ecosystems is change (Levin 1999), yet the essential paradox of ecosystem management is that we seek to preserve that which must change (Pickett & White 1985). While theoretical ecology increasingly recognizes the complex and non-linear nature of ecosystems (Levin 1998; Drake *et al.* 1999; Schulze 2000), the practical management of ecosystems is still largely grounded in the 'command-and-control' approach that assumes problems are well-bounded, clearly defined and generally linear (Holling & Meffe 1996). Command-and-control approaches to ecosystem management frequently result in a reduction in the range of natural variation in systems, to the benefit of minimizing moderate change but at the cost of losing resilience to catastrophic change (Holling & Meffe 1996). In heavily modified and managed systems, such as agriculture and forestry, the simplified structure and reduction of variation imposed exogenously by humans reduces the adaptive responses of the system, making these systems fragile and vulnerable to single stresses (Holling & Meffe 1996; Levin 1998).

Callicott *et al.* (1999) have identified two schools of conservation thought, compositionalism and functionalism. Compositionalists perceive the world as essentially entity-oriented, with an approach to ecology that begins with organisms aggregated into populations. Functionalists perceive the world as essentially process-oriented, with an approach to ecology that begins with solar energy coursing through a physical system that includes but is not limited to the biota. Conservation efforts have generally been structured around the compositionalist world view, with reserves traditionally set aside based on their community-level characteristics (Noss 1990). Yet it is ecosystem processes that are the key players in structuring and maintaining ecosystems: all terrestrial ecosystems are controlled and organized by a small set of key plant, animal and abiotic processes (Holling 1992). The functionalist emphasis on process is perhaps the philosophy better suited to designing and managing landscapes and vegetation in areas that are inhabited and exploited by humans (Callicott *et al.* 1999), with the proviso that both biological and ecological integrity is maintained (Noss 1990).

In this chapter I focus on disturbance as an essential ecosystem process to be addressed in landscape management and design. Disturbance occurs at a variety of spatial and temporal scales, from small and infrequent to rare and large, and thus questions of scale are integral to considerations of the importance of disturbance as an ecosystem process. As we are concerned with managing and designing landscapes – a uniquely human activity – then we must also consider how these questions of scale and disturbance affect the management of ecosystem processes in the socioecological system.

Ecological scales, human fluxes and ecosystem processes

A key property of complex systems is emergence, whereby properties of the ecosystem at large spatial scales result from feedback processes between components occurring at smaller spatial scales (Levin 1998; van de Koppel *et al.* 2005). Larger-scale processes form the context for local dynamics (Lertzman & Fall 1998). Small and fast processes are nested within larger, slower ones in a dynamic hierarchy (Holling 2001). The multiplicity of spatial and temporal scales over which ecosystem processes operate means that there is not one 'correct' scale at which ecosystem processes should be managed (Christensen *et al.* 1996; Lindenmayer & Franklin 2002; Berkes 2004). Yet the scale at which change occurs will determine how that change is evaluated (Hull *et al.* 2002).

Holling (1992) has identified three scale ranges over which a broad class of processes dominate: the microscale, dominated by vegetation processes; the mesoscale, dominated by disturbance and environmental processes; and the macroscale, dominated by geomorphological and evolutionary processes. Mesoscale processes transfer local events into large-scale consequences, and are both abiotic (water, fire, wind) and biotic (dispersal, migration). Maintaining mesoscale processes will maintain landscape connectivity over larger spatial and temporal scales (Soulé *et al.* 2004). This is also the scale at which humans transform and manage landscapes. In Australia, for example, management activities are currently funded and conducted at the catchment (watershed) level. The goals of management are largely similar across a diversity of catchments: (i) conserve biodiversity through the conservation and revegetation of native vegetation and the management of pest animals and invasive weeds; (ii) improve water quality through water management, river and wetland rehabilitation; and (iii) soil conservation. Mostly these management activities operate around a compositionalist philosophy, although processes such as dispersal and migration are addressed by aiming to create corridors between patches of remnant vegetation. As yet, there is little exploration of the idea of managing the 'matrix' as a location where processes such as dispersal and migration can occur (Lindenmayer & Franklin 2002; Manning *et al.* 2004; Soulé *et al.* 2004). Scant emphasis on vegetation outside of high-quality remnants not only neglects the importance of the entire landscape mosaic as habitat (see Chapters 18–20), but also neglects the potential role of lesser quality or 'production' vegetation in maintaining the connectivity of ecosystem processes across all scales.

The fast rate and large spatial extent of land clearing associated with the introduction of industrialized agriculture introduces 'human flux' into the 'natural world' (Pickett & Ostfeld 1995). The introduction of human fluxes into a system generally brings a host of other dimensions – economic, scientific, social and political – beyond the purely ecological ones, creating 'wicked' problems (Rittel & Webber 1973) of ecosystem management. In this socio-ecological context, managing and designing landscapes for conservation is not simply a matter of managing biodiversity and ecosystem processes, but may also involve local and national economics, world trade markets, public policy, political pressures and community expectations. Disturbance is one of the most important mesoscale processes to manage for ecosystem resilience, yet the social and economic impacts of major ecological disturbances such as fire and floods mean that these can rarely be managed as purely ecological problems. Below I discuss the importance of disturbance at small and large scales, and explore some of the opportunities and problems that exist in reinstating such disturbances as essential ecosystem processes.

Disturbance

Disturbances are ubiquitous, inherent and unavoidable. The presence of disturbances across a wide range of spatial and temporal scales and their continuity across all levels of ecological organization is the essence of their importance (White & Jentsch 2001). Disturbance causes change in ecosystems, resulting in both complementarity and redundancy in species' responses that infer ecosystem stability (resilience). In both highly fragmented and largely intact landscapes, disturbances have been removed or dramatically altered in terms of frequency and intensity either directly, as a consequence of the command-and-control approach to ecosystem management, or indirectly, through species extinctions and changes in vegetation (Yates *et al.* 1994).

Small and frequent: Bettongs and bison

Holling (1992) considers the maintenance of vegetation processes that determine plant growth, plant form and soil structure to be the ecosystem processes that structure landscapes at the microscale. While the role of the fauna in pollination, seed dispersal, and nutrient distribution and recycling is widely appreciated, its contribution to the structure of soil and vegetation at the microscale is perhaps less so. Small to medium-sized mammals have largely disappeared from much of the Australian landscape, while it is the large mammals that have been lost from the North American landscape. Yet despite their differences in size, the removal of these mammals from their respective landscapes has had similar implications for the heterogeneity of soil structure, vegetation recruitment and plant diversity at small spatial scales.

Small mammals play an important role in woodland regeneration in southern Australia. While direct activities, such as the seed-caching of brush-tailed bettongs (*Bettongia penicillata*), are crucial to the seed dispersal and regeneration of species with large seeds (Murphy *et al.* 2005), other activities of small mammals may indirectly influence plant regeneration. Sandy soils in southern Australia are commonly water repellent, with a thin, intensely repellent surface layer acting as a barrier to an underlying wettable soil (Garkaklis *et al.* 1998). The diggings of small mammals that feed on hypogeous fungi have lower water repellency than surrounding undisturbed soils (Garkaklis *et al.* 1998), and small mammals were likely to have played an important role in providing microsites with greater water infiltration for seedling establishment. These mammals have all but disappeared over much of their range because of habitat loss and the introduction of feral predators such as European foxes and domestic cats. The loss of small mammals may be a contributing factor to the poor regeneration of woodland trees in southern Australia (Saunders *et al.* 2003). There is no obvious way to reintroduce the small and continuous disturbance to the soil that was achieved by small mammals without reintroducing the small mammals themselves. The reintroduction of small mammals requires dedicated effort to implement breeding programmes and feral animal control, an activity not usually directly linked with the management of vegetation processes.

The distinct grazing pattern of North American bison (*Bos bison*) is a major determinant of the structure of tallgrass prairie plant communities (Hartnett *et al.* 1996; Knapp *et al.* 1999). Preferential grazing of the dominant grasses by bison significantly alters competitive interactions between dominant grasses and subordinate species, leading to increased plant species diversity in grazed areas. Moreover, the wallowing behaviour of bison introduces further spatial heterogeneity and increased plant diversity into prairie ecosystems at both local and regional scales. Bison wallows are small (3–5 m in diameter and 10–30 cm in depth) disturbances that occur when bison paw the ground and roll in exposed soil (Collins & Barber 1985). The hardpan bottom that forms in the wallows often retains moisture in the spring, creating localized habitats that are suitable for ephemeral wetland species, yet in summer the wallows only support plants that can tolerate severe drought (Knapp *et al.* 1999). The vast number of bison that once roamed the Great Plains (30–60 million) meant that these soil depressions were probably abundant and widespread features of the landscape (Knapp *et al.* 1999). Domestic cattle do not display such wallowing behaviour, nor do they show the same grazing preferences (Hartnett *et al.* 1996), and thus the introduction of cattle to these landscapes is unlikely to replicate the ecological role of bison.

Can we overlook such microscale disturbances in either landscape design or the management of vegetation? While it may be the mesoscale processes that maintain landscape connectivity over larger scales (Soulé *et al.* 2004), the loss of microscale processes is likely to lead to some losses in the emergent properties of ecosystems. Although we may not be able to reintroduce these disturbances when implementing a particular landscape conservation strategy, it is important to consider how the absence of such small-scale processes will affect the long-term success of plant regeneration and the maintenance of plant diversity at local and regional scales.

Large and infrequent: Floods

Floodplains experience a highly dynamic disturbance regime that determines the successional trajectory and extent and composition of floodplain vegetation (Johnson 1994). Flow regimes determine the spatial and temporal abundance of various patch types within a riparian mosaic (Scott *et al.* 1997; Richter & Richter 2000). The alteration of flow regimes through river regulation has caused major changes in floodplain vegetation, particularly in semiarid and arid regions, where decreases in flood frequency and magnitude often lead to the replacement of early-successional communities with later-successional communities, often dominated by trees (Johnson 1994). The vegetation of the floodplains in southeastern Australia is associated with characteristic flood frequency ranges, but regulation of the larger river systems has decreased the frequency, duration and extent of flooding (Bren 1988, 1992; Jolly *et al.* 1993). The increased summertime levels and reduced winter-spring levels of the River Murray has allowed the invasion of river red gums (*Eucalyptus camaldulensis*) into extensive natural grasslands in the Barwah-Millewa area, perhaps leading to the development of a new equilibrium state of vegetation (Bren 1992). Conversely, regulation of the River Murray has led to decreased tree health in black box (*E. largiflorens*) and river red gum communities on the Chowilla floodplain downstream of the Barwah-Millewa forests (Jolly *et al.* 1993).

In western North America, infrequent large floods that expose moist sandbars and banks during river fall are important for the establishment of cottonwood (*Populus* spp.), an early pioneer species and the dominant tree species along rivers and streams in semiarid and arid regions (Johnson 1994; Scott *et al.* 1997). Cottonwood woodlands provide structural habitat, particularly for bird species, that is not found within later successional communities (Johnson 1994). As with the Australian example above, river regulation on the Missouri (Scott *et al.* 1997) and Yampa (Richter & Richter 2000) rivers has decreased flood frequency, duration and magnitude, leading to declining populations of cottonwoods and reduced cottonwood regeneration. More stable flow regimes also favour the establishment of invasive species, especially *Tamarix chinensis* (Johnson 1994).

The shift from grasslands to woodlands or from pioneer woodlands to mature forest stands is a natural process, and it may be argued that this is not an inherently undesirable ecological progression. In this context there is no 'right' or 'wrong' ecosystem. Yet the loss of the diversity of patch types within floodplain systems will decrease both species and habitat diversity at regional scales. The diversity of patch types can only be maintained through conserving the natural flood phenomena associated with these rivers. While the restoration of large floods on regulated rivers may be socially unacceptable and economically unfeasible (Richter & Richter 2000), the restoration of floods of adequate magnitude, duration and frequency to maintain the ecological integrity of the river is an essential component of landscape management and design. These are likely to prove challenging objectives for landscape management where there are many (and some powerful) competing interests for water resources, and a lack of integration between policies determining river flows and harvesting allocations and the conservation strategies for floodplain wetlands (Kingsford 2000; Nilsson & Berggren 2000).

Complex and wicked: Fire

Fire is a crucial disturbance in many ecosystems. Inappropriate fire regimes are recognized as detrimental to the maintenance of biological diversity, and the management of fire has been the focus of much research (e.g. van Wilgen *et al.* 1992; Bradstock *et al.* 2002; Veblen *et al.* 2003). Yet fire management remains controversial. Managing fire purely for ecological outcomes is a complex problem – when fire must be managed for societal outcomes, such as the protection of life and property, it becomes a wicked problem (Rittel & Webber 1973). The solutions to wicked problems are not 'true or false', but instead 'better', 'worse' or 'good enough', because stakeholders will have differing views of both the problem and acceptable solutions to the problem, and these views may change over time. For example, the size and severity of the 1988 fires in Yellowstone National Park led many people at the time to conclude that the park had been 'destroyed'. Yet Romme and Turner (2004) argue that, based on Yellowstone's previous fire history and plant responses after the fire, the Park's coniferous forest ecosystems were not degraded or significantly altered by the 1988 fire, and that the fire was within the range of natural variation in that ecosystem.

Although the traditional analytical approaches of science may solve ecological questions related to fire as a disturbance, they will not solve the questions of fire management in the landscape. Living with fire will remain a social issue, even where fire management is well founded upon ecological principles (Dombeck *et al.* 2003). An analytical approach may not even solve the ecological questions. Just as the question 'what is natural?' is currently the focus of much thought in ecological restoration (Callicott 2002), it is an equally appropriate question for fire management (as well as that of other disturbances). Do we manage fire to maximize biological diversity, to mimic pre-European burning regimes, or to return ecological conditions to the Pleistocene? The answer may be 'all of the above', as we aim to manage fire for a limited number of objectives within a range of specific contexts (Bowman 2003; Dombeck *et al.* 2003).

Yet the context for fire-adapted ecosystems in the western USA, southern Australia and Mediterranean Europe is largely urban or rural, and public pressure to suppress fire completely has led to wildfires that have been catastrophic socially and are often perceived as catastrophic ecologically (Cheney 2003; Dombeck *et al.* 2003; Gomes 2006). The exclusion of frequent low-intensity fires in the ponderosa pine (*Pinus ponderosa*) forests in southwestern USA has led to the expansion of trees less adapted to frequent fire, increasing pine biomass (both live and dead) and declining herbaceous productivity.

This change in community composition and structure supports a shift from frequent, low-intensity fires to increasingly larger crown fires (Fulé *et al.* 1997). Conversely, the managed reduction of fuel loads through frequent low-intensity fires in eastern Australia has led to diversity losses and local extinctions, particularly of serotinous obligate seeders, through too frequent disturbance (Keith *et al.* 2002).

Although greater knowledge of species' responses to fire is valuable, the problems inherent in fire management have primarily not been a lack of ecological knowledge, but a lack of institutional and policy sophistication in managing a complex ecological process that also has social repercussions (Cheney 2003; Dombeck *et al.* 2003; Handmer 2003; Tarrant 2003). The conflict between achieving ecological goals through variation of the fire regime in both space and time over a range of scales, and the desire for a command-and-control approach to fire management for preserving life and property is not easily resolved (Gill & Bradstock 2003). The practical realities of our ability to 'control' and 'manage' a process as enigmatic as fire are not to be underestimated either. Hence, moves towards focusing on landscape connectivity need to be further broadened to pursue connectivity between scientific and social settings, and between research, policy and institutional settings (Bradshaw & Bekoff 2001).

Conclusions

The fundamental shift occurring in ecological thinking – that ecosystems are no longer perceived as static identities in equilibrium but are complex systems that are dynamic and unpredictable in space and time – is not yet reflected in environmental policy and planning (Wallington *et al.* 2005). Perhaps because ecological science is currently in transition (Holling 1998; Bradshaw & Bekoff 2001), ecologists have not yet been able to communicate effectively, using compelling evidence, either to policymakers or to the public, that change in ecosystems is both inevitable and desirable. It is now 10 years since Holling and Meffe (1996) stated their 'golden rule' of natural resource management: that natural resource management should strive to retain critical types and ranges of natural variation in ecosystems by facilitating existing processes and variabilities rather than changing or controlling them. While we probably have sufficient scientific and technical knowledge to begin to design landscapes that retain the natural variation in some ecosystem processes across all scales (see Chapter 35), the greater difficulty is in facilitating the policy sophistication and institutional integration required to achieve this. In this context, facilitating disturbance (the greatest driver of change in ecosystems) across all scales is a particular challenge for landscape design and vegetation management. As I have illustrated above, the maintenance of some important small-scale disturbances requires broader ecological management, and the maintenance of larger-scale disturbances is often beholden to objectives considered by many in society to be more important; for example, the provision of water for agriculture in the case of river regulation and flooding, and the protection of life and property in the case of fire. As ecologists, we may design landscapes, but we do not manage them, and for design to be translated into management requires that ecologists engage with policymakers and the public (Meffe 1998; Blockstein 1999; Robertson & Hull 2001).

Key principles for landscape design

1. Aim to maintain the connectivity between, and range of variation of, ecosystem processes across all scales. This may be best achieved by focusing on the maintenance of mesoscale processes such as disturbance, dispersal and migration.
2. Assess, design and manage the landscape mosaic as a whole (rather than as patches within a matrix) to improve the connectivity of processes across scales and increase the likelihood of maintaining the emergent properties of ecosystems.
3. Ensure that the organizations responsible for the management of ecosystem processes have the technical, scientific and financial support to manage processes at the correct scale.
4. Plan for and maintain ecological disturbance so it is integral to landscape design across a range of spatial and temporal scales.
5. Where the management of a particular disturbance has potential impacts on broader society, then landscape design must look to more integrative tools and theories (e.g. systems dynamics, complexity theory; see Bradshaw & Bekoff 2001) to achieve realistic design and management goals.
6. The use of integrative theory in landscape design must be accompanied by greater practical integration between research, policy and management institutions, and increased policy and management sophistication, so that design is translated into practice.

Acknowledgements

I wish to extend my warm thanks to David Lindenmayer and Richard Hobbs for the opportunity to contribute to this book, and to Russell Palmer for alerting me to the wallowing behaviour of bison.

References

Berkes, F. (2004) Rethinking community-based conservation. *Conservation Biology* **18**, 621–630.

Blockstein, D.E. (1999) Integrated science for ecosystem management: an achievable imperative. *Conservation Biology* **13**, 682–685.

Bowman, D. (2003) Bushfires: a Darwinian perspective. In: Cary, G., Lindenmayer, D. & Dovers, S. (eds.) *Australia Burning. Fire Ecology, Policy and Management Issues*, pp. 3–14. CSIRO Publishing, Collingwood, Victoria.

Bradshaw, G.A. & Bekoff, M. (2001) Ecology and social responsibility: the re-embodiment of science. *Trends in Ecology and Evolution* **16**, 460–465.

Bradstock, R.A., Williams, J.E. & Gill, A.M. (2002) *Flammable Australia. The Fire Regimes and Biodiversity of a Continent.* Cambridge University Press, Cambridge.

Bren, L.J. (1988) Effects of river regulation on flooding of a riparian red gum forest on the River Murray, Australia. *Regulated Rivers: Research and Management* **2**, 65–77.

Bren, L.J. (1992) Tree invasion of an intermittent wetland in relation to changes in the flooding frequency of the River Murray, Australia. *Australian Journal of Ecology* **17**, 395–408.

Callicott, J.B. (2002) Choosing appropriate temporal and spatial scales for ecological restoration. *Journal of Bioscience* **27**, 409–420.

Callicott, J.B., Crowder, L.B. & Mumford, K. (1999) Current normative concepts in conservation. *Conservation Biology* **13**, 22–35.

Cheney, P. (2003) Economic rationalism, fear of litigation and the perpetuation of disaster fires. In: Cary, G., Lindenmayer, D. & Dovers, S. (eds.) *Australia Burning. Fire Ecology, Policy and Management Issues*, pp. 150–155. CSIRO Publishing, Collingwood, Victoria.

Christensen, N.L., Bartuska, A.M., Brown, J.H. *et al.* (1996) The report of the Ecological Society of America committee on the scientific basis for ecosystem management. *Ecological Applications* **6**, 665–691.

Collins, S.L. & Barber, S.C. (1985) Effects of disturbance on diversity in mixed-grass prairie. *Vegetatio* **64**, 87–94.

Dombeck, M.P., Williams, J.E. & Wood, C.A. (2003) Wildfire policy and public lands: Integrating scientific understanding with social concerns across landscapes. *Conservation Biology* **18**, 883–889.

Drake, J.A., Zimmerman, C.R., Purucker, T. & Rojo, C. (1999) On the nature of the assembly trajectory. In: Weiher, E. & Keddy, P. (eds.) *Ecological Assembly Rules: Perspectives, Advances, Retreats*, pp. 233–250. Cambridge University Press, Cambridge.

Fulé, P.Z., Covington, W.W. & Moore, M.M. (1997) Determining reference conditions for ecosystem management of southwestern ponderosa pine forests. *Ecological Applications* **7**, 895–908.

Garkaklis, M.J., Bradley, J.S. & Wooller, R.D. (1998) The effects of Woylie (*Bettongia penicillata*) foraging on soil water repellency and water infiltration in heavy textured soils in southwestern Australia. *Australian Journal of Ecology* **23**, 492–496.

Gill, A.M. & Bradstock, R. (2003) Fire regimes and biodiversity: a set of postulates. In: Cary, G., Lindenmayer, D. & Dovers, S. (eds.) *Australia Burning. Fire Ecology, Policy and Management*, pp. 15–25. CSIRO Publishing, Collingwood, Victoria.

Gomes, J.F.P. (2006) Forest fires in Portugal: how they happen and why they happen. *International Journal of Environmental Studies* **63**, 109–119.

Handmer, J. (2003) Institutions and bushfires: fragmentation, reliance and ambiguity. In: Cary, G., Lindenmayer, D. & Dovers, S. (eds.) *Australia Burning. Fire Ecology, Policy and Management Issues*, pp. 139–149. CSIRO Publishing, Collingwood, Victoria.

Hartnett, D.C., Hickman, K.R. & Fischer Walter, L.E. (1996) Effects of bison grazing, fire, and topography on floristic diversity in tallgrass prairie. *Journal of Range Management* **49**, 413–420.

Holling, C.S. (1992) Cross-scale morphology, geometry and dynamics of ecosystems. *Ecological Monographs* **62**, 447–502.

Holling, C.S. (1998) Two cultures of ecology. *Conservation Ecology* **2**, 4 [online] (http://www.consecol.org/vol2/iss2/art4).

Holling, C.S. (2001) Understanding the complexity of economic, ecological and social systems. *Ecosystems* **4**, 390–405.

Holling, C.S. & Meffe, G.K. (1996) Command and control and the pathology of natural resource management. *Conservation Biology* **10**, 328–337.

Hull, R.B., Robertson, D.P., Richert, D., Seekamp, E. & Buhyoff, G.J. (2002) Assumptions about ecological scale and nature knowing best hiding in environmental decisions. *Conservation Ecology* **6**, 12 [online] (http://www.ecologyandsociety.org/vol10/iss1/art15).

Johnson, W.C. (1994) Woodland expansion in the Platte River, Nebraska: Patterns and causes. *Ecological Monographs* **64**, 45–84.

Jolly, I.D., Walker, G.R. & Thorburn, P.J. (1993) Salt accumulation in semi-arid floodplain soils with implications for forest health. *Journal of Hydrology* **150**, 589–614.

Keith, D.A., Williams, J.E. & Woinarski, J.C.Z. (2002) Fire management and biodiversity conservation: key approaches and principles. In: Bradstock, R.A., Williams, J.E. & Gill, A.M. (eds.) *Flammable Australia. The Fire Regimes and Biodiversity of a Continent*, pp. 401–425. Cambridge University Press, Cambridge.

Kingsford, R.T. (2000) Ecological impacts of dams, water diversions and river management on floodplain wetlands in Australia. *Austral Ecology* **25**, 109–127.

Knapp, A.K., Blair, J.M., Briggs, J.M. *et al.* (1999) The keystone role of bison in North American tallgrass prairie. *BioScience* **49**, 39–50.

van de Koppel, J., van der Wal, D., Bakker, J.P. & Herman, P.M.J. (2005) Self-organization and vegetation collapse in salt marsh ecosystems. *The American Naturalist* **165**, E1–E12.

Lertzman, K. & Fall, J. (1998) From forest stands to landscapes: spatial scales and the roles of disturbance. In: Peterson, D.L. & Parker, V.T. (eds.) *Ecological Scale: Theory and Applications*, pp. 339–367. Columbia University Press, New York.

Levin, S.A. (1998) Ecosystems and the biosphere as complex adaptive systems. *Ecosystems* **1**, 431–436.

Levin, S.A. (1999) Towards a science of ecological management. *Ecology and Society* **3**, 6.

Lindenmayer, D.B. & Franklin, J.F. (2002) *Conserving Forest Biodiversity: A Comprehensive Multiscaled Approach.* Island Press, Washington, DC.

Manning, A.D., Lindenmayer, D.B. & Nix, H.A. (2004) Continua and umwelt: novel perspectives on viewing landscapes. *Oikos* **104**, 621–628.

Meffe, G.K. (1998) Conservation scientists and the policy process. *Conservation Biology* **12**, 741–742.

Murphy, M.T., Garkaklis, M.J. & Hardy, G.E.S.J. (2005) Seed caching by woylies *Bettongia penicillata* can increase sandalwood *Santalum spicatum* regeneration in Western Australia. *Austral Ecology* **30**, 747–755.

Nilsson, C. & Berggren, K. (2000) Alteration of riparian systems caused by river regulation. *BioScience* **50**, 783–792.

Noss, R.F. (1990) Can we maintain biological and ecological integrity? *Conservation Biology* **4**, 241–243.

Pickett, S.T.A. & Ostfeld, R.S. (1995) The shifting paradigm in ecology. In: Knight, R.L. & Bates, S.F. (eds.) *A New Century for Natural Resources Management*, pp. 261–278. Island Press, Washington, DC.

Pickett, S.T.A. & White, P.S. (1985) *The Ecology of Natural Disturbance and Patch Dynamics.* Academic Press, San Diego.

Richter, B.D. & Richter, H.E. (2000) Prescribing flood regimes to sustain riparian ecosystems along meandering rivers. *Conservation Biology* **14**, 1467–1478.

Rittel, H. & Webber, M. (1973) Dilemmas in a general theory of planning. *Policy Sciences* **4**, 155–169.

Robertson, D.P. & Hull, R.B. (2001) Beyond biology: toward a more public ecology for conservation. *Conservation Biology* **15**, 970–979.

Romme, W.H. & Turner, M.G. (2004) Ten years after the 1988 Yellowstone fires: Is restoration needed? In: Wallace, L.L. (ed.) *After the Fires: The Ecology of Change in Yellowstone National Park*, pp. 318–361. Yale University Press, New Haven, CT.

Saunders, D.A., Smith, G.T., Ingram, J.A. & Forrester, R.I. (2003) Changes in a remnant of salmon gum *Eucalyptus salmonophloia* and York gum *E. loxophleba* woodland,

1978 to 1997. Implications for woodland conservation in the wheat-sheep regions of Australia. *Biological Conservation* **110**, 245–256.

Schulze, R. (2000) Transcending scales of space and time in impact studies of climate and climate change on agrohydrological responses. *Agriculture, Ecosystems and Environment* **82**, 185–212.

Scott, M.L., Auble, G.T. & Friedman, J.M. (1997) Flood dependency of cottonwood establishment along the Missouri River, Montana, USA. *Ecological Applications* **7**, 677–690.

Soulé, M.E., Mackey, B.G., Recher, H.F. *et al.* (2004) The role of connectivity in Australian conservation. *Pacific Conservation Biology* **10**, 266–279.

Tarrant, M. (2003) Policy, institutions and the law. In: Cary, G., Lindenmayer, D. & Dovers, S. (eds.) *Australia Burning. Fire Ecology, Policy and Management Issues*, pp. 156–161. CSIRO Publishing, Collingwood, Victoria.

Veblen, T.T., Baker, W.L., Montenegro, G. & Swetman, T. (2003) *Fire and Climate Change in Temperate Ecosystems of the Western Americas*. Springer-Verlag, New York.

Wallington, T.J., Hobbs, R.J. & Moore, S.A. (2005) Implications of current ecological thinking for biodiversity conservation: a review of salient issues. *Ecology and Society* **10**, 15 [online] (http://www.ecologyandsociety.org/vol10/iss1/art15).

White, P.S. & Jentsch, A. (2001) The search for generality in studies of disturbance and ecosystem dynamics. *Ecology* **62**, 399–450.

van Wilgen, B.W., Richardson, D.M., Kruger, F.J. & van Hensbergen, H.J. (1992) *Fire in South African Mountain Fynbos: Ecosystem, Community and Species Responses at Swartboskloof*. Springer-Verlag, Berlin.

Yates, C.J., Hobbs, R.J. & Bell, R.W. (1994) Landscape-scale disturbances and regeneration in semi-arid woodlands of southwestern Australia. *Pacific Conservation Biology* **1**, 214–221.

33

Synthesis: Ecosystems and Ecosystem Processes

David B. Lindenmayer and Richard J. Hobbs

Chapters 30 to 32 have focused on ecosystems and ecosystem processes. In his essay, Manning (Chapter 30) dedicates considerable space to carefully defining some of the key attributes of ecosystems and ecosystem processes and then outlining why they are so important. He uses this as a prelude for canvassing the importance of ecosystem management (but see Simberloff, Chapter 26). A particular theme in Manning's essay relates to the implications of global change for ecosystems, ecosystem services and biodiversity loss. He correctly notes that global change entails far more than climate change but includes other key human-derived processes such as landscape modification as well as socioeconomic drivers like globalization (Manning, Chapter 30). Ecosystem management will be exceedingly difficult against a background of rapid and extensive global change. Manning offers a number of approaches for landscape management under such a scenario of change. In particular, he advocates:

- Improved education of the population about the threats to ecosystem services and human populations unless we manage ecosystems better.
- Better protection for the remnants of natural ecosystems. These can still have important ecological roles and provide the 'building blocks' to restore presently degraded ecosystems and landscapes (see also Woinarski, Chapter 10; Mooney, Chapter 31).

Manning (Chapter 30) outlines his deep concern about losses of connectivity in rapidly changing ecosystems and landscapes. He introduces the concept of 'landscape fluidity' and discusses approaches to maintaining it as part of slowing the erosion of connectivity. In the final part of his essay, Manning discusses the urgent need for research and monitoring, particularly long-term work – a critical consideration given anticipated future global change impacts.

The existing paucity of long-term research and monitoring projects to tackle major knowledge gaps is one of the topics that appears in many parts of this book (e.g. Driscoll, Chapter 11). It is also raised in the essays by Lake (Chapter 38) and Mooney in this Section (Chapter 31), particularly in the context of restoration projects and the lack of good data to assess the effectiveness of such efforts.

Manning highlights the importance of ecosystem services, and this theme is further developed by Mooney (Chapter 31), who notes that such services are woefully undervalued (both economically and ecologically) in almost all human development programmes that modify landscapes and ecosystems. Mooney provides a series of stark examples to emphasize this key point and argues that the true values of ecosystem services (as well as the massive costs to restore them if they are lost) need to be calculated as part of any debate on development proposals. In essence, this is a call for landscape ecologists and conservation biologists to engage more intimately in debates on resource development and use, and communicate the ecological and economic values of ecosystems and ecosystem processes far more widely and effectively.

A key plank in Mooney's arguments is that the true ecological and economic values of ecosystem services are sometimes only recognized when attempts are made to restore degraded ecosystems and landscapes. It could be argued that this has been a deficiency of the ecosystem services methodology to date and a problem that reinforces the need for the more proactive stance touted above. Irrespective of the merits of that argument, an increasing recognition of the various values of ecosystems and ecosystem services considerably strengthens the case 'that the best kind of restoration is to maintain what you already have'. This is a key point that has also been made by many authors elsewhere in this book (e.g. Saunders, Chapter 24; Manning, Chapter 30). Mooney further argues that this will be increasingly important in a world subject to major and rapid global change, a point echoed in the essay by Manning (Chapter 30). Maintaining what you already have certainly needs to be a central plank in any conservation programme, but increasingly it may be insufficient on its own to achieve conservation goals, especially where extensive removal and modification of native vegetation has occurred (e.g. see Hobbs, Chapter 43).

Cramer (Chapter 32) takes more of a case study approach to ecosystem management and ecosystem processes in her essay. She illustrates some of the processes operating across a range of spatial scales in southwestern Australia and reinforces ideas of Walker (Chapter 34) about the need to study at a number of scales. Cramer highlights the roles of floods, fires and animal burrowing as differently scaled disturbance regimes that strongly influence many aspects of southwestern Australian landscapes and ecosystems. This work has strong congruence with other chapters in this book that demonstrate the non-static nature of landscapes and ecosystems and the interrelationships between disturbance, vegetation structure and condition, and biotic responses (e.g. Hunter, Chapter 35). Managing these disturbances and other ecological processes is not straightforward because different stakeholders in a landscape have different perspectives on them. The same disturbance event may be seen as good by some but bad by others (Woinarski, Chapter 10). It also may be beneficial for some elements of the biota but highly detrimental to others. Cramer recognizes the management challenges associated with differing perceptions and responses to the array of disturbance regimes that operate (across a range of spatial scales) in all landscapes and ecosystems. She recommends that the management of disturbance should become an integral part of landscape and ecosystem management and suggests that this might be done best by broadening the units for management beyond the individual patch scale to mosaics, landscapes and even regions. This perspective echoes that put forward by many others in this book (Haila, Chapter 3; Mac Nally, Chapter 6; Franklin & Swanson, Chapter 12; Bennett & Radford, Chapter; 18). Additionally, however, she highlights the importance of the need to consider multiple scales, also echoed by, for instance, Walker (Chapter 34). The example of needing regional-scale feral predator control measures to achieve the goal of reintroducing localized animal digging as an important ecosystem process illustrates well the need to consider more than one scale at a time.

Section 9
Disturbance, Resilience and Recovery

34

Disturbance, Resilience and Recovery: A Resilience Perspective on Landscape Dynamics

Brian Walker

Abstract

In the context of managing landscapes, resilience is the amount of disturbance the landscape can experience without shifting to a different regime of function and structure – that is, without changing identity. It places an emphasis on identifying thresholds between such regimes, how to intervene in ways to avoid unwanted regime shifts and how to enhance the resilience of desired regimes. Examples of physical, biological and biophysical threshold effects are presented, leading to a tentative set of 'design' principles. Landscapes are linked social-ecological systems. They are self-organizing systems that are restructured by a few key drivers, and their non-linear dynamics, especially in the social component, call for an adaptive governance approach (Dietz *et al.* 2003). A command-and-control approach to achieving some perceived optimal state is likely to fail.

Keywords: disturbance; landscape; recovery; resilience; scale.

Introduction

A key insight from studies on resilience is that systems cannot be understood or successfully managed by focusing on one scale – the scale of interest. It requires at least three scales – the scale of interest and at least one above and one below (Gunderson & Holling 2002). A perspective on the dynamics of landscapes implicitly focuses on some defined 'landscape' scale, but whatever the landscape that is being considered it is necessary to examine the dynamics of the system above and below that scale as well. Cross-scale interactions strongly influence the dynamics of the landscape as a whole.

With this proviso in mind, my aim in this essay is to present a resilience view of landscape structure, function and dynamics, by examining how they behave in non-linear ways. I begin with a few simple examples and end with some more speculative suggestions about truly complex behaviour that makes landscape management difficult. First, however, it is necessary to explain what is meant by resilience. The essay concludes with some suggested landscape design principles that follow from a resilience perspective.

Resilience

Resilience is the capacity of a system to absorb disturbance and reorganize while undergoing change so as to retain essentially the same function, structure, identity and feedbacks (Walker *et al.* 2004; following Holling 1973). This interpretation places an emphasis on limits to change in a system, focusing on thresholds between alternative 'states' of the system, rather than on the definition of resilience as speed of return to an equilibrium state following a small perturbation from that state (Pimm 1991).

The term 'states' is in quotes because real systems are seldom, if ever, in equilibrium. The notion of alternative stable states, therefore, while strictly speaking correct, is misleading. It is more helpful and informative to consider the alternative stability domains around those notional stable equilibria and the unstable equilibrium that separates them. For this reason it is preferable to refer to and think about alternative system regimes (Scheffer *et al.* 2001; Scheffer & Carpenter 2003), where each regime is a configuration of system states (as defined above) in which the system has essentially the same structure, function and feedbacks, and hence the same 'identity'.

It is important to note that changes in structure and function that occur when a threshold is crossed are induced by the changes in system processes, notably feedbacks, that occur at that threshold. It is the change in feedbacks that demarcates the position of the threshold. Feedback changes are brought about by changes in controlling (usually slowly changing) variables. A well-known example (Carpenter 2003) concerns the alternative regimes in lakes (clear vs. eutrophic, with low and high levels of water phosphate (P), respectively) brought about by feedback changes from slowly increasing levels of P in the sediment. Release of P from the sediment is small under aerobic conditions, but as water P increases so does algal growth and decomposition. When decomposition reduces oxygen below a critical level, the rate of P release from the sediment increases markedly (a feedback change), leading to very high water P, greatly increased algal growth and the eutrophic regime. Work in rangelands (Walker 1993; and see Anderies *et al.* 2002) shows an analogous feedback change in the effects of grass (fast variable) on shrubs (the slowly changing variable), as shown in Fig. 34.1. The feedback is in terms of the effects of fire, and the threshold occurs where there is insufficient grass (due to too many shrubs) to carry a fire of sufficient intensity to kill shrubs. That threshold amount of grass separates a regime of states in which fire is possible from one in which it is ineffective – different function, leading to different structure and different identity.

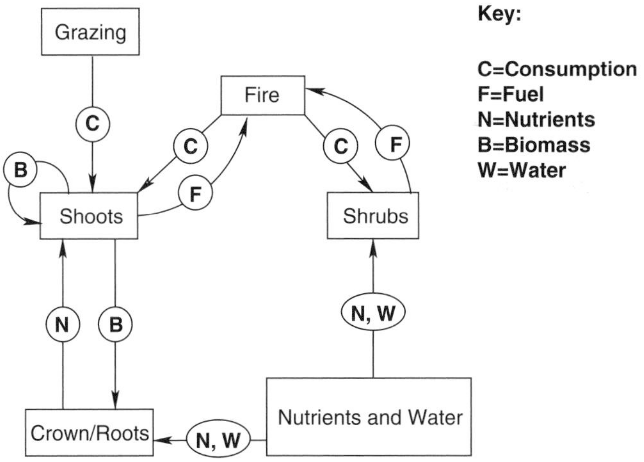

Figure 34.1 **Variables and feedbacks in the dynamics of shrubs and grass in savanna rangelands. The feedback that changes, to cause a regime shift from grassy to shrub-dominated rangeland, is the fire feedback from grass to shrubs. (From Anderies *et al.* 2002, with permission.)**

Not all regime shifts are of the type just described, involving a changing controlling variable. In some cases the controlling variable is a presence or absence phenomenon. The introduction of an exotic pest species, for example or removal of a top predator, can result in changed feedbacks and processes that lead to a different structure and function. The addition or removal of such a variable results immediately in a change in processes that puts the system on a different trajectory – into a different system regime. It may, however, take a long time for the changes in state variables to become evident.

Early studies and examples of resilience changes and the 'flip' from one regime to another (like those described above) were all one-dimensional and mostly biophysical, as illustrated by the hundred or so examples in the thresholds database of the Resilience Alliance (Walker & Meyers 2004). A comparison of several regional scale social-ecological systems (Kinzig *et al.* 2006) suggests, however, that there are a number of interacting threshold effects, at different scales and in different domains (ecological, economic and social). While the single regime shift examples are important, management or policy based solely on them is likely to produce partial solutions – which invariably fail.

The fate of the whole system – its trajectory through time in terms of the combinations of pairs of component alternative regimes – depends on the interactions amongst thresholds and which threshold gets crossed first (dependent on the nature of an external shock, and how close the system is to each threshold). Crossing one threshold may cause an inevitable crossing of another, or preclude it. Depending on the nature of the interactions, a cascade of threshold crossings could theoretically be avoided or promulgated by governance and management. The capacity to manage resilience – to avoid or engineer a threshold crossing, or to change the positions of thresholds (change the dimensions of basins of attraction, the resilience of particular regimes) – is known as adaptability (Walker *et al.* 2004).

Disturbance and resilience at landscape scales

This section presents a few simple examples of resilience in the face of disturbance in landscapes, covering a spectrum of kinds and scales.

A physical example: Salinity in agricultural landscapes

There is a vast literature detailing the changes in landscape hydrology and subsequent salinization, following clearing of native vegetation. A simple account of the dynamics, based on a clear analytical model, is given by Anderies (2005). When agriculture began, water tables in most catchments were well below the rooting zone of the vegetation (e.g. around 30 m in the Goulburn-Broken Catchment, in southeastern Australia). Over millennia, small amounts of salt brought in with rainfall were washed down the profile to below the rooting depth. In most years the vegetation took up all the water – it had come into a dynamic equilibrium with the rainfall regime – and only in very wet years did water go beyond the root zone and recharge the water table. After clearing, the annual crops no longer took up all the water and the excess penetrated to the water table, causing it to rise. As it rose, it brought up the stored salt with it. When the water table reaches c.2 m below the surface (depending on soil type) it crosses a threshold; it is drawn to the surface by capillary action, bringing salt into the top soil layers, reducing plant growth rates (therefore further reducing transpiration and water uptake), altering soil structure (reducing infiltration rates) and effectively bringing about a regime shift to a salinized landscape.

When the water table is 3 m below the surface, the soil is just as productive as when it was 30 m below. But the resilience of the productive soil is much less. A high rainfall 'shock' that previously would have been easily absorbed (the water tables may perhaps have temporarily risen a metre or so) can now cause a regime shift. The resilience of the system can be approximated by the distance between the water table and the c.2-m threshold.

A biological example: Species composition and ecosystem function

It has been shown in rangelands, lakes, forests and coral reefs (Elmqvist *et al.* 2003) that, while functional diversity (species with different functions in the system) confers high performance on a system, response diversity (the diversity of species within a functional group, having different responses to environmental disturbances) confers resilience of that performance. In all cases, as the number of species within a group that is performing some particular function declines, so the vulnerability of that function to some external shock increases. I will use the rangeland example to illustrate how it works. The full details are given in Walker *et al.* (1999).

The rangeland concerned is in southeast Queensland. When in 'good' condition (light grazing pressure) it has some 22 perennial species, with an abundance distribution typical of most ecological communities – that is, with four or five species making up more than 80% of the biomass and a large number of minor species making up the 'tail' of the distribution.

When examined in terms of the functions these species perform in the ecosystem – water uptake, rate of decomposition, maximum size (carbon/nutrient store) – inferred from five functional attributes that could be estimated for each species (see Walker *et al.* 1999 for details) it turns out that the dominant few species that make up the bulk of the biomass are significantly different from each other – that is, they are functionally complementary. However, for most of these dominant species there is one or more of the minor species that is functionally very similar to it.

When a site in the landscape that had been heavily grazed was examined, four of the five most abundant species in the 'good condition' site were found to have either disappeared or been reduced to minor levels, and three of them had been replaced by their functional analogues amongst the minor species. From this we deduce that these minor species, far from being redundant, constitute response diversity in the system, conferring resilience on the functionality of the rangeland.

A biophysical example: Banded landscapes – patterns of vegetation, water and soils driven by topography and rainfall

This is now a well-known phenomenon, adequately described from scales of kilometres, in publications by Pickup (1985), to hundreds of metres (Tongway 1993; Tongway & Ludwig 1997; Ludwig *et al.* 1997), and summarized with work from other countries in the book by Tongway *et al.* (2001). The 'model' is simple: rainfall on an incompletely covered soil (i.e. in arid and semiarid regions) falls on sealed surfaces and begins to run off. As the runoff accumulates, it encounters obstacles to its flow, and eventually one is sufficient to cause increased infiltration of water such that increased grass growth occurs, creating a positive feedback to growth and infiltration. As evidenced in many parts of Africa, Mexico, Australia and India this leads to the 'brousse tigrée' or banded landscape – a self-organizing process at a landscape scale that results in a two-phase pattern with (relatively) narrow bands of dense vegetation across the slope ('runon' phase), separated by wider 'runoff' zones that are sparsely vegetated or bare. Ludwig *et al.* (1994) provide a model for calculating the equilibrium band widths based on rainfall, soil and topography.

The dynamics are interesting. In dry years insufficient runoff penetrates to the lower parts of the vegetated bands, and they die off. In wet years, the increased flow through the zone washes the accumulated topsoil from the bottom of this now-dead part of the band down to the top of the next vegetated band, resulting in an episodic uphill migration of the vegetation bands. The system is self-organizing in this way, provided the level of grazing remains below a critical level. Above that threshold, runoff arriving at a banded zone begins to run through it fast and in small channels (there is insufficient impediment to surface flow), rather than seeping through and infiltrating. The result is a loss of the banded pattern and net loss of water, die-off of the remaining grass cover (that received too little infiltrating water), and a shift from a productive, albeit banded, landscape to an unproductive landscape with little pattern and a significantly reduced 'rain use efficiency' (Le Houerou 1984), retaining little of the rainfall (Ludwig *et al.* 1999). The resilience of the system is the amount of a rainfall 'shock' required to exceed the capacity of the bands to stop the runoff and induce infiltration. The level of the shock required depends on the cover of grass in the vegetated bands, which in turn depends on the amount of rainfall and level of grazing pressure.

A variant of this model, invoking a hierarchical view of landscape processes, places more emphasis on the catena-type dynamics involving changes in hydrological processes through incision leading to a drying out of the soil profile and subsequent soil erosion (Pringle & Tinley 2003). It explains many cases of 'degradation' in at least Western Australia and southern Africa.

A further example

The three examples above all involve feedback processes in ecosystems that are either irreversible or reversible with hysteretic effects, resulting in alternative stability domains for the same level of a controlling variable. Another kind of threshold effect is one where there is a fully reversible sharp decline in a variable of interest (such as the amount of biodiversity) over a narrow range of a controlling variable. An example is the often quoted drop in persistence of number of species in a landscape below some level of vegetation cover (e.g. Andren 1994; Bennett & Ford 1997; Walpole 1999).

Combined regimes

Extending the reasoning of the multiple regime shifts (see Kinzig *et al.* 2006) we can conceive of landscapes in which there is: (i) a threshold effect in vegetation cover associated with the maintenance of ecosystem function in terms of hydrology and nutrient retention and (ii) a threshold effect in animal activity providing the nucleus for vegetation patches by bringing in seeds, in bush clumps (e.g. in the Gascoyne-Murchison region of Western Australia, K. Tinley, personal communication). If the first threshold (vegetation cover) is exceeded, ecosystem function declines to the point that bush clumps cannot survive, and bird populations also disappear. If bird populations decline (for other reasons) below the activity threshold required to maintain the clumps, it then leads to the first threshold being crossed. Attention by management to only one threshold is insufficient.

'Specified' vs. 'general' resilience: A note of warning

Much of the developing literature on getting from theory into practice is concerned with the resilience 'of what to what' (see Carpenter *et al.* 2001). This is specified resilience. If we know what the possible alternative regimes are, and something about their relative values to society, and we can estimate how likely it is that a regime shift will occur (how close to the threshold we are), we can get some idea of how much effort it is worth expending to avoid the regime shift. This approach can lead to putting much effort into enhancing resilience of a particular regime to one or more particular shocks, for example the resilience of the grassy (vs. thicket) regime of a rangeland to drought and grazing pressure. The danger is that focusing on this can become a narrow form of optimization. Some work in the Robustness Program of the Santa Fe Institute sheds light on this ('robustness' can be equated to resilience for our purposes here).

In examining robustness in different systems (vegetation, the internet), Carlson and Doyle (2002) argue that the amount of robustness in such systems is conserved, and that they are necessarily 'robust yet fragile' – that is, the more robust you make them in one way, the more fragile they necessarily become in others. They call this the 'HOT' model (Highly Optimized Tolerance). Whether resilience is a conserved property of ecosystems is beyond the scope of this essay, and it doesn't alter the important message: trying to increase the resilience of a system in one way may well decrease its resilience in others. In an analogous way, increasing resilience at one scale can decrease it at another. General resilience is more difficult to analyse, but attributes of systems that have been shown to confer resilience, generally, include such things as diversity, tight feedbacks, modularity, well-developed networks and 'memory' (seed banks, local knowledge) (see Levin 1999 for a useful discussion).

Key landscape design principles

I hesitate to label what follows as 'principles', but the key take-home messages from this perspective of landscape dynamics are:

1 Landscapes are self-organizing systems with limits to the amount of change they can experience before losing the capacity to remain in the same system 'regime'.
2 Structure and function in landscape systems are determined by a few key drivers – almost always including the redistribution pattern of water and nutrients. Landscapes have evolved to absorb normal fluctuations in these drivers, and they are restructured by exceptional episodic events. The corollary is that systems that are event-driven need to be event-managed; this includes being prepared to take advantage of such events (in a positive way) and increasing resilience to cope with such events.
3 All managed landscapes are linked social-ecological systems that have non-linear dynamics, and the non-linear dynamics of the social domains of such systems are more difficult to understand than the ecological dynamics. The combined dynamics are driven largely by the governance (institutional arrangements) of the system, and instituting an evolving form of adaptive governance (see Dietz *et al.* 2003) is a key 'design' principle.
4 A top-down, command-and-control approach to achieving some perceived optimal landscape state is likely to fail. In contrast to such an approach, resilience management and governance aims to understand the possible alternative regimes the system can be in – in terms of individual regime shifts and the overall trajectory of the system. The objective is twofold: First, assess if alternative regimes might exist and, if possible, their probabilities. For this set, identify the thresholds associated with each, and the feedback processes involved. Second, manage the feedbacks associated with each threshold so as to either avoid crossing it (or engineer a crossing to the original regime, if the system has already undergone a regime shift), or to change the positions of the threshold.
5 Do not fall into the trap of becoming too narrowly focused on optimization for some particular goal(s). While enhancing resilience with regard to known threatening disturbances, do not lose sight of the need to maintain 'general' resilience – with no identified regime shifts and without specifying which kinds of shocks. The difficulty with promoting general resilience is that it is even harder to value than specified resilience.

6 Resilience costs: maintaining or increasing resilience is done at the expense of foregone extra profits from a highly 'efficient' resource use system. The challenge to scientists is to be able to contrast the costs and benefits of 'efficient' production with the costs and benefits of maintaining resilience. In both cases the costs are short-term. The benefits for 'efficient' management are immediate; for resilience management they are generally longer-term. Part of any operational landscape design must include some assessment of the benefits and costs of alternative 'design' options.
7 As a final comment, the word 'design' is problematic. It raises images of an optimal state of a landscape. There is no such thing for a complex adaptive system. It is like asking what will be the final optimal state of biological evolution; a nonsense question – it is an ongoing process, with branch points, just like an evolving landscape. 'Design' for landscape management has to be a dynamic process of governance and management centred on principles 4 and 5.

Principles

1 Landscapes are self-organizing systems with limits to the amount of change they can experience before losing the capacity to remain in the same system 'regime'.
2 **Structure and function in landscape systems are determined by a few key drivers – almost always including the redistribution pattern of water and nutrients – and they are restructured by exceptional episodic events.** Systems that are event-driven need to be event-managed.
3 **All managed landscapes are linked social-ecological systems that have non-linear dynamics, driven largely by the governance (institutional arrangements) of the system.** Instituting an evolving form of adaptive governance is a key 'design' principle.
4 **A top-down, command-and-control approach to achieving some perceived optimal landscape state is likely to fail.** As opposed to such an approach, resilience management and governance aims to understand and manage the possible alternative regimes the system can adopt.

5 While enhancing resilience with regard to known threatening disturbances, do not lose sight of the need to maintain 'general' resilience.
6 Part of any operational landscape design must include some assessment of the benefits and costs of 'efficient' production (of a state of the system, or particular species, etc.) with benefits and costs of maintaining resilience.
7 **There is no such thing as an optimal state of a landscape.** 'Design' for landscape management has to be a dynamic process of governance and management centred on principles 4 and 5.

References

Anderies, J.M. (2005) Minimal models and agro-ecological policy at the regional scale: An application to salinity problems in southeastern Australia. *Regional Environmental Change* **5**, 1–17.

Anderies, J.M., Janssen, M.A. & Walker, B.H. (2002) Grazing management, resilience and the dynamics of a fire driven rangeland. *Ecosystems* **5**, 23–44.

Andren, H. (1994) Effects of habitat fragmentation on birds and mammals in landscapes with different proportions of suitable habitat: a review. *Oikos* **71**, 355–366.

Bennett, A. & Ford, L. (1997) Land use, habitat change and the distribution of birds in fragmented rural environments: a landscape perspective from the Northern Plains, Victoria, Australia. *Pacific Conservation Biology* **3**, 244–261.

Carlson, J.M. & Doyle, J. (2002) Complexity and robustness. *Proceedings of the National Academy of Sciences of the USA* **99**, 2538–2545.

Carpenter, S.C., Walker, B.H., Anderies, M. & Abel, N. (2001) From metaphor to measurement: Resilience of what to what? *Ecosystems* **4**, 765–781.

Carpenter, S.R. (2003) *Regime Shifts in Lake Ecosystems.* Ecology Institute, Oldendorf/Luhe, Germany.

Dietz, T., Ostrom, E. & Stern, P.C. (2003) The struggle to govern the commons. *Science* **302**, 1907–1912.

Elmqvist, T., Folke, C., Nystrom, M. *et al.* (2003) Response diversity, ecosystem change and resilience. *Frontiers in Ecology and Environment* **1**, 488–494.

Gunderson, L. & Holling, C.S. (eds.) (2002) *Panarchy: Understanding Transformations in Human and Natural Systems.* Island Press, Washington, DC.

Holling, C.S. (1973) Resilience and stability of ecological systems. *Annual Review of Ecological Systems* **4**, 1–23.

Janssen, M.A., Anderies, J.M., Stafford Smith, M. & Walker, B.H. (2002) Implications of spatial heterogeneity of grazing pressure on the resilience of rangelands. In: Janssen, M.A. (ed.) *Complexity and Ecosystem Management: The Theory and Practice of Multi-agent Systems*, pp. 103–123. Edward Elgar Publishers, Cheltenham/Northampton.

Kinzig, A.P., Ryan, P., Etienne, M., Allyson, H., Elmqvist, T. & Walker, B.H. (2006) Resilience and regime shifts: assessing cascading effects. *Ecology and Society* **11**, 20 [online] (http://www.ecologyandsociety.org/vol11/iss1/art20/).

Le Houerou, H.N. (1984) Rain-Use Efficiency, a unifying concept in arid land ecology. *Journal of Arid Environments* **7**, 1–35.

Levin, S. (1999) *Fragile Dominion: Complexity and the Commons*. Perseus Books, Reading, MA.

Ludwig, J.A., Tongway, D.J. & Marsden, S.G. (1994) A flow-filter model for simulating the conservation of limited resources in spatially heterogeneous, semi-arid landscapes. *Pacific Conservation Biology* **1**, 209–215.

Ludwig, J., Tongway, D., Freudenberger, D., Noble, J. & Hodgkinson, K. (eds.) (1997) *Landscape Ecology Function and Management: Principles from Australia's Rangelands*. CSIRO, Melbourne.

Ludwig, J.A., Tongway, D.J. & Marsden, S.G. (1999) Stripes, strands or stipples: modeling the influence of three landscape banding patterns on resource capture and productivity in semi-arid woodlands, Australia. *Catena* **37**, 257–273.

Pickup, G. (1985) The erosion cell – a geomorphic approach to landscape classification in range assessment. *Australian Rangeland Journal* **7**, 114–121.

Pimm, S.L. (1991) *The Balance of Nature?* University of Chicago Press, Chicago, IL.

Pringle, H. & Tinley, K. (2003) Are we overlooking critical determinants of landscape change in Australian rangelands? *Ecological Management and Restoration* **4**, 180–186.

Scheffer, M. & Carpenter, S.R. (2003) Catastrophic regime shifts in ecosystems: linking theory to observation. *Trends in Ecology and Evolution* **18**, 648–656.

Scheffer, M., Carpenter, S.R., Foley, J.A., Folke, C. & Walker, B. (2001) Catastrophic shifts in ecosystems. *Nature* **413**, 591–596.

Scholes, R.J. & Walker, B.H. (1993) *An African Savanna: Synthesis of the Nylsvley study*. Cambridge University Press, Cambridge.

Tongway, D.J. (1993) Functional analysis of degraded rangelands as a means of defining appropriate restoration techniques. In: Gaston, A., Kernick, M. & Le Houerou, H. (eds.) *Proceedings of the Fourth International Rangeland Congress. Montpellier, France, 22–26 April 1991*, pp. 166–168. Service Central d'Information Scientifique et Technique, Montpellier.

Tongway, D. & Ludwig, J. (1997) The conservation of water and nutrients within landscapes. In: Ludwig, J., Tongway, D., Freudenberger, D., Noble, J. & Hodgkinson, K. (eds.) *Landscape Ecology, Function and Management: Principles From Australia's Rangelands*, pp. 13–22. CSIRO, Melbourne.

Tongway, D.J., Valentin, C. & Seghieri, J. (2001) Banded vegetation patterning in arid and semi-arid environments: Ecological processes and consequences for management. *Ecological Studies no. 149*, Springer Verlag, New York.

Walker, B.H. (1993) Rangeland ecology: Understanding and managing change. *Ambio* **22**(2–3), 80–87.

Walker, B. & Meyers, J.A. (2004) Thresholds in ecological and social-ecological systems: A developing database. *Ecology and Society* **9**, 3 [online] (http://www.ecologyandsociety.org/vol9/iss2/art3).

Walker, B.H., Kinzig, A. & Langridge, J. (1999) Plant attribute diversity, resilience, and ecosystem function: The nature and significance of dominant and minor species. *Ecosystems* **2**, 1–20.

Walker, B.H., Holling, C.S., Carpenter, S.C. & Kinzig, A.P. (2004) Resilience, adaptability and transformability. *Ecology and Society* **9**, 3 [online] (http://www.ecologyandsociety.org/vol9/iss2/art5).

Walpole, S.C. (1999) Assessment of the economic and ecological impacts of remnant vegetation on pasture productivity. *Pacific Conservation Biology* **5**, 28–35.

35

Core Principles for Using Natural Disturbance Regimes to Inform Landscape Management

Malcolm L. Hunter Jr.

Abstract

The natural dynamics of a landscape, particularly its disturbance regimes, can inform managers how to design anthropogenic disturbances (e.g. logging and controlled burning) that are more ecologically sensitive than conventional management. Although perfect emulation of a natural disturbance is impossible, the effects of an anthropogenic disturbance may be ameliorated to some degree if it resembles the natural disturbances to which species have evolved various adaptations. For example, the spatial patterns that characterize a natural disturbance (e.g. size, shape and juxtaposition of disturbed patches) or the temporal patterns (e.g. return interval between disturbances, or seasonality of events) can be emulated to some degree. Similarly, the patterns of legacy materials left after a disturbance (e.g. abundance and distribution of logs and snags) also merit emulation. Although landscape management based on natural patterns has intuitive appeal, it should not become a ball and chain to constrain management; it should simply be a dominant technique in a diverse repertoire, and a catalyst for creative, ecologically sensitive thinking.

Keywords: anthropogenic disturbances; fire management; forest management; landscape management; natural disturbance regimes.

Introduction

Many people have at least a rudimentary grasp of the fact that landscapes change over time, either by witnessing the after-effects of a major disturbance or by viewing historic photos of a familiar area. The average person may not welcome such change for, as individuals, we tend to be quite conservative about what we consider to be our 'natural habitat', even though as a species *Homo sapiens* is a habitat generalist. Of course ecologists seek to have a deeper understanding of landscape dynamics, often viewing them as multifaceted phenomena that merit close study. Natural resource managers share this interest, particularly because landscape dynamism often represents a challenge for them, something to be controlled so that they can provide a steady stream of natural resources, including relatively stable habitat for people.

In recent years, the work of natural resource managers has grown more challenging because they are expected not only to produce a sustainable supply of resources for people, but also to maintain habitat for myriad other species, a goal implicitly encapsulated in the word 'biodiversity'. This has led to the idea of viewing landscape dynamics, especially natural disturbance regimes, as a set of processes that can inform natural resource management rather than be avoided or controlled. More specifically, many ecologists have suggested that natural disturbance regimes can provide models that natural resource managers should strive to emulate in their manipulations of ecosystems (Perera *et al.* 2004). For example, timber-harvesting regimes can be designed partly to resemble natural disturbances that kill large numbers of trees. The fundamental premise here is that species have evolved adaptations to respond to natural disturbances and thus, to the extent that human-induced disturbances can resemble natural disturbances, then species may be better able to cope with our landscape manipulations.

The idea of emulating natural disturbance regimes in resource management has been particularly prominent in boreal forests, one of the Earth's most dynamic landscapes, where stand-replacing crown fires often cover tens of thousands of hectares and have a return interval of only 50–200 years. Northern hemisphere ecologists began thinking about boreal forest fires as models for forest harvesting in the 1990s (Hunter 1993; Haila 1994; Bunnell 1995; Angelstam 1998) and their interest has not abated (see 22 papers in Perera *et al.* 2004). Similar ideas were first expressed by Australian ecologists at about the same time (Attiwill 1994a, 1994b) and interest also continues, albeit with less focus on managing native forests for timber and more attention directed towards prescribed burning in forests that are now national parks. In this essay I will explore some of the core principles associated with the idea of emulating natural disturbance regimes, drawing heavily upon work done in boreal forests and Australia. I will primarily use fire, forest and forestry examples, but most of the principles are more widely relevant – for example, to how managers control river floods and grassland fire regimes.

The essay begins with two sections that highlight important caveats regarding two common assumptions: first, that emulating natural disturbances can resolve all the problems associated with anthropogenic disturbance regimes; and second, that one can take a 'black-box' approach to interspecific differences in responses to disturbances. The following three sections focus on three main attributes of disturbance regimes: temporal patterns, spatial patterns and the legacies left after disturbances. The last three sections briefly consider the integration of natural and anthropogenic disturbances, the sources of information about natural disturbance regimes and the integration of natural disturbance emulation with an alternative model.

Emulating natural disturbance regimes has limitations

Anthropogenic disturbances cannot be perfect analogues of natural disturbances, and thus emulating natural disturbances is not a panacea; it is merely a model for making natural resource management more ecologically sensitive. This is especially true if anthropogenic disturbances are driven by a goal of extracting natural resources. To take an obvious example, logging involves removal of significant numbers of boles whereas windstorms merely uproot trees and fires leave many charred snags and logs (Foster *et al.* 1997). Furthermore, terrestrial resource extraction usually requires roads, which have major concomitant environmental problems (Forman *et al.* 2003). While it is unrealistic to expect that emulating natural disturbances can mitigate all the negative consequences of natural resource exploitation, it certainly can be an improvement over management that essentially ignores natural disturbance regimes, as in the agricultural paradigm of plantation forestry or management that seeks to work around them. The latter approach is exemplified in the 1962 edition of the classic US silviculture textbook. Under the heading 'Silviculture as an Imitation of Nature', Smith (1962) wrote:

> The fact that [the forester] must know the course of natural succession does not indicate that he should necessarily allow it to proceed. Economic factors ultimately decide the silvicultural policy on any given area; the objective is to operate so that the value of benefits derived . . . exceeds by the widest possible margin the value of efforts expended.

The quote may be over 40 years old but the perspective is still quite common. In situations where the goal is strongly focused on emulating a natural disturbance, for example setting a prescribed fire in a national park, it is reasonable to expect a much closer approximation, although it will still be imperfect because of differences in fire intensity and other parameters (Gill 1999). In particular, risks to human health and infrastructure will almost always constrain the use of prescribed fires.

Species differ in their responses to disturbances

Not surprisingly, ecologists have positive feelings towards nature and thus they tend to see natural disturbances as a 'good' thing. Overlooking a burned forest landscape, many ecologists probably feel a faint sense of intellectual superiority because we can see beauty in the aftermath of a fire while the average person sees only ugly destruction. While it is understandable that ecologists tend to think holistically about ecosystems or communities, there are some risks in believing that natural disturbances are 'good' for all species. Obviously, some species, notably early successional species, may thrive in the aftermath of a disturbance, but many other species resort to mechanisms, some better than others, merely to survive a disturbance. For example, survival of an animal population might depend on a few lucky individuals that find refugia during the disturbance and sufficient food and cover afterwards. Alternatively, population survival may require considerable vagility, that is, individuals that can move beyond the boundary of the disturbance, find suitable habitat, and return later. Or, the population may survive only in the sense that a metapopulation structure facilitates recolonization of a site later in succession.

To be explicit, I consider something to be beneficial for a species if it fosters a relatively large, persistent population. In some cases, a disturbance may have a short-term negative effect, but be beneficial in the longer term – for instance, an event that kills mature individuals but leads to better habitat for a large, new generation. Many Australian plants benefit from competition-limiting fires by having fire-resistant seeds (e.g. some *Acacia* and *Banksia* species) or being able to resprout from epicormic buds, roots or lignotubers (e.g. many *Eucalyptus* species) (Gill 1999; Whelan *et al.* 2002). By extension, some Australian animals presumably benefit from disturbances because of their association with post-disturbance vegetation and environments (e.g. certain species of *Pseudomys* mice and various invertebrates) (York 1999; Whelan *et al.* 2002). However, the icon of Australia, the koala (*Phascolarctos cinereus*), and many other species are harmed by fires (Martin & Handasyde 1999). Indeed, Woinarski (1999) lists 51 species of threatened Australian birds for which 'inappropriate fire regimes' are believed to be a significant problem. In most cases, this refers to fires, often human-set fires, that are too frequent and intense. It may be helpful to think of species in functional groups based on their responses to disturbance, but their diversity belies simple polar constructs such as 'disturbance-sensitive' versus 'disturbance-adapted' (Gill *et al.* 2002).

Because species are affected by disturbances in different ways and to different degrees, emulating a natural disturbance regime does not absolve managers from paying attention to the habitat needs of species that merit special management, for example an endangered species or economically important species. To take one challenging example, populations of the endangered ground parrot (*Pezoporus wallicus*) are reduced by fire, but they may also decline if their habitat remains unburned for too long. Thus their management requires determination of an optimal fire interval, and this has proven elusive because it varies among vegetation types and across the species' geographic range (Woinarski 1999; Keith *et al.* 2002a).

Temporal patterns

Landscapes are dynamic at two broad temporal scales. First, and ultimately most important, are the geological processes that uplift and erode mountain ranges, mould the paths of rivers and glaciers, shift continental plates, and so on over millions of years. By shaping topography, climate, bedrock and soils these processes have a fundamental role in determining how biota are distributed and how these distributions change over long periods. Second are the relatively short-term cycles of natural disturbance and succession that are measured in decades, centuries and sometimes millennia. Disturbances with a meteorological or biotic origin come to mind first – fires, windstorms, floods and outbreaks of insects – but some geological processes such as volcanoes and landslides happen so rapidly that they are also disturbances that initiate succession (Franklin & McMahon 2000).

Understandably, natural resource managers focus on short-term cycles because they are much more synchronized with human timescales and they set the clock for ecological succession. A key question managers need to consider is: are the intervals between natural disturbance events different from the intervals between anthropogenic disturbances? For example, foresters commonly schedule the logging of forests to occur when annual growth falls below long-term average growth, and this is usually a shorter interval than the period between natural events that kill trees at a given site (i.e. the return interval). In some forests, these numbers are not extremely disparate – roughly 50–200 years for the fire return interval in boreal forests vs. 40–60 years for logging (Harvey *et al.* 2002), or 75–150 years for intense fires in mountain ash (*Eucalyptus regnans*) forests of Victoria vs. 50–80 years for clearfell logging (Mackey *et al.* 2002). In contrast, the interval between natural stand-replacing events in the sub-boreal forests of northeastern North America has been estimated at 800–14,000 years vs. 50–80 years for a typical clearfelling rotation, and such a disparity clearly demands a partial-cutting approach designed to emulate partial disturbances (Seymour *et al.* 2002). Disturbing a forest prematurely may prevent it from developing crucial features such as large trees, hollows and an uneven-age structure (Harvey *et al.* 2002; Lindenmayer & McCarthy 2002). Matching a logging rotation to the mean return interval does not mean aligning two numbers. Variation in return intervals and other parameters needs to be recognized and accommodated (Seymour & Hunter 1999). Operating across the natural range of variability will also help to provide habitat for a wider range of species (Bunnell 1995).

Two short-term temporal issues – time of year and time of day – can affect the impact of disturbances (e.g. dry season fires and midday fires are relatively intense, and breeding season disturbances kill relatively more animals) and consequently managers consider time of year and day when undertaking prescribed burning (Bradstock *et al.* 2002). The importance of time of year in determining resource extraction is less clear; for example, in boreal forests most people would favour winter logging when snow and frozen ground limit soil damage even though fires occur mainly in summer and autumn.

Prescribed burning raises another temporal issue: the prospect of setting frequent low-intensity fires (e.g. every 4–5 years) that will remove fuel and lower the probability of having infrequent intense fires. This is a complex and controversial topic; suffice to say here that some species are much better adapted to infrequent fires than frequent ones (e.g. obligate-seeder plants that regenerate from seeds after fires and that require 10 or more years for adults to reach maturity) (Bradstock *et al.* 1997; Whelan 2002; Andersen *et al.* 2005).

Finally, standing on the brink of what is likely to be a period of dramatic climate change, we need to consider the fact that natural disturbance regimes change over time (Clark 1988; Cary 2002; Mackey *et al.* 2002). In particular, warmer, drier climates lead to more frequent fires while periods of climatic instability may be characterized by more storms. This should not be an excuse for inaction. However, it does mean that we are aiming at a moving target and should not be accelerating the pace of change that will challenge species' ability to adapt (Chambers *et al.* 2005).

Spatial patterns

Ultimately landscape dynamics are the primary driver of spatial patterns. In some landscapes gradients are a dominant feature reflecting the geological processes that generate topography (with all of its effects on climate and hydrology) and other environmental attributes. In other landscapes, relatively sharp boundaries between different types of environment and/or vegetation generate a landscape mosaic, often construed as comprising patches, matrix and corridors (see Chapter 4). My impression, as a human who usually flies at 10,000 metres, is that mosaics are more common than gradients. This may reflect the fact that disturbances, both natural and anthropogenic, typically generate fairly sharp, visible edges, as do some geological processes (e.g. those that form lakes and bedrock intrusions). It is less clear if many other species perceive the landscape as a mosaic at the scales at which they use habitat; for example, a lizard might perceive a forest-grassland edge to be a climate gradient (Schlaepfer & Gavin 2001; Fischer & Lindenmayer 2006).

Natural disturbances can provide spatial models for resource managers in at least three respects: where do disturbances occur on the landscape, and what is their typical size and shape? Small disturbances will generate patches in the larger matrix while large disturbances will constitute a matrix within which it is important to consider the location, size and shape of undisturbed patches, often called skips (e.g. unburned patches on steep slopes, in riparian areas or on wet soils; Hunter 1993; DeLong & Kessler 2000; Mackey *et al.* 2002). The importance of patch size, shape and spatial arrangement (isolation vs. clustering) is well documented in the extensive literature on fragmentation and edge effects (e.g. Lindenmayer & Fischer 2006; see also Chapter 4). Suffice to say here that some species fare best in highly heterogeneous landscapes, while others select more coarsely grained landscapes; emulating natural disturbances will favour whatever seems most natural and appropriate for a given landscape. For example, the spatial patterns of boreal forest fires have led to proposals for variably sized clearcuts (5–100 ha) aggregated into large clusters that cover thousands of hectares with the uncut areas designed to emulate skips (Hunter 1993). Similarly, in Australia the spatial patterning of wildfires may, at least sometimes, inform the strategic placement of fuel-reduction burns and fire breaks (Keith *et al.* 2002b).

Disturbance legacies

Although fires burn organic material and floods and winds can redistribute it, the aftermath of disturbances is characterized by an enormous abundance and diversity of legacies. Residual living material may consist of whole plants (both individuals or skipped patches) that escaped destruction as well as seeds and underground parts. Dead material will range from leaf litter to entire snags and logs. Obviously the living material is key to regeneration and secondary succession after stand-replacing disturbances and development of an uneven-age structure after partial disturbances. However, it is only quite recently that the full importance of the dead material has been appreciated. Snags, logs and smaller material provide cover for animals, substrates for seedlings, nutrient reservoirs, soil protection and more (Franklin *et al.* 2002; Lindenmayer & McCarthy 2002).

Foresters have long focused on the living trees, seedlings and seeds left after logging as the foundation for future stand development, and it should be easy to broaden this thinking to view these structures as crucial ecosystem elements writ large (e.g. 'life boats' that allow other species to persist and recolonize the post-logging forest). It has been somewhat more difficult for foresters to see all the dead material remaining after a disturbance as important structures rather than wasted commodities, but this perspective is growing steadily.

Inevitably there will be less legacy material after a resource extraction operation than a natural disturbance (e.g. because some logs have gone to the mill); the best that advocates of ecosystem integrity can do is to push for narrowing the difference. In response, foresters will ask ecologists: 'So how much do we need to leave behind?' Unfortunately, in the current state of research, this is often a difficult question to answer beyond a general 'the more the better'. In some cases it is reasonable to offer an answer based on the habitat requirements of studied species; an example is the work of Lindenmayer *et al.* (1991) on the relationship between the density of living and dead trees with hollows and the species richness and abundance of arboreal marsupials. However, making a comprehensive statement about the broad ecological benefits (i.e. across the whole range of biota and ecological processes) derived from leaving 5 snags/ha vs. 10/ha vs. 15/ha is not possible at this time. Incidentally, one conspicuous difference between the legacies of fires and logging – the burning of litter and charring of logs – can be reduced by burning debris after a logging operation.

The interaction of natural and anthropogenic disturbances

Initiating anthropogenic disturbances based on natural disturbances will usually not replace natural disturbance regimes in a compensatory fashion and thus managers have to be concerned about the additive effects of having both disturbance types. Under some special circumstances, natural disturbances might be managed to some degree: for example, firebreaks can limit the spread of fires; prescribed burns and thinning stands from below can reduce fuel loads; and growing relatively short trees in dense stands may make them less susceptible to windthrow than growing tall, isolated trees (Brown & Davis 1973). However, with the possible exception of low-intensity frequent ground fires, our understanding of how to control natural disturbances is very limited (Cary & Bradstock 2003).

One way to deal with this issue is to let natural disturbances determine the time and place of disturbance events and then have the resource extraction follow; 'salvage logging' is the forestry term for this. There is some logic to this approach but serious problems can arise given the way salvage logging is usually practised (Lindenmayer *et al.* 2004; Schmiegelow *et al.* 2006). First, it is common for salvage logging to remove all merchantable material, thus ignoring the value of legacies. Second, natural disturbances and the prospect of salvage logging can provide an excuse to permit wood extraction from national parks and other areas set aside for their ecological value, and to build roads into formerly remote areas. Third, salvage logging can produce a surfeit of timber in one time and place, while mills that are far from the salvage site will continue to demand their wood supply, and thus the salvage timber becomes additive to the regular wood supply. These problems are manageable to some extent – for example, salvage logging can be forced to leave legacy structure just as regular logging does – but salvage logging should not be allowed in areas designated for their ecological value because its effects are profoundly different from those of a natural disturbance alone.

Investigating disturbance regimes

Information about natural disturbance regimes is often incomplete because natural landscapes where these processes can still unfold are uncommon in most parts of the world. Nevertheless, the temporal patterns of disturbances can often be reconstructed from historic records, dendrochronology, and the stratigraphy of pollen and charcoal. These same sources can also provide some idea of spatial patterns of past disturbances by searching for temporal concordance among different places, but it is difficult to obtain an adequate sample size to accomplish this with both fine resolution and a large scale. To understand the legacies of disturbances, it is probably adequate simply to survey some recent disturbances if they are reasonably representative. Lack of information may constrain emulating natural disturbances, but it should not derail it. For example, there will almost always be enough information to determine that significant legacies should be left after anthropogenic disturbances.

Note that I have skirted the issue of what constitutes 'natural' (Hunter 1996). Is 40,000 years of aboriginal burning long enough to assume that species have adapted to it? What about 200 years of European activity? What about the respective periods for the American continents, roughly 10,000 and 500 years? The distinction may not be crucial because ultimately long-term disturbance regimes are more of an inspiration for ecologically sensitive thinking than a sacrosanct model that must be followed slavishly.

An alternative model

Although I have long advocated emulating natural disturbance regimes (Hunter 1990, 1999), most recently under the rubric of 'Using nature's template' (M.L. Hunter & F.K.A. Schmiegelow, in preparation), I do not believe that this is always the optimum model for natural resource management for four reasons: (i) the desirability of addressing the habitat needs of particular species under some circumstances (i.e. endangered species or economically important species including *Homo sapiens*); (ii) the likelihood that global climate change will profoundly alter natural disturbance regimes; (iii) uncertainty about details of the historic, natural disturbance regime; and (iv) ambiguity over what is 'natural', especially in a place like Australia with its history of human disturbances dating back 40,000 years or more.

In many circumstances the 'Using nature's template' model should be integrated with an alternative model, 'Diversity begets diversity'. The simple idea here is that to provide habitat for an entire biota we should manage ecosystems in a diverse manner – 'not put all our eggs into one basket' to express it in terms of folk wisdom. At its coarsest level, taking a diverse approach to landscape management constitutes a 'triad model' with three main components: (i) protected natural ecosystems where natural disturbance regimes would prevail; (ii) ecosystems managed for natural resources while trying to emulate natural disturbances to some degree; and (iii) ecosystems managed for production of commodities under an agricultural paradigm (Hunter & Calhoun 1995; Seymour & Hunter 1999). At a finer level, 'diversity begets diversity' means using the emulation of natural disturbances as a catalyst for diverse, creative and ecologically responsive thinking, not as an impediment; it thus becomes an essential element for adaptive management.

Key landscape design principles

1. **Emulating natural landscape dynamics during anthropogenic disturbances such as logging may ameliorate the impact of human activities.** This idea is based on the assumption that species have evolved adaptations in response to natural disturbance regimes and are better able to respond to anthropogenic disturbances that resemble natural disturbances to some degree.
2. **This amelioration will never be perfect for it is not possible for an anthropogenic disturbance to replicate exactly a natural one**, especially if the anthropogenic disturbance is motivated by resource extraction.
3. **Both natural disturbances and anthropogenic disturbances will benefit some species (i.e. lead to larger, more persistent populations) and**

harm others to differing degrees and in diverse ways. This means that emulating natural disturbances requires a balancing act that may be weighted towards species of conservation or economic concern.

4 **Temporal patterns, especially the return intervals between natural disturbances, can provide a model for the periodicity of anthropogenic disturbances.** Typically humans disturb ecosystems more frequently than natural disturbances so emulation of natural patterns will usually require slowing the pace of human activities.

5 **Over time, natural disturbance regimes will change in response to global climate change, and this will require natural resource managers to be sensitive to these changes and nimble in their responses.**

6 **The location, size, shape and other spatial features of natural disturbances can inform the design of anthropogenic disturbances.** For example, emulating natural disturbances could mean modifying anthropogenic disturbances to generate smaller and less regularly shaped patches that are clustered in certain parts of the landscape.

7 **Natural disturbances leave enormous amounts of living and dead legacies (e.g. live trees, snags, logs and more, individually and in skips) and although resource extraction necessitates removing some of this material, it is imperative that significant quantities be left behind for its diverse values.**

8 **Anthropogenic and natural disturbances typically have additive effects on the landscape because it is generally not possible to control natural disturbances.** Salvaging resources after a disturbance can mitigate this phenomenon, but must be carefully undertaken to avoid exacerbating impacts, and it should not be allowed in areas that have been reserved for their ecological values.

9 **Information about natural disturbance regimes is best gathered from ecosystems that have been set aside as scientific benchmarks.** Historic records, dendrochronology, and the stratigraphy of pollen and charcoal can also help, especially to establish temporal patterns.

10 **Emulating natural disturbances should be considered a dominant, but not exclusive, management model, to be used with others that maintain a diverse environment.** Its use should be particularly scrutinized if it could generate a critical shortage of habitat for some species. Using emulation of natural disturbances as an inspiration for creative, ecologically sensitive thinking will integrate well with adaptive management.

References

Andersen, A.N., Cook, G.D., Corbett, L.K. *et al.* (2005) Fire frequency and biodiversity conservation in Australian tropical savannas: implications from the Kapalga fire experiment. *Austral Ecology* **30**, 155–167.

Angelstam, P.K. (1998) Maintaining and restoring biodiversity in European boreal forests by developing natural disturbance regimes. *Journal of Vegetation Science* **9**, 593–602.

Attiwill, P.M. (1994a) The disturbance of forest ecosystems: The ecological basis for conservative management. *Forest Ecology and Management* **63**, 247–300.

Attiwill, P.M. (1994b) Ecological disturbance and the conservative management of eucalypt forests in Australia. *Forest Ecology and Management* **63**, 301–346.

Bradstock, R.A., Tozer, M. & Keith, D.A. (1997) Effects of high frequency fire on floristic composition and abundance in a fire-prone heathland near Sydney. *Australian Journal of Botany* **45**, 641–655.

Bradstock, R.A., Williams, J. & Gill, A.M. (eds.) (2002) *Flammable Australia: the Fire Regimes and Biodiversity of a Continent.* Cambridge University Press, Cambridge.

Brown, A.A. & Davis, K.P. (1973) *Forest Fire: Control and Use*, 2nd edn. McGraw-Hill, New York.

Bunnell, F.B. (1995) Forest-dwelling vertebrate faunas and natural fire regimes in British Columbia: patterns and implications for conservation. *Conservation Biology* **9**, 636–644.

Cary, G. (2002) Importance of a changing climate for fire regimes in Australia. In: Bradstock, R.A., Williams, J.E. & Gill, A.M. (eds.) *Flammable Australia: the Fire regimes and Biodiversity of a Continent*, pp. 26–48. Cambridge University Press, Melbourne.

Cary, G. & Bradstock, R. (2003) Sensitivity of fire regimes to management. In: Cary, G., Lindenmayer, D.B. & Dovers, S. (eds.) *Australia Burning. Fire Ecology, Policy and Management Issues*, pp. 65–81. CSIRO Publishing, Melbourne.

Chambers, L.E., Hughes, L. & Weston, M.A. (2005) Climate change and its impact on Australia's avifauna. *Emu* **105**, 1–20.

Clark, J.S. (1988) Effect of climate change on fire regimes in northwestern Minnesota. *Nature* **334**, 233–235.

DeLong, S.C. & Kessler, W.B. (2000) Ecological characteristics of mature forest remnants left by wildfire. *Forest Ecology and Management* **131**, 93–106.

Fischer, J. & Lindenmayer, D.B. (2006) Beyond fragmentation: the continuum model for fauna research and conservation in human-modified landscapes. *Oikos* **112**, 473–480.

Forman, R.T.T., Sperling, D., Bissonette, J.A. *et al.* (2003) *Road Ecology*. Island Press, Washington, DC.

Foster, D.R., Aber, J.D., Melillo, J.M., Bowden, R.D. & Bazzaz, F.A. (1997) Forest response to disturbance and anthropogenic stress. *Bioscience* **47**, 437–445.

Franklin, J.F. & MacMahon, J.A. (2000) Ecology – messages from a mountain. *Science* **288**, 1183–1185.

Franklin, J.F., Spies, T.A., van Pelt, R. *et al.* (2002) Disturbances and the structural development of natural forest ecosystems with some implications for silviculture. *Forest Ecology and Management* **155**, 399–423.

Gill, A.M. (1999) Biodiversity and bushfires: an Australia-wide perspective on plant-species changes after a fire event. In: Gill, A.M., Woinarski, J. & York, A. (eds.) *Australia's Biodiversity – Responses to Fire: Plants, Birds and Invertebrates.* Environment Australia Biodiversity Technical Paper 1, pp. 9–53. Environment Australia, Canberra, Australia.

Gill, A.M., Bradstock, R.A. & Williams, J.E. (2002) Fire regimes and biodiversity: Legacy and vision. In: Bradstock, R.A., Williams, J. & Gill, A.M. (eds.) *Flammable Australia: the Fire Regimes and Biodiversity of a Continent*, pp. 429–446. Cambridge University Press, Cambridge.

Haila, Y., Hanski, I.K., Niemala, J. *et al.* (1994) Forestry and the boreal fauna–matching management with natural forest. *Annales Zoologici fennici* **31**, 187–202.

Harvey, B.D., Leduc, A., Gauthier, S. & Bergeron, Y. (2002) Stand-landscape integration in natural disturbance-based management of the southern boreal forest. *Forest Ecology and Management* **155**, 369–385.

Hunter, M.L. Jr. (1990) *Wildlife, Forests, and Forestry: Principles of Managing Forests for Biological Diversity*. Prentice-Hall, Englewood Cliffs, NJ.

Hunter, M.L. Jr. (1993) Natural fire regimes as spatial models for managing boreal forests. *Biological Conservation* **65**, 115–120.

Hunter, M.L. Jr. (1996) Benchmarks for managing ecosystems: Are human activities natural? *Conservation Biology* **10**, 695–697.

Hunter, M.L. Jr. (ed.) (1999) *Maintaining Biodiversity in Forest Ecosystems.* Cambridge University Press, Cambridge.

Hunter, M.L. Jr. & Calhoun, A. (1995) A triad approach to land use allocation. In: Szaro, R. & Johnston, D. (eds.) *Biodiversity in Managed Landscapes*, pp. 447–491. Oxford University Press, New York.

Keith, D., McCaw, W.L. & Whelan, R.J. (2002a) Fire regimes in Australian heathlands and their effects on plants and animals. In: Bradstock, R., Williams, J. & Gill, A.M. (eds.) *Flammable Australia: the Fire Regimes and Biodiversity of a Continent*, pp. 199–237. Cambridge University Press, Cambridge.

Keith, D., Williams, J. & Woinarski, J. (2002b) Fire management and biodiversity conservation: key approaches and principles. In: Bradstock, R., Williams, J. & Gill, A.M. (eds.) *Flammable Australia. The Fire Regimes and Biodiversity of a Continent*, pp. 401–425. Cambridge University Press, Cambridge.

Lindenmayer, D.B. & Fischer, J. (2006) *Landscape Change and Habitat Fragmentation: an Ecological and Conservation Synthesis.* Island Press, Washington, DC.

Lindenmayer, D.B. & Franklin, J.F. (2002) *Conserving Forest Biodiversity.* Island Press, Washington, DC.

Lindenmayer, D. & McCarthy, M.A. (2002) Congruence between natural and human forest disturbance: a case study from Australian montane ash forests. *Forest Ecology and Management* **155**, 319–335.

Lindenmayer, D.B., Cunningham, R.B., Tanton, M.T., Nix, H.A. & Smith, A.P. (1991) The conservation of arboreal marsupials in the montane ash forests of the Central Highlands of Victoria, south-east Australia. III. The habitat requirements of Leadbeater's possum, *Gymnobelideus leadbeateri* McCoy and models of the diversity and abundance of arboreal marsupials. *Biological Conservation* **56**, 295–315.

Lindenmayer, D.B., Foster, D.R., Franklin, J.F. *et al.* (2004) Salvage harvesting policies after natural disturbances. *Science* **303**, 1303.

Mackey, B.G., Lindenmayer, D.B., Gill, A.M., McCarthy, M.A. & Lindesay, J. (2002) *Wildlife, Fire and Future Climate.* CSIRO Publishing, Melbourne.

Martin, R. & Handasyde, K. (1999) *The Koala.* UNSW Press, Sydney.

Perera, A.H., Buse, L.J. & Weber, M.G. (2004) *Emulating Natural Forest Landscape Disturbances.* Columbia University Press, New York.

Schlaepfer, M.A. & Gavin, T.A. (2001) Edge effects on lizards and frogs in tropical forest fragments. *Conservation Biology* **15**, 1079–1090.

Schmiegelow, F.K., Stepnisky, D.P., Stambaugh, C.A. & Koivula, M. (2006) Reconciling salvage logging of boreal forests with a natural disturbance management model. *Conservation Biology* **20**, 971–983.

Seymour, R.S. & Hunter, M.L Jr. (1999) Principles of ecological forestry. In: Hunter, M.L. Jr. (ed.) *Maintaining Biodiversity in Forest Ecosystems*, pp. 22–64. Cambridge University Press, Cambridge.

Seymour, R.S., White, A.S. & deMaynadier, P.G. (2002) Natural disturbance regimes in northeastern North America: evaluating silvicultural systems using natural scales and frequencies. *Forest Ecology and Management* **155**, 357–367.

Smith, D.M. (1962) *The Practice of Silviculture*, 7th edn. Wiley and Sons, New York.

Whelan, R.J. (2002) Managing fire regimes for conservation and property protection: an Australian response. *Conservation Biology* **16**, 1659–1661.

Whelan, R., Rodgerson, L., Dickman, C.R. & Sutherland, E.F. (2002) Critical life cycles of plants and animals: developing a process-based understanding of population changes in fire-prone landscapes. In: Bradstock, R.A., Williams, J.E. & Gill, A.M. (eds.) *Flammable Australia. The Fire Regimes and Biodiversity of a Continent*, pp. 94–124. Cambridge University Press, Melbourne.

Woinarski, J.C.Z. (1999) Fire and Australian birds: A review. In: Gill, A.M., Woinarski, J.C. & York, A. (eds.) *Australia's Biodiversity – Responses to Fire: Plants, Birds and Invertebrates.* Environment Australia Biodiversity Technical Paper 1, pp. 55–112. Environmen Australia, Canberra, Australia.

York, A. (1999) Long-term effects of repeated prescribed burning on forest invertebrates: Management implications for the conservation of biodiversity. In: Gill, A.M., Woinarski, J.C. & York, A. (eds.) *Australia's Biodiversity – Responses to Fire: Plants, Birds and Invertebrates.* Environment Australia Biodiversity Technical Paper 1, pp. 181–259. Environmen Australia, Canberra, Australia.

36

Synthesis: Disturbance, Resilience and Recovery

David B. Lindenmayer and Richard J. Hobbs

The essay by Hunter (Chapter 35) focuses largely on the potential use of natural disturbance regimes as a template to guide human disturbances in landscapes. In contrast, most of the essay by Walker (Chapter 34) revolves around the concepts of thresholds and resilience. Some of the key points from these contributions are presented below, together with some of the insights derived from provocative discussions that followed the two presentations in this session.

The use of natural disturbances as a guide to manage human disturbance regimes (such as logging and grazing) is a concept that has been proposed for more than a decade (Hunter, Chapter 35). The underlying premise for this approach to management is that species are likely to be best adapted to disturbance regimes with which they evolved. In contrast, they may be susceptible to the effects of novel kinds of disturbance. Hence, the goals of biodiversity conservation (and also some of the goals of landscape management) might be better achieved by increasing levels of congruence between human disturbance and natural disturbance. Hunter (Chapter 35) concludes that while this approach to managing human disturbance regimes has considerable merit, it is not without its limitations. Some of these include:

- Human disturbance can never be a perfect analogue for natural disturbance.
- The needs of particular taxa may not be met by such an approach, and highly targeted additional management actions may be required to conserve such taxa (see Simberloff, Chapter 26).
- It can be very difficult to determine what is 'natural' and hence what is an appropriate natural disturbance regime.
- Benchmarks are needed to guide the development of appropriate disturbance regimes and help assess consistency with, and departures from, desired states of vegetation structure and condition (see also Franklin & Swanson,

Chapter 12). However, developing these benchmarks is a far from straightforward task (Woinarski, Chapter 10).

- Human and natural disturbances are not independent processes in the landscape. Rather, there are potential cumulative effects resulting from both of them occurring in the same broad area. Post-natural disturbance salvage harvesting is an example.

Hunter (Chapter 35) provides some guidance on the use of human disturbance regimes given these limitations. First, while there is merit in trying to bring the impacts associated with human disturbance closer to those of natural disturbance, there will be occasions when this is not the appropriate management approach. Other strategies may be needed to achieve specific objectives (e.g. the targeted manipulation of the habitat of a particular taxon). Second, avoid doing the same thing everywhere – in other words, given incomplete knowledge and uncertainty, use a risk-spreading approach. Third, be aware of the cumulative problems that can arise from coupling human and natural disturbances. For example, in some cases (such as landscapes and ecosystems already heavily disturbed by humans), it may be nonsensical to add natural disturbance as well.

Like Hunter, Walker (Chapter 34) also acknowledges the crucial importance of disturbance in shaping landscapes and influencing biota. He discusses the potential for disturbances to produce non-linear changes, in particular threshold responses in which critical change points are crossed. Major changes (e.g. in ecological processes, species interactions, population sizes and chances of persistence) may ensue in which there is a sudden switch from an apparently stable state to a markedly different state. Discussions of real and potential thresholds appear in many other chapters including those on aquatic ecosystems (e.g. Calhoun, Chapter 37; and Cullen, Chapter 39), ecosystems and ecosystem processes (e.g. Manning, Chapter 30), and the total amount of vegetation cover (Bennett & Radford, Chapter 18; Fischer & Lindenmayer, Chapter 20). Walker (Chapter 34) describes evidence for a large number of documented threshold responses. However, a key issue remains about how to identify thresholds well before they might occur and, in turn, help ensure that management practices do not drive ecosystems, species and ecological processes close to critical change points. Unfortunately, most research on thresholds to date seems to be theoretical or descriptive – describing threshold responses after they have occurred. A predictive and anticipatory capacity is crucial for any aspect of natural resource management, particularly if breaching a threshold results in effects that are irreversible or extremely difficult or very expensive to reverse. The desirability of developing this capacity has to be matched against the prevailing institutional ethos in many places of ignoring potential problems until they become critical and then instigating crisis management. This type of ethos virtually ensures that thresholds will not be recognized and acted on until they have already been crossed. An additional issue is that while threshold responses undoubtedly exist, there are also many kinds of response trajectories that do not include thresholds characterized by critical breakpoints. Differentiating the kinds of ecosystems, landscapes, species and ecological processes that are prone to threshold responses from those that exhibit other kinds of responses remains a major research and methodological challenge.

Resilience is the other key theme of the essay by Walker (Chapter 34). He describes some of the various characteristics that provide resilience to ecosystems, landscapes and species assemblages. He also outlines ways that it might be possible to manage for increased resilience, but warns that improving resilience in one dimension may reduce resilience in another – a sobering conclusion for on-ground resource management and biodiversity conservation. Many of the concepts and approaches associated with exploring resilience in ecosystems, landscapes and species assemblages are in their infancy and this emerging arena of research seems likely to be a rich one to explore in the coming years.

Although the two preceding chapters are remarkably different, there are some areas for possible convergence. Both recognize that disturbance makes ecosystems and landscapes complex and dynamic entities. Perhaps also through tailoring human disturbance regimes to match more closely natural disturbance regimes (Hunter, Chapter 35), it may be possible to confer greater resilience on landscapes and ecosystems (or at least not diminish it) and hence avoid them approaching or crossing thresholds or critical change points.

Section 10
Aquatic Ecosystems and Integrity

37

Principles for Conserving Wetlands in Managed Landscapes

Aram J.K. Calhoun

Abstract

We have an opportunity to reverse trends of wetland ecosystem degradation and loss by embracing management guidelines based on the best available science, engaging stakeholders in conservation decisions, and conducting research to inform conservation practices. Clarity and consistency in wetland definitions, classification, and inventory methods are essential for assessing the wetland resource to be conserved. Stronger wetland regulations, attention to wetland functions (especially of smaller wetlands), and consideration of the linkages among wetlands and between aquatic and terrestrial systems are crucial components in ensuring the integrity of wetland resources. Conservation theory can be put into practice when we have an educated citizenry, scientists who are active in conservation policy, interdisciplinary teams working towards creative solutions, and stakeholders involved in practical conservation efforts.

Keywords: aquatic; management principles; wetland classification; wetland conservation; wetland inventory; wetlands.

Introduction

'Bigger is better, wetter is better': this has been the mantra for wetland conservation priorities globally, and as a result the integrity of aquatic resources has been greatly compromised through direct loss, fragmentation and degradation. This lack of acknowledgment of small, ephemeral wetlands, or vegetated wetlands that are never flooded, reflects a greater human bias against wetlands. Historically, wetlands were seen as the sources of disease and noxious animals, and hence they were systematically drained or filled thereby 'reclaiming' them for agriculture and development (Dahl 2006). By contrast, it was recently estimated that although wetlands occupy less than 6% of the Earth's surface, they provide 40% of the Earth's renewable services (i.e., hydrological, water quality, biogeochemical cycling) (Zedler 2003a). From a biodiversity perspective, they provide habitat for a much richer biota than predicted based on area alone (De Meester *et al.* 2005). Now that over 50% of global wetlands have probably been degraded or lost (Finlayson & Spiers 1999; Millennium Ecosystem Assessment 2005), and this loss has been linked to unprecedented flooding events and species declines (Lehtinen *et al.* 1999; Dahl 2006), our attention is once again on wetlands. Now we are asking, what and how much should we conserve? And how should we do it?

These questions can be answered through focusing on core principles for managing wetlands from a landscape perspective. Specifically, the goal of this chapter is to provide background on three key issues relevant to improving wetland conservation strategies: wetland regulation; wetland classification and inventory; and geographically isolated wetlands and connectivity. I also suggest a general approach to moving conservation theory into practice followed by specific guidance for direct application of these principles to wetland management. A list of management principles designed to address failures and gaps in current approaches to wetland management is provided at the end of the essay.

The distinction between wetland and aquatic systems is often blurred and varies from country to country. For the purposes of this chapter, I define wetlands as any ecosystem (with the exception of the open ocean and other deepwater habitats) in which the flora or fauna are dependent upon water as habitat for any part of their life history or are adapted to living in inundated conditions.

Wetland protection in the regulatory arena

The integrity of both aquatic and terrestrial ecosystems is threatened by regulatory approaches that treat aquatic and terrestrial systems as isolated and discrete management units. Although regulatory protection of wetlands is a blunt-instrument approach to conserving functions, it is a fundamental first step that raises the profile of wetlands and sets a minimum standard of protection to be emulated by other entities (from local through to higher levels of government).

The ecological and hydrological connections between uplands and wetlands are only now being recognized as the cascading effects of mismanagement have forced us into a 'crisis research' and 'crisis response' mode, for example as in the aftermath of Hurricane Katrina in the southeastern USA. We all know ecosystems do not exist in a vacuum, but ecological principles seem to fall by the wayside during strategic planning. If we are to slow the degradation of wetland ecosystems, we need to take a serious look at how we perceive and manage aquatic and terrestrial resources.

Wetland definition

We must have definitions in order to regulate. Definitions vary among countries, states and regulating agencies, and thus regulations may be inconsistent even within the same geographical region. A clear definition of wetland, hopefully one that transcends political boundaries, is the foundation upon which to build a conservation strategy. The definition will reflect the cultural perception of wetlands (often as articulated by regulators or natural resource managers). For example, a widely used definition in Australia states: 'Wetlands are quirks and aberrations of the hydrologic cycle' (Paijmans *et al.* 1985), suggest that wetlands are more of a problem than a national treasure. I would argue that a definition that is neutral and simply recognizes the unique functions of lands subject to periods of saturation or inundation would reflect our modern understanding of wetland values. Such a broad definition (as defined for this chapter) would cover the full spectrum of wetlands from ephemeral pools on decadal flood cycles to permanent water bodies.

Inadequate protection of wetland functions

Most wetland regulations were developed with the intent of conserving water quality. For example, in the USA there is no federal legislation specifically designed to conserve wetlands. Rather, wetlands were added as 'navigable waters' to the Clean Water Act in 1977 and hence fell under the jurisdiction of the Army Corps of Engineers, an agency at that time dedicated to draining wetlands for agriculture. Declines in migratory waterfowl were the primary trigger that broadened regulation to address wildlife habitat functions. The Ramsar Convention, convened in Iran in 1971, focused on conserving wetlands of international importance and initially targeted large marshes important for waterfowl (and hence sporting recreation) (Davis 1994). Water quality and game species habitat are worth conserving, but regulations still fail to address functions (at multiple scales) that are equally important including transfer of energy and propagules among aquatic systems and terrestrial systems (through biotic and hydrological connections), biogeochemical cycling and biodiversity (Mitsch & Gosselink 2000).

Lack of consideration of context and cumulative impacts

Wetlands are typically regulated as discrete, isolated units. Regulations typically fail to address wetland integrity beyond the scale of the individual wetland (context) and with regard to multiple impacts over time (cumulative impacts). For example, decisions authorising activities that will impact on a given wetland are made without considering the status of the resource in the watershed. Is the wetland the last example of a certain type of shrub swamp? Will its removal substantially increase inter-wetland distances in the watershed? Many wetland functions (e.g., habitat for fauna with biphasic life histories, species organized in metapopulations, nomadic species) are intimately tied to the adjacent terrestrial systems and other wetlands but such contextual considerations are rarely considered in permitting for impacts.

Similarly, regulators often do not base decisions on assessments of historical impacts. For example, a single wetland may legally be subjected to multiple impacts if over time the impacts are attributed to different projects. The loss of an individual wetland is often not weighed against regional wetland losses. Such cumulative losses over time in a watershed degrade the landscape-scale ecosystem services provided by wetlands.

Wetland science as a discipline is quite young (the international Society of Wetland Scientists was established in 1981), and many wetland regulations do not reflect our current understanding of wetland structure and functions. Amendments to existing regulations based on our current understanding of wetland functions are hard to pass and even harder to implement. *The Wetlands Policy of the Commonwealth Government of Australia* (Environment Australia 1997) is an example of an attempt to redress these failures, but policy development is proving to be easier than implementation.

Concept of no-net-loss

A number of countries have a policy of no-net-loss of wetlands that requires mitigation when wetlands are destroyed or degraded (Marsh *et al.* 1996). The type of mitigation required is based on wetland functional assessment protocols that typically place lower value on smaller wetlands and wooded wetlands (National Research Council 2001). As a result, in many cases, diverse wetlands are replaced with open-water pools ringed with common marsh plants (e.g., *Typha* spp.) (National Research Council 2001). In some cases, preservation or 'enhancement' counts as mitigation for loss. Thus, in practice, current policies lead to a net loss in both area and function (Brown & Lant 1999). Decisions regarding wetland mitigation or restoration should take into account the historical wetland landscape: what types of wetlands were present and at what density, proportion and spatial configuration (Windham *et al.* 2004)? Concomitantly, in the current landscape, we need to know which wetlands are vulnerable, or experiencing greatest losses, or are irreplaceable.

Wetland classification and inventory

Classification schemes

A shared classification system creates a common language that allows people to discuss complex phenomena. Wetland classification, coupled with inventory, will provide conservation planners with a powerful tool for assessing the resource and creating management strategies. Because many countries have not yet developed, or are in the process of developing, classification systems, there is opportunity to learn from the strengths and weaknesses of classification systems already in place.

In the USA, Australia, and Canada formal wetland classification was initiated to identify and classify aquatic habitat for waterfowl (Martin *et al.* 1953; Shaw & Fredine 1956; Pressey & Adam 1995). Survey and classification of wetlands was well underway in many countries by the 1970s, but a classification system that went beyond waterfowl habitat was not developed until 1979, when the classification of Cowardin *et al.* (1979) was developed to describe all wetland systems, not simply waterfowl marshes. The classification is primarily based on broad setting (lake, river, ocean) and more finely based on dominant vegetation or substrate structure (greater than 30% cover; e.g., deciduous forest, shellfish bed). This system has served as a template for the current Ramsar wetland classification scheme (Davis 1994), and systems for the Mediterranean (Hecker *et al.* 1996), South Africa (Dini & Cowan 2000) and western Europe (Davies & Moss 2002). Wetland inventory based on a vegetation structure, and to some degree composition, is a valuable tool for gathering baseline data on wetland type and distribution, assessing wildlife habitat, and for tracking changes in the resource base over time. However, this system does not address other wetland functions. For example, a deciduous forested wetland situated on a headwater slope would receive the same classification as a deciduous forested wetland in a groundwater depression adjacent to a river, but their landscape positions dictate differences in water regime, chemistry and to some extent biota (i.e., species influenced equally by hydrology and dominant vegetation type).

Systems based on hydrogeomorphy have been developed in the USA, Australia, and Canada to address these issues of wetland hydrological, biogeochemical and landscape-scale functions (Brinson 1993; Zoltai & Vitt 1995; Semeniuk & Semeniuk 1997). Using this approach, wetlands are classified by landscape position (headwater, riverine, depressional, perched and so on; see Brinson 1993 for details), dominant hydrological inputs (groundwater, surface water runoff, precipitation), geology, and landform. These considerations are particularly relevant for assessing wetland function in the face of global climate change. Classifying wetlands according to landscape position is not unlike guidance given for reserve design that suggests more focus on capturing altitudinal gradients and landforms and less on community composition (see Hunter *et al.* 1988; Groves 2003). Researchers have presented these concepts in diverse frameworks (hydrogeomorphic classes, Brinson 1993; wetland templates, Bedford 1996; wetland continua, Euliss *et al.* 2004; hydrological landscapes, Winter 2001), but they all address the need for incorporating temporal, hydrological, and climatic components into wetland classification schemes. This approach, coupled with a vegetation community classification typical of the earlier systems, will give a better assessment of wetland functions and will greatly improve wetland assessment for mitigation.

Inventory

A wetland inventory is only as strong as the classification system adopted. If a classification has been adopted that encompasses the range of wetland attributes (vegetation structure and hydrogeomorphic setting), an inventory can provide the baseline data needed to assess the current distribution of wetlands and to track changes. However, neither classification nor inventory methods are standardized. A review of the status of wetland inventories was the topic of a special issue of *Vegetatio* (vol. 118, 1995), based on a book by Finlayson and van der Valk (1995), in which these issues, and the fact that many countries lack inventories, was brought to the forefront. Since then, the Ramsar Convention on Wetlands initiated a Global Review of Wetland Resources and Priorities for Wetland Inventory (GRoWI; see Finlayson & Spiers 1999; Finalyson 2003). This study similarly reports that most inventories are not comprehensive, methods are not well defined, and often the goal of the inventory was not to collect comprehensive baseline data.

Isolated wetlands and landscape connectivity

The lack of protection for the drier-end wetlands (i.e., can't float a duck or sail a boat in them year round) and smaller or geographically isolated wetlands (those not directly associated with aquatic systems such as lakes, streams/rivers, including small ponds) highlights a shortsightedness. In particular, these wetlands are often crucial to wetland connectivity and thus conservation of wetland landscapes will not be effective unless they are fully considered. Recent use by scientists of the term 'isolated wetland' for smaller or wooded wetlands is the result of a landmark US Supreme Court decision in 2001 (*Solid Waste Agency of Northern Cook County v. U.S. Army Corps of Engineers*, 531 U.S. 159; Downing *et al.* 2003) that removed wetlands not adjacent to open-water aquatic resources (i.e., 'isolated') from federal protection. This setback in wetland conservation precipitated a flurry of research on 'isolated' wetland functions; two international journal special issues were published to highlight the value of so-called 'isolated' wetlands (*Wetlands* vol. 23, 2003; and *Wetlands Ecology and Management*, vol. 13, 2005).

These publications reflect a significant body of research supporting the important role of 'isolated' wetlands in conserving aquatic integrity at the landscape scale (see Leibowitz 2003). 'Isolated' wetland functions – specifically the transfer of energy, matter (water, sediment), nutrients and organisms between isolated wetlands and other aquatic and terrestrial systems, often mediated by intermittent hydrological connections – argue against the label 'isolated' (Pringle 2001). A few examples are provided below to highlight the most recent findings that should guide future management policies.

Ecological connectivity of 'isolated' wetlands has been well documented in the amphibian, reptile and bird literature. Many herptiles, especially amphibians with biphasic life histories, use multiple habitats during their lives including isolated wetlands, open water bodies, forested wetlands and terrestrial habitats (Gibbs 2000; Marsh & Trenham 2001; Semlitsch & Bodie 2003; Trenham & Shaffer 2005; Regosin *et al.*, in press).

Smaller, geographically isolated wetlands serve as stepping stones between wetlands and more aquatic systems. Local populations of non-wetland breeding species including birds, small mammals and herptiles, may be threatened by loss of wetland stepping stones (Gibbs 1993, 2000), which may also provide amphibians with refugia from predators. A variety of species that are dependent on larger wetlands for breeding, hibernation or aestivation only occupy these wetlands if they are located in high-density-wetland landscapes – for example, Blanding's turtle (*Emydoidea blandingi*), spotted turtle (*Clemmys guttata*) and black tern (*Chlidonias niger*) (Haig *et al.* 1998; Joyal *et al.* 2001; Miller *et al.* 2001; Naugle *et al.* 2001). In arid landscapes where wetlands are temporally dynamic (e.g., western North America, inland Australia), wetland density and distribution must be considered to maintain an array of waterbirds (Plissner *et al.* 2000; Roshier *et al.* 2002). There is also growing anecdotal evidence of use of geographically isolated wetlands by small and large mammals as moist refugia during dry seasons, as resting areas in winter (less snow cover in temperate and boreal regions) and as foraging areas during the growing season (A. Calhoun, unpublished).

Documentation of the contribution of 'isolated' wetland to regional biodiversity is widely available in the wetland literature (Zedler 2003b; De Meester *et al.* 2005; Scheffer & van Geest 2006). Seasonal or ephemeral wetlands (e.g., arid wetlands including freshwater and saltwater lakes and claypans, prairie potholes, vernal pools, playa lakes and woodland pools) often support a biota that is adapted to life in temporary waters found in no other wetland types (Colburn 2004), and in situations where these wetlands are truly hydrologically isolated from other wetlands, may support a genetically distinct biota (De Meester *et al.* 2005).

Isolated wetlands are also not necessarily hydrologically isolated. Indeed, the degree of hydrological connectivity is variable among wetlands, with some being linked to regional aquifers while others are intermittently connected with other wetlands and aquatic resources (Cable Rains *et al.* 2006). Fewer are truly hydrologically isolated from other systems (Pringle 2001; Leibowitz & Vining 2003; Winter & LeBaugh 2003; Euliss *et al.* 2004). The concept of an isolation-connectivity continuum has been suggested that recognizes the dynamic nature of hydrological connections among these 'isolated' wetlands by considering spatial and temporal aspects of connection: for example, depending on the climate (flood vs. drought), geographically isolated wetlands may be physically connected to other wetlands annually or only every 10 years (see Leibowitz & Vining 2003).

The role of geographically isolated wetlands in maintaining water quality and enhancing water supply is not well researched. The position and density of isolated wetlands in a watershed strongly suggest that they will play a role in water storage, filtration and nutrient transformations that may influence the water quality of other aquatic resources.

A landscape perspective that sees wetlands in a broad climatic and hydrological perspective fosters thinking that will allow managers to link 'isolated' wetland functions to off-site effects on open waters. All these functions suggest a level of biotic and abiotic connectivity that needs to be integrated into land-use policy.

Translating conservation theory into practice

The take-home message presents a difficult challenge: it is that all systems are inextricably intertwined and hence landscape integrity is as strong as its weakest link. Isn't this a basic principle of ecology? Why are we just now beginning to understand that aquatic and terrestrial systems should be managed together? Part of the answer lies in the complexity of translating theory into practice. We have to look to human culture to see where conservation theory and practice part ways.

- We are organized by social and political units, not by ecological units.
- We are wired for thinking in the here and now (instant gratification) and less comfortable with sacrificing in the present for the future.
- We have trouble valuing things we cannot see, hold or understand (hence the denitrification functions of forested wetlands do not roll off most people's lips as a wetland value).
- Concepts of private land (where in many cases the majority of the resources lie) collide with conserving ecosystem services for the public good, particularly in Western cultures or in developing countries where resources are exploited by outside interests.
- The tenets of a basic education include reading, writing and maths. Ecology is not a required topic for formally educated citizens.

I believe these broad generalizations provide the guidance that conservation scientists and planners need to rectify a trend of ecosystem degradation in general and to redress negative trends in their discipline in particular. Specifically, give some time and focus to:

- Education. It is incumbent upon conservation scientists to translate complex ecological principles into terms that are meaningful for the general public and to provide guidelines for practitioners. This can be done through involvement in community conservation initiatives, publications written for non-scientists and public presentations. Strive to make abstract functions palpable to the public and to link seemingly remote functions (such as biogeochemical cycling) to human benefit.
- Activism. This is not to be confused with chaining yourself to a tree or lying between a wetland and the bulldozer. It means taking the extra step to provide input at some level to policy decisions, to work for public education, and to take the extra step to respond to poor management decisions in a constructive, positive and very public fashion. Bring the public into your orbit and do not get lost in academia.
- Integrated problem solving. Work across disciplines. Moving conservation theory into practice requires taking into consideration socioeconomic issues, understanding human motivations for decisions (psychology) and providing incentives. We need creative approaches, and this can only be done by merging science and humanities, by putting our science into a human context.
- Involve citizens in practical conservation and management plans. There is a plethora of literature on the importance of stakeholder involvement in effective land management and policy initiatives (Theobald & Hobbs 2002; Savan & Sider 2003; Moore & Koontz 2003; Berkes 2004). An educated citizen with a strong sense of place is a powerful conservation tool.

Key steps in applying core principles of conservation planning to practical wetland management

Wetland conservation starts at home. Assume that the management guidelines outlined above have been embraced and a classification and inventory that takes into account wetland form and function is in place. What should wetland managers do next?

Technical approach (the easy part)

1. Identify crucial wetland resources, using the wetland inventory (e.g., vulnerable, rare or exemplary).
2. Identify hydrological linkages among wetlands and aquatic resources. Consider hydrological linkages that may change through time based on climate or annual precipitation patterns.
3. Identify species of concern and consider their within-season as well as annual movements and habitat needs, including the special needs of nomadic species. These will make clear linkages among resources that need to be maintained through corridors or other tools.
4. Be comprehensive in assessments of threats to wetland ecosystems. Exurban and suburban development should be considered in risk assessments.
5. Identify regions at high risk of loss based on population trends and, more importantly, build out scenarios based on current levels of protection (see Theobald & Hobbs 2002 for a case study/model of this approach).
6. Identify low-hanging fruits from a conservation perspective: intact wetlands that abut public lands or that may be easily conserved, or areas on degraded lands that may be relatively inexpensive to purchase and have great restoration potential.
7. Overlay wetland data with data layers for other crucial ecological functions to identify areas that meet diverse needs.
8. Track the efficacy of conservation strategies through evaluation and monitoring plans and provide mechanisms for adaptive management.

Sociopolitical context (the hard part)

1. Articulate the issues and the choices clearly (based on your above groundwork) to local stakeholders and involve them at the outset.
2. Develop goals for conservation: do you want to restore wetlands to the distribution and density of pre-development eras? Do you want representation of all the natural wetland types? Are you focusing only on rare and vulnerable types in your watershed? Do you want to conserve an array of wetland functions or only specific functions?
3. Ask if current wetland regulation or conservation strategies meet these goals. If they don't, you will need to work towards making those changes or complementing them.
4. Do you want to expand your work to include other watersheds or similar contiguous ecoregions?

Research gaps

Even if we have all the data necessary to make conservation decisions, it is a challenge to define our goals and decide in the present how we want a landscape to look and function in the future. Still, there are major gaps in our knowledge. Key things that need to be addressed to advance wetland conservation include:

- Thresholds. We need to know thresholds for collapse of function or populations at multiple scales (Groffman *et al.* 2006). For example, what proportion of headwater streams or riparian areas need to be conserved in a watershed to maintain hydrological functions? What density of wetlands is needed in a region to maintain biological and hydrological connectivity?
- Exurban/suburban impacts. We need information about the effects of low-density housing and development on wildlife (Miller & Hobbs 2002; Theobald 2005; Rubbo & Kiesecker 2005).
- Carbon and nutrient exchanges. We need to quantify exchanges of carbon and nutrients among wetlands and between wetlands and terrestrial systems.

Wetland management principles

1 Wetland regulations should be designed to conserve an array of wetland functions, not limited to water quality, waterfowl habitat and recreation. They should address cumulative impacts, and connectivity of wetland, aquatic and terrestrial resources, and be comprehensive enough to protect both individual wetlands and the overall integrity of landscapes in which wetlands occur.
2 Wetland mitigation should address both wetland function and area and should focus on maintaining wetland classes that naturally occur in a region (i.e., creation of simple ponds across a landscape that does not naturally have ponds is not responsible mitigation).
3 Wetland resources should be classified using a system that puts them in a climatic and hydrological framework in order to capture their dynamism and their role in landscape-scale processes. Vegetation-based classifications should be melded with those based on hydrogeomorphology.

4 Wetland classification and inventory systems should be comprehensive and standardized to facilitate information exchange globally. This could be done at a basic level by adopting a core template of information to be collected and through clarifying definitions and methods of data collection. This template could reflect minimal data standards, with variations on that template reflecting regional needs.
5 The spatial and temporal distribution of geographically isolated wetlands must be considered when evaluating the integrity of aquatic and terrestrial resources.
6 Conservation theory can be put into practice through an educated citizenry, by scientists who are active in conservation policy, by interdisciplinary teams working towards creative solutions and by stakeholders involved in practical conservation efforts.
7 Effective wetland conservation begins locally. Local efforts should strive to fill gaps left by governmental regulations through working with regional stakeholders to conserve wetland resources.
8 Tailor research to meet research gaps that hinder wetland conservation planning. Ideally, wetland policy is based on the best available science.

References

Bedford, B.L. (1996) The need to define hydrologic equivalence at the landscape scale for freshwater wetland mitigation. *Ecological Applications* **6**, 57–68.

Berkes, F. (2004) Rethinking community-based conservation. *Conservation Biology* **18**, 621–630.

Brinson, M.M. (1993) *A Hydrogeomorphic Classification for Wetlands.* US Army Corps of Engineers, Waterways Experiment Station, Vicksburg.

Brown, P.H. & Lant, C. (1999) The effect of wetland mitigation banking on the achievement of no-net-loss. *Environmental Management* **23**, 333–345.

Cable Rains, M., Fogg, G.E., Harter, T., Dahlgren, R.A. & Williamson, R.A. (2006). The role of perched aquifers in hydrological connectivity and biogeochemical processes in vernal pool landscapes, Central Valley, California. *Hydrological Processes* **20**, 1157–1175.

Colburn, E.A. (2004) *Vernal Pools: Natural History and Conservation.* McDonald and Woodward Publishing, Virginia, VA.

Cowardin, L.M., Carter, V., Golet, F.C & LaRoe, E.T. (1979) *Classification of Wetlands and Deepwater Habitats of the United States.* United States Fish and Wildlife Service, Washington, DC.

Dahl, T.E. (2006) Status and trends of wetlands in the conterminous United States 1998 to 2004. US Department of the Interior, Fish and Wildlife Service, Washington, DC.

Davies, C.E. & Moss, D. (2002) *EUNIS Habitat Classification. Final Report to the European Topic Center on Nature Protection and Biodiversity*, p. 125. European Environment Agency, Copenhagen.

Davis, T.J. (1994) *The Ramsar Convention Manual: a Guide to the Convention of Wetlands of International Importance Especially as Waterfowl Habitat.* Ramsar Convention Bureau, Gland, Switzerland.

De Meester, L., Declerck, S., Stoks, R. *et al.* (2005) Ponds and pools as model systems in conservation biology, ecology, and evolutionary biology. *Aquatic Conservation: Marine and Freshwater Ecosystems* **15**, 715–725.

Dini, J.A. & Cowan, G.I. (2000) *Classification System for the South African Wetland Inventory*, Second Draft. South African Wetlands Conservation Programme, Department of Environmental Affairs and Tourism, Pretoria.

Downing, D.M., Winer, C. & Wood, L.D. (2003) Navigating through Clean Water Act jurisdiction: a legal review. *Wetlands* **23**, 475–493.

Environment Australia (1997) The Wetlands Policy of the Commonwealth Government of Australia [online]. URL: http//www.environment.gov.au/water/wetlands/publications/pubs/policy.pdf

Euliss, N.H. Jr., LaBaugh, J.W., Fredrickson, L.H. *et al.* (2004) The wetland continuum: a conceptual framework for interpreting biological studies. *Wetlands* **24**, 448–458.

Finlayson, C.M. (2003) The challenge of integrating wetland inventory, assessment, and monitoring. *Aquatic Conservation, Marine and Freshwater Ecosystems* **13**, 281–286.

Finlayson, C.M. & Spiers, A.G. (eds.) (1999) *Global Review of Wetland Resources and Priorities for Wetland Inventory.* Supervising Scientist Report 144, Supervising Scientist, Canberra.

Finlayson, C.M. & van der Valk, A.G. (eds.) (1995) *Classification and Inventory of the World's Wetlands.* Advances in Vegetation Science 16, Kluwer Academic Publishers, Dordrecht.

Gibbs, J.P. (1993) Importance of small wetlands for the persistence of local populations of wetland-associated animals. *Wetlands* **13**, 25–31.

Gibbs, J.P. (2000) Wetland loss and biodiversity conservation. *Conservation Biology* **14**, 314–317.

Groffman, P.M., Baron, J.S., Blett, T. *et al.* (2006) Ecological thresholds: the key to successful environmental management or an important concept with no practical application? *Ecosystems* **9**, 1–13.

Groves, C. (2003) Drafting a conservation blueprint. Island Press, Washington, DC.

Haig, S.M., Mehlman, D.W., & Oring, L.W. (1998) Avian movements and wetland connectivity in landscape conservation. *Conservation Biology* **12**, 749–758.

Hecker, N., Costa, L.T., Farinha, J.C. & Tomas Vives, P. (1996) *Mediterranean Wetlands Inventory: Data Recording*, vol. 2. MedWet/Wetlands International, Slimbridge, UK/Instituto da Concervaco da Nautreza, Lisbon.

Hunter, M.L. Jr., Jacobson, G. & Webb, T. (1988) Paleoecology and the coarse-filter approach to maintaining biological diversity. *Conservation Biology* 2, 375–385.

Joyal, L.A., McCollough, M. & Hunter, M.L. Jr. (2001) Landscape ecology approaches to wetland species conservation: a case study of two turtle species in southern Maine. *Conservation Biology* **15**, 1755–1762.

Lehtinen, R.M., Galatowitsch, S.M. & Tester, J.R. II (1999) Consequences of habitat loss and fragmentation for wetland amphibian assemblages. *Wetlands* **19**, 1–12.

Leibowitz, S.G. (2003) Isolated wetlands and their functions: an ecological perspective. *Wetlands* **23**, 517–531.

Leibowitz, S.G. & Vining, K.C. (2003) Temporal connectivity in a prairie pothole complex. *Wetlands* **23**, 13–25.

Marsh, D.M. & Trenham, P.C. (2001) Metapopulation dynamics and amphibian conservation. *Conservation Biology* **15**, 40–49.

Marsh, L.L., Porter, D.R. & Salvesen, D.A. (1996) *Mitigation Banking: Theory and Practice*. Island Press, Washington, DC.

Martin, A.C., Hotchkiss, N., Uhler, F.M. & Bourn, W.S. (1953) *Classification of Wetlands of the United States*, p. 14. US Fish and Wildlife Service, Special Scientific Report – Wildlife No. 20, Washington, DC.

Millennium Ecosystem Assessment (2005) *Ecosystems and Human Well-Being: Wetlands and Water Synthesis*. World Resources Institute, Washington, DC.

Miller, J.R. & Hobbs, R.J. (2002) Conservation where people live and work. *Conservation Biology* **16**, 300–337.

Miller, J.R., Fraterrigo, J.M., Hobbs, N.T., Theobald, D.M. & Wiens, J.A. (2001) Urbanization, avian communities, and landscape ecology. In: Marzluff, J.M., Bowman, R. & Donnelly, R. (eds.) *Avian Ecology and Conservation in an Urbanizing World*, pp. 117–137. Kluwer Academic Publishers, New York.

Mitsch, W.J. & Gosselink, J.G. (2000) *Wetlands*, 3rd edn. John Wiley and Sons, New York.

Moore, E.A. & Koontz, T.M. (2003) A typology of collaborative watershed groups: citizen-based, agency-based, and mixed partnerships. *Society and Natural Resources* **16**, 451–460.

National Research Council (2001) *Compensating for Wetland Losses under the Clean Water Act*. National Academy Press, Washington, DC.

Naugle, D.E., Johnson, R.R., Estey, M.E. & Higgins, K.F. (2001) A landscape approach to conserving wetland bird habitat in the prairie pothole region of eastern South Dakota. *Wetlands* **21**, 1–17.

Paijmans, K., Galloway, R.W., Faith, D.P. *et al.* (1985) *Aspects of Australian Wetlands*. Division of Water and Land Resources Technical Paper no. 44. CSIRO, Canberra.

Plissner, J.H., Haig, S.M., & Oring, L.W. (2000) Postbreeding movements of American avocets and implications for wetland connectivity in the western Great Basin. *The Auk* **117**, 290–298.

Pressey, R.L.& Adam, P. (1995) A review of wetland inventory and classification in Australia. *Vegetatio* **118**, 81–101.

Pringle, C.M. (2001) Hydrologic connectivity and the management of biological reserves. A global perspective. *Ecological Applications* 11, 981–998.

Regosin, J.V., Windmiller, B.S., Homan, R.N. & Reed, J.M. Variation in terrestrial habitat use by four pool-breeding amphibian species. *Journal of Wildlife Management* **69**,(4), 1481–1493.

Roshier, D.A., Robertson, I.A. & Kingsford, R.T. (2002) Responses of waterbirds to flooding in an arid region of Australia and implications for conservation. *Biological Conservation* **106**, 399–411.

Rubbo, M.J. & Kiesecker, J.M. (2005) Amphibian breeding distribution in an urbanized landscape. *Conservation Biology* **19**, 504–511.

Savan, B. & Sider, D.(2003) Contrasting approaches to community-based research and a case study of community sustainability in Toronto, Canada. *Local Environment* **8**, 303–316.

Scheffer, M. & van Geest, G.J. (2006) Small habitat size and isolation can promote species richness: second-order effects on biodiversity in lakes and ponds. *Oikos* **112**, 227–231.

Semeniuk, V. & Semeniuk, C.A. (1997) A geomorphic approach to global classification for natural inland wetlands and rationalization of the system used by the Ramsar Convention – a discussion. *Wetlands Ecology and Management* **5**, 145–158.

Semlitsch, R.D. & Bodie, J. R. (2003) Biological criteria for buffer zones around wetlands and riparian habitats for amphibians and reptiles. *Conservation Biology* **17**, 1219–1228.

Shaw, S.P. & Fredine, C.G. (1956) *Wetlands of the United States*, p. 67. U.S. Fish and Wildlife Service, Circular 39.

Theobald, D. (2005) Landscape patterns of exurban growth in the USA from 1980 to 2020. *Ecology and Society* **10**, 32.

Theobald, D.M. & Hobbs, N.T. (2002) A framework for evaluating land use planning alternatives: protecting biodiversity on private land. *Conservation Ecology* **6**, 5 [online] (http://www.consecol.org/vol6/iss1/art5).

Trenham, P.C. & Shaffer, H.B. (2005) Amphibian upland habitat use and its consequences for population viability. *Ecological Applications* **15**, 1158–1168.

Windham, L., Laska, M.S. & Wollenberg, J. (2004) Evaluating urban wetland restorations: case studies for assessing connectivity and function. *Urban Habitats* **2**, 130–145.

Winter, T.C. (2001) The concept of hydrologic landscapes. *Journal of the American Water Resources Association* **37**, 335–349.

Winter, T.C. & LaBaugh, J.W. (2003) Hydrologic considerations in defining isolated wetlands. *Wetlands* **23**, 532–540.

Zedler, J.B. (2003a) Wetlands at your service: reducing impacts of agriculture at the watershed scale. *Frontiers in Ecology and Environment* **1**, 65–72.

Zedler, P.H. (2003b) Vernal pools and the concept of 'isolated wetlands.' *Wetlands* **23**, 597–607.

Zoltai, S.C. & Vitt, D.H. (1995) Canadian wetlands: Environmental gradients and classification. *Vegetatio* **118**, 131–137.

38

Flowing Waters in the Landscape

P.S. Lake

Abstract

Flowing waters are an essential component of landscapes, although only recently has this integration come to be fully accepted. Streams or lotic systems are not networks, but are unidirectional ramifications with a dynamic patchiness. This patchiness is largely regulated by flow-generated forces, especially those of flow-generated disturbance – floods and droughts. Connectivity – longitudinal, lateral and vertical – is a vital attribute of lotic systems and as such has been disrupted or severed by human intervention in many cases. Riparian zones are crucial transition zones between streams and their catchments. They contribute substantially to catchment biodiversity and serve to mediate the lateral movements of biota, chemicals and sediments between streams and their catchments. Floodplains are riparian zones and with flooding they are highly productive systems. However, human activities have degraded many floodplains by curtailing or preventing natural floods. Stream restoration is now gaining momentum. Many projects are small, short-term and uncoordinated, when the need is for them to be long-term and integrated at the catchment scale. Projects need to be rigorously designed and evaluated so that knowledge is acquired in order progressively to improve stream restoration.

Keywords: connectivity; patchiness; restoration; riparian zones; streams.

Introduction

In the early development of landscape ecology, rivers and riparian zones were seen as corridors with sharp, linear boundaries and with little internal structure. Linked with the corridor approach has been the obvious and somewhat persistent discrepancy between the spatial and functional attributes of flowing waters (lotic systems) both internally and in relation to their surrounding landscapes.

This essay focuses on the scape ecology of flowing waters, their riparian zones and their catchments. I will deal with the three essential attributes of lotic systems, namely the structural–functional nature of flowing waters, the crucial property of connectivity and the effects of disturbance. Along the way, I hope to bring together concepts in landscape and aquatic-scape ecology and to highlight the importance of terrestrial-lotic linkages in influencing both terrestrial and lotic ecosystems within catchments.

Flowing waters worldwide have been extensively degraded, mostly by human activities directly altering the essential attributes of lotic systems and terrestrial-lotic linkages. Such activities have threatened the ecological sustainability of flowing waters and have impaired the sustainable maintenance of lotic ecosystems (Lake 2005). Many of the changes may be irreversible whereas others may be alleviated, even reversed, by restoration.

Catchments and flowing waters

Flowing waters have been, and are, the major force shaping many catchments. Thus, much of the patchiness in the terrestrial environment of catchments has been created by flowing water. Divides between catchments, although hydrological, affect the distribution of freshwater biota and may also influence terrestrial biota.

Independently of landscape ecology, a comprehensive understanding of the links between hydrology, fluvial geomorphology and ecology has been steadily developed. Major steps forward were the recognition of the biogeochemical linkages between catchments and streams (e.g. Likens *et al.* 1977), and that catchment land use can govern stream condition (Hynes 1975). Building on this was the discovery that catchment detritus inputs were important in stream metabolism (Cummins 1974). This led to the river continuum concept (Vannote *et al.* 1980), whereby catchment inputs and longitudinal changes down rivers strongly influence the structure of biotic assemblages and ecological processes. Subsequently, disturbance, especially by floods, was recognized as being influential in structuring streams (e.g. Resh *et al.* 1988). Concurrently, there was recognition of the importance of riparian zones to stream condition and of flooding to sustain floodplain ecosystems (Junk *et al.* 1989).

Changes in perception of lotic systems in landscapes

Wiens (2002) pointed out that the land-focused approach of landscape ecology had viewed streams and their riparian zones simply as corridors and that streams were seen as lacking internal structure. Stream ecologists themselves are partly to blame for this situation as they have long investigated streams by concentrating on the channel itself rather than the bidirectional linkages between the channel and its terrestrial surrounds (Fisher *et al.* 2001). Clearly, flowing water systems have patchiness (Wiens 2002) and mosaics of patchiness generate streamscapes. The spatial scale of patches varies, for example, from decomposing leaves (Palmer *et al.* 2000), to patches of stones and gravel (Barmuta 1989), to log jams in big rivers (Sedell & Froggatt 1984). Patchiness in flowing waters is dynamic and as in terrestrial systems, patch duration appears to be scaled with size (Wiens 1989) – a sunken leaf may last days, a big log jam for years. The incorporation of landscape ecology into fluvial ecology has been most marked in studies on the terrestrial-aquatic transitions that occur with the expansion and contraction of floods (Stanley *et al.* 1997; Amoros & Bornette 2002; Ward *et al.* 2002). The view of streamscapes being made up of dynamic mosaics of patches has generally been accepted (Townsend 1989), although the patchiness is better understood in terms of physical structure and inhabitants, than in terms of ecological processes. Crucial to maintaining the dynamic patchiness are the two major components of connectivity and the disturbance regime.

Flowing waters and the axes of connectivity

Streams and rivers have been envisaged as networks – structures that are meshed together as in a net. However, streams are basically unidirectional, branched structures or ramifications that with the decline in altitude progressively coalesce. They are usually portrayed as linear two-dimensional systems, but they are three-dimensional structures; streams are complex ramifications with steep sections of high flow energy and gentle slopes of low energy. Although ecological processes are poorly understood at present in a space-specific context, key processes, such as detritus and nutrient retention, are clearly linked to the three-dimensional rather than the two-dimensional structure of streams. Stream ecologists have mostly worked at the levels of sites and have neglected a crucial property of ramifications, namely the junctions or nodes (Fisher *et al.* 2004). Such junctions vary considerably dependent on the relative size of the two streams. Junctions as sites of mixing may have high heterogeneity (physically, chemically and biologically) and may be important hotspots in stream metabolism and biodiversity.

In any scape, connectivity between patches and between scapes can be important (Lake 2003). In the case of lotic systems, water-driven connectivity is essential. Longitudinal connectivity has long been appreciated from the times of zone definition in rivers to more recent concepts, such as the river continuum concept. Following the longitudinal division of rivers into upland headwaters, constrained upland valley sections and finally lowland floodplain sections, the movements of water, sediments, chemicals (including nutrients), detritus and biota vary in levels of magnitude, retention and processing. Nutrients, such as nitrogen (N) and phosphorus (P), may be taken up by biota, metabolized and released back into the water column as they move downstream – a process encapsulated in the nutrient spiralling concept (Elwood *et al.* 1983). In streams, as opposed to many other systems, the transport of chemicals may dominate over their involvement in biological processing.

Longitudinal connectivity is bidirectional, with the downstream axis dominating the upstream one. However, some forms of upstream connectivity are crucial, such as fish migrations. Anadromous fish, such as salmon, undergo lengthy migrations to spawn; their carcasses may not only fertilize the stream, but through wasteful predation may also fertilize the forest (e.g. Willson *et al.* 1998). Anthropogenic barriers, ranging from small weirs to large dams, have greatly disrupted longitudinal connectivity.

Headwater streams are vital to a river's integrity as they may comprise up to 75% of total stream length and capture up to 80% of the water. In terms of research and management they have been neglected. Indeed, management has concentrated on the valley and floodplain sections of rivers (the gauged and audited sections). As headwater streams comprise the major source not only of water, but also of sediments and nutrients, it is essential that they are intact and not 'drains' from damaged catchments. It is illuminating to look at headwater streams in a natural forest, and then go out to cleared land and see the changes to headwater streams – with, for example, loss of riparian zones, human-made barriers, channels straightened and associated wetlands drained. As the classic study of Likens *et al.* (1970) showed, clear-felling catchments can cause a hitherto retentive system to leak, lifting chemical concentrations in headwater streams.

Lateral connectivity occurs along two major routes, from the catchment to the stream and from the river to the floodplain. From the upland catchment to the stream, water carries into the stream sediment and soluble chemicals, ranging from inorganic ions to complex forms of dissolved organic matter. The latter fuels much of the microbially driven ecological processing of nutrients in streams. As Fisher *et al.* (2004) indicate, flowpaths from the catchment to the stream occur both on and below the land surface, and their spatial extent, duration and number in a catchment depend on precipitation levels, vegetation cover and geology. Thus, an Order-1 stream may only be an occasional flowpath. In pursuing the nature of flowpaths, it becomes clear that the terrestrial and aquatic components of a catchment blend into each other (Fisher *et al.* 2004). Furthermore, there is the difference between perceived surface patchiness and the spatial design of ecological processes.

Lowland rivers naturally flood their floodplains. Floodplain width may be only hundred of metres, but can be hundreds of kilometres (e.g. Puckridge *et al.* 2000). The advent of floodwaters stimulates high levels of production – a production pulse that makes floodplains one of the most productive ecosystems in the world (Tockner & Stanford 2002). Biodiversity is increased by floods triggering hatching, breeding and succession and creating hubs for immigration. By inundating the floodplain, floods generate a complex mosaic of habitats or patches, that change with succession and with disconnection as floods recede (Ward *et al.* 2002). Thus, for both landscape design and river condition, maintaining intact floodplains that are naturally flooded is vital. Yet in innumerable instances, floods have been curtailed and floodplains have been settled, exploited and alienated from their rivers. For example, over 90% of the floodplain of the mid- and lower Mississippi is leveed (Sparks *et al.* 1998).

The connectivity axis of channel bottom to subsurface to hyporheic zone to groundwater is bidirectional – to and from the surface. This zone can be particularly important for processing of nutrients, such as nitrogen (Fisher *et al.* 2004). The hyporheic zone harbours a distinctive fauna – the hyporheos. In some rivers it can extend considerable distances laterally from the stream channel (Boulton *et al.* 1998).

Boundaries and riparian zones

As indicated by flowpaths, terrestrial-aquatic boundaries are both diffuse and ever-changing. While surface linkages, such as temporary rills and runnels, are easily discernible, groundwater linkages are difficult to trace. Yet these links are vital. For example, groundwater flows provide the baseflow of streams. As flow in a stream fluctuates, so do the terrestrial-aquatic boundaries (Stanley *et al.* 1997), and these changes are relatively rapid with floods and drawn-out with drought (Boulton 2003). On floodplains, with the expansion and contraction of floods, there is a rich and dynamic tapestry of boundary changes. Thus, instead of rivers and streams being linear and sharply bounded structures, they actually have blurred and ever-changing boundaries.

Riparian zones have been regarded as ecotones (Naiman & Décamps 1997) mediating transactions between streams and their hinterland. In landscape ecology they have been largely regarded as corridors for wildlife and as habitats rich in diversity. Indeed, the latter is startlingly shown by Sabo *et al.* (2005), who revealed that riparian zones may not contain a higher number of species than their hinterland, but support a significantly different assemblage of species that may add more than 50% to regional biodiversity. Riparian zones supply streams with detritus (both particulate organic matter and dissolved organic matter) and coarse wood – a valuable component of instream habitat. They shade streams, thereby moderating water temperatures, and regulate sediment inputs. Very importantly, riparian zones moderate the inputs of chemicals, particularly in groundwater flowpaths. In a landscape sense, they are a valuable component in spite of occupying only a small area – at least in constrained streams.

Riparian zones are usually defined by their vegetation. In low-order upland streams, the riparian zone width may vary considerably and may even extend almost up to the ridgeline. As this area is the water-gathering part of rivers, this is not surprising. However, the extent of this area is not currently recognized by management in many cases. Riparian zone boundaries are complex and not the sharply defined linear boundaries that we impose on these zones in order to 'protect' or even restore them. In many places, riparian zones have been fragmented and narrowed, greatly impeding their capacity to protect streams and wetlands (Houlahan & Findlay 2004).

Floodplains in the natural state are covered with riparian vegetation (Ward *et al.* 2002; Johnson 2002), and thus the riparian zone can be very wide. Unfortunately, such riparian zones have been poorly managed and have been degraded by extensive clearing, agriculture and river regulation. Lack of regular flooding due to regulation can lead to major losses of riparian vegetation (Johnson 2002) along with losses in biodiversity and production (Tockner & Stanford 2002).

Trophic subsidies

Trophic subsidies are linked with riparian zones. The importance of detritus from riparian zones fuelling in-stream food webs is well known (Cummins 1974; Allan 1995). More recently, the input of riparian arthropods into streams to augment food for top predators (fish) has emerged as a major component of in-stream food webs (Baxter *et al.* 2005). When floodplains are flooded, there is a boom in the production of aquatic fauna that, in turn, is preyed upon by terrestrial fauna. Insects emerging from floodplain water bodies are consumed by terrestrial predators, including odonates, birds and bats. The nature, magnitude and trophic effects of this floodplain-terrestrial subsidy are poorly known (Ballinger & Lake 2006).

Present and past disturbances

Disturbances are forces that can greatly change, if not damage, ecosystems and can occur as pulses, or rise to a constant level as presses or steadily build in strength as ramps (Lake 2000). The responses to these disturbances can also be pulses, presses or ramps (Lake 2000). Streams are naturally subject to disturbance, and they may be regarded as being disturbance-dependent ecosystems. Streams are subject to many forms of disturbance, from landslides to deoxygenation, but the major ones are flow-generated – floods and droughts. Disturbances can interact; for example, floods can create massive sedimentation.

From studies across a variety of streams, we know that floods can damage and create habitat, deplete biota and change stream morphology; we also know that the stream biota, through the use of refugia, its high mobility and colonization capacity, can recover relatively rapidly from such disturbance (Lake 2000). Floods in upland streams exert a major influence on patchiness. Large floods, such as one in 100-year floods, can markedly reconfigure both the river channel and the adjoining landscape (Parsons *et al.* 2005).

Floods on floodplains are powerful generators of aquatic habitat and patchiness. As floods expand, waterbodies are connected with each other and with the channel. As floodwaters recede, complex patterns of disconnections and of patchiness occur. The elimination of regular and large floods by river regulation, water extraction and flood protection constitutes a major form of anthropogenic disturbance (Tockner & Stanford 2002).

Droughts arise slowly; indeed, it can be difficult to determine when a drought starts. There are many definitions of drought, but hydrological drought occurs when surface water levels drop to non-normal levels for an extended period of time (McMahon & Finlayson 2003). It becomes evident after meteorological and agricultural drought and is defined in respect of normally expected flows. Droughts are large scale and lengthy disturbances. As they develop, flow in rivers decreases and habitats are disconnected and may dry out (Stanley *et al.* 1997). Flow can cease creating pools, which undergo deleterious changes in water quality, such as low oxygen levels, and an increase in biotic interactions, such as predation (Lake 2003). Droughts do not change channel morphology, but may leave long lags, such as changes in the distribution of the biota, even creating local extinctions (Boulton 2003; Lake 2003). Human activities have exacerbated droughts; river regulation and water extraction accelerate water loss, refugia may be depleted, and changes in catchment surfaces, such as by land clearance, can reduce water infiltration and storage in catchments and hence the provision of base flows.

Past disturbances generated by land settlement may operate as legacies that affect stream condition. European settlement with grazing on the granitic Strathbogie Ranges near Euroa, Victoria (combined with episodic floods after drought), generated massive and irreversible sedimentation, leading to 'sand slugs' in downstream creeks (Davis & Finlayson 2000). Such sand slugs exacerbate the effects of both floods and droughts. Dryland salinization is another example of a very negative disturbance legacy (Hart *et al.* 2003).

Worthwhile restoration

In many parts of the world streams and rivers and their catchments are in a degraded, ecologically unsustainable condition. There have been many projects to restore flowing waters, but overall their effectiveness has been limited. Most projects are site-specific, poorly designed and uncoordinated with other projects. The majority have not been adequately monitored or monitored at all. There are fortunately some exceptions, such as the very extensive $500 million restoration of the Kissimmee River in Florida with monitoring of 60 performance measures (Whalen *et al.* 2002).

The NCEAS Working Group on River Restoration Science and Synthesis drew up a list of essential ingredients for ecologically successful restoration (Palmer *et al.* 2005).

- The design of a river restoration project should be based on a specific guiding image of a more dynamic, healthy river.
- The river's ecological condition must show measurable improvement.
- The river system must be more self-sustaining and resilient to external perturbations, so that only minimal follow-up maintenance is needed.
- During the construction phase, no lasting harm should be inflicted on the ecosystem.
- Both pre- and post-assessments must be completed and data made publicly available.

To be effective, planning and execution of a river restoration project should be conducted at the catchment level. A preliminary step should be to audit the disturbances (past and present), determine and rank those that are affecting the catchment and its streams, and then design a scape-based plan for the restoration work. Restoration must be pitched at the large spatial extent; small, isolated restoration projects can be ineffective (Roni 2005). With large-scale restoration interventions, restoration of catchments and streams can synergistically create ecological benefits (Roni 2005).

Conclusions

Flowing waters, temporary and permanent, are an essential component of, and a major force in, landscapes. Streams are ramifications rather than networks, and as water flow is gravity-driven, streams may be more insightfully depicted in three dimensions than two. For a long time, streams and their riparian zones were regarded by landscape ecologists as corridors and channels, and stream ecologists mainly investigated in-channel phenomena. Patchiness in flowing waters is dynamic and is substantially driven by flow-generated disturbance. Lotic biota have adapted to this dynamism as shown by their distinctive traits. As for landscapes, patchiness in streamscapes is determined by structural and biotic attributes. Whether such patchiness is reflected in biogeochemical transactions and ecological processes remains uncertain.

Connectivity is a vital attribute of lotic systems. There are three axes of connectivity – vertical, longitudinal and lateral. Floods increase the extent and levels of all three axes – with longitudinal connectivity being magnified in constrained sections and lateral connectivity being magnified in unconstrained floodplain sections. Droughts, especially those of long duration, disrupt connectivity (both lateral and longitudinal).

From an across-scape perspective, covering a catchment and its streams, the 'terrestrial-aquatic dichotomy is a false one' (Fisher *et al.* 2004). This applies not only to the biota (e.g. trophic subsidies) but also to biogeochemical transactions.

Design principles applied to flowing waters in landscapes

1 For the effective management of water bodies, with either flowing or standing waters and either temporary or perennial, it is advisable to plan and manage at the catchment level.
2 A vital attribute of flowing waters is connectivity, along three axes – longitudinal, lateral and vertical; disruption of the axes of connectivity alters stream condition.
3 **Flowing waters have a dynamic patchiness across a range of spatial extents and this patchiness is maintained by flow variability and natural disturbance.** Maintaining natural levels of flow variability and disturbance is essential for sustaining biotic assemblages and ecological processes.
4 **Riparian zones are vital transition zones that mediate transactions between catchments and their flowing waters.** Both depletion and fragmentation of riparian zones lessen their aquatic and terrestrial biodiversity and impair their function in maintaining and protecting stream condition.
5 As restoration of flowing waters is rapidly becoming more common, it is essential that restoration is coordinated at the catchment level and is rigorously designed and evaluated, so that our understanding of the restoration ecology of streams can be progressively improved.

Acknowledgements

I am grateful to Land and Water Australia for a Senior Research Fellowship whilst writing this essay.

References

Allan, J.D. (1995) *Stream Ecology. Structure and Function of Running Waters.* Chapman & Hall, London.

Amoros, C. & Bornette, G. (2002) Connectivity and biocomplexity in waterbodies of riverine floodplains. *Freshwater Biology* **47**, 761–776.

Ballinger, A. & Lake, P.S. (2006) Energy and nutrient fluxes from rivers and streams into terrestrial food webs. *Marine and Freshwater Research* **57**, 15–28.

Barmuta, L.A. (1989) Habitat patchiness and macroinvertebrate community structure in an upland stream in temperate Victoria, Australia. *Freshwater Biology* **21**, 223–236.

Baxter, C.V., Fausch, K.D. & Saunders, W.C. (2005) Tangled webs: reciprocal flows of invertebrate prey link streams and riparian zones. *Freshwater Biology* **50**, 201–220.

Boulton, A.J. (2003) Parallels and contrasts in the effects of drought on stream macroinvertebrate assemblages. *Freshwater Biology* **48**, 1173–1185.

Boulton, A.J., Findlay, S., Marmonier, P., Stanley, E.H. & Valett, H.M. (1998) The functional significance of the hyporheic zone in streams and rivers. *Annual Review of Ecology and Systematics* **29**, 59–81.

Cummins, K.W. (1974) Structure and function of stream ecosystems. *BioScience* **24**, 631–641.

Davis, J.A. & Finlayson, B. (2000) *Sand Slugs and Stream Degradation: The Case of the Granite Creeks, North-east Victoria.* Technical Report 7/2000. Cooperative Research Centre for Freshwater Ecology, Canberra.

Elwood, J.W., Newbold, J.D., O'Neill, R.V. & Van Winkle, W. (1983) Resource spiraling: an operational paradigm for analyzing lotic ecosystems. In: Fontaine, T.F.D. III & Bartell, S.M. (eds.) *Dynamics of Lotic Ecosystems*, pp. 3–42. Ann Arbor Science, Ann Arbor, MI.

Fisher, S.G., Welter, J., Schade, J. & Henry, J. (2001) Landscape challenges to ecosystem thinking: creative flood and drought in the American Southwest. *Scientia Marina* **65** (Suppl. 2), 181–192.

Fisher, S.G., Sponseller, R.A. & Heffernan, J.B. (2004) Horizons in stream biogeochemistry: flowpaths to progress. *Ecology* **85**, 2369–2379.

Hart, B.T., Lake, P.S., Webb, J.A. & Grace, M.R. (2003) Ecological risk to aquatic systems from salinity increases. *Australian Journal of Botany* **51**, 689–702.

Houlahan, J.E. & Findlay, C.S. (2004) Estimating the 'critical' distance at which adjacent land-use degrades wetland and sediment quality. *Landscape Ecology* **19**, 677–690.

Hynes, H.B.N. (1975) The stream and its valley. *Verhandlungen der Internationalen Vereinigung für Theoretische und Angewandte Limnologie* **19**, 1–5.

Johnson, W.C. (2002) Riparian vegetation diversity along regulated rivers: contribution of novel and relict habitats. *Freshwater Biology* **47**, 749–759.

Junk, W.J., Bayley, P.B. & Sparks, R.E. (1989) The flood pulse concept in river-floodplain systems. *Canadian Special Publication of Fisheries and Aquatic Sciences* **106**, 110–127.

Lake, P.S. (2000) Disturbance, patchiness, and diversity in streams. *Journal of the North American Benthological Society* **19**, 573–592.
Lake, P.S. (2003) Ecological effects of perturbation by drought in flowing waters. *Freshwater Biology* **48**, 1161–1172.
Lake, P.S. (2005) Perturbation, restoration and seeking ecological sustainability in Australian flowing waters. *Hydrobiologia* **552**, 109–120.
Likens, G.E., Bormann, F.H., Johnson, N.M., Fisher, S.W. & Pierce, R.S. (1970) Effects of forest cutting and herbicide treatment on nutrient budgets in Hubbard Brook watershed-ecosystem. *Ecological Monographs* **40**, 23–47.
Likens, G.E., Bormann, F.H., Pierce, R.S., Eaton, J.H. & Johnson, N.M. (1977) *Biogeochemistry of a Forested Ecosystem*. Springer-Verlag, New York.
McMahon, T.A. & Finlayson, B.L. (2003) Droughts and anti-droughts: the low flow hydrology of Australian rivers. *Freshwater Biology* **48**, 1147–1160.
Naiman, R.J. & Décamps, H. (1997) The ecology of interfaces: riparian zones. *Annual Review of Ecology and Systematics* **28**, 621–658.
Palmer, M.A., Swan, C.M., Nelson, K., Silver, P. & Alvestad, R. (2000) Streambed landscapes: Evidence that stream invertebrates respond to the type and spatial arrangement of patches. *Landscape Ecology* **15**, 563–576.
Palmer, M.A., Bernhardt, E.S., Allan, J.D. *et al.* (2005) Standards for ecologically successful river restoration. *Journal of Applied Ecology* **42**, 208–217.
Parsons, M., McLoughlin, C.A., Kotschy, K.A., Rogers, K.H. & Rountree, M.W. (2005) The effects of extreme floods on the biophysical heterogeneity of river landscapes. *Frontiers in Ecology and the Environment* **3**, 487–494.
Puckridge, J.T., Walker, K.F. & Costelloe, J.F. (2000) Hydrological persistence and the ecology of dryland rivers. *Regulated Rivers: Research and Management* **16**, 385–402.
Resh, V.H., Brown, A.V., Covich, A.P. *et al.* (1988) The role of disturbance in stream ecology. *Journal of the North American Benthological Society* **7**, 433–455.
Roni, P. (ed.) (2005) *Monitoring Stream and River Restoration*. American Fisheries Society, Bethesda, MD.
Sabo, J.L., Sponseller, R., Dixon, M. *et al.* (2005) Riparian zones increase regional species richness by harboring different, not more, species. *Ecology* **86**, 56–62.
Sedell, J.R. & Froggatt, J.L. (1984) Importance of streamside forests to large rivers: The isolation of the Willamette River. Oregon, U.S.A., from its floodplain by snagging and streamside forest removal. *Verhandlungen der Internationalen Vereinigung für Theoretische und Angewandte Limnologie* **22**, 1828–1834.
Sparks, R.E., Nelson, J.C. & Yin, Y. (1998) Naturalization of the flood regime in regulated rivers. *BioScience* **48**, 706–720.
Stanley, E.H., Fisher, S.G. & Grimm, N.B. (1997) Ecosystem expansion and contraction in streams. *BioScience* **47**, 427–435.
Tockner, K. & Stanford, J.A. (2002) Riverine flood plains: present state and future trends. *Environmental Conservation* **29**, 308–330.

Townsend, C.R. (1989) The patch dynamics concept of stream community ecology. *Journal of the North American Benthological Society* **8**, 36–50.

Vannote, R.L., Minshall, G.W., Sedell, J.R. & Cushing, C.E. (1980) The river continuum concept. *Canadian Journal of Fisheries and Aquatic Sciences* **37**, 130–137.

Ward, J.V., Tockner, K., Arscott, D.B. & Claret, C. (2002) Riverine landscape diversity. *Freshwater Biology* **47**, 517–540.

Whalen, P.J., Toth, L.A., Koebel, J.W. & Strayer, P.K. (2002) Kissimmee River restoration: a case study. *Water Science and Technology* **45**, 55–62.

Wiens, J.A. (1989) Spatial scaling in ecology. *Functional Ecology* **3**, 385–398.

Wiens, J.A. (2002) Riverine landscapes: taking landscape ecology into the water. *Freshwater Biology* **47**, 501–515.

Willson, M.F., Gende, S.M. & Marston, B.H. (1998) Fishes and forest: expanding perspectives on fish-wildlife interactions. *BioScience* **48**, 455–462.

39

Water in the Landscape: The Coupling of Aquatic Ecosystems and their Catchments

Peter Cullen

Abstract

The health of a river is directly influenced by its catchment. Soils and topography are important influences on the relative contributions of surface runoff and groundwater to river flow. Vegetation patterns determine runoff and influence what proportion of rainfall runs off or soaks into the soil and may enter groundwater. If deep-rooted vegetation is replaced by shallow-rooted plants waterlogging may occur after a time of 20–50 years as water moves past the root zone and enters groundwater. If there is salt in the landscape, dryland salinity may result. As native vegetation is replaced by agricultural plants there is commonly an increase in the sediment and nutrients entering the river from the catchment and this will impair river health, especially when agricultural activity takes place on more than 30% of the catchment. The riparian zone mediates the influence of the overall catchment on river health, and consequently is a primary target in efforts to restore or protect river health. It is extreme events, such as droughts, floods and fires, that cause profound and long-lived impacts on river health.

Keywords: connectivity; nutrients; riparian; water.

Introduction

Landscapes are never stable, but are constantly reacting to the environment as they respond to climate and various biological processes such as invasions and predation. Climate, and water in particular, are powerful factors in landscape formation. It is the availability of water in the soil that determines the vegetation cover, and in turn this vegetation determines the pulses of water and materials from the catchment that drive the aquatic ecosystems.

Rivers comprise both the channel that carries low flows, and the floodplain that accommodates larger flows. The floodplain may contain wetlands that mediate the flow of water and materials. The river will be fringed with riparian vegetation, dependent on the availability of water and itself affecting the aquatic environment in various ways. Rivers are underlain by groundwater, which may either supply or remove water from the surface features. The end of a river can be either in a terminal wetland or in an estuary.

Rivers integrate the pulses of water and other materials draining from the catchment, and hence are direct reflections of their catchments. Freshwater ecologists have long sought to understand the role of catchments in driving aquatic ecosystems (Hasler 1975; Likens *et al.* 1977). It is only more recently, however, that the discipline of landscape ecology has developed and started to connect with freshwater ecology (Wiens 2002; Allan 2004).

Humans extract water from rivers to support urban communities and agricultural activities. They store water in dams for drier periods, and seek to reduce the natural variability of river flow. The variability is, however, a key driver of the health of the aquatic ecosystem and so a tension arises between maintaining natural processes and modifying them for human needs. The extraction of water, the reduction of variability and the return of contaminated water have all damaged the rivers. As communities seek to restore the health of rivers, they need to understand the connections between catchment activities and the health of the rivers.

Water pathways

Rain that falls on land runs off into streams and infiltrates into the soil, or is evaporated from leaves, soils and puddles. Water entering the soil may be either taken up by plant roots and returned to the atmosphere as evapotranspiration, or it enters the groundwater.

In dry weather, streamflow comes from groundwater (often called baseflow) and following rainfall there are pulses of runoff from the soil surface (often called quickflow). The proportion of each flow type is a function of vegetation, soils and geology, and the intensity of rainfall.

Both the surface and subsurface flows will carry dissolved and particulate materials to the river. The velocity of water movement will determine what particulate material is mobilized from the soil surface and transported to the river, and it will be deposited as the water velocity drops. It may later be resuspended by subsequent runoff events and again redeposited as water velocity drops.

Not all of a catchment necessarily contributes runoff to a stream, and in many catchments only a limited area may contribute to surface runoff. The soil and vegetation characteristics of these contributing areas are those that determine the amount of water and materials that enter the river and so are much more important than other parts of the catchment. The riparian vegetation along the riverbank also mediates the connection between the catchment and the river.

The connection of a river to the groundwater system is also not always a simple one-directional movement. Lamontagne *et al.* (2006) report that the Murray River in Australia lost water to the groundwater beneath the floodplain during baseflow conditions, but this was reversed following floods. The Chowilla floodplain discharged saline groundwater from the floodplain to the river for 18 months following a medium flood.

Catchment vegetation and water pathways

The vegetation of a catchment not only determines the rate at which water runs off a catchment, but also determines how much of a year's rainfall actually enters the groundwater.

In Australia, native vegetation is adapted to drought. Perennial vegetation commonly has deep roots that extract the available water that falls each year. These systems were in long-term equilibrium, with the annual rainfall being used by plants and a fairly stable groundwater depth. When the deep-rooted native vegetation was cleared for agriculture and replaced with shallow-rooted annual plants, more water entered the groundwater and it consequently rose, leading to waterlogging. This 'leakage' of water into groundwater commonly occurred at rates between 15 and 150 mm/year. Once the groundwater rises to within 1–2 m of the soil surface, capillary action takes over and evaporation from the soil surface leads to the deposition of salt (Cullen 2003). This is an example of a threshold being reached where an apparently stable situation suddenly switches to another state that may be very difficult or impossible to reverse (Folke *et al.* 2004).

When trees are replaced with grass there is a reduction in evapotranspiration, an increase in runoff and rising groundwater. These changes have been observed in small plot studies and there is some evidence that the outcomes continue at the basin scale. This effect occurs relatively quickly, especially when the vegetation canopy is closed. The reduction in flow is significant in high-rainfall zones (>700 mm/year). In low- to medium-rainfall zones (400–700 mm/year) the reduction is often small, ambiguous and difficult to predict reliably (see O'Loughlin & Nambiar 2001; Hairsine & Polglase 2004). Zhang *et al.* (1999) brought together datasets from around the world, and the results are consistent with preliminary studies in the Comet catchment in Queensland, which indicate runoff may have increased by around 35% following clearing of about 50% of large vegetation in the mid-1960s (T. McMahon, personal communication). There are many documented examples of decline in water quality, aquatic habitat and the river biota as forest is cleared and replaced with pasture (e.g. Townsend *et al.* 2004).

The reverse process, where plantation forests are introduced into a catchment, can cause a decline in stream flow (Vertessy *et al.* 2003). Planting of deep-rooted trees is commonly recommended to reduce dryland salinity problems, because they transpire water and cause less to enter groundwater. Consequently, less groundwater may be available to provide baseflow for streams (Scott 2005).

Hydrological disturbances: Droughts and floods

Stream ecosystems are rarely in equilibrium but are responding to the pulses of water that come from runoff events. Floods and droughts are events that may reset the system in various ways. This is most obvious in arid area aquatic ecosystems that experience boom and bust cycles (Roshier *et al.* 2001).

Lake (2003) explored the impacts of droughts on stream communities. Droughts disrupt the hydrological connectivity of the aquatic systems, causing pools to be disconnected and drying out of some reaches. Droughts have marked effects on the densities and size or age structure of populations, on community composition and diversity, and on ecosystem processes. Organisms can resist the effects of drought by the use of refugia, and the availability of such refugia may strongly influence the capacity of the biota to recover from droughts.

Collier and Quinn (2003) explored the impacts of a 1-in-28-year flood on invertebrate communities in streams draining pasture and forest catchments in Waikato, New Zealand. The number of taxa and their density declined markedly at the forested site after the flood, but there was a delayed response at the pasture site, reflecting greater initial resistance to this pulse disturbance. Community composition was less stable at the pasture site, where the percent abundance of taxa was highly variable prior to the flood and over the 2-year post-flood sampling period. Both sites showed some recovery within 5–7 months, but community stability at the pasture site had not returned to pre-flood composition within 2 years. This study indicates that the magnitude and duration of responses to major pulse disturbances (the flood) can depend on the presence or absence of an underlying press disturbance (conversion of forest to pasture).

Water is also an important mechanism for seed dispersal along river corridors, and onto floodplains. Merritta and Wohl (2006) estimated that as many as 120 million seeds were transported along free-flowing reaches of two Rocky Mountain (USA) streams in a single growing season. Renofalt *et al.* (2005) introduced seeds of an alien species into headwaters of the free-flowing Vindel River in northern Sweden and found significant invasion of adjacent riparian areas.

Hydrological disturbances: Horizontal connectivity with the floodplain

The lateral exchange of water between the river channel and floodplain during floods is an important ecological process for both the terrestrial and aquatic systems (Ward *et al.* 2002). The intermittent connection drives the exchange of organic matter and inorganic nutrients (Amoros & Bornette 2002). Fish move out onto the flooded floodplain to forage. The building of dams reduces the frequency and duration of small and medium floods with consequential ecological impacts (Walker & Thoms 1993; Roberts & Marston 2000).

Much of Australia has been in drought during the 2000–2005 period, and this, on top of the 'man-made' drought caused by excessive extraction of irrigation water has seen the loss of many river redgums along the Murray River (Jolly 1996). This is another example of a threshold effect that was not anticipated, was sudden and is probably irreversible. The frequency of flooding may be a critical disturbance required for the survival of particular plants in riparian areas (Wolfert 2002).

Because floodplains are fertile, humans often seek to settle on them. They then seek to protect their developments from the flooding processes that made the floodplain fertile, and build levee banks to try to keep floodwaters away. Gergel *et al.* (2002) showed that if levees were set well back from the river they had only a minor impact on flood height and overbank flood velocities, and that floodplain communities within the levee banks were little altered. Levees had a greater impact on flood height and overbank flood velocities of larger-magnitude events.

Land use and nutrient exports

Many studies have shown a strong relationship between land use and nutrient exports from various sized catchments (e.g. Carpenter *et al.* 1998; Harris 2001). The runoff characteristics of a catchment, determined by slope, soil type and vegetation cover, impact on nutrient exports from catchments (e.g. Wickham *et al.* 2005). Nutrient export coefficients in terms of kg/ha/year are commonly used to provide estimates of nitrogen and phosphorus loadings from catchments to water bodies (Cullen & O'Loughlin 1982). Typical coefficients for phosphorus in kg P/ha/year are: forests <0.1; poor-quality grazing 0.2; high-quality grazing 0.6; and urban areas >1.2. These are annual estimates and are useful for crude catchment assessments of likely nutrient loading as land uses are changed.

The bulk of material moves during runoff events and is commonly in particulate form. The importance of high-flow events are even more apparent from Table 39.1 showing some data on phosphorus exports found in Berri Berri Creek draining into Lake Burrogarang near Sydney in southeastern Australia. Table 39.1 shows the total phosphorus coming off the catchment over the entire 314-day study, and the amount of phosphorus that came off in selected high-flow events. These results show that 61% of the phosphorus and 41% of the water moved down the Creek in 1% of the time during the study. Had this single 3-day high-flow event not been measured, then the nutrient export would have been underestimated by 60%.

These large flood flows also transport to streams sizable woody debris that provides habitat for a range of organisms. In some catchments human activities have developed point sources of nutrients, frequently sewage treatment plants, which may discharge at a more constant rate and commonly have large amounts of dissolved materials.

Table 39.1 **Influence of flow on exports of phosphorus, Berri-Berri Creek 1988–9 (P. Cullen, unpublished data).**

Sampling period	No. of days	% of time	Flow (ML)	% of flow	Phosphorus (kg)	Phosphorus (%)
Entire	314	100	52,994	100	2874	100
Flows over:						
300 ML/day	25	8	35,984	68	2334	81
600 ML/day	10	3	29,772	56	2123	74
1200 ML/day	6	2	26,407	50	1989	69
2400 ML/day	3	1	21,672	41	1764	61

Fire and catchment processes

Fires are a common element in many landscapes, and can have significant impacts on water quantity and quality for many years (Zierholz *et al.* 1995). Langford (1976) measured the change in runoff from mountain ash forests in southern Australia following fire. He demonstrated reduced water yield for a 10–20-year period as regrowth increased leaf area and hence evapotranspiration, before recovery to pre-fire conditions after perhaps 30 years or more. Kuczera (1987) reported reductions in water yield following a bushfire in an ash-mixed species eucalypt forest in southern Australia. These factors can have significant impacts for water planning. English *et al.* (2005) explored sediment and nutrient redistribution in the eucalypt forests of the Nattai National Park near Sydney following major bushfires of 2001–2002 and found large amounts of sediment and phosphorus mobilization in storms that followed the fires. High-intensity fires cause a range of biological and physical changes to soils that will impact on infiltration and runoff processes.

Vieira *et al.* (2004) examined the impact of post-fire flash floods that brought considerable sediment to the stream on stream insect communities in New Mexico in the USA. The first major flood reduced total insect density and taxon richness to near zero, although density returned rapidly to pre-fire levels because of colonization by taxa that were generalist feeders with strong larval dispersal. Taxon richness did not recover until 4 years after the fire, and community composition in the burned stream still differed from pre-fire and reference stream compositions after 6 years.

Role of riparian vegetation

Riparian vegetation is an important boundary between the terrestrial and aquatic ecosystems and may ameliorate some of the impacts of catchment land use on river health. An intact riparian community acts as a filter for particulate material coming from the catchment. Considerable work has been done on the design of such filter strips to protect stream values (Herron & Hairsine 1998; Prosser & Karssies 2001). The riparian tree community can add leaf litter and large woody debris to the stream, providing food and habitat. The canopy provides shade to the aquatic ecosystem, reducing stream temperature and affecting fish behaviour (Pusey & Arthington 2003).

Jansen and Robertson (2001) have explored the impacts of domestic livestock grazing of riparian areas, and commonly the first step in stream restoration is to exclude grazing directly from stream banks. Rios and Bailey (2006) examined the influence of riparian vegetation on macroinvertebrate community structure in streams and found that taxon richness of the macroinvertebrate community increased with increased tree cover in the riparian zone at the reach scale. They found no relationship between the macroinvertebrate community and land cover at the whole-basin scale.

The coupling of catchments and rivers

There are a number of mechanisms by which catchments determine the ecological integrity of a river system, and the study of these processes is not a simple one due to covariation between factors, non-linear responses and long-term legacy effects (Allan 2004). Nevertheless, a number of major mechanisms are now recognized.

In studies of nested catchments in Maryland, USA, Wickham *et al.* (2005) found that changes in nitrogen and phosphorus exports were a function of instream decay, subcatchment land-cover composition and subcatchment stream length. Stream decay refers to uptake by plants and by particulates settling within the stream. These materials may be remobilized during higher flows, causing what is called nutrient spiralling as they move some distance downstream with each freshet of water, before again settling.

Vondracek *et al.* (2005) examined 425 stream reaches from three North American ecoregions and found that both riparian and catchment-scale land use explained significant variation in water quality, channel morphology, and fish distribution and density. Fish and macroinvertebrate assemblages can be positively affected by increasing the extent of perennial riparian and upland vegetation.

The impact of land-use change on river health

The mechanisms listed in Table 39.2 interact in various ways. As a catchment is converted from forest to an agricultural landscape, a number of changes can be observed (after Allan 2004). Landscape metrics such as the proportion of agriculture in the catchment and the amount of trees in the riparian zone explain many of the observed changes.

Table 39.2 **Mechanisms of catchment-river coupling.**

Process	**Drivers**	**Consequences to river system**
Sedimentation	Clearing of vegetation Drought Fire impacts Agricultural tillage Overgrazing	Deposition of sand in river, creating sand slugs, filling of deeper pools and loss of refugia Deposition of finer materials interfering with benthic organisms and fish spawning areas Increased turbidity with impacts on light penetration and hence plant communities Scouring by particulates impacting on habitat and or organisms directly
Water storage and delivery	Dams Flow changes for water delivery	Remove small-medium floods, reduce frequency of overbank flooding and hence connectivity with floodplain Loss of flood pulses that provide triggers for fish breeding Cold water releases may impact on biota Release of irrigation water may give seasonal reversal of flow patterns Unnaturally high summer flows will impact on floodplain vegetation
Overextraction of surface water	Overallocation of water or failure to manage within sustainable extraction levels	Loss of in-stream habitat due to reduced flows Loss of native fish Loss of connected wetlands and bird breeding events Algal blooms in slow-moving weir pools Blockage of estuarine connection

Water returns	Surface drainage from irrigation Stormwater and waste disposal from urban communities	Altered flow patterns Increased contaminant loads
Land clearing	Increased runoff and contaminant mobilization Loss of large woody debris	Increased contaminant loads Reduced woody debris in river that provides substrate and habitat for various organisms and alters hydraulic character of stream
Riparian damage	Grazing or clearing of riparian vegetation	Reduced shading causing increase in stream temperature Decrease in bank stability Reduction in leaf fall hence reduced food sources for biota Reduction in filtering effects means more sediment, nutrients and contaminants reach waterway
Extraction of groundwater	Extraction beyond sustainable yield	Reduced stream flow after lag of perhaps 20–50 years
Levee banks	To protect development from flooding	Isolation of river from floodplain and its wetlands Reduced exchange of nutrients and organic matter Loss of feeding habitat for fish
Nutrient enrichment	Agricultural fertilizers Sewage disposal Urban stormwater	Encourages growth of filamentous and planktonic algae, which replace macrophytes
Contaminant pollution	Agricultural chemicals Mining	Various toxicants in water column and sediment may directly affect biota and human uses of water

As trees are removed from the riparian zone, then woody debris in the stream is reduced providing less habitat for aquatic biota, stream temperatures increase due to lack of shading, light levels in the water column increase and nutrient inputs increase. These cause filamentous algae to dominate benthic production and, in extreme cases, blooms of planktonic algae develop in slow-moving sections. Initially this leads to an increase in macroinvertebrates, especially grazers, due to the increased organic matter as a food source.

As agricultural activity in the catchment increases, there are occasional significant pulses of sediment, nutrients and agricultural chemicals. There have been examples where pesticides have removed grazing organisms from the water, leading to excessive growth of algae.

The clearing of native vegetation increases runoff, causing soil and bank erosion, leading to degradation of in-stream habitat (Downes *et al.* 2006). Cultivated land can provide large amounts of sediment should rainfall follow cultivation. Storm flow commonly increases in magnitude and frequency causing changes to the hydrological regime that impact on biota in various ways.

Allan (2004) reports various studies and suggests that commonly river health remains good until agricultural activity involves 30–50% of the catchment area. A New Zealand study reported that aquatic invertebrate populations were undamaged until agricultural land use reached 30%, but beyond this pollution-tolerant species dominated (Quinn 2000).

Principles

A number of broad principles arise from this review of catchment–river interactions.

1 **Landscapes are not static**, but are in a state of constant change as they respond to climate and other factors such as grazing pressure and invasive species. Commonly these changes are slow and hardly perceptible.
2 From time to time **extreme events occur that can lead to dramatic changes.** Droughts, floods, fires and invasions of pest species can trigger dramatic and sudden change that may be effectively irreversible. These more extreme events are key drivers of landscape change and push a landscape across a change threshold that is commonly not obvious until after it has been reached. These events need to be the focus of attention in catchment management.
3 Catchment **land use is an important driver of river health**, and river health seems commonly to be impaired when agricultural activities cover more than 30% of a catchment area, although distance from the river, and the health of the riparian vegetation can mediate these effects.
4 The **removal of native vegetation can markedly alter the hydrological behaviour** of a catchment. As trees are removed, runoff commonly increases, as does the amount of water flowing below the root zone and

entering groundwater, and this may lead to consequent waterlogging and salinity, although there may be a lag of 20–50 years before these impacts are obvious. Conversely, introducing forests into a catchment will reduce streamflow downstream.

5 Maintaining **connectivity between a river channel and its floodplain is important** to river health, and the resilience of an aquatic ecosystem to recover from events that may change it. Many systems in semiarid areas are adapted to cycles of wetting and drying. Particular areas like deeper pools and wetlands may be important refuges that allow organisms to survive extreme high- or low-flow events.

6 **The riparian zone has a profound impact on stream ecosystem processes.** An intact riparian zone provides shade and reduces stream temperature, provides leaf litter and woody debris, protects riverbanks from erosion, and filters nutrients and materials from the catchment. The restoration of degraded rivers commonly starts with protection and restoration of riparian zones, and this can mitigate some of the impacts of land uses further up the catchment.

7 **Nutrient export from a catchment increases with intensification of land use**, and much of this moves as particulate material during episodic high-runoff events. Dissolved material commonly increases in importance with fertilizer use.

8 **Groundwater and surface water are connected unless shown otherwise.** Overallocation of groundwater will lead to a reduction in river flow after a lag period of perhaps 20–40 years (Evans 2005). The depth to groundwater will also affect riparian vegetation (Horton *et al.* 2001).

References

Allan, J.D. (2004) Landscapes and riverscapes: the influence of land use on stream ecosystems. *Ann. Rev. Ecol. Evol. Syst.* **35**, 257–284.

Amoros, C. & Bornette, G. (2002) Connectivity and biocomplexity in waterbodies of riverine floodplains. *Freshwater Biology* **47**, 761–776.

Carpenter, S.R., Caraco, N.F., Correll, D.L., Howarth, R.W., Sharpley, A.N. & Smith, V.H. (1998) Nonpoint pollution of surface waters with phosphorus and nitrogen. *Ecological Applications* **8**, 559–568.

Collier, K.J. & Quinn, J.M. (2003) Land-use influences macroinvertebrate community response following a pulse disturbance. *Freshwater Biology* **48**, 1462–1481.

Hairsine, P.B. & Polgose, P. (2004) Maximizing the benefits of new tree plantations in the Murray-Darling Basin. *A Joint Statement by CSIRO Forestry and Forest Products and CSIRO Land and Water* [online]. URL: http://www.clw.csiro.au/staff/hairsinep/FFP–LW19052004.pdf.

Cullen, P. (2003) Salinity. In: Attiwill, P. & Wilson, B. (eds.). *Ecology, An Australian Perspective*, pp. 474–488. Oxford University Press, Melbourne.

Cullen, P.W. & O'Loughlin, E.M. (1982) Non-point sources of pollution. In: O'Loughlin, E.M. & Cullen, P.W. (eds.) *Prediction in Water Quality*, 437–453. Australian Acadamy of Science, Canberra.

Downes, B.J., Lake, P.S., Glaister, A. & Bond, N.R. (2006) Effects of sand sedimentation on the macroinvertebrate fauna of lowland streams: are the effects consistent? *Freshwater Biology* **51**, 144–160.

English, P., Wallbrink, P., Humphreys, G. *et al.* (2005) Impacts of water quality by sediments and nutrients released during extreme bushfires. In: *Report 2: Tracer Assessment of Post-Fire Sediment and Nutrient Redistribution on Hillslopes, Nattai National Park, NSW*. Sydney catchment authority–CSIRO Land & Water [online]. URL: http://www.clw.csiro.au/publications/consultancy/2005/SCA-Report2.pdf.

Evans, R. (2005) Double accounting of surface water and groundwater resources: the tyranny of the time lag. Outlook 2005: Proceedings ABARE Conference. Canberra, 1–2 March. pp. 1–7.

Folke, C., Carpenter, S., Walker, B. *et al.* (2004) Regime shifts, resilience, and biodiversity in ecosystem management. *Annual Review in Ecology, Evolution and Systematics* **35**, 557–581.

Gergel, S.E., Dixon, M.D. & Turner, M.G. (2002) Consequences of human-altered floods: levees, floods, and floodplain forests along the Wisconsin River. *Ecological Applications* **12**, 1755–1770.

Harris, G.P. (2001) Biogeochemistry of nitrogen and phosphorus in Australian catchments, rivers and estuaries: effects of land use and flow regulation and comparison with global patterns. *Marine and Freshwater Research* **52**, 139–149.

Hasler, A.D. (ed.) *Coupling of Land and Water Systems*. Ecological Studies 10. Springer-Verlag, New York.

Herron, N.F. & Hairsine, P.B. (1998) A scheme for evaluating the effectiveness of riparian zones in reducing overland flow to streams. *Australian Journal of Soil Research* **36**, 683–698.

Horton, J.L., Kolb, T.E. & Hart, S.C. (2001) Physiological response to groundwater depth varies among species and with river flow regulation. *Ecological Applications* **11**, 1046–1059.

Jansen, A. & Robertson, A.I. (2001) Relationships between livestock management and the ecological condition of riparian habitats along an Australian floodplain river. *Journal of Applied Ecology* **38**, 63–75.

Jolly, I.D. (1996) The effects of river management on the hydrology and hydroecology of arid and semi-arid floodplains. In: Anderson, M.G., Walling, D.E. & Bates, P.D. (eds.) *Floodplain Processes*, pp. 577–609. Wiley, New York.

Kuczera, G. (1987) Prediction of water yield reductions following a bushfire in ash-mixed species eucalypt forest. *Journal of Hydrology* **94**, 215–236.

Lake, P.S. (2003) Ecological effects of perturbation by drought in flowing waters. *Freshwater Biology* **48**, 1161–1172.

Lamontagne, S.B., Leaney, F.W. & Herczeg, A.L. (2006) Patterns in groundwater nitrogen concentration in the riparian zone of a large semi-arid river (River Murray, Australia). *River Research and Applications* **22**, 39–54.

Langford, K.J. (1976) Change in yield of water following a bushfire in a forest of *Eucalyptus regnans. Journal of Hydrology* **29**, 87–114.

Likens, G.E., Borman, F.H., Pierce, R.S., Eaton, J.S. & Johnson, N.M. (1977) *Biogeochemistry of a Forested Ecosystem*. Springer-Verlag, New York.

Merritta, D.M. & Wohl, E.E. (2006) Plant dispersal along rivers fragmented by dams. *River Research and Applications* **22**, 1–26.

O'Loughlin, E. & Nambiar, E.K.S. (2001) *Water and Salinity issued in Agroforestry. No. 8. Rural Industries Research and Development Corporation, Canberra.*

Prosser, I. & Karssies, L. (2001) Designing filter strips to trap sediment and attached nutrients. *Riparian Land Management Technical Guideline 1*. Land and Water Australia, Canberra.

Pusey, B.J. & Arthington, A. (2003) Importance of the riparian zone to the conservation and management of freshwater fish. A review. *Marine and Freshwater Research* **54**, 1–16.

Quinn, J.M. (2000) Effects of pastoral development. In: Collier, K.J & Winterbourn, M.J. (eds.) *New Zealand Stream Invertebrates: Ecology and Implications for Management*, pp. 208–229. Caxton, Christchurch.

Renofalt, B.M., Jansson, R. & Nilsson, C. (2005) Spatial patterns of plant invasiveness in a riparian corridor. *Landscape Ecology* **20**, 165–176.

Rios, S.L. & Bailey, R.C. (2006) Relationship between riparian vegetation and stream benthic communities at three spatial scales. *Hydrobiologia* **553**, 153–160.

Roberts, J. & Marston, M. (2000) *Water Regime of Wetland and Floodplain Plants in the Murray Darling Basin: A Source Book of Ecological Knowledge*. CSIRO Land and Water Technical Report. CSIRO, Canberra.

Roshier, D.A., Robertson, A.I., Kingsford, R.T. & Green, D.G. (2001) Continental-scale interactions with temporary resources may explain the paradox of large populations of desert waterbirds in Australia. *Landscape Ecology* **16**, 547–556.

Scott, D.F. (2005) On the hydrology of industrial timber plantations. *Hydrological Processes* **19**, 4203–4206.

Townsend, C.R., Downes, B.J., Peacock, K. & Arbuckle, C.J. (2004) Scale and the detection of land-use effects on morphology, vegetation and macroinvertebrate communities of grassland streams. *Freshwater Biology* **49**, 448–462.

Vertessy, R.A., Zhang, L. & Dawes, W.R. (2003) Plantations, river flows and river salinity. *Australian Forestry* **66**, 55–61.

Vieira, N.K., Clements, M.W.H., Guevara, L.S. & Jacobs, B.F. (2004) Resistance and resilience of stream insect communities to repeated hydrologic disturbances after a wildfire. *Freshwater Biology* **49**, 1243–1259.

Vondracek, B., Blann, K.L., Nerbonne, J.F. *et al.* (2005) Land use, spatial scale, and stream systems: lessons from an agricultural region. *Environmental Management* **36**, 775–791.

Walker, K.F. & Thoms, M.C. (1993) Environmental effects of flow regulation on a semi-arid lowland river: the River Murray, Australia. *Regulated Rivers: Research & Management* **8**, 103–119.

Ward, J.V., Tockner, K., Arscott, D.B. & Claret, C. (2002) Riverine landscape diversity. *Freshwater Biology* **47**, 517–539.

Wickham, J.D., Riitters, K.H., Wade, T.G. & Jones, K.B. (2005) Evaluating the relative roles of ecological regions and land-cover composition for guiding establishment of nutrient criteria. *Landscape Ecology* **20**, 791–798.

Wiens, J.A. (2002) Riverine landscapes: taking landscape ecology into the water. *Freshwater Biology* **47**, 501–515.

Wolfert, H.P., Hommel, P.W.F.M., Prins, A.H. & Stam, M.H. (2002) The formation of natural levees as a disturbance process significant to the conservation of riverine pastures. *Landscape Ecology* **17** (Suppl. 1), 47–57.

Zhang, L., Dawes, W.R. & Walker, G.R. (1999) Predicting the effect of vegetation changes on catchment average water balance. Cooperative Research Centre for Catchment Hydrology, Technical Report 99/12. 35. URL: http://www.catchment.crc. org.au/pdfs/technical.

Zierholz, C., Hairsine, P.B. & Booker, F. (1995) Runoff and soil erosion in bushland following the Sydney bushfires. *Australian Journal of Soil and Water Conservation* **8**, 28–37.

40

Synthesis: Aquatic Ecosystems and Integrity

David B. Lindenmayer and Richard J. Hobbs

Many books on landscape ecology and conservation biology are terrestrially focused. Aquatic ecosystems are ignored or paid only very limited attention. While we have not given the 'wet parts' of landscapes the full attention they deserve, we have done our best to ensure that they have not been treated as an afterthought. Unfortunately, terrestrial landscape ecology and aquatic ecology appear to be largely parallel literatures with relatively few individuals attempting to engage in what is undoubtedly useful cross-fertilization. Indeed, it is instructive to recognize the similarities in some of the key features that characterize both terrestrial and aquatic ecosystems – examples are connectivity, patchiness (and the existence of mosaics), dynamism, structural complexity, habitat amount, habitat loss, scale and multiscaled effects. Landscapes and streamscapes also share common problems concerning variations in classification, with consequent implications for legislation and resource-use practices (Calhoun, Chapter 37, cf. Chapters 2–5). They also share the problem that they can be easy to disrupt but very difficult to restore (see also Woinarski, Chapter 10).

Landscapes and streamscapes are clearly more integrated than has often been recognized, particularly by terrestrial ecologists; Lake (Chapter 38) notes that only relatively recently has the intersection of landscape ecology and streamscape ecology begun to be better developed. Calhoun (Chapter 37) laments the fact that wetlands are often entirely overlooked in land-use planning and decision-making – many do not even appear in inventories. This is despite the enormous ecosystem service role played by wetlands.

All the authors of Chapters 37–39 highlight the importance of the integration of terrestrial and aquatic ecosystems. For example, floods and droughts strongly connect land and aquatic ecosystems (Lake, Chapter 38). Similarly, Cullen (Chapter 39) briefly describes work from a range of locations where vegetation clearance instigates salinization and declines in river health, often with significant negative effects, as when agricultural development exceeds 30–50% of the catchment area. Calhoun (Chapter 37) emphasizes the fact that many species have life histories dependent not only on a range of aquatic environments but also on adjacent terrestrial ones.

Unfortunately, land managers often fail to recognize the intimate and complex interrelatedness of terrestrial and aquatic ecosystems and, in turn, do not fully appreciate the need to co-manage terrestrial and aquatic ecosystems (Calhoun, Chapter 37). Lake (Chapter 38) discusses this problem in a forestry context, particularly in terms of the shortcomings of riparian buffer systems. Cullen (Chapter 39) highlights the impacts of agricultural development on aquatic ecosystems. Yet better recognition of such interrelationships and the impacts of land-use practices on aquatic ecosystems is one of the factors pivotal to more effective streamscape restoration. Monitoring is another pivotal factor. It is impossible to gauge progress on any management action without monitoring, and impossible to determine what went right and what did not. This has been well known for many decades and applies to all aspects of landscape and conservation management – for example, see Driscoll (Chapter 11) and Luck (Chapter 16) – and is not confined to river and aquatic ecosystem restoration. Despite this, the current record on successful sustained monitoring programmes remains worse than ordinary in almost all jurisdictions. Improvements in management will be extremely difficult without dramatically improving that record.

There are two final issues associated with managing aquatic ecosystems that were touched on in this latter section. The first is that extreme events such as floods and droughts have profound impacts on aquatic ecosystems. Their effects can be significantly magnified when they occur in combination with human impacts such as land clearing, intensive livestock grazing in riparian areas and streamscape modifications (e.g. dams, canalization, heavy regulation of water flows). Managing extreme events can be both extremely difficult to predict and extremely difficult to manage. Rather than allowing events to drive management responses, wherever possible it may be better to anticipate extreme events and plan contingencies before they occur. This equates to building general resilience into systems (Walker, Chapter 34). The second issue is that some ecological processes and responses to human modification in aquatic ecosystems do not follow linear trajectories. Rather, all three authors noted the occurrence of thresholds (see also Walker, Chapter 34), where aquatic ecosystems suddenly switch from an apparently stable state to a markedly different state. Cullen (Chapter 39) describes rising groundwater and subsequent capillary action drawing salt through the soil, as an example of a threshold. Importantly, crossing some of these thresholds may produce states that are either irreversible or extremely difficult to reverse, making a better understanding of thresholds crucial for aquatic ecosystem management as well as essential for attempts to instigate effective river and wetland restoration programmes.

Section 11
"Bringing It All Together"

41

Does Conservation Need Landscape Ecology? A Perspective from Both Sides of the Divide

John A. Wiens

Abstract

As the emphasis of conservation has shifted from protecting species to include entire ecological systems or 'functional landscapes', the need for closer linkages between conservation and landscape ecology has become obvious. Several emerging principles of landscape ecology can inform conservation decisions about which places to protect, how to implement that protection, how to manage or restore the places once they are protected and how to balance human activities with the protection of biodiversity. Doing this, however, requires closing the gap that exists between these disciplines by establishing shared goals, addressing inconsistencies of scale and changing institutional cultures to reconcile the desire of academics for 'more research' with the 'just do it' attitude of conservationists.

Keywords: biodiversity; conservation; landscape; scale; The Nature Conservancy.

Introduction

Conservation is about the protection, preservation and management of the Earth's biodiversity. This biodiversity is rapidly being eroded, and the major cause of biodiversity loss is the transformation of the Earth by human actions. As the Earth's population, economies and demands for resources continue to grow, changing land use will pose ever greater challenges to biodiversity and those who seek to conserve it. Creating more reserves and protected areas by itself will not meet these challenges – conservation must look as well to the places where people live and work (Redford & Richter 1999; Miller & Hobbs 2002), to entire landscapes, writ large. Landscape ecology would seem to have much to offer towards addressing these challenges.

Yet, despite several valiant attempts (notably Bissonette 1997; Gutzwiller 2002; Lindenmayer & Franklin 2002), landscape ecology has not become part of the mainstream of conservation biology and has had little direct impact on how conservation is actually done. While there are always gaps between science and the practices that could most benefit from that science, this gap is particularly disconcerting because of the urgency of conservation and the seeming relevance of landscape ecology.

But is it relevant? Does landscape ecology really have much to offer conservation practice? I'm drawn to consider these questions because (to paraphrase Joni Mitchell) I've looked at them from both sides now. I spent decades in academia pursuing the twists and turns of spatial ecology, which sometime during the 1980s metamorphosed into landscape ecology. I published lots of papers and directed lots of students, with the goal of increasing our understanding of how spatial patterns influence ecological processes, secure in the belief that surely somehow, somewhere, this knowledge would be relevant to the world at large. Now I'm on the other side, working to infuse scientific knowledge and practices into the work of the world's largest conservation organization. And what do I see?

To address this question it is useful to consider some of the current issues and prevailing themes of conservation and how conservation is actually done. If landscape ecology is to be relevant to conservation, it must contribute to conservation thinking and practice. It is unrealistic to expect landscape ecology to be equally relevant to all aspects of conservation, however. My intention here is to frame the conceptual issues and describe the elements of conservation practice, then outline the primary themes of landscape ecology, and then consider where these sets are most likely to intersect. I'll use the work of The Nature Conservancy (TNC) as an example because I know it best, but it is generally representative of the approach taken by many conservation organizations and government agencies.

The conservation side

Issues and emerging themes in conservation biology

Conservation is not a conceptually unified endeavour. While there may be general agreement about 'preserving the Earth's biodiversity', people and organizations differ in their perceptions of how this translates into specific goals. Much of the focus is on the protection of species, particularly rare, endemic or threatened ones (and, inordinately, plant and vertebrate ones; see Ponder & Lunney 1999). In the USA this focus is formalized in the Endangered Species Act. Internationally, it is recognized in, for example, the IUCN Red List, CITES, and the Convention on Biological Diversity. Papers dealing with demography, population viability, metapopulation structure, genetics and single-species habitat requirements dominate the scientific journals in conservation biology.

A different approach, embodied by 'ecosystem management' or 'habitat conservation', emphasizes the goal of protecting entire functioning systems or habitat types. These may be local (e.g. fens, vernal pools) or regional (e.g. boreal coniferous forest). The focus is on habitats and the ecological functions and species they contain rather than on the species alone. The debate over whether conservation efforts should be concentrated on 'hotspots' of biodiversity (species richness and endemism) or should consider a broader array of habitats that represent the full spectrum of biodiversity ('coldspots' as well as hotspots; Kareiva & Marvier 2003) is a reflection of the difference between species goals and habitat goals for conservation.

There are also differences in the scales at which conservation is targeted. While it is arguably true that all conservation is ultimately local, conservation goals and strategies are expressed over multiple scales, from very local (that fen or vernal pool) to global (e.g. the Convention on Biological Diversity, the Millennium Ecosystem Assessment). It is not always obvious how efforts at different scales can be linked together, and the scales at which resource management are implemented often do not coincide with the functional scales of the resources they are aimed at managing (Turner *et al.* 2002; Wiens *et al.* 2002). Moreover, conservation efforts at a given scale generally do not recognize that the species or systems of interest operate at multiple, different scales. For example, both migratory warblers (*Dendroica* spp.) and sedentary manakins (*Pipra* spp.) may occupy the same tropical forest habitat in Costa Rica; protection measures that work for the manakins may offer little benefit to the warblers.

Two broader issues are increasingly pervading conservation. The first concerns dynamics. Despite the hopeful belief of the public (and too many ecologists) that natural systems are normally in a relatively stable state (the 'balance of nature') or at least can be treated as if they are, evidence abounds that they are not. Long-term trends are superimposed on short-term variations, and both may involve natural and anthropogenic changes that often interact to produce 'surprising' (from a stable nature perspective) dynamics. Conservation must deal with the combined effects of climate disruption, which is driven by global forces with some local tuning, and land-use change, which is driven by local practices with a powerful influence of global economics. The difficulties of combining these factors are formidable, the more so because the dynamics may often be non-linear (Lindenmayer & Luck 2005; Groffman *et al.* 2006), and assessing these changes requires future projections. Such projections are inevitably couched in uncertainty – uncertainty that exceeds the comfort level of many politicians, resource managers and members of the public.

The second issue concerns people. Over the past 50 years, conservation has often come to be viewed as nature versus people. Parks, wilderness areas, wildlife refuges and nature reserves were established in which human activities and resource extraction were forbidden or sharply curtailed. Many in the environmental movement still wish for places in which nature is untouched by human hands. But there are few places on Earth that do not bear the human footprint, so conservation in the absence of humans simply will not succeed in protecting the Earth's biodiversity. Conservation must be couched in terms of conservation *with* people, recognizing that the places where people live and work can contribute to biodiversity protection. The challenge is in determining which human activities, at what levels, are compatible with conservation objectives. It also involves reframing the value of biodiversity protection, not simply in terms of ethics or beauty or spiritual values but in ways that can be incorporated into economic valuation systems. This is the thinking behind the push to make 'ecosystem services' a major focus of conservation efforts (Daily & Ellison 2002; Millennium Ecosystem Assessment 2005).

Conservation in practice

The Nature Conservancy aims to preserve biodiversity by protecting places – the lands and waters that organisms and natural communities need to survive.

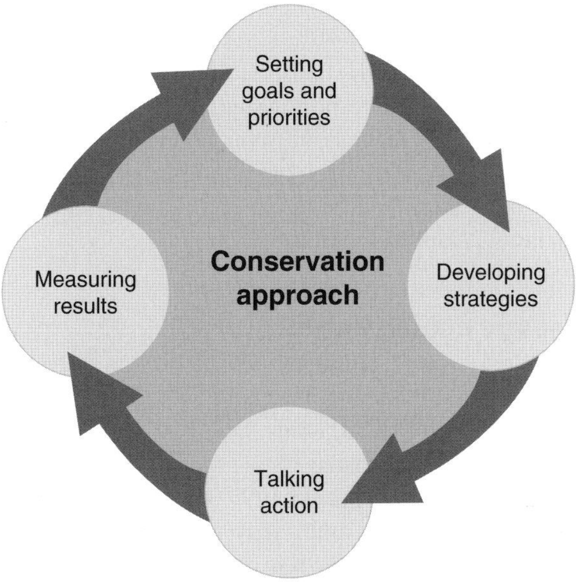

Figure 41.1 **The basic elements of the conservation approach practised by TNC.**

The focus is on habitats and ecosystems as well as species. TNC's approach includes four components (Fig. 41.1). First, conservation cannot be done everywhere and should not be done just anywhere. It is necessary to **set priorities** among places meriting conservation attention in order to identify those areas where conservation work should be concentrated to achieve the greatest results (Groves 2003). TNC uses iterative site-selection algorithms to define a complementary set of conservation areas that, if protected or managed appropriately, have a high potential to protect the biodiversity characteristic of an entire **ecoregion** – a large area of land or water delineated by climate and landform and containing geographically distinct natural communities.

This ecoregional planning tells one **where** to work, but not **how** to work there. Should one purchase targeted areas, create conservation easements, broker the establishment of government preserves, or use some other means to place lands under protection? And once a place is 'protected', what then? Because the threats influencing conservation targets – species, communities or ecosystems – within a place are often varied and complex, conservation efforts will be most effective if they are focused on abating or managing those factors that most directly impact biodiversity. Conservation **strategies** must be developed. These, in turn, dictate the conservation **actions** that may enhance the status or persistence of conservation targets. Finally, it is important to determine whether all this planning, strategizing and action has actually produced the desired results – ultimately, preserving biodiversity. This requires **measuring success** (or outcomes). Ideally, the results of this evaluation are then used to adjust the initial priorities or strategies.

The landscape ecology side

There are two ways to think about landscape ecology and its potential to contribute to conservation. First, although there has been ongoing debate about whether 'landscape' can legitimately be viewed as a level of ecological organization or a spatial scale (e.g. Forman 1995; King 2005), the fact is that many people do view it in these ways, and it may be useful to forego academic arguments momentarily to accept this view (even though I do not agree with it). Second, landscape ecology has generated a distinctive set of concepts ('principles') and tools. If the discipline is to have much relevance to conservation, it must be through these concepts and tools.

Landscape as a scale

Forman (1995) defined landscapes as mosaics of land elements ('patches') occurring over a kilometre-wide scale. The concepts of landscape ecology, however, may apply over multiple scales, from that of a beetle traversing a grassland 'microlandscape' mosaic (which from a beetle's worldview may be every bit a 'landscape') to that of a bison undertaking (in former times) movements over hundreds of kilometres of the same grasslands. Yet there is some comfort in relating landscape to scales of human perception and practicality in the sense that this accords with the scales of land tenure and management. Used in this way, it immediately suggests that the domain of applicability of landscape ecology to conservation is limited to scales between, say, a few dozen hectares to perhaps tens or hundreds of square kilometres. This perspective is embodied in TNC's focus on 'functional landscapes' as targets in conservation planning (Poiani *et al.* 2000; Groves 2003).

Landscape concepts

The essential message of landscape ecology is that not all places in a landscape are the same, and the ways in which they differ and their locations relative to one another affect what goes on within and among those places. Several authors (notably Hobbs 1994) have defined the central concepts or principles of landscape ecology as they relate to landscape structure, function and change. (Liu and Taylor (2002) suggest adding 'landscape integrity' as a measure of landscape 'health', but in my view this opens a Pandora's box of debate about what 'health' is and how to measure it.) Landscapes have a spatial structure – places have explicit locational relationships to other places. They have functional properties – the spatial structure affects how organisms, materials, disturbances or information move across space and therefore how things (species, communities, nutrients, etc.) relate to one another locationally as well as functionally. And these relationships change over time – landscapes are **dynamic**, both structurally and functionally.

While it is useful to partition landscapes in this way, I find it more helpful to consider six essential features of landscapes ('principles') that combine structure, function and change. Here they are:

1 **Landscape elements ('patches') differ in quality.** Most visualizations of landscapes portray a mosaic of patches that differ from one another – this is what creates landscape structure. But simply mapping this structure is cartography, not ecology. The differences among patches that enable mapping of a mosaic also imply differences in ecological conditions among the patches that determine their 'quality' to the organisms occupying them. Places in a landscape differ with regard to factors influencing individual reproduction and survival, interactions among species that may affect community composition, and biophysical properties that determine the retention, release and exchanges of materials and nutrients.
2 **Patch boundaries influence what goes on within and among landscape elements.** The boundaries of landscape elements serve not only to define these patches structurally, but may determine how organisms, materials, nutrients, energy or disturbances move across a landscape. Movements within patches differ among different types of patch – termed 'patch viscosity'. Patch boundaries therefore represent transitions from one set of movement parameters to another. But there is more to it than that. Boundaries differ in their 'permeability' to movements, and the permeabilities differ among organisms,

materials, etc. The arrangement and characteristics of boundaries in a landscape can therefore exert a powerful influence on how things are distributed among patches and how these distributional dynamics change over time.

3 **Patch context matters.** The combination of intrinsic differences among patches with the influence of patch boundaries on movements through a landscape means that how patches are configured in the landscape can have major consequences. The context of a given patch – the characteristics of neighbouring patches – influences what goes on within that patch as well as its interactions and exchanges with its neighbours. One immediate result is that, unless one is comparing different landscapes, metrics such as 'average' patch quality for a landscape as a whole have little meaning. Measures such as population density, habitat suitability, survival probability and so on are not just spatially dependent, but spatially dependent in ways that relate to the structural configuration of the landscape.

4 **Connectivity is an important feature of landscapes.** How the arrangement of patches in a landscape and characteristics of patch boundaries affect movement is reflected in the connectivity of the landscape. Connectivity is a functional attribute of landscapes, one that depends on the properties of interest – one species versus another, different kinds of disturbance, and so forth. A landscape that offers few impediments to movement of a habitat generalist may be quite disconnected to a habitat specialist. But connectivity also has a structural component. Corridors – bands of habitat that join otherwise separated patches of similar habitat in a mosaic – may facilitate the movement not only of organisms but of pathogens, predators and disturbances.

5 **Scale is also important.** It is becoming almost trite in ecology to observe that everything depends on scale. We have little difficulty in acknowledging that our perception of the structural configuration of a landscape changes with changes in the spatial scale of resolution – this is what different map scales are all about. Patches and linkages that are evident at one scale disappear and are replaced by different patterns at another scale. But scale also relates to the functional attributes of a landscape. Different organisms move through landscapes and perceive landscape patterns at different scales, so what may be a highly heterogeneous mosaic of patches of vastly different quality to one organism may be one homogeneous patch to another. Moreover, neither landscape structure nor the ecological responses to landscapes change smoothly and continuously with scale. There are scaling non-linearities – patterns and processes may change gradually over some range of scales, then abruptly change to something quite different with a small additional change in scale (Wiens 1989; Groffman et al. 2006).

6 **Landscapes are dynamic.** Although maps give the impression that landscape patterns are frozen in time, both the patterns and the functions they engender are not static. Any farmer knows well the seasonal and annual changes in the landscape, whether due to natural or anthropogenic forces. Remote sensing indicates clearly how landscapes change on multiple scales of time and space. These changes create a shifting stage on which ecological processes are played out. The distributions and abundances of species across the landscape change, pathways of structural and functional connectivity change, patch qualities change, species interactions change. The landscape is (to borrow a metaphor from Horn & MacArthur 1972) like a multicoloured harlequin engaged in an improvised dance.

Seeing both sides: Where do conservation and landscape ecology intersect?

Let's stick with the premise that conservation aims to preserve species and functional ecological systems by protecting the places they need to survive. Conservation planning at broad global or ecoregional scales can help to identify areas or regions in which the payoff for conservation efforts, in terms of the 'biobang for the buck', is likely to be greatest. Efforts to influence policy at similarly broad scales can be effective in establishing an overarching context for conservation efforts or abating widespread threats. But if, indeed, the major threats to conservation are land use and land-use change, then a good deal of conservation action must be directed at the scale of land use, through policy (e.g. land-use zoning restrictions), establishing protected areas, ecological restoration, designation of human activities that are compatible with biodiversity protection and the like. This is where the landscape rubber meets the conservation road.

For place-based conservation, the major decisions are which places to protect, how to implement that protection, and how to manage or restore the places once they are under some form of protection. The primary message from landscape ecology is that these decisions cannot be made about a place in isolation from its location and relationships in a broader landscape. Because temporal changes affect both landscapes and conservation, the decisions also need to be based not just on how things are now, but on probable future conditions. A place that has high biodiversity value now may not in the future, and even if it does, the surrounding landscape may change in ways that affect the sustainability of this value.

It should be obvious that effective conservation cannot be based on an island-like perception of protected areas (i.e. 'no park is an island'; Janzen 1983). These areas are parts of landscapes that can enhance or erode the sanctity of a reserve, depending on how the landscape is used. To the extent that the 'principles' of landscape ecology hold true, effective conservation must be directed towards entire landscapes – managing the matrix (Lindenmayer & Franklin 2002). But these broader landscapes usually involve multiple ownerships, many with competing priorities for land use, and the various land uses have different compatibilities with the objectives of biodiversity conservation. Managing the matrix entails working with various stakeholders to balance their priorities, while at the same time aiming for some combination of human uses of the landscape that, within those constraints, minimizes the impacts on biodiversity. All in a dynamic and uncertain world. It is easy to see why conservationists are sometimes reluctant to embrace the complexities of landscapes and instead focus on 'protected areas' to the exclusion of their surroundings.

Bringing the sides together: Bridging the gap between science and practice

Clearly, landscape ecology does offer principles, concepts and insights that are needed if the focus of conservation is to be expanded beyond threatened species or hotspots of species richness to include not just 'functioning ecosystems' but the array of spatial relationships and interactions that affect a species or community in a particular place. Despite this relevance, there remains a wide gap between the science of landscape ecology and the reality of conservation practice and action (Fig. 41.2). Why does this gap persist, and what can be done to bridge it?

Monica Turner and her colleagues (2002) identified several sources of division between landscape ecology and natural resource management. I will focus on three of these. First, there are differences in **goals**. Landscape ecology strives to understand the causes and consequences of the spatial heterogeneity of landscapes, whereas conservation is aimed at preserving biodiversity. Framed this way, little common ground is immediately apparent. But if the goal of landscape ecology is extended to include using that understanding to enhance conservation and natural-resource management, and that of conservation is directed to include functioning landscapes, then the congruence is obvious. If these goals can be expressed in operational terms, so that progress towards

Figure 41.2 **Bridging the gap between the worlds of academic science and conservation practice. (Modified from *The Economist*.)**

meeting the goals can be evaluated, then it may be possible to define a goal held in common.

Second, there are several **incongruities of scale**. The 'places' of place-based conservation may range from a few hectares to tens or hundreds of square kilometres, and land ownership and management may vary over a similar range of scales. The everyday perception of 'landscape' falls somewhere between these extremes. Yet the experimental tests and studies that add validity to (or cast doubt on) the principles of landscape ecology are generally conducted at much finer spatial scales, in the order of metres to tens of metres (Kareiva & Andersen 1988; Gardner *et al.* 2001). Conservationists, land managers, the general public and landscape scientists all wear blinders that lead them to focus on some scales and ignore others. As a consequence, they see places and landscapes differently and end up talking about different things and proposing practices that may be effective at some scales but not others. A solution to this difficulty would be for multiscale perspectives to become common to all of these approaches. This is much easier said than done, although the increasing availability of remote-sensing imagery and analysis over a broad range of scales provides the needed platform.

Third, differences in **institutional cultures** tend to maintain or widen the science–practice gap. In academic science, the emphasis is on research that will further our understanding of how landscape structure influences ecological phenomena, and the reward system strongly favours those who obtain grants and publish. In management and conservation, the emphasis is on actions that will produce quick results. In the face of rapid land-use change, managers and conservationists do not have the luxury of waiting for 'more research' and the additional knowledge that it brings; 'just do it' is the prevailing mantra. Yet managers and conservationists often regard science, even science that is done without 'more research', as producing answers to particular, narrow questions.

How much scientific certainty is needed to make conservation decisions, however, is an unresolved question. Scientists need to understand the limitations of their science and recognize that answers for which P is greater than the conventional 0.05 are often useful for management. At the same time, managers should recognize that such answers may carry substantial uncertainty and learn to accept uncertainty into their decision-making.

Conclusions

So where does this leave us? Certainly landscape ecology does have insights and methods that can enhance conservation efforts. But these insights do not apply to all conservation issues or at all scales. Conservation targeted at ensuring the persistence of a particular species, for example, may benefit from an understanding of how the habitat of that species is affected by its surroundings, but the primary knowledge needed will come from basic ecology and population biology. Efforts aimed at developing regional or global priorities for conservation, whether based on 'hotspots' or on some form of representation, generally don't need the detail that comes from a landscape perspective. Aside from local land-use policy, most policy approaches to conservation do not entail the sorts of spatial issues that landscape ecology is so well suited to address. But there is a large component of habitat- or place-based conservation that focuses at scales where the knowledge and tools of landscape ecology are not only relevant, but are crucial to achieving the desired conservation results. We can no longer proceed as if our goals will be met by protecting and managing priority biodiversity areas independent of their landscape surroundings. They won't be.

Some landscape principles

If landscape ecology has something significant to offer to natural-resource management and conservation, it is through the following principles:

1 The essence of a landscape is that it is not all the same – places in the landscape differ; the landscape is spatially heterogeneous. This means that:

- Landscape elements (patches) differ in their internal properties ('quality') – conditions or resource levels that influence the survival or reproduction of organisms, the accumulation of nutrients, productivity, the spread of disturbances and the like. This means that different places in a landscape may require different management in order to maintain or enhance patch quality.
- The boundaries of landscape elements affect the exchanges of organisms, materials or disturbances among adjacent patches. This means that management of patch boundaries may be used to facilitate or impede flows across a landscape (e.g. individual dispersal, invasive spread).
- The landscape context of individual patches influences what goes on within and among patches in a landscape. This means that it may be important to manage not just the overall composition of a landscape, but how different elements are arranged relative to one another.
- Landscape connectivity is a measure of the ease or difficulty with which organisms, materials or disturbances can move across an entire landscape. Because it is determined by characteristics of what is moving as well as what it is moving through, landscape management must consider the characteristics of the organisms, materials, or disturbances of interest as well as the structural configuration of the landscape.

2 How these features of a landscape are expressed, and how organisms, materials or disturbances respond to landscape structure, differs with the scale on which the landscape is viewed. This means that management conducted at one scale may have the desired effects only for features operating at that scale; broad-based landscape management requires a multiscale perspective to be effective.

3 Neither landscape structure nor landscape processes are static in time. Landscapes and the factors that affect organisms, materials and disturbances change, and they change in different ways at different scales. This means that management must be flexible – practices that are effective at one time may be ineffective or counterproductive at another.

References

Bissonette, J.A. (ed.) (1997) *Wildlife and Landscape Ecology: Effects of Pattern and Scale.* Springer-Verlag, New York.

Daily, G.C. & Ellison, K. (2002) *The New Economy of Nature.* Island Press, Washington, DC.

Forman, R.T.T. (1995) *Land Mosaics: The Ecology of Landscapes and Regions.* Cambridge University Press, Cambridge.

Gardner, R.H., Kemp, W.M., Kennedy, V.S. *et al.* (eds.) (2001) *Scaling Relations in Experimental Ecology.* Columbia University Press, New York.

Groffman, P.M., Baron, J.S., Blett, T. *et al.* (2006) Ecological thresholds: the key to successful environmental management or an important concept with no practical application? *Ecosystems* **9**, 1–13.

Groves, C.R. (2003) *Drafting a Conservation Blueprint.* Island Press, Washington, DC.

Gutzwiller, K.J. (ed.) (2002) *Applying Landscape Ecology in Biological Conservation.* Springer-Verlag, New York.

Hobbs, R.J. (1994) Landscape ecology and conservation: moving from description to application. *Pacific Conservation Biology* **1**, 170–176.

Horn, H.S. & MacArthur, R.H. (1972) Competition among fugitive species in a Harlequin environment. *Ecology* **53**, 749–752.

Janzen, D.H. (1983) No park is an island: increase in interference from outside as park size decreases. *Oikos* **41**, 402–410.

Kareiva, P. & Andersen, M. (1988) Spatial aspects of species interactions: the wedding of models and experiments, In: Hastings, A. (ed.) *Community Ecology*, pp. 35–50. Springer-Verlag, New York.

Kareiva, P. & Marvier, M. (2003) Conserving biodiversity coldspots. *American Scientist* **91**, 344–351.

King, A.W. (2005) Hierarchy theory and the landscape . . . level? Or, words do matter. In: Wiens, J. & Moss, M. (eds.) *Issues and Perspectives in Landscape Ecology*, pp. 29–35. Cambridge University Press, Cambridge.

Lindenmayer, D.B. & Franklin, J.F. (2002) *Conserving Forest Biodiversity.* Island Press, Washington, DC.

Lindenmayer, D.B. & Luck, G. (2005) Synthesis: thresholds in conservation and management. *Biological Conservation* **124**, 351–354.

Liu, J. & Taylor, W.W. (2002) Coupling landscape ecology with natural resource management: paradigm shifts and new approaches. In: Liu, J. & Taylor, W.W. (eds.) *Integrating Landscape Ecology into Natural Resource Management*, pp. 3–19. Cambridge University Press, Cambridge.

Millennium Ecosystem Assessment (2005) *Ecosystems and Human Well-Being: Synthesis.* Island Press, Washington, DC.

Miller, J.R. & Hobbs, R.J. (2002) Conservation where people live and work. *Conservation Biology* **16**, 330–337.

Poiani, K.A., Richter, B.D., Anderson, M.G. *et al.* (2000) Biodiversity conservation at multiple scales: functional sites, landscapes, and networks. *BioScience* **50**, 133–146.

Ponder, W. & Lunney, D. (eds.) (1999) *The Other 99%: The Conservation and Biodiversity of Invertebrates*. Transactions of the Royal Zoological Society of New South Wales, Mosman, NSW.

Redford, K.H. & Richter, B.D. (1999) Conservation of biodiversity in a world of use. *Conservation Biology* **13**, 1246–1256.

Turner, M.G., Crow, T.R., Liu, J. *et al.* (2002) Bridging the gap between landscape ecology and natural resource management. In: Liu, J. & Taylor, W.W. (eds.) *Integrating Landscape Ecology into Natural Resource Management*, pp. 433–460. Cambridge University Press, Cambridge.

Wiens, J.A. (1989) Spatial scaling in ecology. *Functional Ecology* **3**, 383–397.

Wiens, J.A., Van Horne, B. & Noon, B.R. (2002) Integrating landscape structure and scale into natural resource management. In: Liu, J. & Taylor, W.W. (eds.) *Integrating Landscape Ecology into Natural Resource Management*, pp. 23–67. Cambridge University Press, Cambridge.

42

What Are We Conserving? Establishing Multiscale Conservation Goals and Objectives in the Face of Global Threats

J. Michael Scott and Timothy H. Tear

Abstract

The conservation movement is entering a new era defined by increasing uncertainty and complexity, driven primarily by large-scale human-induced threats impacting conservation efforts at global and local scales. In this context, how is it possible to establish the necessary long-term goals and objectives at ecologically relevant scales to chart our conservation future? We call for a dramatic shift in current operating practices, which demands significant increases in collaboration and investment, particularly in evidence-based conservation assessments, citizen science projects, alternative scenario building and establishing dynamic, multiscale goals and objectives. For this change to occur, it will require that landscape ecologists and conservation biologists work together and become far more engaged in societal decision-making processes than in the past to build more support and understanding for conservation issues in society at large. Establishing ecologically defensible hierarchical goals and objectives based on sound science emerges as a crucial part of this process. We identify six key principles for landscape ecologists and conservation biologists to follow in establishing shared goals and objectives. Business as usual is insufficient if science is more explicitly to guide societal decisions to address the growing intensity and complexity of threats that local and global conservation efforts now face.

Keywords: conservation goals; evidence-based conservation; global climate change; science and society.

Introduction: A new era in conservation

Conservation is entering a new era, one defined by increasing uncertainty and complexity. As human-induced threats increase in magnitude and scope, conservation faces challenges on unprecedented spatial and temporal scales, and as the degree of their interactions are largely unknown, it is ultimately more difficult to demonstrate that conservation strategies now in place are working. One of the most comprehensive studies to support this broad conjecture was the Millennium Ecosystem Assessment (2005). To conduct their global assessment, the authors identified five 'drivers' of ecosystem service impoverishment (i.e. habitat change, climate change, invasive species, overexploitation and pollution), and concluded that most of these drivers are expected to increase rapidly in their impact, leaving none of the 13 ecosystems they evaluated untouched. In particular, climate change and pollution were expected to have rapidly increasing impacts across all ecosystems, and in only one ecosystem (temperate forests) for one driver (habitat change), were the direct impacts expected to decrease.

Consequently, altered climatic patterns and biogeochemical cycles are likely to change in ways that decrease biological diversity and ecosystem services of benefit to people. Evidence is accumulating that global influences are impacting local conservation efforts beyond the capabilities of local managers to control. We can no longer look the other way and say it is someone else's problem. The expression 'think globally – act locally' is now more relevant, because 'local' is increasingly the expression of 'global'.

In this new context of increasing global threats and local impacts, it becomes even more imperative to establish appropriate conservation goals and objectives. Our ability to achieve success in conservation at any scale now demands that we, landscape ecologists and conservation biologists, chart and direct our conservation futures together. Conservation practitioners are being pushed to tackle issues that stretch traditional norms and demand that decisions are made in the context of increasing complexity and greater uncertainty. We must embrace and engage conservation planning at broader spatial scales, confront environmental threats of greater scope and magnitude, and address increasingly complex ecological relationships – a practice that will take us well beyond the traditional and well-established concepts of protected areas (Newmark 1987; Carroll *et al.* 2004; Scott *et al.* 2004). It is in this context that landscape ecology, and those who practise it, have an important and much-needed role to play.

Establishing a foundation for goal and objective setting in conservation

The terms 'goals', 'objectives' and 'targets' are often used interchangeably but they differ in important ways. Conservation **goals** are intended to draw attention to a cause, to attract supporters and resources, and inspire participants to achieve the goal, whether globally broad, such as launching transboundary conservation efforts, or more narrowly local, such as sustaining open spaces in a single watershed (Table 42.1). Conservation **objectives**, in contrast, identify the outcomes of conservation actions that impact specific areas, populations or individuals that must be managed to meet goals. Objectives are more quantifiable and measurable. Goals provide the broad context in which more specific measurable objectives of conservation projects are implemented (Table 42.2). The term **conservation targets** describes the collective group of biodiversity elements – species, ecological communities and broad-scale ecosystems – that are selected as surrogates to represent the biological diversity.

Conservation biologists have established a strong theoretical foundation for achieving conservation goals and objectives (Pressey *et al.* 1993, 2003; Margules & Pressey 2000; Davis *et al.* 2006). Tear *et al.* (2005) identified five basic principles for conservation planning: (i) state clear goals; (ii) define measurable objectives; (iii) separate science from feasibility; (iv) follow the scientific method; and (v) anticipate change (Table 42.2). In the light of the current 'implementation crisis' (Knight *et al.* 2006) and the growing impacts of global threats (Millennium Ecosystem Assessment 2005), we would add a sixth: achieve political buy-in from stakeholders, implementers and affected parties. The success of any conservation effort is dependent upon support from the individuals and institutions that will be impacted by the effort. The reintroduction of the grey wolf (*Canis lupus*) provides an excellent example of the importance of 'stakeholder' buy-in to the successful accomplishment of a conservation project (Fisher 1995; Phillips & Smith 1996).

Establishing goals and measurable objectives in conservation planning has too often failed to match the scale of environmental problems. If we are to

Table 42.1 **Mission or goal statements for conservation organizations spanning global to local emphases.**

Hierarchical scale	**Organization**	**Mission or goal**	**Reference**
Global	Conservation International	To conserve the Earth's living natural heritage, our global biodiversity, and to demonstrate that human societies are able to live harmoniously with nature	http://www.conservation.org
	The Nature Conservancy	To preserve plants, animals and natural communities that represent the diversity of life on Earth by protecting the lands and waters they need to survive	http://www.nature.org
Hemispheric	Audubon	To conserve and restore natural ecosystems, focus on birds, other wildlife and their habitats for the benefit of humanity and the Earth's biological diversity	http://www.audubon.org
	Ducks Unlimited	Conserves, restores and manages wetlands and associated habitats for North American Waterfowl	http://www.ducks.org/
National	Australia	A pattern of development that improves the total quality of life both now and in the future in a way that maintains the ecological processes on which life depends	Environmental Advisory Panel 1996
Regional	Northwest Forest Plan	To produce a predictable and sustainable level of timber sales and non-timber resources	Thomas *et al.* 2006
	Sonoran Desert Plan	Protect the heritage and natural resources of the West	Pima County 2001
Local	Natural Heritage Land Trust	Dedicated to conservation of natural areas and open spaces in Dane and the surrounding counties	http://www.nhlt.org
	Palouse Land Trust	Conserve the open space, wildlife habitat, water quality and scenery of the Palose	http://palouselandtrust.org

Table 42.2 **The fundamental principles for establishing credible goals and measurable objectives in conservation planning (modified from Tear *et al.* 2005, reproduced with permission of the American Institute of Biological Sciences).**

Principle	**Description**
State clear goals	Well-defined, unambiguous statements that are brief, yet visionary, and are used as the basis for more specific objective setting
Define measurable objectives	Measurable by some standard scale (e.g. number or percent) over time (e.g. months or years) and space (e.g. for a political or ecological region like a state or ecoregion)
Separate science from feasibility	Science alone must drive the process for setting objectives. Feasibility may then be considered to evaluate the likelihood of achieving the stated objectives
Follow the scientific	Build on previous knowledge, method conduct and document a transparent and repeatable process, document assumptions, quantify sources of error, and subject findings to peer review. In addition, thoroughly document sources of information, highlight weaknesses/information gaps, and suggest ways to improve through further research or improving the process in subsequent iterations
Anticipate change	As objective setting is a science, expect objectives to change as knowledge and science change, and employ the concepts of adaptive management
Achieve buy-in from stakeholders	Buy-in from stakeholders will reduce barriers to implementation and increase possible sources of funding

meet the challenge that global threats pose to the world's species, ecosystems, ecological processes and biological phenomena, the scale of our goals and objectives must match the scale of the challenge. Efforts to reduce threats to these conservation targets must be structured to match the scale of the threat. This will demand a more active linkage between local conservation efforts and global threat-reduction efforts. Success in achieving goals and objectives may be more successfully achieved if the general public and the corporate world are engaged in framing and implementing conservation plans.

There is a growing need for landscape ecology principles and practitioners in this new era of increased scale, scope and complexity. As landscape ecologists become more engaged in establishing bolder conservation goals and associated objectives, they should heed the following points:

1 **Place project-specific conservation planning goals and objectives in a hierarchy from local to global.** The pervasiveness of large-scale threats demands that even small conservation projects consider their global context. Landscape ecologists can contribute knowledge of hierarchy theory and spatially explicit tools to help bring data and information into local conservation planning. For example, The Nature Conservancy attempts to draw direct links between their local conservation actions and their global mission, which can be a complicated process (Table 42.3).
2 **Identify the stakeholders and implementers of conservation goals and objectives at all scales, and engage them early in the planning process.** In planning, the process can be as, or more, important than the product. Early engagement and active involvement of all parties is a fundamental tenet of successful planning. As goal-setting is mostly about societal choice, and establishing measurable objectives is the means to achieve the goals, broader engagement and agreement are crucial components of success. Landscape ecologists need to be a part of these discussions to ensure that the principles of landscape ecology are incorporated into the decision-making process.
3 **Establish mutually agreed goals and objectives that capture the full range of variation across the priority conservation targets.** Establishing broad goals and measurable conservation objectives that are biologically relevant should consider the historic range and variation of the conservation target. Society's choices for establishing viability can range broadly depending upon the degree of exploitation or security that is desired (Table 42.4). Such decisions should consider not only the abiotic and habitat-related components, but also the genetic and behavioural expressions of ecological variation to ensure that redundancy and resilience have received adequate

Table 42.3 **An example of the hierarchical structure necessary to make direct links from global to local conservation efforts. Based on the Catskill Mountain conservation project in New York, the organizational mission of The Nature Conservancy was linked to on-the-ground actions by a series of well-defined goals, objectives and actions. This example is illustrative of a growing trend for articulating distinct global to local connections, and the need to stratify and prioritize in order to cross spatial scales and describe these connections.**

Organizational hierarchy	Description	Organizational responsibilities
Organizational mission	To preserve plants, animals and natural communities that represent the diversity of life on Earth by protecting the lands and waters they need to survive	Long-term, organization-wide, global goal
Organizational goal	By 2015, The Nature Conservancy will work others to ensure the effective conservation of places that represent at least 10% of every Major Habitat Type on Earth	Shorter-term, organization-wide responsibility to achieve the mission stratified by Major Habitat Type, specified by time period and expressed by a measurable outcome
Organizational goal stratification: by Major Habitat Type and realm	Effective conservation of places that represent at least 10% of the Temperate Mixed Deciduous Forest in each realm	Organization-wide goal expressed via a single Major Habitat Type
Major Habitat Type stratification: example goal for one realm	Effective conservation of places that represent at least 10% of the Temperate Broadleaf and Mixed Forest in the Nearctic Realm	Each Major Habitat Type goal is stratified by biophysical realm. TNC Political Regions are responsible for each Major Habitat Type goal in their realm and to prioritize efforts by crisis and opportunity ecoregions
Realm stratification and prioritization:	In the High Allegheny Plateau Ecoregion, conserve all major forest types by protecting	Within each realm and priority ecoregion, the Conservancy's associated Region and

example objectives for one ecoregion	multiple (>1) high-quality examples of large, relatively unfragmented blocks (>25,000 acres) of each forest type within each ecoregional subsection where they occur	operating units (in this case the Eastern US Conservation Region and the NY, PA and NJ Chapters) are responsible for effective conservation of the Major Habitat Type, ideally expressed and stratified across an individual ecoregion Quantifiable goals are established for each conservation target across the ecoregion, ideally expressed by an abundance measure and some stratification across the ecoregion
Ecoregion stratification and prioritization: example selection of projects/places	In the Catskill Mountain project area, conserve the high-quality examples of the Northern Hardwood Forest system identified in the High Allegheny Ecoregional plan	The operating unit (e.g. Eastern NY Chapter) and the project team (e.g. the Catskill Mountains conservation project) have the responsibility for contributing to the effective conservation of the ecoregional targets and their occurrences in the project area
Project/place conservation target prioritization and designation: example of aligning project and ecoregion priority conservation targets	Ensure the effective conservation of all ecoregionally significant Northern Hardwood Forest blocks necessary to meet the ecoregional goal and the six contiguous ecoregional priority forest blocks in the Catskill Mountains	Across the ecoregion, and within the conservation project boundary, efforts are undertaken to effectively conserve the conservation target(s) selected as priorities
Strategies to ensure the long-term persistence of the priority	In the Beaverkill Forest Block: *Objective 1*: Reduce or maintain the 12 priority invasive plant species to <1% cover within the designated weed-free area of the Beaverkill and	Measurable objectives are identified by the project team to ensure the long-term persistence of each priority conservation target and its associated occurrences. This is carried

(*Continued*)

Table 42.3 **(continued)**

Organizational hierarchy	**Description**	**Organizational responsibilities**
conservation targets: examples of measurable objectives from one project/place	Panther Mountain forest blocks and adjacent to priority ecoregional streams by 2024 *Objective 2*: Reduce nitrogen deposition in the Catskills to <7 kg/ha/yr by 2020 *Objective 3*: Reduce the risk of forest core fragmentation by achieving permanent protection status on 15% of the acres not permanently protected by 2010	out by assessing the viability, threats and conservation management status (both protection status and management effectiveness), and identifying the measurable objectives most important to ensure persistence. Ideally, monitoring programmes are established to gather the data necessary to track progress towards each measurable objective (i.e. strategy effectiveness measures), and where possible also track the status of targets or threats where no action is being taken as early warning status measures to signal if strategies should change
Strategic actions: examples of some initial steps necessary to evaluate if measurable objectives are appropriate and achievable	*Objective 1 Action*: By 2007, develop and complete an invasive plant species inventory of the Beaverkill and Panther Mountain forest core areas *Objective 2 Action*: Assemble science on impact of atmospheric deposition to the Catskills Northern Hardwood Forest and other embedded biodiversity conservation targets within the forest system *Objective 3 Action*: By 2006, meet with all significant landowners within the Beaverkill block and discuss land protection opportunities	The project team (e.g. Catskill Mountain project team) identifies a series of strategic actions to clarify the steps necessary to achieve each measurable objective Multiple objectives combine to ensure the effective conservation of each priority conservation target within the project boundary

Table 42.4 **Different choices for establishing viability goals to conserve individual populations. Each goal could result in significantly different numbers of individuals as the associated measurable objective to evaluate the effectiveness of conservation efforts (modified from Tear *et al.* 2005, reproduced with permission of the American Institute of Biological Sciences.)**

Viability goals	**Description**
Minimum viable population	The smallest isolated population that has a specified statistical chance of remaining extant for a given period of time in the face of foreseeable demographic, genetic and environmental stochasticity and natural catastrophes (Meffe & Carroll 1994, p. 562; see also Soulé & Wilcox 1980; Shaffer 1981; Beissinger & Westphal 1998)
Ecologically viable population	Population that maintains critical interactions (e.g. behavioural, ecological, genetic) and thus helps ensure against ecosystem degradation. In general, these populations sizes much larger than estimated simply to require population persist over time (see Peery *et al.* 2003; Soulé *et al.* 2003)
Recreationally viable population	Population that supports recreational activities under specified conditions for a specified period of time (similar to criteria for minimum viable population above). Classic examples would be hunted populations, such as waterfowl and big game, and sport fisheries. For these populations, the impact of hunting or fishing can be manipulated through the use of different definitions of take, such as daily and seasonal bag limits, to reduce the number of individuals taken, size limits (e.g. not allowing removal of fish larger than 18 inches to protect brood stock), or catch-and-release permits to limit incidental mortality resulting from no-take fishing (see Barnhart & Roelofs 1989)
Commercially viable populations	Population that supports commercial activities under specified harvest levels for specified periods of time. Concepts such as maximum sustainable and optimally sustainable yields traditionally have been used to predict commercially permitted harvest levels (see Lovejoy 1996)

consideration within the conservation plan. Providing spatially explicit analyses, and easily understood mapped information to guide planning decisions are just some of the skills and products that landscape ecologists could provide to improve ecological assessments and their incorporation into decision-making. Landscape ecologists could also help to increase the capacity for stakeholders to consider multiple sets of differing objectives to achieve the same broadly stated goal, as this can help to increase the likelihood of consensus among varied stakeholders. Finally, spatially explicit modelling to promote scenario building is a much-needed tool that landscape ecologists can contribute to advance these discussions.

4 **Establish goals and measurable objectives while anticipating change.** Conservation decision-making always occurs in the absence of complete information. Multidisciplinary teams must commit not only to making decisions with limited information, but also to a process for revisiting decisions as new information becomes available. In particular, monitoring efforts should be directed to track progress towards measurable objectives, which must be an integral component of anticipating change. Landscape ecologists can help in crafting the appropriate set of indicators that will be sensitive to management actions. Given the increasing size and scale of threats to conservation, landscape-scale metrics derived from remotely sensed information will become increasingly important to track changes in status of conservation targets and their threats. In addition, landscape ecologists can facilitate the use of global, national and regional datasets at local scales, and in turn, use local data to feed back into larger-scale modelling processes.

5 **Establish standards for goal and objective setting that facilitate effective evaluation of the products and the process.** There are many differing perspectives on what goals and objectives are, how they should be set, what a credible process should look like, and how best to determine if conservation efforts are effective. Conservation biologists and landscape ecologists need to work together to establish the appropriate standards necessary to ensure a credible goal- and objective-setting process, and to put in place monitoring programmes to determine the effectiveness of conservation actions.

Expanding the conservation toolbox to achieve better results

One of the most important roles for science and scientists in this age of increasing uncertainty and complexity is to help society choose its future by bringing the latest tools and knowledge so that we can better define how long-term visions might be achieved. Traditionally, scientists have remained largely removed from such decision-making processes. Our call for a more engaged scientific community demands that scientists maintain their credibility by describing the relationship of science to policy without being prescriptive about policy options or using value-laden language (Haseltine 2006). Changing from the current 'business as usual' model is necessary if scientists are to be major players in shaping policies that promote conservation efforts in the service of society. There are several ways to do this that involve expanding the toolbox of conservation information and services to achieve better results.

Increase the use of evidence-based conservation science

Evidence-based science, in which all available evidence that is applicable to a specific cleanly stated problem is rigorously reviewed, has had a powerful influence in other disciplines, particularly medicine, but has been meagrely deployed in conservation. Using this approach could move conservation beyond practices where actions are often based on anecdotal evidence, hearsay, common sense, personal experience or 'expert opinion' to a systematic appraisal of past practices based on an objective review (Pullin and knight 2001; Sutherland *et al.* 2004). The Centre for Evidence-Based Conservation (http://www.cebc.bham.ac.uk/) has produced several systematic reviews that 'summarize, appraise and communicate the results of a large quantity of research and information' on topics as diverse as prescribed fire impacts on bogs or the effectiveness of control programmes for ragwort and muskrats. But review alone is insufficient to be effective the information must be transferred to, and used by, decision-makers. Communicating the results of policy- and management-relevant research will increase the visibility and credibility of conservation science and scientists in the public policy domain. Increasing our collective investment in evidence-based conservation, including information systems that allow streamlined application of information to decision-making, could result in more effective conservation, reduced costs and increased respect for landscape ecologists and conservation biologists by policy-makers and managers.

Encourage more citizen science projects

As human populations shift towards urban centres people are becoming more disconnected from nature and natural resources. The resulting erosion of knowledge and appreciation for nature and its benefits to society is likely to lead to greater apathy. Overall, the conservation movement has failed to make the connections between the targets we endeavour to conserve, the protected areas that provide their essential habitat, and the ecosystem services they provide to people. The relationships between economic and social wealth and human well-being, and natural resources must be more evident. Only then can society more fully understand the true impacts of human-induced, and globally distributed threats to biodiversity. Engaging more people in citizen conservation efforts is a necessary strategy to facilitate increased support and understanding for conservation issues in society at large – a society that must ultimately choose conservation goals (Whitfield 2001). Finding ways to balance ecologically defensible conservation goals and objectives with competing economic and social pressures is essential if conservation issues are not to disappear from the public mind. Increased citizen involvement is essential if we are to move beyond the current 'culture of emergency and conflict' (Redford & Sanjayan 2003) to one of engagement and problem-solving.

Increase multidisciplinary conservation partnerships

It is one thing to call for more 'citizen science', but we recognize that such efforts alone will not be sufficient dramatically to change society's perceptions. These efforts must be matched with broader multidisciplinary constituencies and partnerships. For example, more numerous and more sizable public and private partnerships are needed to combine the resources and expertise necessary if we want to reverse many of the global threats degrading biological diversity and essential ecosystem services provided to society at large. Kareiva *et al.* (2006) have called for much more explicit, formalized partnerships between non-governmental conservation organizations and the federal government to advance protection efforts under the Endangered Species Act. These partnerships would illustrate how greater collaboration could result in improved protection for critical biodiversity habitat, more effective goal-setting processes and a more equitable distribution of costs.

Expand the use of multiple scenarios

Reaching consensus with diverse stakeholder groups is no easy task. Scientists have an important role to play in the process of setting objectives by helping to craft multiple scenarios (Schwartz 1991). Alternative scenarios have been an important factor in resolving recovery efforts for wide-ranging species in complicated biological and socioeconomic systems. Scenarios have also helped to establish legally defensible, public mandates for statewide plans that directly support local action (Tear *et al.* 2005). Landscape ecologists can play an important role in creating and assessing the crucial spatial framework in which global threats can be linked with potential conservation solutions for reducing the impact of the threats. For example, there is an urgent need to identify spatially explicit options for improving the connectivity among reserve networks to enhance the movement of species in response to changing climate. Similarly, assessment of the risk from a single threat or multiple threats across large landscapes by determining the sensitivity of biodiversity elements and priority conservation areas is needed. In these cases, like many others, spatially explicit information and analysis backed by a landscape ecology perspective can help not only to improve the assessment process, but also ultimately the prioritization of scarce resources.

Establish dynamic and hierarchical goals and objectives

Establishing appropriate conservation goals and objectives has always been challenged by a lack of information. Our understanding of the direct and indirect relationships that exist between human activities and ecological systems is in its infancy. A lack of strong predictive relationships, a high degree of variability, large error terms, significant data uncertainty and the unpredictability of the human behavior all help to explain why we were able to put a man on the Moon over three decades ago but can't be sure that our grandchildren will live in a world that includes polar bears in the wild. Yet we must be able to establish appropriate conservation goals to sustain the natural resources we depend upon and determine the necessary actions to achieve these goals. However, we know that achievement of a conservation goal or objective at one scale is no guarantee that it will be met at others (Scott *et al.* 2004). Landscape ecologists and conservation biologists must work together on conservation projects that span local to global issues, as these require specific skills and knowledge about hierarchical relationships. It is also necessary to establish methods for gathering and incorporating new evidence that help not only in initially setting goals and objectives, but also in implementing a dynamic process that ensures appropriate revision as more evidence becomes available.

Conclusions

In summary, it is time that conservation departs from 'business as usual' in order to address the increasing severity and intensity that global threats pose to all conservation efforts. In this world of increasing uncertainty and complexity arising from global threats, it is essential to establish the mechanisms by which science and scientists can more explicitly contribute to decision-making by society about appropriate goals for conservation efforts. Clear, direct links between conservation actions and societal benefits are needed. Landscape ecology and ecologists should play a central and integral role in this transition period, helping to shape not only the information to strengthen these links, but to contribute the skills and knowledge to communicate the importance and relevance of sound conservation practices for everyone's benefit.

Key principles for goal and objective setting in landscape ecology

1 Place project-specific conservation planning goals and objectives in a hierarchy from local to global.
2 Identify the stakeholders and implementers of conservation goals and objectives at all scales, and engage them early in the planning process.
3 Establish mutually agreed goals and objectives that capture the full range of variation across the priority conservation targets.
4 Establish goals and measurable objectives while anticipating change.
5 Establish standards for goal and objective setting that facilitate effective evaluation of the products and the process before the process begins.

References

Barnhart, R.A. & Roelofs, T.D. (1989) Catch and release fishery: a Decade of experience. California Cooperative Fisheries Unit Arcata California.

Beissinger, S.R. & Westphal, M.I. (1998) On the use of demographic models of population viability in endangered species management. *Journal of Wildlife Management* **62**, 821–841.

Carroll, C., Noss, R.F., Paquet, P.C. & Schumaker, N. (2004) Extinction debt of protected areas in developing landscapes. *Conservation Biology* **18**, 1110–1112.

Davis, F.D., Costello, C. & Stoms, D. (2006) Efficient conservation in a utility-maximization framework. *Ecology and Society* **11**(1), 33 [online]. URL: http://www.ecologyandsociety.org/vol11/iss1/art33/

Fisher, H. (1995) *Wolf Wars: The Remarkable Inside Story of the Restoration of Wolves to Yellowstone*. Falcon Press, Helena, MT.

Haseltine, S.D. (2006) Scientists should help frame the question. *Bioscience* **56**, 289–290.

Kareiva, P.M., Andelman, S., Doak, D.F. *et al.* (1998) Using science in habitat conservation plans. National Center for Ecological Analysis and Synthesis (http://www.aibs.org/books/resources/hcp-1999-01-14.pdf).

Kareiva, P.M., Tear, T.H., Solie, S., Brown, M.L., Sotomayor, L. & Yuan-Farrell, C. (2006) Nongovernment organizations. In: Goble, D.D., Scott, J.M. & Davis, F.W. (eds.) *The Endangered Species Act at Thirty: Renewing the Conservation Promise*, pp. 176–192. Island Press, Covelo, CA.

Knight, A.T., Cowling, R.M. & Campbell, B.M. (2006) An operational model for implementing conservation action. *Conservation Biology* **20**, 408–419.

Lovejoy, T.E. (1996) Beyond the concept of sustainable yield. *Ecological Applications* **6**, 363.

Margules, C.R. & Pressey, R.L. (2000) Systematic conservation planning. *Nature* **405**, 243–253.

Meffe, G.K. & Carroll, C.R. (1994) *Principles of Conservation Biology*. Sinauer Associates, Sunderland, MA.

Millennium Ecosystem Assessment (2005) *Our Human Planet: Summary Report for Decision Makers*. Island Press, Covelo, CA.

Newmark, W.C. (1987) A land bridge island perspective on mammalian extinctions of large mammals in Western North American parks. *Nature* **325**, 430–432.

Peery, C.A., Kavanagh, K.L. & Scott, J.M. (2003) Pacific salmon: setting ecologically defensible recovery goals. *BioScience* **53**, 622–623.

Phillips, M.K. & Smith, D.W. (1996) *The Wolves of Yellowstone: The Inside Story*. Voyager Press, Stillwater, MN.

Pima County (2001) *Sonoran Desert Conservation and Comprehensive Land Use Plan*. Pima County, Tucson, AZ.

Pressey, R.L., Humphries, C.J., Margules, C.R., Vanewright, R.I. & Williams, P.H. (1993) Beyond opportunism: key principles for systematic reserve selection. *Trends in Ecology and Evolution* **8**, 124–128.

Pressey, R.L., Cowling, R.M. & Rouget, M. (2003) Formulating conservation targets and biodiversity pattern and process in the Cape Floristic Region, South Africa. *Biological Conservation* **112**, 99–127.

Pullin, A. & Knight, T. (2001) Effectiveness of conservation practice: pointers from medicine and public health. *Conservation Biology* **15**, 50–54.

Redford, K.H. & Sanjayan, M.A. (2003) Retiring Cassandra. *Conservation Biology* **17**, 1463–1464.

Schwartz, P. (1991) *The Art of the Long View*. Currency Doubleday, New York.

Scott, J.M., Loveland, T., Gergelly, K., Strittholt, J. & Staus, N. (2004) National Wildlife Refuge System: ecological context and integrity. Natural; *Resources Journal* **44**, 1041–1066.

Shaffer, M.L. (1981) Minimum population sizes for species conservation. *BioScience* **31**, 131–134.

Soulé, M.E. & Wilcox, B.A. (1980) *Conservation Biology. An Evolutionary-Ecological Perspective*. Sinauer Associates, Sunderland, MA.

Soulé, M.E., Estes, J.A., Berger, J., *et al.* (2003) Ecological effectiveness: conservation goals for interactive species. *Conservation Biology* **17**, 1238–1250.

State of the Environment Advisory Council. (1996) Australia's State of the Environment. Commonwealth Minister for the Environment Department of the Environment and Water Resources.

Sutherland, W.J., Pullin, A.S., Dolman, P. & Knight, T.M. (2004) The need for evidence based conservation. TREE 305–309.

Svancara, L.K., Brannon, R. & Scott, J.M. (2005) Policy driven versus evidence-based conservation: a review of political targets and biological needs. *Bioscience* **55**, 989–995.

Tear, T.H., Kareiva, P., Angermeier, P.L. *et al.* (2005) How much is enough? The recurrent problem of setting quantitative objectives in conservation. *Bioscience* **10**, 835–849.

Thomas, J.W., Franklin, J.F., Gordon, J. & Johnson, K.N. The northern forest plan: origin, components, implementation, experience and suggestions for change. *Conservation Biology* **20**, 277–287.

Whitfield, J. (2001) The budding amateurs. *Nature* **414**, 578–579.

43

Goals, Targets and Priorities for Landscape-Scale Restoration

Richard J. Hobbs

Abstract

Given that resources available for conservation management are always likely to be limiting, it becomes essential that management actions are prioritized and directed towards explicitly stated goals and targets. Often conservation goals are poorly enunciated and activities carried out on a more-or-less *ad hoc* basis. Landscape-scale conservation and restoration activities need to be set within the context of the particular landscape being examined and can be prioritized on the basis of the current landscape condition, which includes consideration of extent and condition of remaining native vegetation and the characteristics of the matrix (particularly in relation to connectivity). In addition, some consideration of the relative value of different conservation assets, the degree of threat that they are under, and the relative likelihood of effective management treatments can help assign priorities and indicate the degree of intervention required in each case.

Keywords: goals; landscape context; prioritization; restoration.

Introduction

What are we trying to achieve when we set about managing and designing landscapes for conservation? The answer may seem obvious – for instance, we can say that we are aiming to conserve the native biota of the region. However, despite the fact that this aim may derive from legislation, policy and societal expectations at a variety of levels from local to national, it hides a wealth of detail and does not necessarily give clear guidance as to what needs to be done in particular places or under particular situations. What does 'conserve the native biota of the region' mean in practical terms? Here are some of the questions to be considered in this context.

1 Which elements of the biota are we concerned with – all the biota, selected groups, species that are rare, species that are important as ecosystem engineers or drivers, or charismatic species that people are particularly engaged with or concerned about?
2 How do we define the region – on biogeographic boundaries, on jurisdictional boundaries, at a relatively fine scale or at a broad scale?
3 What does conserving the biota entail – 'habitat' reservation, management and restoration, individual species management, control of undesirable species, managing ecosystem processes or all of these?

Much conservation activity is undertaken without a clear understanding of how it contributes to any goals and objectives. Sometimes the goals and objectives are tacit and assumed, and often they are couched in such generalized language as to mean very little. In this case, any conservation management activity can be justified regardless of whether it is a cost-effective, urgent or useful thing to do or not. However, where time, funds and resources are limiting, as is usually the case, it becomes important to consider what are the most useful things to do initially, and how to get the best bang for one's buck. In order to be able to do this, a clear understanding of the objectives is essential, and these objectives need to be set in the context of the current state of the conservation resources, the armoury of conservation actions possible in that context, and the potential for effective action. In this chapter, I examine the need to prioritize conservation activities, particularly in relation to ecological restoration carried out in a conservation context. The chapter is partially based on material that originally appeared in Hobbs (2005). Restoration and conservation management are increasingly viewed as complementary activities, with restoration often now forming an important element of conservation management (Dobson *et al.* 1997; Young 2000).

Why restoration?

Ecological restoration can be carried out for a number of reasons. Sometimes the aim is to restore highly disturbed, but localized sites, such as mine sites. In this case restoration often entails amelioration of the physical and chemical characteristics of the substrate and ensuring the return of vegetation cover. In other cases the aim is to improve the productive capability in degraded production lands. Restoration in these cases aims to return the system to a sustainable level of productivity by, for instance, reversing or ameliorating soil erosion or salinization problems in agricultural or range lands.

More pertinent to the topic of this book, however, is the use of restoration to enhance nature conservation values in protected landscapes by reversing the impacts of various degrading forces, for example, by removing an invasive animal or plant from a protected area. In addition to this patch-based restoration, broader-scale restoration may also be required to restore ecological processes over landscape-scale or regional areas. This will include the reinstatement of broad-scale connectivity, disturbance regimes and flows of biota, water and nutrients (e.g. Soulé *et al.* 2004).

There is an increasing recognition that protected areas alone will not conserve biodiversity in the long term, and that production and protection lands are linked by landscape-scale processes and flows (e.g. hydrology, movement of biota). Methods of integrating conservation and productive use are thus required, as for instance in the biosphere reserve and core-buffer-matrix models (Hobbs 1993; Noss & Cooperrider 1994; Morton *et al.* 1995; Craig *et al.* 2001). Management in this case entails: (i) maintaining and restoring conservation value in protected areas; (ii) returning conservation value to portions of the productive landscape, preferably through an integration of production and conservation values; and/or (iii) ensuring that land uses within a region do not have adverse impacts on the region's ecological processes.

Ecological restoration thus occurs along a continuum – from the rebuilding of totally devastated sites to the limited management of relatively unmodified sites (Hobbs & Hopkins 1990) – and at a variety of scales, from small, localized efforts to restore particular sites to landscape- or regional-scale programmes that aim to reinstate lansdscape flows and processes; for example, restoration of the Florida Everglades involves extensive restoration activities over much of southern Florida (South Florida Water Management District 1996). The specific goals of restoration and the techniques used will obviously differ from case to case. In general terms, however, restoration aims to return the degraded system to something that is protective, productive, aesthetically pleasing, or valuable in a conservation sense.

Within these broad general aims, more specific goals are required to guide the restoration process. Ecosystem characteristics that may be considered when considering restoration goals include structure, species composition, pattern and heterogeneity, function, species interactions and resilience (Hobbs & Norton 1996). This set of characteristics is complex, and often individual components are considered as primary goals. For instance, restoration of a mine site may aim to replace the complement of plant species present prior to disturbance, while other situations may have the restoration of particular ecosystem functions as a primary aim (e.g. bioremediation of eutrophication in lakes or the manipulation of vegetation cover to modify water use).

Unfortunately, restoration goals are often poorly defined or stated in general terms relating to the return of the system to some pre-existing condition. The definition of the characteristics of this condition has proved problematic, because it assumes a static situation. Ecologists increasingly consider that natural systems are dynamic, that they may exhibit alternative (meta-)stable states (Beisner *et al.* 2003; Suding *et al.* 2004), and that the definition of what is the 'natural' ecosystem in any given area may be difficult (Sprugel 1991). Indeed, the concept of 'naturalness' has itself been the subject of much recent debate, especially in relation to landscapes with long histories of human habitation (Brunson 2000; Oliver *et al.* 2002; Povilitis 2002; see also Chapters 10 and 35).

Landscape-scale restoration

Most of the information and methodologies concerning ecological restoration centre on individual sites. However, site-based restoration has to be placed in a broader context, and is often insufficient on its own to deal with large-scale restoration problems (Hobbs & Norton 1996; Hobbs & Harris 2001). Landscape- or regional-scale processes are often either responsible for ecosystem degradation at particular sites or alternatively have to be restored to achieve restoration goals. Hence restoration is often needed both within particular sites and at a broader landscape scale.

How are we then to go about restoration at a landscape scale? What are the relevant aims? What landscape characteristics can we modify to reach these aims, and do we know enough to be able confidently to make recommendations on priorities and techniques?

There are a number of steps in the development of a programme of landscape-scale restoration, which can be outlined as follows (from Hobbs 2005):

1 Assess whether there is a problem that requires attention; for instance:
   - changes in biotic assemblages (e.g. species loss or decline, invasion);
   - changes in landscape flows (e.g. species movement, water and/or nutrient fluxes);
   - changes in aesthetic or amenity value (e.g. decline in favoured landscape types).

2 Determine the causes of the perceived problem; for instance:
   - removal and fragmentation of native vegetation;
   - changes in pattern and abundance of vegetation/landscape types;
   - cessation of historic management regimes.

3 Determine realistic goals for restoration; for instance:
   - retention of the existing biota and prevention of further loss;
   - slowing or reversal of land or water degradation processes;
   - maintenance or improvement of productive potential;
   - integrated solutions tackling multiple goals.

4 Develop cost-effective planning, prioritization and management tools for achieving agreed goals:
   - determining priorities for action in different landscape types and conditions;
   - spatially explicit solutions;
   - acceptance and 'ownership' by managers and landholders;
   - an adaptive approach that allows course corrections when necessary.

This list is easy to construct but more difficult to deal with in detail. For instance, the initial assessment of whether there is a problem or not requires the availability of a set of readily measurable indicators of landscape 'condition' or 'health'. There have been attempts to use the concept of ecosystem health as an effective means of discussing the state of ecosystems (Costanza *et al.* 1992; Cairns *et al.* 1993; Shrader-Frechette 1994), although such attempts have been criticized as not being very useful in practice (Lancaster 2000). Ecosystem health can be characterized by examining the system's vigour (or activity, production), organization (or the diversity and number of interactions between system components) and resilience (the system's capacity to maintain structure and function in the presence of stress (Rapport *et al.* 1998). Attempts have also been made to produce readily measurable indices of ecosystem health for a number of different ecosystems, although there is still debate over whether these are useful or not. In Australia there are ongoing attempts to arrive at easily obtained measures that provide sensible information on the current condition and value of particular areas of vegetation (e.g. Parkes *et al.* 2003, 2004; Freudenberger *et al.* 2004; McCarthy *et al.* 2004; Oliver 2004; Williams 2004).

In the same way, there have been attempts to develop a set of measures of landscape condition; for instance Aronson and Le Floc'h (1996) present three groups of what they term 'vital landscape attributes', which aim to encapsulate: (i) landscape structure and biotic composition; (ii) functional interactions among ecosystems; and (iii) degree, type and causes of landscape fragmentation and degradation. While their list of 16 attributes provides a useful start for thinking about these issues, it fails in its attempt to provide a practical assessment of whether a particular landscape is in need of restoration, and if so, what actions need to be taken. Considerable progress is being made in this, at least for landscape flows, for instance in the Landscape Function Analysis approach developed for Australian rangelands (see Ludwig *et al.* 2004; Tongway 2004).

Once a problem has been perceived, the correct diagnosis of its cause and prescription of an effective treatment is by no means simple. The assumption underlying landscape ecology is that landscape processes are in some way related to landscape patterns. Hence, by determining the relationship between pattern and process, one is better able to predict what will happen to the processes in which one is interested (biotic movement, metapopulation dynamics, system flows, etc.) if the pattern of the landscape is altered in particular ways. Thus, we are becoming increasingly confident that we can, for instance, predict the degree of connectivity in a landscape from the proportion of the landscape in different cover types. As proportion of a particular cover types decreases, a threshold value may be reached at which physical connectivity rapidly decreases (Pearson *et al.* 1996; Wiens 1997; With 1997). Similarly, as landscapes become more fragmented, there may be thresholds or breakpoints where relatively large numbers of species drop out. Hobbs and Harris (2001) have argued that there may be different types of thresholds at the landscape scale, with some being biotically driven (in the case of connectivity-related processes) and others being abiotically driven (in the case of physical changes such as altered hydrology). The possibility of the existence of different types of thresholds means that clear identification of the primary driving forces is essential before restoration is attempted. There will be little point in trying to deal with biotic issues before treating abiotic problems.

A number of other important questions have to be asked in terms of restoration. First, does the threshold work the same way on the way up as it did on the way down, or is there a hysteresis effect? In other words, in a landscape where habitat area is being increased, will species return to the system at the same rate as they dropped out when habitat was being lost? Second, what happens when pattern and process are not tightly linked? For instance, studies in central Europe have illustrated the important role of traditional management involving seasonal movement of sheep between pastures in dispersing seeds around the landscape (Bakker *et al.* 1996; Fischer *et al.* 1996; Poschlod *et al.* 1996). The long-term viability of some plant species may be threatened by the cessation of this process, and restoration in this case will not involve any modification of landscape pattern – rather it will entail the reinstatement of a management-mediated process of sheep movement. Hence, correct assessment of the problem and its cause and remedy require careful examination of the system and its components rather than generalized statements of prevailing dogma.

What to do: Where and in what order?

Given the considerations above, how do you then go about determining how to conduct restoration at a landscape scale? Here, I relate what we have been thinking about in the context of the agricultural regions of Australia, where landscape fragmentation and habitat modification have caused numerous and extensive problems of land degradation and biodiversity decline. We have been examining the question of what remedial measures can be taken to prevent further loss of species and assemblages in these altered landscapes. A set of general principles, derived from island biogeography theory, suggest that bigger patches are better than small patches, connected patches are better than unconnected, and so on. For fragments in agricultural landscapes, such principles can be translated into the need to retain existing patches, especially large ones, and existing connections, and to revegetate in such a way as to provide larger patches and more connections (Hobbs 1993). Ryan (2000) indicated clearly the lack of evidence that carrying out such revegetation will actually do anything useful, although more recent work is providing useful pointers that differing types of revegetation will have varying degrees of impact in terms of conservation outcomes (e.g. Hobbs *et al.* 2003a; Lindenmayer & Hobbs 2004).

Nevertheless, important questions remain concerning what sort of landscape-level management and revegetation is appropriate for different landscapes. Given that funds, time and resources are always likely to be limiting, we will never be able to do everything that would be desirable. In that case, we need to try to ensure that we undertake activities that are most likely to achieve desirable outcomes in the most cost-effective and efficient way. We also need to set priorities because it is likely that time and resources will not permit the achievement of all the necessary actions.

If we can accept that priority actions involve, first, the protection of existing fragments; second, their effective management; and third, restoration and revegetation, where do we go from there? Which are the priority areas to retain? How do we decide which bits are more important and hence should be afforded a higher priority? Should we concentrate on retaining the existing fragments or on revegetation, and relatively how many resources (financial, personnel, etc.) should go into each? How much revegetation is required, and in what configuration? When should we concentrate on providing corridors versus additional habitat? If we are to make a significant impact in terms of conserving remaining fragments and associated fauna, these questions need to be addressed in a strategic way.

McIntyre and Hobbs (1999, 2000) attempted to examine these questions in the context of different types of landscape, with these types being defined by the extent of vegetation removal and modification. They identified four broad types of landscapes – intact and relictual landscapes at the two extremes, and two intermediate states, variegated and fragmented. In variegated landscapes, the habitat still forms the matrix, whereas in fragmented landscapes, the matrix comprises 'destroyed habitat'. Each of their four levels is associated with a particular degree of habitat destruction. Habitat modification acts to create a layer of variation in the landscape over and above the straightforward spatial patterning caused by vegetation destruction. There is a tendency for habitats to become progressively more modified with increasing levels of destruction, owing to the progressively greater proportion of edge in remaining habitats. Knowing what type of landscape you are dealing with can assist in deciding where on the landscape to allocate greater and lesser efforts towards different management actions (McIntyre & Hobbs 2000). Further to this, figuring out how far it is possible to generalize findings and management guidelines from one landscape type to another and among different biogeographic zones would allow for more confidence that guidelines and prescriptions will actually be effective (Hobbs & McIntyre 2005).

Three types of action could be applied to habitats for their conservation management:

1 **Maintain** the existing condition of habitats by removing and controlling threatening processes. It is generally much easier to avoid the effects of degradation than it is to reverse them.
2 **Improve** the condition of habitats by reducing or removing threatening processes. More active management may be needed to initiate a reversal of condition (e.g. removal of exotic species, reintroduction of native species) in highly modified habitats.
3 **Reconstruct** habitats where their total extent has been reduced below viable size using replanting and reintroduction techniques. As this is so difficult and expensive, it is a last-resort action that is most relevant to fragmented and relictual landscapes. We have to recognize that restoration will not come close to restoring habitats to their unmodified state, and this reinforces the wisdom of maintaining existing ecosystems as a priority.

The next stage is to link these activities to specific landscape components (matrix, connecting areas, buffer areas, fragments) in which they would be most effective and to determine priorities for management action in different landscape types. A general approach might be to build on strengths of the remaining habitat by filling in gaps and increasing landscape connectivity, increasing the availability of resources by rehabilitating degraded areas, and expanding habitat by revegetating to create larger blocks and restore poorly represented habitats. The first priority is the maintenance of elements that are currently in good condition. This will be predominantly the vegetated matrix in intact and variegated landscapes, and the remnants that remain in good condition in fragmented landscapes. There may well be no remnants left in good condition in relict landscapes. Maintenance will involve ensuring the continuation of population, community and ecosystem processes that result in the persistence of the species and communities present in the landscape. Note that maintaining fragments in good condition in a fragmented system may also require activities in the matrix to control landscape processes, such as hydrology.

The second priority is the improvement of elements that have been modified in some way. In variegated landscapes, buffer areas and corridors may be a priority, while in fragmented systems, improving the surrounding matrix to reduce threatening processes will be a priority, as indicated above. In relict landscapes, improving the condition of fragments will be essential for their continued persistence. Improvement may involve simply dealing with threatening processes such as stock grazing or feral predators, or may involve active management to restore ecosystem processes, improve soil structure, encourage regeneration of plant species, or reintroduce flora or fauna species formerly present there (Hobbs & Yates 1997, 2000; Yates *et al.* 2000).

Reconstruction is likely to be necessary only in fragmented and relict areas. Primary goals of reconstruction will be to provide buffer areas around fragments, to increase connectivity with corridors and to provide additional habitat (Hobbs 1993). While some basic principles of habitat reconstruction have been put forward, the benefits of such activities have rarely been quantified. Questions remain about which characteristics of 'natural' habitat are the most important to try to incorporate into reconstruction, and what landscape configurations are likely to be most effective. While there have been recent attempts to provide guidance on this, at least in relation to some aspects of landscape management (e.g. Lindenmayer & Hobbs 2004; Salt *et al.* 2004), as Vesk and Mac Nally (2006) point out, we still have a long way to go to be able to move confidently from putting lines on maps to putting effective habitat back on the ground.

All of this still needs to be placed within a prioritization framework that allows decisions on what to do, where and when. Attempts to develop prioritization frameworks are underway in a variety of arenas in Australia, particularly in relation to the disbursement of government funds for conservation activities by regional Natural Resource Management groups. Hobbs *et al.* (2003b) and Hobbs & Kristjanson (2003) have discussed potential ways of arriving at priorities for action, based on a consideration of the relative value of particular landscape elements, the degree of threat these elements are under and the likelihood of success of management intervention (Table 43.1 and Fig. 43.1). These schemes relate mainly to the problem of maintaining conservation areas in salinizing landscapes – that is, landscapes in which significant areas of land are becoming saline due to rising saline water tables resulting from broadscale clearing of native vegetation (George *et al.* 1995; Cramer & Hobbs 2002). A modified version has appeared in policy documents relating to prioritizing public investment (Department of Environment 2003). In addition to allowing some prioritization of actions, these schemes also indicate the appropriate level of intervention required in each case. While such schemes go some way to providing a rational basis for making decisions on priorities, many problems still remain: for instance, how are relative values allocated, and how is the likelihood of success estimated? In addition, only the threat from salinization is considered in this scheme: when other threats are added in, the dimensionality of the process increases greatly and it becomes immensely complex.

Table 43.1 **Landscape management intervention grid based on level of threat and probability of long-term persistence or system recovery (from Cramer & Hobbs 2002).**

Probability of long-term persistence or system recovery			
High	**C. Threat minimization and prevention** Low disturbance, amelioration of hydrological processes feasible *Large remnant, protected location, high heterogeneity, slow rate of change*	**F. Prompt protection and restoration** Low disturbance, amelioration of hydrological processes feasible *Large/small remnant, unprotected location, high heterogeneity, fast rate of change*	**I. Urgent protection or restoration** Low disturbance, amelioration of hydrological processes feasible *Small remnant, unprotected location, low heterogeneity, fast rate of change*
Medium	**B. Low-level management of threats** Some disturbance, some amelioration of hydrological processes feasible *Large remnant, protected location, high heterogeneity, slow rate of change*	**E. Active threat reduction** Some disturbance, some amelioration of hydrological processes feasible *Large/small remnant, unprotected location, high heterogeneity, fast rate of change*	**H. Fast-tracked management intervention** Some disturbance, some amelioration of hydrological processes feasible *Small remnant, unprotected location, low heterogeneity, fast rate of change*
Low	**A. No immediate management action** High disturbance, amelioration of hydrological processes not feasible *Large remnant, protected location, high heterogeneity, slow rate of change*	**D. Long-term low-level management** High disturbance, amelioration of hydrological processes not feasible *Large/small remnant, unprotected location, high heterogeneity, fast rate of change*	**G. Transition to a new system** High disturbance, amelioration of hydrological processes not feasible *Small remnant, unprotected location, low heterogeneity, fast rate of change*
	Low	*Medium*	*High*
		Level of threat	

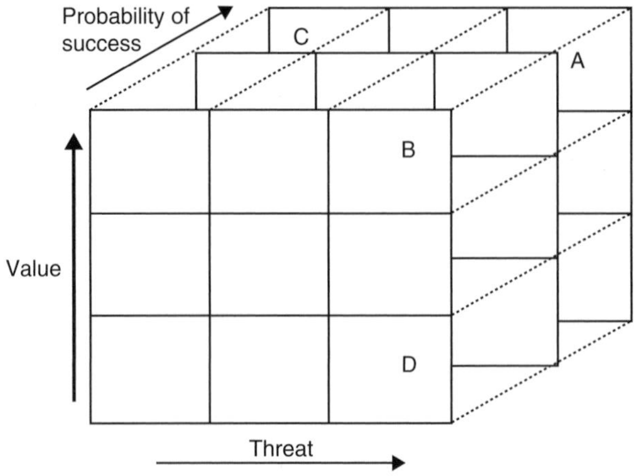

Figure 43.1 **Three-dimensional scheme for priority setting based on level of threat, value of resource and probability of success of management or restoration (a three-dimensional expansion of Table 43.1). 'A' represents a high-value asset at high risk, for which there is a good probability that urgent management/restoration measures will be successful. 'B' represents a high-value asset at high risk, for which the probability of management/restoration success is low. 'C' represents a high-value asset at low risk, for which preventative measures are likely to be successful. 'D' represents a relatively low-value asset at high risk, with a low probability that management/restoration measures will be successful. (From Hobbs *et al.* 2003b, with permission.)**

Principles

Because every landscape is unique and generalized principles have to be translated into landscape-specific priorities, options and actions, perhaps a set of generic principles has limited value. Instead, I suggest a set of principles for deciding 'what to do where' may be more useful.

1 Conservation management is likely to be context-specific and will vary depending on the objectives set, current condition of the landscape and resources available.
2 Management decisions should be based on a prior assessment of landscape characteristics and which biotic elements are to be considered.
3 Set objectives based on this assessment and establish a prioritization of actions that are most likely to achieve these objectives.

References

Aronson, J. & Le Floc'h, E. (1996) Vital landscape attributes: missing tools for restoration ecology. *Restoration Ecology* **4**, 377–387.

Bakker, J.P., Poschlod, P., Strykstra, R.J., Bekker, R.M. & Thompson, K. (1996) Seed banks and seed dispersal: important topics in restoration ecology. *Acta Botanica Neerlandica* **45**, 461–490.

Beisner, B.E., Haydon, D.T. & Cuddington, K. (2003) Alternative stable states in ecology. *Frontiers in Ecology and Environment* **1**, 376–382.

Brunson, M.W. (2000) Managing naturalness as a continuum: setting limits of acceptable change. In: Gobster, P.H. & Hull, R.B. (eds.) *Restoring Nature: Perspectives from the Social Sciences and Humanities*, pp. 229–307. Island Press, Washington, DC.

Cairns, J.J., McCormick, P.V. & Niederlehner, B.R. (1993) A proposed framework for developing indicators of ecosystem health. *Hydrobiologia* **263**, 1–44.

Costanza, R., Norton, B.G. & Haskell, B.D. (1992) *Ecosystem Health: New Goals for Environmental Management*. Island Press, Washington, DC.

Craig, J.L., Mitchell, N. & Saunders, D.A. (2001) *Nature Conservation 5: Nature Conservation in Production Landscapes*. Surrey Beatty & Sons, Chipping Norton, NSW.

Cramer, V.A. & Hobbs, R.J. (2002) Ecological consequences of altered hydrological regimes in fragmented ecosystems in southern Australia: impacts and possible management responses. *Austral Ecology* **27**, 546–564.

Department of Environment (2003) *Salinity Investment Framework Interim Report – Phase I, 2003, Department of Environment, Salinity and Land Use Impacts Series No. SLUI 32*. Department of Environment, Perth.

Dobson, A.P., Bradshaw, A.D. & Baker, A.J.M. (1997) Hopes for the future: restoration ecology and conservation biology. *Science* **277**, 515–522.

Fischer, S.F., Poschlod, P. & Beinlich, B. (1996) Experimental studies on the dispersal of plants and animals on sheep in calcareous grasslands. *Journal of Applied Ecology* **33**, 1206–1222.

Freudenberger, D., Harvey, J. & Drew, A. (2004) Predicting the biodiversity benefits of the Saltshaker Project, Boorowa, NSW. *Ecological Management and Restoration* **5**, 5–14.

George, R.J., McFarlane, D.J. & Speed, R.J. (1995) The consequences of a changing hydrologic environment for native vegetation in south Western Australia. In: Saunders, D.A., Craig, J.L. & Mattiske, E.M. (eds.) *Nature Conservation 4: The Role of Networks*, pp. 9–22. Surrey Beatty & Sons, Chipping Norton, NSW.

Hobbs, R.J. (1993) Can revegetation assist in the conservation of biodiversity in agricultural areas? *Pacific Conservation Biology* **1**, 29–38.

Hobbs, R.J. (2005) Restoration ecology and landscape ecology. In: Wiens, J.A. & Moss, M.R. (eds.) *Issues and Perspectives in Landscape Ecology*, pp. 217–229. Cambridge University Press, Cambridge.

Hobbs, R.J. & Harris, J.A. (2001) Restoration ecology: repairing the Earth's ecosystems in the new millennium. *Restoration Ecology* **9**, 239–246.

Hobbs, R.J. & Hopkins, A.J.M. (1990) From frontier to fragments: European impact on Australia's vegetation. *Proceedings of the Ecological Society of Australia* **16**, 93–114.

Hobbs, R.J. & Kristjanson, L.J. (2003) Triage: how do we prioritize healthcare for landscapes? *Ecological Management and Restoration* **4** (Suppl.), S39–S45.

Hobbs, R.J. & McIntyre, S. (2005) Categorizing Australian landscapes as an aid to assessing the generality of landscape management guidelines. *Global Ecology and Biogeography* **14**, 1–15.

Hobbs, R.J. & Norton, D.A. (1996) Towards a conceptual framework for restoration ecology. *Restoration Ecology* **4**, 93–110.

Hobbs, R.J. & Yates, C.J. (1997) Moving from the general to the specific: remnant management in rural Australia. In: Klomp, N. & Lunt, I. (eds.) *Frontiers in Ecology: Building the Links*, pp. 131–142. Elsevier, Amsterdam.

Hobbs, R.J. & Yates, C.J. (2000) *Temperate Eucalypt Woodlands in Australia: Biology, Conservation, Management and Restoration.* Surrey Beatty & Sons, Chipping Norton, NSW.

Hobbs, R.J., Catling, P.C., Wombey, J.C., Clayton, M., Atkins, L. & Reid, A. (2003a) Faunal use of bluegum (*Eucalyptus globulus*) plantations in southwestern Australia. *Agroforestry Systems* **58**, 195–212.

Hobbs, R.J., Cramer, V.A. & Kristjanson, L.J. (2003b) What happens if we can't fix it? Triage, palliative care and setting priorities in salinising landscapes. *Australian Journal of Botany* **51**, 647–653.

Lancaster, J. (2000) The ridiculous notion of assessing ecological health and identifying the useful concepts underneath. *Human and Ecological Risk Assessment* **6**, 213–222.

Lindenmayer, D.B. & Hobbs, R.J. (2004) Fauna conservation in plantation forests – a review. *Biological Conservation* **119**, 151–168.

Ludwig, J.A., Tongway, D.J., Bastin, G.N. & James, C.D. (2004) Monitoring ecological indicators of rangeland functional integrity and their relation to biodiversity at local to regional scales. *Austral Ecology* **29**, 108–120.

McCarthy, M.A., Parris, K.M., van der Ree, R. *et al.* (2004) The habitat hectares approach to vegetation assessment: an evaluation and suggestions for improvement. *Ecological Management and Restoration* **5**, 24–27.

McIntyre, S. & Hobbs, R.J. (1999) A framework for conceptualizing human impacts on landscapes and its relevance to management and research. *Conservation Biology* **13**, 1282–1292.

McIntyre, S. & Hobbs, R.J. (2000) Human impacts on landscapes: matrix condition and management priorities. In: Craig, J., Saunders, D.A. & Mitchell, N. (eds.) *Nature Conservation 5: Nature Conservation in Production Environments*, pp. 301–307. Surrey Beatty & Sons, Chipping Norton, NSW.

Morton, S.R., Stafford Smith, D.M., Friedel, M.H., Griffin, G.F. & Pickup, G. (1995) The stewardship of arid Australia; ecology and landscape management. *Journal of Environmental Management* **43**, 195–217.

Noss, R.F. & Cooperrider, A.Y. (1994) *Saving Nature's Legacy: Protecting and Restoring Biodiversity*. Island Press, Washington, DC.

Oliver, I. (2004) A framework and toolkit for scoring the biodiversity value of habitat, and the biodiversity benefits of land use change. *Ecological Management & Restoration* **5**, 75–77.

Oliver, I., Smith, P.L., Lunt, I. & Parkes, D. (2002) Pre-1750 vegetation, naturalness and vegetation condition: what are the implications for biodiversity conservation? *Ecological Management and Restoration* **3**, 176–178.

Parkes, D., Newell, G. & Cheal, D. (2003) Assessing the quality of native vegetation: the 'habitat hectares' approach. *Ecological Management & Restoration* **4**, S29–S38.

Parkes, D., Newell, G. & Cheal, D. (2004) The development and *raison d'être* of 'habitat hectares': a response to McCarthy *et al.* (2004). *Ecological Management and Restoration* **5**, 28–29.

Pearson, S.M., Turner, M.G., Gardner, R.H. & O'Neill, R.V. (1996) An organism-based perspective of habitat fragmentation. In: Szaro, R.C. & Johnston, D.W. (eds.) *Biodiversity in Managed Landscapes: Theory and Practice*, pp. 77–95. Oxford University Press, New York.

Poschlod, P., Bakker, J., Bonn, S. & Fischer, S. (1996) Dispersal of plants in fragmented landscapes. In: Settele, J., Margules, C., Poschlod, P. & Henle, K. (eds.) *Species Survival in Fragmented Landscapes*, pp. 123–127. Kluwer Academic Publishers, Dordrecht.

Povilitis, T. (2002) What is a natural area? *Natural Areas Journal* **21**, 70–74.

Rapport, D.J., Costanza, R. & McMichael, A.J. (1998) Assessing ecosystem health: Challenges at the interface of social, natural and health sciences. *Trends in Ecology and Evolution* **13**, 397–402.

Ryan, P. (2000) The use of revegetated areas by vertebrate fauna in Australia: a review. In: Hobbs, R.J. & Yates, C.J. (eds.) *Temperate Eucalypt Woodlands in Australia: Biology, Conservation, Management and Restoration*, pp. 318–335. Surrey Beatty & Sons, Chipping Norton, NSW.

Salt, D., Lindenmayer, D.B. & Hobbs, R.J. (2004) *Trees and Biodiversity: A Guide for Improving Biodiversity Outcomes in Tree Plantings on Farms*. RIRDC, Canberra.

Shrader-Frechette, K.S. (1994) Ecosystem health: a new paradigm for ecological assessment. *Trends in Ecology and Evolution* **9**, 456–457.

Soulé, M.E., Mackey, B.G., Recher, H.F. *et al.* (2004) The role of connectivity in Australian conservation. *Pacific Conservation Biology* **10**, 266–279.

South Florida Water Management District (1996) *Discover a Watershed: The Everglades*. The Watercourse, Montana State University, Bozeman, MT.

Sprugel, D.G. (1991) Disturbance, equilibrium, and environmental variability: what is 'natural' vegetation in a changing environment? *Biological Conservation* **58**, 1–18.

Suding, K.N., Gross, K.L. & Houseman, G.R. (2004) Alternative states and positive feedbacks in restoration ecology. *Trends in Ecology and Evolution* **19**, 46–53.

Tongway, D. J. & Hindley, N.L. (2004) *Landscape Function Analysis: Methods for Monitoring and Assessing Landscapes, with Special Reference to Minesites and Rangelands.* CSIRO Sustainable Ecosystems, Canberra.

Vesk, P.A. & Mac Nally, R. (2006) The clock is ticking – Revegetation and habitat for birds and arboreal mammals in rural landscapes of southern Australia. *Agriculture, Ecosystems and Environment* **112**, 356–366.

Wiens, J.A. (1997) Metapopulation dynamics and landscape ecology. In: Hanski, I.A. & Gilpin, M.E. (eds.) *Metapopulation Biology: Ecology, Genetics, and Evolution*, pp. 43–62. Academic Press, New York.

Williams, J. (2004) Metrics for assessing the biodiversity values of farming systems and agricultural landscapes. *Pacific Conservation Biology* **10**, 145–163.

With, K.A. (1997) The theory of conservation biology. *Conservation Biology* **11**, 1436–1440.

Yates, R.J., Hobbs, R.J. & Atkins, L. (2000) Establishment of perennial shrub and tree species in degraded *Eucalyptus salmonophloia* remnant woodlands: effects of restoration treatments. *Restoration Ecology* **8**, 135–143.

Young, T.P. (2000) Restoration ecology and conservation biology. *Biological Conservation* **92**, 73–83.

44

A Contribution to the Development of a Conceptual Framework for Landscape Management: A Landscape State and Transition Model

Peter Cale

Abstract

Bringing together our understanding of landscape structure and function to improve our capacity to design sustainable landscapes for the future requires a theoretical framework for landscape ecology. This framework must contain a classification of landscape types, a definition of the drivers of change within landscapes, an explanation of the interaction between landscape pattern and function, and a consideration of how these processes differ between scales. I propose a conceptual model, in the form of a Landscape State and Transition Model, that provides these four components in a simple structure. The effect of change on emergent properties of landscapes (i.e. ecosystem health and resilience) is not likely to be linear, but will vary depending on the biotic and abiotic regime shifts that a given landscape has undergone. I explore various consequences of change within different states, and suggest that this approach may enhance the design of landscapes by focusing efforts on the most important consequences instead of relying on generic designs. There are important aspects of landscape ecology still to be resolved that will restrict our capacity to produce effective landscape designs. However, developing conceptual models, albeit incomplete ones, can improve landscape design by forcing managers to address these issues in a clear and transparent manner.

Keywords: disturbance; regime shift; scale; State and Transition Model; threshold.

Introduction

Landscape design should ultimately aim to incorporate the accumulated knowledge about landscape pattern and function and their interactions, and thereby assist in directing landscape management. This goal requires an effective theoretical framework for landscape ecology, something acknowledged by others to be lacking (Wiens 1995; McIntyre & Hobbs 2000).

Landscapes comprise mosaics, and one of the greatest hurdles in developing effective landscape designs is addressing the complexity of these mosaics. Even simple mosaics of a few patch types have an almost unlimited number of configurations. This makes universal generalization unlikely, and instead we should be seeking generalizations appropriate to a particular type of landscape (Wiens 1995). An effective classification of landscape types is an essential part of such an approach. This classification needs to be defined not solely on landscape pattern but also on function, because the former does not necessarily group similar landscapes when the responses of species are considered (Cale & Hobbs 1994).

I present here a conceptual model of landscapes as my attempt at providing a framework that will bring together our understanding of landscapes. This model is deliberately simple in structure, because it needs to be understood by the range of stakeholders who must be engaged to ensure effective landscape management. It includes a general classification of the drivers of landscape change and how they relate at the landscape and patch scales. I consider the likely effect of landscape change on important landscape properties (i.e. ecosystem health and resilience) and explore various consequences of change within different states. I finish by discussing some aspects of developing a theoretical framework for landscape ecology that are still to be resolved, and detailing principles for landscape design derived from the model.

Processes of landscape change

The first requirement of a conceptual model for landscapes is a description of the major drivers of landscape change and how they interact to influence change. Landscapes change through the imposition of disturbances, or changes in disturbance regimes (Harris & Hobbs 2001). These disturbances can be either **endogenous** or **exogenous** (McIntyre & Hobbs 2000). Endogenous disturbances are those to which the system has been exposed during its evolution and they generate the system's 'natural' dynamics. Exogenous disturbances are either changes to, or the loss of, endogenous disturbances, or novel disturbances generated by new elements introduced to the system (e.g. new patch types or introduced species).

Drivers of change can result from the introduction or removal of biotic or abiotic components of the system at either the landscape or patch scale. However, they take different forms and alter the system in different ways at these two scales (Fig. 44.1). At the landscape scale they alter landscape pattern (composition and structure), resulting in changes to landscape function, which in turn alters landscape biotic and abiotic processes (Table 44.1). At the patch scale, drivers of change indirectly alter landscape pattern and function through cumulative changes to patch-scale biotic and abiotic processes (Table 44.1).

Many phenomena have been identified as landscape-scale processes, and it has been argued that the plethora of candidates reflects the poor theoretical basis for landscape ecology (Ehrenfeld 2000). However, all of these processes are controlled by the connectivity between patches, or the dynamics across patch boundaries. These two parameters therefore represent landscape function. The choice of processes to be considered will depend on the landscape and issues being considered.

One of the most fundamental changes to landscapes induced by humans is the shift in the relative importance of landscape- and patch-scale drivers of change. Landscape-scale disturbances (e.g. vegetation clearance, introduction of new patch types, altered fire and hydrological regimes) now dominate landscape change (Adamson & Fox 1982; Ehrlich 1993; Dobson *et al.* 1997), and often do so at the short timescales associated with patch-scale disturbances (De Leo & Levin 1997; Dobson *et al.* 1997). The relative importance of landscape- and patch-scale disturbances is now a distinctive characteristic of landscapes (Table 44.2). For example, Australian rangelands are strongly influenced by patch-scale drivers (i.e. introduction of livestock grazing), although fire regimes can also be important (Westoby *et al.* 1989). In contrast, agricultural landscapes of southern Australia are dominated by broad-scale vegetation clearance (Saunders *et al.* 1991; Dobson *et al.* 1997). Although patch-scale disturbances are frequently important in these landscapes, they are strongly influenced by this single landscape disturbance (Saunders *et al.* 1991). The relative mix of patch and landscape drivers of change is therefore an important

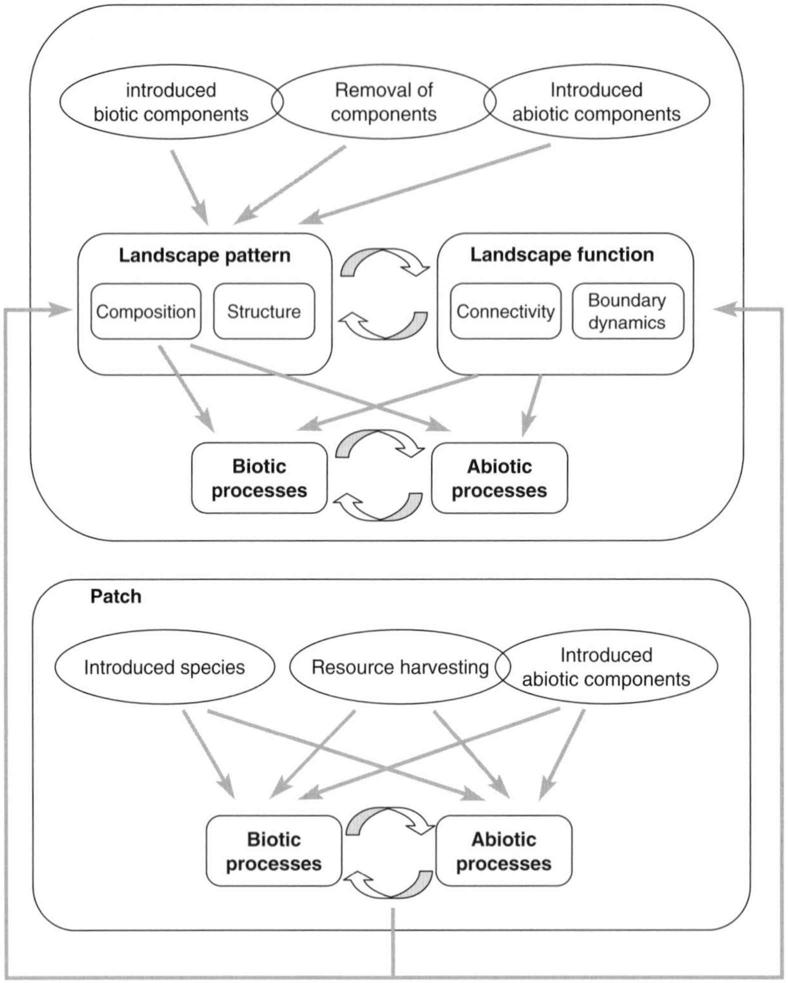

Figure 44.1 **Drivers of change operate differently at the landscape and patch scales. Landscape-scale processes are altered by changes in landscape composition and through altered landscape function resulting from changes to landscape pattern. Patch-scale processes are altered directly by drivers of change and through their cumulative effects they result in changes to landscape pattern and function. Landscape composition is a reflection of patch type and quality, while landscape structure is a reflection of patch configuration (size and disposition).**

Table 44.1 **Examples of disturbances, and biotic and abiotic processes at the landscape and patch scales.**

	Scale	
	Patch	**Landscape**
Disturbances		
Introduced biotic components	Introduced grazers (livestock, rabbits) Introduced predators (foxes) Weeds Pathogens	Patch types (paddocks, urban blocks)
Removal of components	Forestry Mining Firewood collection Hunting	Vegetation clearance
Introduced abiotic components	Fire Fertilizers Pesticides and herbicides	Fire regimes Hydrology
Processes		
Biotic	Primary productivity Herbivory Predation Mutualism	Population dynamics Dispersal
Abiotic	Energy cycling Nutrient cycling Water flux Soil dynamics	Energy movement* Material movement*

*Movement at the landscape scale represents shifts across patch boundaries.

characteristic to consider in classifying landscapes, because it influences the way landscapes behave and therefore how they should be managed.

A Landscape State and Transition model

Alternative State models have developed as an alternative paradigm to traditional Clementsian succession for describing the response of systems to disturbance

Table 44.2 **The relative importance of landscape- and patch-scale processes as drivers of landscape change differs between landscape types. The dominant scale is represented in bold for each landscape type.**

Landscape type	Scale of process	
	Patch	Landscape
Urban	**Dominated by introduced components**	**Vegetation clearance and introduced systems dominate**
Intensive agriculture	**Introduced biotic and abiotic components generally important**	**Vegetation clearance and introduced systems dominate**
Extensive agriculture	Varies in importance and extent throughout the landscape	**Dominant process is vegetation clearance with varying levels of introduced systems**
Rangelands	**Introduced biotic and abiotic components dominate change**	Less important except introduced abiotic components (e.g. fire regimes)
Forestry	**Resource harvesting dominant process**	Less important, though fire regimes sometimes an important exception
Mining	**Resource harvesting dominant process**	Less important, though hydrology sometimes an important exception

(Adamson & Fox 1982; Hobbs & Norton 1996; Scheffer & Carpenter 2003; Suding *et al.* 2004). This paradigm recognizes that systems do not always follow linear trajectories back to their original state following a disturbance. However, it is important to note it does not preclude this process and therefore non-linear responses are simply one end of the spectrum of possible responses (Scheffer & Carpenter 2003).

Alternative states represent regime shifts in which positive feedbacks make the new state resilient to a return to the original state. This difference between forward and backward shifts between states is known as **hysteresis** (Scheffer *et al.* 2001; Scheffer & Carpenter 2003), and the non-linear response to a disturbance that results in a regime shift has generally been referred to as a **threshold** (Hobbs & Norton 1996; Suding *et al.* 2004). This use of the term 'threshold' should not be confused with the proposed thresholds in vegetation loss that result in accelerated levels of species loss (Luck 2005). Although related, the latter use of 'threshold' is a description of a collection of phenomena associated with the particular, albeit commonly faced, disturbance of vegetation clearance for which the evidence is still inconsistent (Parker & Mac Nally 2002; Lindenmayer *et al.* 2005; Radford *et al.* 2005).

One type of Alternative State model proposed as the basis for a theoretical framework for landscape restoration is the State and Transition model (Hobbs & Norton 1996; Whisenant 1999). These models have been successfully developed to address management at the patch scale (Westoby *et al.* 1989; Yates & Hobbs 1997), but have not yet been used at the landscape scale.

Description of a generic model

Figure 44.2 represents a generic Landscape State and Transition model consisting of seven states with patch- and landscape-scale thresholds reduced to single transitions. In real landscapes multiple transitions and thresholds associated with different disturbances at both scales would be expected (see Fig. 44.3). As with any abstraction of reality, these models reduce complexity by simplifying the range of disturbances considered. The decision about what level of simplification is appropriate depends on the purpose of the model, and therefore should be determined on a case-by-case basis.

The transitions from State 1 to State 5 (S1 to S5 in Fig. 44.2) represent the modification of a landscape through patch-scale changes. Landscapes of structurally modified patches (S2) will continue to change until function is altered (S5), or they will revert to their original state (S1) depending on whether or not the existing disturbances are maintained. State 2 therefore represents a 'transitional state' (*sensu* Westoby *et al.* 1989). Transitional states are likely to represent conditions of low resilience where the system's trajectory is sensitive to new disturbances.

The cumulative effect of functional changes within patches can ultimately result in changes to landscape function (Fig. 44.1). Therefore, State 5 represents a change in state at the landscape scale despite the lack of landscape-scale drivers. The relative importance of patch-scale drivers of change to landscape function depends on the spatial extent of the changes and/or the range of patch types that are affected.

It is likely that landscape pattern can be changed to some degree without substantial changes in landscape function (De Leo & Levin 1997). Therefore, removing the processes driving these changes will allow the landscape to

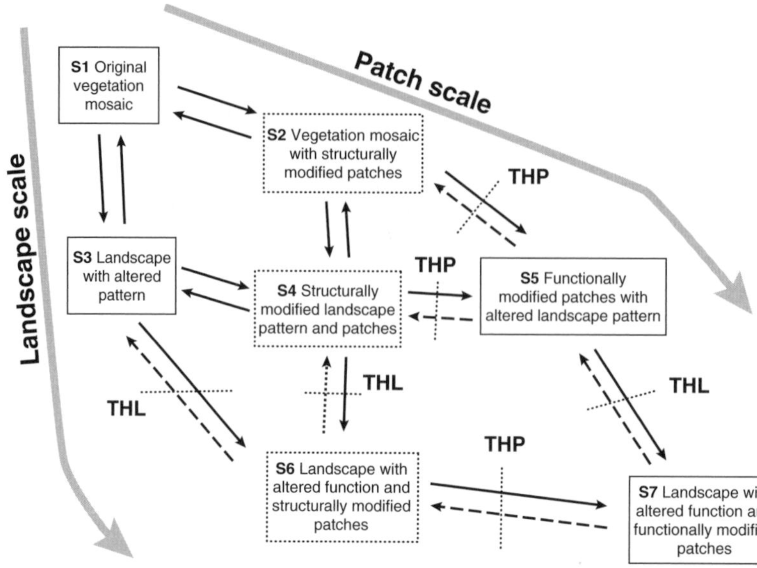

Figure 44.2 **A generic Landscape State and Transition model. Patch and landscape changes are represented in the direction of their respective arrows. Boxes with solid outlines represent states; boxes with dotted outlines represent transitional states. Solid arrows represent transitions between states. Dashed arrows represent those transitions that require an external input to occur, because the system has crossed a patch-scale threshold (THP) or a landscape-scale threshold (THL). See text for explanation.**

recover its original state (S3 to S1), although elements of its pattern may have changed irreversibly. It is unlikely that real landscapes would be subject to landscape drivers of change and not patch-scale drivers. Therefore, State 3 is likely to be theoretical, while State 4 represents real landscapes. The fundamental changes to landscapes ensuing from crossing landscape thresholds will inevitably result in changes to patch function (Saunders *et al.* 1991). Therefore, State 6 represents a landscape in transition from structurally to functionally modified patches (S7).

This Landscape State and Transition model provides a framework for comparing landscapes of a similar type (i.e. agricultural or rangelands) and providing potential pathways for management of these landscapes. Alternatively, the model can document historical change in a particular landscape and thereby provide the basis for determining relationships between drivers of

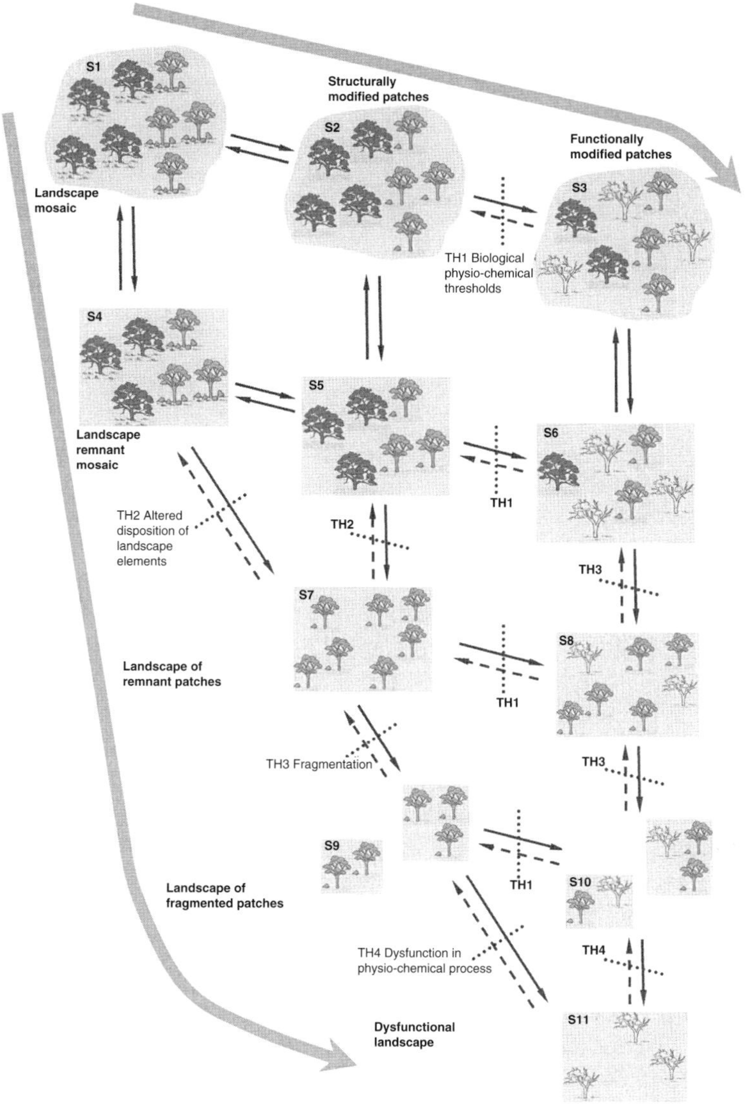

Figure 44.3 **An example of a Landscape State and Transition model for the agricultural landscapes of southern Australia. Solid arrows represent transitions between states. Dashed arrows represent those transitions that require an external input to occur, because the system has crossed a threshold. See text for explanation.**

change and changes in biodiversity or function. Landscape types in a particular region often face a common pattern of development, and therefore may be comparable at this generic level. However, a limitation of this model is that it does not explicitly address history. Landscapes can have the same broad drivers of change, but be subject to different histories (e.g. climatic events, single catastrophes), which are important to the ontogeny of the landscape's current state (Scheffer & Carpenter 2003).

This simple model identifies a number of unanswered questions with respect to landscape change. What level of landscape or patch change results in the transition from altered landscape pattern to altered landscape function? Further, does altering landscape pattern change the position of thresholds associated with changing patch function (S2–S5 vs. S4–S5)? These are difficult questions to answer, being specific to landscapes or landscape types. However, considering the potential for such interactions could have a profound influence on the approach to managing different landscapes.

Example of the model for southern Australian agricultural landscapes

The Landscape State and Transition model for agricultural landscapes dominated by broad-scale vegetation clearance presented in Fig. 44.3 is not intended to be comprehensive, but serves to demonstrate the potential use of this conceptual framework. The states and transitions S1 to S3 represent a simplified description of landscapes dominated by patch-scale disturbances. The associated thresholds are those traditionally identified in vegetation restoration, including: the loss of plant species or microsymbionts associated with grazing (Yates & Hobbs 1997; Spooner *et al.* 2002); introduction of new species (Adamson & Fox 1982; Hobbs 2001); and altered soil properties (Nulsen 1993; Whisenant 1999).

Vegetation clearance is not a random process because vegetation communities can be preferentially cleared depending on their value for agriculture. This can result in a decline in the diversity of the landscape mosaic, and ultimately in a shift from remnants containing vegetation mosaics to ones dominated by single vegetation patches (i.e. S1 to S7 in Fig. 44.3). From a faunal perspective, these preferentially cleared vegetation communities are often important, because they represent areas of high productivity (Paton *et al.* 2004).

Landscapes in States 9 and 10 have crossed a second landscape threshold and are functionally fragmented (TH3). Percolation Theory suggests, depending on the assumptions used, that there is a threshold in patch connectedness at between 40% and 70% loss of vegetation (Andrén 1994; With 1997). Whether this structural fragmentation manifests itself as functional fragmentation (i.e. inhibition of dispersal of species or their propagules) is dependent on the species and landscape in question (Lindenmayer *et al.* 2005; Radford *et al.* 2005). However, fragmentation is not solely about changes in connectivity; it also represents changes in the shape and size of landscape patches. The fragmentation threshold also represents the stage when exogenous disturbances become more important than endogenous ones, as remnants become too small to maintain endogenous disturbance regimes and adjacent agricultural patch types generate new exogenous disturbances (Saunders *et al.* 1991).

Native vegetation is a major driver of many physicochemical processes, such as hydrology, nutrient cycling and erosion (Naiman & Décamps 1997; Whisenant 1999). Once this vegetation is replaced by agriculture these processes are changed and can result in landscape-scale dysfunction (e.g. secondary salinity). Such changes probably result in crossing biological and/or physicochemical thresholds at the patch scale (TH1 in Fig. 44.3). Therefore, recovery from this state, if possible at all, will require addressing thresholds at multiple scales (i.e. S11 to S9 via S10).

Landscape- and patch-scale disturbances can produce landscapes that are functionally similar, but structurally different. For instance, fragmentation may result from vegetation clearance that induces a decline in connectedness resulting in a reduced frequency of dispersal (e.g. S7 to S9 or S8 to S10 in Fig. 44.3), or from patch-scale disturbances that reduce the production of potential dispersers (e.g. S2 to S3, S5 to S6 or S7 to S8 in Fig. 44.3). Therefore, addressing the decline in functional connectivity requires an understanding of its causes followed by management at the appropriate scale(s) to mitigate these particular problems.

Predicted changes to landscape properties

Ecosystem health is defined by the system's vigour, organization and resilience (Rapport *et al.* 1998; Harris & Hobbs 2001). Harris and Hobbs (2001) proposed that the concept of ecosystem health could be developed into the 'diagnostic toolbox' for managing ecosystems. As such it represents a concept that is valuable in describing the collective changes within a system in terms of degradation, which is essential for determining the goals of landscape design.

Ecosystem health will vary within the identified states of this Landscape State and Transition model, because patch-scale drivers of change do not

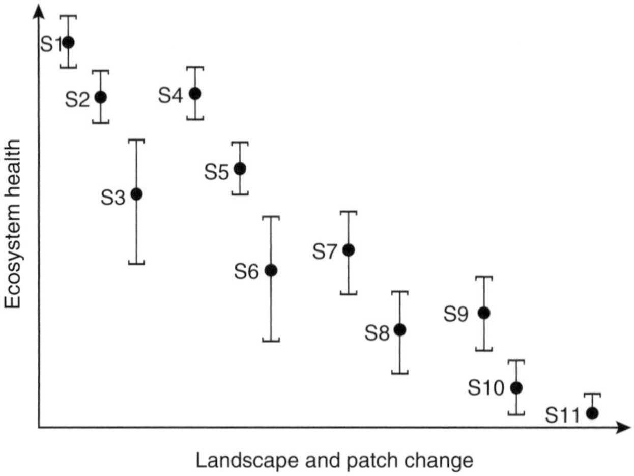

Figure 44.4 **Differences in ecosystem health within states subject to different levels of landscape and patch change. The points represent the likely ecosystem health of the states identified in Fig. 44.3. Variation in ecosystem health within these states is indicated by the bars.**

necessarily occur evenly across the landscape (Fig. 44.4). A second source of variation is an artefact of this particular model that has simplified the patch-scale drivers to a single transition. Therefore, ecosystem health will vary depending on the number of different patch functions that have changed and the spatial extent of these changes. The level of variation in ecosystem health will be greatest where patch-scale drivers dominate landscape change (e.g. States 3 and 6), but as landscape thresholds are crossed, variation should decline as the landscape as a whole becomes more dysfunctional.

The Landscape State and Transition model enables an assessment of the constraints on the level of improvement that can be achieved through action. These constraints are controlled by the current state of the landscape and the particular disturbances that are being addressed. Therefore, it provides a means of assessing the achievability of particular management goals or the consequences of ignoring existing disturbances.

Walker (1995) proposed that biological diversity could be conserved through the management of ecosystem resilience. This idea has also expanded into the realm of managing social–ecological systems where human activity and ecological processes are acknowledged to be inextricably linked (Gunderson 2000; Carpenter *et al.* 2005; Cumming *et al.* 2005). So an interesting question to ask is what might happen to the resilience of a landscape as that landscape shifts between different states? Changes in the resilience of landscapes to new disturbances when subject to vegetation clearance are of particular significance from a landscape restoration perspective. It is unlikely that changes in landscape resilience will be linear, because of correlations in species' responses to major disturbances. For example, functionally similar species are often lost as a group (e.g. ground-foraging insectivorous birds; Cale 1994; Reid 1999). This exacerbates the resultant decline in resilience by reducing the system's capacity to provide particular functions (Walker 1995) – a key component of resilience known as response diversity (see Chapter 20).

The loss of landscape mosaic elements (TH2 in Fig. 44.3) also represents a significant decline in resilience through a decline in what Dunning *et al.* (1992) termed landscape complementation and supplementation. In addition, the selectively cleared patch types are often refugia for species during periods of stress (e.g. drought or fire), and therefore represent important components of the system's resilience (Morton 1990). Fragmentation exacerbates these losses by disconnecting the remaining elements of the landscape, thereby reducing their functional diversity (Dunning *et al.* 1992), and also by reducing the size of individual remnants, thus increasing the likelihood of single events causing catastrophic loss. At this stage of landscape degradation remnants represent discontinuous components of the system, with each maintaining differing levels and aspects of the system's function due to their individual histories and to chance. Therefore, remnants in combination represent the collective resilience of the landscape, but due to fragmentation, this resilience may not manifest itself.

These additional impacts of fragmentation provide a slightly different emphasis on the relative importance of habitat loss and fragmentation in degraded landscapes. Habitat loss has generally been considered the most important problem in modified landscapes, but the influence of fragmentation has generally been restricted to consideration of the loss of landscape connectivity for species. The influence of fragmentation in driving the balance between endogenous and exogenous disturbances and the disruption of emergent properties of landscapes (e.g. resilience) may make it more important than generally thought. This will depend on the landscape and the types of changes that occur within it. These complex, non-linear changes in the relative importance of habitat loss and fragmentation, and consequently endogenous and exogenous disturbances, mean that using standard blueprints for landscape design will not always address the most important problems in a particular landscape. A focus on determining the specific importance of these phenomena and their interactions within a landscape will result in designs that produce more successful and cost-effective management.

Issues yet to be addressed

A fundamental principle of landscape ecology is that landscape components interact (Wiens 1995). However, we tend to study landscapes with a reductionist approach, expecting landscapes to equal the sum of their structures and functions, which is not the case (Dunning *et al.* 1992). Identifying and managing emergent properties of landscapes, such as resilience, may partly address this problem (Gunderson 2000; Walker *et al.* 2002). However, resilience cannot be easily measured, and so practical application of this concept reverts to measuring surrogates with uncertain relationships to the property they try to represent (Bennett *et al.* 2005; Carpenter *et al.* 2005; Cumming *et al.* 2005). Surrogates for resilience therefore face the same problems and uncertainties encountered when using threatened species as surrogates for guiding restoration (Lambeck 1997; Simberloff 1998).

Another issue that remains unresolved is how to deal with the intrinsic dynamics of landscapes when identifying changes and setting goals. Properties such as fragmentation and endogenous disturbances are interwoven with landscape dynamics. When does an endogenous disturbance become an exogenous one through a change in regime? When does a landscape shift from one state to another and therefore represent the extent of resilience in the system? These questions are difficult to answer in practice because we do not know what the intrinsic variability within a system is.

Although the concept of thresholds has been valuable in advancing the theoretical basis for landscape restoration, in practice the identification and management of thresholds remains poorly understood. As it is difficult to be sure if and where thresholds exist until they have been crossed (Carpenter *et al.* 2005), their management is reduced to mitigating their effects after the event. Further, our inability to distinguish intrinsic dynamics from directional changes means that thresholds will not easily be identified until the changes have well exceeded the threshold or have persisted for a long time. In either situation, the capacity to recover from such changes is likely to be reduced.

These unresolved problems relate to the transfer of our theoretical understanding of how landscapes might work to the practical application of this theory to real landscape problems. This problem is certainly not restricted to landscape restoration and management, and some are now arguing that such problems may be ultimately unresolvable (Funtowicz *et al.* 1999; Ludwig 2001). This means that ecologists must acknowledge the limitations of their science and identify where it is most appropriately applied. Landscape designs represent intellectual constructs from one perspective, and although they can assist in deciding on an appropriate direction for management, there are alternative perspectives of equal legitimacy (Funtowicz *et al.* 1999). Where science should play its strongest role is in assessing the outcomes of the decisions made by the community and thereby refining future decisions through improved understanding of the system.

Principles for landscape management

General principles about the behaviour of complex systems such as landscapes are difficult to enunciate without the common criticism of the 'exception to the rule'. This means that principles are often reduced to 'motherhood' statements with little practical value, or are abandoned completely for the description of specific case studies. Both of these extreme responses are unsatisfactory if a conceptual framework for landscape management is to be achieved. An alternative approach is to have a hierarchy of principles contingent on the conditions met in the higher levels of the hierarchy. The first level in this hierarchy defines the approach to landscape design and therefore addresses the issues of 'design of what for what?' These principles are not solely issues of ecology, but must also consider social issues.

- Not all of the species and processes within a landscape can be managed by a single set of design principles, and therefore effective landscape design is impossible without a clear set of objectives.
- The objectives define the landscape(s) in question.
- Management objectives have ecological and social components that ultimately cannot be addressed separately.
- Uncertainties in the response of landscape properties to change mean that landscape designs that emphasize diversity in approaches and are implemented within an experimental framework are less risky than 'one approach fits all' designs.

Addressing these broad principles results in a clearly defined restriction in the domain of the management problem. The principles identified at the next level in the hierarchy are now contingent on this problem domain. Four principles regarding landscape design and management are evident from the conceptual model presented here.

Landscape design principles

1 The interactions between landscape- and patch-scale disturbances drive changes in landscape pattern and function – that is, the landscape is more than the sum of its individual parts.
2 The relative importance of landscape- and patch-scale disturbances within a landscape is an important descriptor of different landscape types, because it influences the way such landscapes behave and therefore how they should be managed.
3 Thresholds in biotic and abiotic processes, when crossed, result in regime shifts that represent alternative landscape states, which are resistant to returning to their original state.
4 Functional connectivity can result from landscape-scale changes to connectedness or through patch-scale changes to the productivity of propagules. Therefore, management of connectivity must address potential changes at both scales to be effective.

References

Adamson, D.A. & Fox, M.D. (1982) Change in Australasian vegetation since European settlement. In: Smith, J.M.B. (ed.) *A History of Australasian Vegetation*, pp. 109–160. McGraw Hill, Sydney.

Andrén, H. (1994) Effects of habitat fragmentation on birds and mammals in landscapes with different proportions of suitable habitat: a review. *Oikos* **71**, 355–366.

Bennett, E.M., Cumming, G.S. & Peterson, G.D. (2005) A systems model approach to determining resilience surrogates for case studies. *Ecosystems* **8**, 945–957.

Cale, P.G. (1994) The effects of landscape fragmentation on the bird community of the Kellerberrin district of the Western Australian wheatbelt. Masters Thesis, University of Western Australia, Perth.

Cale, P.G. & Hobbs, R.J. (1994) Landscape heterogeneity indices: Problems of scale and applicability, with particular reference to animal habitat description. *Pacific Conservation Biology* **1**, 183–193.

Carpenter, S., Westley, F. & Turner, M.G. (2005) Surrogates for resilience of social-ecological systems. *Ecosystems* **8**, 941–944.

Cumming, G.S., Barnes, G., Perz, S. *et al.* (2005) An exploratory framework for the empirical measurement of resilience. *Ecosystems* **8**, 975–987.

De Leo, G.A. & Levin, S. (1997) The multifaceted aspects of ecosystem integrity. *Conservation Ecology* **1**, article 3 [online] (http://www.ecologyandsociety.org/vol1/iss1/art3).

Dobson, A.P., Bradshaw, A.D. & Baker A.J.M. (1997) Hopes for the future: Restoration ecology and conservation biology. *Science* **277**, 515–522.

Dunning, J.B., Danielson, B.J. & Pulliam, H.R. (1992) Ecological processes that affect populations in complex landscapes. *Oikos* **65**, 169–175.

Ehrenfeld, J.G. (2000) Defining the limits of restoration: The need for realistic goals. *Restoration Ecology* **8**, 2–9.

Ehrlich, P.R. (1993) The scale of the human enterprise. In: Saunders, D.A., Hobbs, R.J. & Ehrlich, P.R. (eds.) *Nature Conservation 3: Reconstruction of Fragmented Ecosystems*, pp. 3–8. Surrey Beatty & Sons, Chipping Norton, NSW.

Funtowicz, S., Martinez-Alier, J., Munda, G. *et al.* (1999) *Information Tools for Environmental Policy under Conditions of Complexity*. European Environmental Agency, Copenhagen [online] (http://reports.eea.eu.int/ISSUE09/en/envissue09.pdf).

Gunderson, L.H. (2000) Ecological resilience – in theory and application. *Annual Review of Ecology and Systematics* **31**, 425–439.

Harris, J.A. & Hobbs, R.J. (2001) Clinical practice for ecosystem health: the role of ecological restoration. *Ecosystem Health* **7**, 195–202.

Hobbs, R.J. (2001) Synergisms among habitat fragmentation, livestock grazing, and biotic invasions in southwestern Australia. *Conservation Biology* **15**, 1522–1528.

Hobbs, R.J. & Norton, D.A. (1996) Towards a conceptual framework for restoration ecology. *Restoration Ecology* **4**, 93–110.

Lambeck, R.J. (1997) Focal species: a multi-species umbrella for nature conservation. *Conservation Biology* **11**, 849–856.

Lindenmayer, D.B., Cunningham, R.B. & Fischer, J. (2005) Vegetation cover thresholds and species responses. *Biological Conservation* **124**, 311–316.

Luck, G.W. (2005) An introduction to ecological thresholds. *Biological Conservation* **124**, 299–300.

Ludwig, D. (2001) The era of management is over. *Ecosystems* **4**, 758–764.

McIntyre, S. & Hobbs, R. (2000) Human impacts on landscapes: matrix condition and management priorities. In: Craig, J.L., Mitchell, N. & Saunders, D.A. (eds.) *Nature Conservation 5: Nature Conservation in Production Environments: Managing the Matrix*, pp 301–307. Surrey Beatty & Sons, Chipping Norton, NSW.

Morton, S.R. (1990) The impact of European settlement on the vertebrate animals of arid Australia: a conceptual model. *Proceedings of the Ecological Society of Australia* **16**, 201–213.

Naiman, R.J. & Décamps, H. (1997) The ecology of interfaces: Riparian zones. *Annual Review of Ecology and Systematics* **28**, 621–658.

Nulsen, R.A. (1993) Changes in soil properties. In: Hobbs, R.J. & Saunders, D.A. (eds.) *Reintegrating Fragmented Landscapes: Towards Sustainable Production and Nature Conservation*, pp. 107–145. Springer-Verlag, New York.

Parker, M. & Mac Nally, R. (2002) Habitat loss and the habitat fragmentation threshold: an experimental evaluation of impacts on richness and total abundances using grassland invertebrates. *Biological Conservation* **105**, 217–229.

Paton, D.C., Rogers, D.J. & Harris, W. (2004) Birdscaping the environment: restoring the woodland systems of the Mt Lofty Region, South Australia. In: Lunney, D. (ed.) *Conservation of Australia's Forest Fauna*, pp. 331–358. The Royal Zoological Society of NSW, Sydney.

Radford, J.Q., Bennett, A.F. & Cheers, G.J. (2005) Landscape-level thresholds of habitat cover for woodland-dependent birds. *Biological Conservation* **124**, 317–337.

Rapport, D.J., Costanza, R. & McMichael, A.J. (1998) Assessing ecosystem health. *Trends in Ecology and Evolution* **13**, 397–402.

Reid, J. (1999) *Threatened and Declining Birds in the New South Wales Sheep-Wheat Belt: I. Diagnosis, Characteristics and Management*. Consultancy report to NSW National Parks and Wildlife Service. CSIRO Wildlife & Ecology, Canberra.

Saunders, D.A., Hobbs, R.J. & Margules, C.R. (1991) Biological consequences of ecosystem fragmentation: a review. *Conservation Biology* **5**, 18–32.

Scheffer, M. & Carpenter, S.R. (2003) Catastrophic regime shifts in ecosystems: linking theory to observation. *Trends in Ecology and Evolution* **18**, 648–656.

Scheffer, M., Carpenter, S.R., Foley, J.A., Folke, C. & Walker, B. (2001) Catastrophic shifts in ecosystems. *Nature* **413**, 591–596.

Simberloff, D. (1998) Flagships, umbrellas, and keystones: Is single-species management passé in the landscape era? *Biological Conservation* **83**, 247–257.

Spooner, P., Lunt, I. & Robinson, W. (2002) Is fencing enough? The short-term effects of stock exclusion in remnant grassy woodlands in southern NSW. *Ecological Management and Restoration* **3**, 117–126.

Suding, K.N., Gross, K.L. & Houseman, G.R. (2004) Alternative states and positive feedbacks in restoration ecology. *Trends in Ecology and Evolution* **19**, 46–53.

Walker, B. (1995) Conserving biological diversity through ecosystem resilience. *Conservation Biology* **9**, 747–752.

Walker, B., Carpenter, S., Anderies, J. *et al.* (2002) Resilience management in social-ecological systems: a working hypothesis for a participatory approach. *Conservation Ecology* **6**, article 14 [online] (http://www.ecologyandsociety.org/vol6/iss1/art14).

Westoby, M., Walker, B. & Noy-Meir, I. (1989) Opportunistic management for rangelands not at equilibrium. *Journal of Range Management* **42**, 266–274.

Whisenant, S.G. (1999) *Repairing Damaged Wildlands. A Process-Oriented, Landscape-Scale Approach*. Cambridge University Press, Cambridge.

Wiens, J.A. (1995) Landscape mosaics and ecological theory. In: Hansson, L. Fahrig, L. & Merriam, G. (eds.) *Mosaic Landscapes and Ecological Processes*, pp. 1–26. Chapman & Hall, London.

With, K.A. (1997) The application of neutral landscape models in conservation biology. *Conservation Biology* **11**, 1069–1080.

Yates, C.J. & Hobbs, R.J. (1997) Woodland restoration in the Western Australian wheatbelt: A conceptual framework using a state and transition model. *Restoration Ecology* **5**, 28–35.

45

Principles of Landscape Design that Emerge from a Formal Problem-Solving Approach

Hugh P. Possingham and Emily Nicholson

Abstract

For landscape and conservation ecology to become useful branches of applied ecology, they must be able to inform policy and management decisions, and help solve real problems within the constraints of economic reality. This means applying problem-solving tools used commonly in economics, applied mathematics and engineering. The crucial role of landscape ecologists in conservation is to provide relationships between biodiversity and attributes of the landscape that can be changed by management or policy, such as the configuration of protected or restored areas. We set out seven principles of landscape design for biodiversity based on a formal problem-solving approach. We illustrate these principles by providing a general formulation of the spatial conservation resource allocation problem and with some examples from the recent literature. We argue that good management decisions can only be made using a formal decision-theory approach, where the objectives are clearly stated, constraints such as finances are included, and the ecological information required to formulate the problems are transparent. In this way decisions can be made with full acknowledgement of the assumptions and uncertainties in the process, and we can learn from past successes and failures through a process of active adaptive management.

Keywords: conservation resource allocation; decision theory; management objectives; problem formulation; process models.

Introduction

If landscape and conservation ecology are to become useful arms of applied ecology, then they must borrow problem-solving concepts from economics, applied mathematics and engineering. Economists, applied mathematicians and engineers pose and solve real problems within the constraints of economic reality. They use rational decision-making processes, where the objectives, constraints and underlying assumptions and uncertainties are explicit. This **decision-theory approach** can and should be applied to landscape design and environmental management (Shea *et al.* 1998; Possingham *et al.* 2001). Unfortunately, landscape and conservation ecologists expend much of their effort trawling for ecological patterns and trying to disprove null hypotheses that are invariably false (Gerber *et al.* 2005). Worse still, some conservation biologists devise crude decision-support schemes that have no clear objective, ignore costs and/or rely on arbitrary weighting and scoring methods to find solutions. This risks bringing the field into disrepute and renders conservation efforts inefficient. Instead, applied landscape ecologists should strive to provide problem-solvers with statistical and/or process-based relationships between biodiversity and attributes of the landscape that can be changed by management or policy. We need to be clear that **science informs policy and management; it does not make policy and management decisions.**

In this essay we discuss a framework for posing and solving problems in landscape design for the conservation of biodiversity. Using a formal decision-theory approach, objectives are clearly stated, finances are included, and the ecological information required to formulate the problems is transparent. We provide a new and general description of the Spatial Conservation Resource Allocation Problem (SCRAP) to illustrate our point, where the question posed is: how should conservation resources be distributed in time and space to different activities to minimize the loss of biodiversity in a region? We propose seven principles of landscape design, summarized at the end of the chapter. These ideas are illustrated throughout in the context of the formulation of the spatial conservation resource allocation problem and with some examples from the recent literature.

What is the goal?

Defining the precise goal of a conservation problem is often the most important and difficult step in a rational decision-making process (Shea *et al.* 1998; Possingham & Wilson 2005), and is probably the point at which most attempts falter. Here are two examples, from the management of fire regimes and the prioritization of habitats for conservation.

First, agencies will argue about whether the goal of fire management is to have more or less frequent fires in a conservation park. This implies that the action is the goal rather than the outcome, which is a common mistake, especially where governments are involved. A better goal might be to maintain a fraction of the park in each of several post-fire succession states, or manage the park to minimize the number of extinctions. We have posed and solved a simple version of this problem (Richards *et al.* 1999) where the objective was to maximize the amount of time that each of three successional states covers 20% of the park, and where fire management decisions were the actions.

Second, in landscapes where we are attempting to prevent the destruction of the most important pieces of habitat, a frequent response by agencies is to assign each habitat patch a score based on a variety of interdependent attributes like patch size, patch shape, number of threatened species in the patch, species richness of the patch, patch location relative to other patches and so forth; an example is the Common Nature Conservation Classification System that is widely used in Queensland (Chenoweth EPLA 2000). The amalgamated score, which invariably has no well-defined physical dimensions, is then used to rank the patches forming an order of merit (e.g. EPA 2002; Root *et al.* 2003; Parkes *et al.* 2004). The score/merit of each patch can lead to multiple responses – acquire for conservation, protect from development, manage for biodiversity – in the absence of any stated goal (Mace *et al.* 2006). This kind of approach, which has been used in decision-making from wetland trading (Stein *et al.* 2000) to landscape prioritization for conservation and restoration (Fattorini 2006; Twedt *et al.* 2006), ignores two things. First it ignores the vast literature on conservation planning dating back to the mid-1980s, which attempts to formulate such problems properly (Cocks & Baird 1989). Second, the scoring approach fails to recognize that individual patches (or actions) cannot be ranked in isolation because their value depends on other decisions made in the landscape – **the whole is more than the sum of the parts.** Whether dealing with weed incursions or the dispersal of wildlife between habitat patches, the management of the whole landscape influences the value of an action on a particular land parcel.

Broadly speaking there are two kinds of approaches to systematic landscape design. The first is to meet a variety of biodiversity targets while minimizing net expected costs – for example, the minimum set problem in conservation planning (Cocks & Baird 1989; Pressey *et al.* 1989; Possingham *et al.* 1993; Meir *et al.* 2004). The second is maximizing biodiversity benefit within a fixed budget – for example, the maximal coverage problem of conservation planning (Underhill 1994; Nicholson & Possingham 2006; Wilson *et al.* 2006). Both have the advantage that they do not attempt to mix the two currencies, biodiversity benefit and money. Amalgamating biodiversity benefit and money into a single currency requires the use of either valuation methods (Bandara & Tisdell 2004), or explicit political trade-offs, both of which are contentious.

Both the minimum set and maximal coverage problems include an implicit trade-off between money and biodiversity. The implicit trade-off in the minimum set problem is the level of the biodiversity target (e.g. 30% of the distribution of every species must be conserved), while it is the size of the budget in the case of the maximal coverage problem. Either way, **inclusion of finances is essential**, whether through the objective (minimize total cost) or through a resource constraint (annual budget). Ignoring money in landscape design is like advising someone how to do their weekly shopping on the assumption that every item in the supermarket costs the same amount.

In this chapter we will focus our discussion on problems where we maximize biodiversity benefit within a fixed budget. The thinking is illustrated with a fairly general formulation that we call the Spatial Conservation Resource Allocation Problem (SCRAP).

The Spatial Conservation Resource Allocation Problem (SCRAP)

Consider a landscape that is divided into planning units (polygons that cover the entire two-dimensional region of interest). In each planning unit in the landscape we can take a variety of actions; for example, acquire land for conservation, acquire some component of the development rights, fence, control feral predators and/or reduce the probability of fire.

We will assume that there is a budget of B_t dollars available per year, t, of the management time horizon T. Let the amount of money spent on action k

in planning unit i in year t be b_{ikt}. Each year the total cost of all actions must be less than or equal to the budget. Assume that the set of these actions across all the planning units, $\mathbf{b}_{kt}$, changes the state of the system in year t, $\mathbf{y}_t$, and both the set of actions and the system state can affect the chance of persistence of any species, j, in the system. Given these definitions a general formulation of the SCRAP with species persistence as the prime objective is

$$\text{Maximize} \qquad \sum_{i} P_{jT}(\mathbf{y}_0, \mathbf{b}_{k1}, \ldots, \mathbf{b}_{kT}) \tag{45.1}$$

where P_{jT} is the probability that species j is extant by year T (a terminal time at which our plan is evaluated), given the initial state of the system and the sequence of actions each year. The sum of persistence probabilities gives the expected number of species surviving at the end of the management time horizon.

This objective is maximized subject to a series of constraints and equations describing the dynamics of the system (ecosystem and population models). The **constraints** include the annual budget,

$$\sum_{i} \sum_{j} b_{ikt} \leq B_t \quad \text{for all years } t = 1, \ldots, I, \tag{45.2}$$

and constraints on the resources allocated to what actions can occur in different planning units (e.g. selective logging could be ruled out in estuaries).

The **system dynamics** are equations describing the way in which the **system state**, $\mathbf{y}_t$, changes due to actions, system parameters and external stochastic forces. For example, with our fire management problem (Richards *et al.* 1999), the system state is the fraction of the park in each successional state, and this state changes stochastically depending on the process of succession and random fire events, both modified by management actions. The system state each year, $\mathbf{y}_t$, can include not just attributes of the population of every species (including its genetic structure) but also aspects of the environment in which the species live.

The SCRAP assumes that the overall objective is to maximize the number of species that persist over some long timeframe. We may choose to generalize this objective by including other aspects of the environment that people would like to conserve, like some measures of water flow, and provide rewards for every year that these services (including species existence services) are provided. The possibilities are endless, although adding complexities then requires knowledge about how the actions and random events affect ecosystem services, and interactions between all these services. **Describing and predicting how the state of an ecosystem determines outcomes like extinction probabilities and the flows of ecosystem services is the realm of scientists** – this is the task to which landscape ecologists should apply themselves, a point we return to below.

While formulating the general resource allocation problem is intellectually engaging, it is of little practical value unless we can be more specific. Here we provide specific examples of the SCRAP.

Example 1: The classic 'maximum coverage' conservation planning problem

The classical conservation planning problem (Kirkpatrick 1983; Cocks & Baird 1989) and its relatives represent the only widespread and generally accepted use of decision theory in conservation biology. The simplest version of the objective is to ensure that a fixed fraction of a variety of conservation features (e.g. habitat types and species) is conserved in protected areas for the minimum total cost. The related problem is the maximal coverage problem where, given a fixed budget, we try to maximize biodiversity gains, such as ensuring that as many features as possible meet a fixed conservation target. These problems are time-independent, ignoring both the dynamics of the species and the landscape during the process of acquisition, and only one action is possible: reserve acquisition. The performance of conservation action is evaluated solely on the final reserve system.

Overall this is a well-posed and sensible approach. Two concerns with this version of the SCRAP are the lack of dynamics and the fact that the performance of the system is solely a function of the number of species and/or habitat that meet a prespecified target. In reality we have no guarantee that meeting an arbitrarily set target ensures persistence. Most targets are based on political pragmatism rather than ecological principles (Kirkpatrick 1998; Cabeza & Moilanen 2001; Reyers *et al.* 2002). However, with some thought we are getting better at defining 'adequate' targets; for example, setting species-specific targets for the area or population size required for viability, adjusting targets according to levels of threat faced and considering large-scale movement and evolutionary dynamics (e.g. Cowling *et al.* 1999; Nicholls 1999; Burgman *et al.* 2001; Cowling & Pressey 2001; Verboom *et al.* 2001; Pressey *et al.* 2003). Ignoring time simplifies the problem but leaves the manager unclear about how to prioritize between options and when to take actions. Meir *et al.* (2004) describe a dynamic version of the problem where planning units can be destroyed and the acquisition process needs to meet the constraints of an annual budget.

Example 2: The optimal habitat restoration problem

The optimal habitat restoration problem is a minor variant on the maximal coverage problem, but now the expenditure is not on acquisition but restoration. If we know the 'cost' (where cost may have an economic, social and political dimension) of restoring any planning unit and the benefits of those planning units to meeting species targets then we can find optimal plans for restoring habitat (Westphal *et al.* 2003; Westphal & Possingham 2003).

Ideally we would know how landscapes affect viability, but in the absence of this information, we may rely instead on statistical models relating landscapes variables to habitat suitability – for example planning unit i has an x% chance of having species j if it is restored to a certain habitat quality and it is embedded in landscape y. This sits comfortably with many ecologists as it utilizes the methods of statistical habitat models. The problem is that we do not know how statistical habitat models translate into species viability.

Example 3: The optimal patch protection problem

In an attempt to include species persistence in landscape planning, we used models of metapopulation viability to find optimal reserve systems that maximize the long-term persistence of multiple species (Nicholson & Possingham 2007; Nicholson *et al.* 2006). The objective relates directly to extinction risk, a function of the ecology of the species and the amount, quality and spatial configuration of habitat in the landscape (Frank & Wissel 2002). Although able to deal with some spatial dynamics, the current formulation in Nicholson *et al.* (2006) does not include the temporal dynamics of the problem described in the SCRAP formulation. Furthermore, the choice of objective function – for example, maximize number of species that persist, versus minimize the probability any species goes extinct – can alter the outcome, and therefore needs to be carefully considered (Nicholson & Possingham 2006).

Huge challenges remain in furthering this approach, in particular for more complex landscapes and combinations of species. The metapopulation models for species viability are approximations of idealized systems. We have no simple way of converting more complex landscapes into probabilities of persistence for use in a reserve optimization framework. Even when using a metapopulation approach, with its relatively simple view of the landscape and species, we rarely have sufficient information on species to put values to many of the parameters. Solving these challenges relies on moving from statistical relationships of patterns of presence/abundance into process-based viability models. In these mechanistic models the fundamental population processes of birth, death and movement will be complex functions of the landscape – easy to say but daunting to address empirically (Tyre *et al.* 2001).

What use is a landscape ecologist?

A discussion of finding better functional forms relating management actions to biodiversity benefits leads us back to the role of landscape ecology. Good objectives are directly related to biodiversity outcomes – for example, extinction probabilities. Bad objectives are vague and uncertain surrogates for biodiversity – for example, landscape metrics or combinations of landscape metrics. The reason people don't always use the good objectives is because they are hard to calculate and parameterize. Crude surrogates, like the 'habitat hectares' metric (McCarthy *et al.* 2004; Parkes *et al.* 2004) that represents a rough measure for the condition and extent of habitat, are easier to measure and optimize. Many functions are available for describing the relationship between biodiversity benefit and increasing cost or effort (Hof & Raphael 1993; Arponen *et al.* 2005). The difficulty lies in relating this benefit function to meaningful outcomes. When the benefit of an action is estimated without a basis in ecological theory, and when we use a rough surrogate for persistence, such as the amount of habitat or a landscape metric, we run the risk of optimizing something that is not directly related to the outcomes we want, such as species persistence.

We believe that the primary task of a useful landscape ecologist is to determine the relationships between measurable surrogates and real biodiversity outcome measures – for example, turn landscape pattern metrics and habitat condition metrics into probabilities of presence (e.g. Lindenmayer *et al.* 2002; Wintle *et al.* 2005), and, better still, probabilities of persistence. To be succinct, **landscape ecologists should be trying to define the functional forms of *P*,** or equally well-defined and logical objectives (eqn. 45.1) in the SCRAP.

Solving the problem

Although the problems that we have formulated may seem very large and complex, finding a range of good solutions poses few difficulties. The SCRAP is technically solvable using stochastic dynamic programming, although simplifying assumptions are required for exact solutions to be found due to computational constraints. Alternatively, approximate near-optimal solutions may be obtained using modern optimization tools, such as genetic algorithms and simulated annealing. Many of these search algorithms have already been used to solve conservation planning problems (e.g. Moilanen & Cabeza 2002; Westphal *et al.* 2003; Nicholson *et al.* 2006). Finding the very best solution to these problems is fortunately irrelevant for most real-world problems. Despite the unrealistic calls for more exact optimization (Rodrigues & Gaston 2002; Fischer & Church 2005), there is so much uncertainty in the models, data and politics of any real problem that a variety of good alternatives is far more useful to decision-makers and managers. Future challenges include devising good methods to find and communicate good solutions to the SCRAP, in particular robust and accessible rules of thumb (Regan *et al.* 2006; Wilson *et al.* 2006).

Dealing with uncertainty

Practitioners often assume that large amounts of uncertainty about parameters and processes make a formal decision-making process useless. Ironically it is under great uncertainty that decision-making tools can provide the most substantial advantages over raw human intuition. By borrowing methods from applied mathematics, economics and engineering, we are continually expanding the toolbox for dealing with ecological uncertainties (e.g. Drechsler 2004; Regan *et al.* 2005; Nicholson & Possingham 2007). It is worth noting that when considering uncertainty our objective may change. For example, rather than maximizing the expected outcome (like maximizing expected profit when investing in the stock market), we may wish to minimize the uncertainty about the outcome (Milner-Gulland *et al.* 2001), or maximize robustness to uncertainty by choosing the management strategy that delivers a particular performance outcome subject to the greatest amount of uncertainty (Ben-Haim 2001).

Learning and active adaptive management

Merely maximizing some biodiversity objective ignores one final issue – learning. Given the uncertainty about many parameters and processes embedded in the problem formulation, there must be merit in reducing some of that uncertainty through learning. If the problem is properly formulated we can simultaneously learn and act to achieve optimal outcomes. This is the concept of active adaptive management, a process for learning while doing.

Active adaptive management contrasts with passive adaptive management, where learning occurs serendipitously (Shea *et al.* 1998). While the concept of active adaptive management has been around for a long time (Walters 1986), the idea is rarely applied to conservation problems (Gerber *et al.* 2005). Active adaptive management is further complicated by cultural and social issues surrounding the experimental management of national parks and state forests. Specifically the approach embraces uncertainty, which does not suit people, politicians and agencies, who may prefer a clear and invariant plan. A great deal more research is required in this area, in particular in the development of robust rules-of-thumb that allow managers to allocate sufficient effort to monitoring outcomes to enable learning.

Final words

It is a truism to say that you cannot simultaneously maximize two different objectives. Win–win decision-making is wonderful but rare. However, a decision-theory approach, with explicit costs and objectives, allows us to examine the trade-offs between alternative currencies and find transparent compromise solutions.

Land managers and management agencies will usually consider other issues in addition to economics and biodiversity. For example, regions interested in riparian restoration will consider the impact of actions, such as fencing, habitat restoration and altering agricultural practices, on the amount of nutrients and sediments flowing into streams, estuaries and water storage systems. Biodiversity itself, being such a broad concept, implies multiple objectives that cannot be amalgamated into a single objective function. So far we have focused on the case of minimizing the expected loss of species. Other biodiversity objectives can include: (i) maximize the opportunity for the continued evolution of existing species and the evolution of new species; (ii) maximize the delivery of ecosystem services, such as salinity mitigation and carbon sequestration; and (iii) maximize the number of different ecological communities that are 'viable'.

While constructing better spatially explicit process models of ecosystems will challenge us technically for a good time, considering multiple criteria concurrently requires even more careful thought about objectives and constraints in the problem formulation. Where there are several objectives that cannot easily be turned into the same currency (e.g. soil loss, nutrient pollution and biodiversity) there are a variety of well-known decision-making methods, like multicriteria decision analysis, that can provide objective and repeatable advice to decision-makers (Drechsler 2004; Mendoza & Martins 2006). Researchers have already produced software for supporting such decision-making (e.g. Janssen *et al.* 2005; Hajkowicz 2006). The situation is further confused by the fact that the objectives interact (e.g. some ecosystem services, like pollination, may be part and parcel of ensuring species' persistence and future evolutionary potential). Developing sensible, reliably quantifiable and more sophisticated biodiversity objectives is a challenging topic for future research. The associated socioeconomic problem is equally intricate. If we choose to include a complex suite of objectives, such as species' persistence, ecosystem services and landscape aesthetics, we need to be able to construct an appropriately weighted objective function – for example, how much is the persistence of the orange-bellied parrot for another year worth relative to more natural flow regimes in a river?

At this point the astute reader might ask the question – surely there is a great deal of subjectivity about how problems are formulated? Does this mean that our approach is just as arbitrary as the scoring and rule-based methods we have criticized so vehemently earlier in this chapter? It is true that problem formulation is a complex social construct. By necessity it is a translation of the aspirations of, and constraints on, society into a suite of mathematical equations. Once done it is transparent. It is not modelling – it is translation (like translating English into French). By using a formal decision-theory approach we can test the sensitivities of alternative goals and measures of the benefits of management actions; this is impossible when there is no explicit statement of the goals. Our challenge is to translate into mathematics as best we can our hopes, dreams and desires for landscapes that are functional and biodiverse. There are innumerable papers that fail to achieve this because they simply don't bother to pose their problem properly using objectives, constraints, decision variables and state dynamics.

Acknowledgements

The chapter contains ideas and labour shamelessly stolen from innumerable colleagues in the spatial ecology lab (http://www.uq.edu.au/spatialecology/) and elsewhere. We also thank Peter Baxter and David Lindenmayer for their valuable comments on earlier drafts.

Seven principles for managing and designing landscapes for biodiversity conservation

1. To inform directly landscape design one must use a decision-theory approach with clear and quantifiable goals.
2. Ecological principles inform practice but do not alone determine practice.
3. In landscape design the whole is more than the sum of the parts.
4. Landscape design problems should accommodate financial costs of different actions.
5. Solving the landscape design problem, once properly posed, requires decision support tools that are readily available from economists, mathematicians and engineers.
6. Uncertainty about parameters and processes is no excuse for not using formal decision-making tools and concepts.
7. Active adaptive management is management with a plan for learning, integrating research with management and monitoring, and enabling us to learn from our successes and failures.

References

Arponen, A., Heikkinen, R.K., Thomas, C.D. & Moilanen, A. (2005) The value of biodiversity in reserve selection: representation, species weighting, and benefit functions. *Conservation Biology* **19**, 2009–2014.

Bandara, R. & Tisdell, C. (2004) The net benefit of saving the Asian elephant: a policy and contingent valuation study. *Ecological Economics* **48**, 93–107.

Ben-Haim, Y. (2001) *Information-Gap Decision Theory: Decisions Under Severe Uncertainty*. Academic Press, London.

Burgman, M.A., Possingham, H.P., Lynch, A.J.J. *et al.* (2001) A method for setting the size of plant conservation target areas. *Conservation Biology* **15**, 603–616.

Cabeza, M. & Moilanen, A. (2001) Design of reserve networks and the persistence of biodiversity. *Trends in Ecology and Evolution* **16**, 242–248.

Chenoweth EPLA (2000) *Common Nature Conservation Classification System, Version 99709*. Western Subregional Organisation of Councils (WESROC), Brisbane, Queensland.

Cocks, K.D. & Baird, I.A. (1989) Using mathematical programming to address the multiple reserve selection problem: an example from the Eyre Peninsula, South Australia. *Biological Conservation* **49**, 113–130.

Cowling, R.M. & Pressey, R.L. (2001) Rapid plant diversification: Planning for an evolutionary future. *Proceedings of the National Academy of Sciences of the USA* **98**, 5452–5457.

Cowling, R.M., Pressey, R.L., Lombard, A.T., Desmet, P.G. & Ellis, A.G. (1999) From representation to persistence: requirements for a sustainable system of conservation areas in the species-rich mediterranean-climate desert of southern Africa. *Diversity and Distributions* **5**, 51–71.

Drechsler, M. (2004) Model-based conservation decision aiding in the presence of goal conflicts and uncertainty. *Biodiversity and Conservation* **13**, 141–164.

EPA (2002) *Biodiversity Assessment and Mapping Methodology (BAMM)*. Environmental Protection Agency, Brisbane, Queensland.

Fattorini, S. (2006) A new method to identify important conservation areas applied to the butterflies of the Aegean Islands (Greece). *Animal Conservation* **9**, 75–83.

Fischer, D.T. & Church, R.L. (2005) The SITES reserve selection system: a critical review. *Environmental Modeling and Assessment* **10**, 215–228.

Frank, K. & Wissel, C. (2002) A formula for the mean lifetime of metapopulations in heterogeneous landscapes. *American Naturalist* **159**, 530–552.

Gerber, L.R., Beger, M., McCarthy, M.A. & Possingham, H.P. (2005) A theory for optimal monitoring of marine reserves. *Ecology Letters* **8**, 829–837.

Hajkowicz, S. (2006) Multi-attributed environmental index construction. *Ecological Economics* **57**, 122–139.

Hof, J.G. & Raphael, M.G. (1993) Some mathematical-programming approaches for optimizing timber age-class distributions to meet multispecies wildlife population objectives. *Canadian Journal of Forest Research* **23**, 828–834.

Janssen, R., Goosen, H., Verhoeven, M.L., Verhoeven, J.T.A., Omtzigt, A.Q.A. & Maltby, E. (2005) Decision support for integrated wetland management. *Environmental Modelling and Software* **20**, 215–229.

Kirkpatrick, J.B. (1983) An iterative method for establishing priorities for the selection of nature reserves: an example from Tasmania. *Biological Conservation* **25**, 127–134.

Kirkpatrick, J.B. (1998) Nature conservation and the Regional Forestry Agreement process. *Australian Journal of Environmental Management* **5**, 31–37.

Lindenmayer, D.B., Cunningham, R.B., Donnelly, C.F. & Lesslie, R. (2002) On the use of landscape surrogates as ecological indicators in fragmented forests. *Forest Ecology and Management* **159**, 203–216.

McCarthy, M.A., Parris, K.M., van der Ree, R. *et al.* (2004) The habitat hectares approach to vegetation assessment: An evaluation and suggestions for improvement. *Ecological Management and Restoration* **5**, 24–27.

Mace, G.M., Possingham, H.P. & Leader-Williams, N. (2006) Prioritizing choices in conservation, In: Macdonald, D.W. & Service, K. (eds.) *Key Topics in Conservation Biology*, pp. 17–34. Blackwell Publishing, Oxford.
Meir, E., Andelman, S. & Possingham, H.P. (2004) Does conservation planning matter in a dynamic and uncertain world? *Ecology Letters* 7, 615–622.
Mendoza, G.A. & Martins, H. (2006) Multi-criteria decision analysis in natural resource management: A critical review of methods and new modelling paradigms. *Forest Ecology and Management* **230**, 1–22.
Milner-Gulland, E.J., Shea, K., Possingham, H., Coulson, T. & Wilcox, C. (2001) Competing harvesting strategies in a simulated population under uncertainty. *Animal Conservation* **4**, 157–167.
Moilanen, A. & Cabeza, M. (2002) Single-species dynamic site selection. *Ecological Applications* **12**, 913–926.
Nicholls, A.O. (1999) Integrating population abundance, dynamics and distribution into broad-scale priority setting. In: Mace, G.M., Balmford, A. & Ginsberg, J.R. (eds.) *Conservation in a Changing World*, pp. 251–272. Cambridge University Press, Cambridge.
Nicholson, E. & Possingham, H.P. (2006) Objectives for multiple species conservation planning. *Conservation Biology* **20**, 871–881.
Nicholson, E. & Possingham, H.P. (2007) Making conservation decisions under uncertainty for the persistence of multiple species. *Ecological Applications* **17**, 251–265.
Nicholson, E., Westphal, M.I., Frank, K. *et al.* A new method for conservation planning for the persistence of multiple species. *Ecology Letters* **9**, 1049–1060.
Parkes, D., Newell, G. & Cheal, D. (2004) The development and raison d'être of 'habitat hectares': A response to McCarthy *et al.* (2004). *Ecological Management and Restoration* **5**, 28–29.
Possingham, H.P. & Wilson, K.A. (2005) Biodiversity: Turning up the heat on hotspots. *Nature* **436**, 919–920.
Possingham, H.P., Day, J., Goldfinch, M. & Salzborn, F. (1993) The mathematics of designing a network of protected areas for conservation. In: Sutton, D., Cousins, F. & Pierce, C. (eds.) *12th Australian Operations Research Conference*, pp. 536–545. University of Adelaide, Adelaide.
Possingham, H.P., Andelman, S.J., Noon, B.R., Trombulak, S. & Pulliam, H.R. (2001) Making smart conservation decisions. In: Soulé, M.E. & Orians, G.H. (eds.) *Conservation Biology: Research Priorities for the Next Decade*, pp. 225–244. Island Press, Washington, DC.
Pressey, R.L., Nicholls, A.O. & Margules, C.R. (1989) Application of a numerical algorithm to the selection of reserves in semi-arid New South Wales. *Biological Conservation* **50**, 263–278.
Pressey, R.L., Cowling, R.M. & Rouget, M. (2003) Formulating conservation targets for biodiversity pattern and process in the Cape Floristic Region, South Africa. *Biological Conservation* **112**, 99–127.

Regan, H.M., Ben-Haim, Y., Langford, B. *et al.* (2005) Robust decision making under severe uncertainty for conservation management. *Ecological Applications* **15**, 1471–1477.

Regan, T.J., McCarthy, M.A., Baxter, P.W.J., Panetta, F.D. & Possingham, H.P. (2006) Optimal eradication: when to stop looking for an invasive plant. *Ecology Letters*, **9**, 759–766.

Reyers, B., Fairbanks, D.H.K., Wessels, K.J. & van Jaarsveld, A.S. (2002) A multicriteria approach to reserve selection: addressing long-term biodiversity maintenance. *Biodiversity and Conservation* **11**, 769–793.

Richards, S.A., Possingham, H.P. & Tizard, J. (1999) Optimal fire management for maintaining community diversity. *Ecological Applications* **9**, 880–892.

Rodrigues, A.S.L. & Gaston, K.J. (2002) Optimisation in reserve selection procedures – why not? *Biological Conservation* **107**, 123–129.

Root, K.V., Akcakaya, H.R. & Ginsberg, L. (2003) A multispecies approach to ecological valuation and conservation. *Conservation Biology* **17**, 196–206.

Shea, K., Amarasekare, P., Mangel, M. *et al.* (1998) Management of populations in conservation, harvesting and control. *Trends in Ecology and Evolution* **13**, 371–374.

Stein, E.D., Tabatabai, F. & Ambrose, R.F. (2000) Wetland mitigation banking: A framework for crediting and debiting. *Environmental Management* **26**, 233–250.

Twedt, D.J., Uihlein, W.B. & Blaine Elliott, A. (2006) A spatially explicit decision support model for restoration of forest bird habitat. *Conservation Biology* **20**, 100–110.

Tyre, A.J., Possingham, H.P. & Lindenmayer, D.B. (2001) Inferring process from pattern: Can territory occupancy provide information about life history parameters? *Ecological Applications* **11**, 1722–1737.

Underhill, L.G. (1994) Optimal and sub-optimal reserve selection algorithms. *Biological Conservation* **70**, 85–87.

Verboom, J., Foppen, R., Chardon, P., Opdam, P. & Luttikhuizen, P. (2001) Introducing the key patch approach for habitat networks with persistent populations: an example for marshland birds. *Biological Conservation* **100**, 89–101.

Walters, C.J. (1986) *Adaptive Management of Renewable Resources*. Macmillan, New York.

Westphal, M.I. & Possingham, H.P. (2003) Applying a decision-theory framework to landscape planning for biodiversity: follow-up to Watson *et al. Conservation Biology* **17**, 327–330.

Westphal, M.I., Pickett, M., Getz, W.M. & Possingham, H.P. (2003) The use of stochastic dynamic programming in optimal landscape reconstruction for metapopulations. *Ecological Applications* **13**, 543–555.

Wilson, K.A., McBride, M., Bode, M. & Possingham, H.P. (2006) Prioritising global conservation efforts. *Nature* **440**, 337–340.

Wintle, B.A., Elith, J. & Potts, J. (2005) Fauna habitat modelling and mapping in an urbanising environment; A case study in the Lower Hunter Central Coast region of NSW. *Austral Ecology* **30**, 729–748.

46

From Perspectives to Principles: Where to From Here?

Richard J. Hobbs and David B. Lindenmayer

Science may be described as the art of systematic over-simplification.
(Karl Popper)

These models are bridges of snow spanning crevasses of ignorance.
(Cohen & Stewart 1994)

Introduction

This book contains 34 perspectives from a wide range of different authors plus 11 synthesis pieces from us as editors. After all this, are we any further forward than before? As stated in Chapter 1, we aimed to gather together a range of perspectives on the various issues currently being debated within conservation biology, landscape ecology and restoration ecology that have a bearing on how landscapes are planned and managed for conservation. From these perspectives, we hoped to be able to extract a number of principles that could be usefully carried forward into the practical application of the science in real-world conservation management.

To what extent have we succeeded in this? Certainly, during the workshop at Bowral it became evident that actually nailing such principles to the table was difficult. Capturing meaningful principles reminded us of watching a kitten trying to catch sunflecks on the ground: you think you have your paw on it and then it flickers off somewhere else. There was a definite problem with the level of generality that could be obtained, because building sufficient generality into a principle usually consigned it to the wasteheap of motherhood statements and truisms – nobody could disagree with them, but at the same time they were next to useless.

A key lesson from this is that ecological principles have to be contextual. All ecological systems are ultimately unique and their composition and processes are functions of where they are, what the physical and chemical environment is like, what their spatial context and surroundings are, and what their history and current level and type of human use is. The search for generalizable principles (Popper's systematic oversimplification) thus has to be conducted in recognition of this contextual nature of ecological systems, and this undoubtedly means that the same set of principles will not be appropriate everywhere. This has emerged from the different perspectives presented in this book: in each section, the different authors writing on ostensibly the same theme and following the same guidelines have, almost without exception, produced entirely different chapters. Even the literature cited in the different chapters is almost completely non-overlapping. This diversity in what has been produced undoubtedly reflects the diversity of viewpoints and experience garnered by the authors in their respective parts of the world and in their individual ecological, social and political contexts. It has been said of economists that if you laid them all out end to end they still would never reach a conclusion, and from the diversity observed here, the same charge could perhaps be levelled at ecologists and conservation biologists. Of course, the participants at the workshop were chosen for their contrasting and non-overlapping areas of interest and expertise, and so the result is not perhaps surprising.

Do we need to throw our hands in the air in despair at the prospect of never being able to find commonality? Fortunately, we believe that this is not the case. A number of key ideas appeared repeatedly, from the discussions at the workshop, the exercise where participants were asked to produce their own personal list of principles emerging from the workshop and the chapters produced for this book. These have already been aired in many of the chapters, in the lists of principles at the end of each chapter and in the synthesis pieces at the end of each section. In this chapter, we present a brief overview of these emergent themes. These are, we believe, a set of generic principles that apply regardless of the location or type of system being managed, and these represent overarching considerations concerning visions, goals, approaches and procedures on one hand and key elements of how landscapes function on the other.

Key principles

1 **Develop long-term shared visions and from these, quantifiable objectives and constraints.** Much conservation management is undertaken without consideration of the ultimate objective and whether this is achievable given the ecological, social and economic constraints in place. Clear objectives

need to be derived from a broader vision of what people want from landscapes in the future. What should they contain? What should they look like? What services do we want from them? Hence there is a need for better problem definition, and for priority setting because not all goals are equal. We need to be able to identify the best management options to achieve a particular goal and minimize the risk of unacceptable failure. To do this we need to establish the relationships between management actions and the state of the system and hence the best management option(s) to apply. Current goal setting is often neither well focused (some goals are too abstract, for instance 'maintaining ecosystem health') whereas others are not holistic enough (often cross-agency cooperation will be needed for better holistic goal setting). Continuous ecological improvement could be one of the goals (linked to experimental management and monitoring to help assess outcomes relative to goals). Multiple plausible scenarios need to be assessed for achieving goals – there often will be no single best plan for a landscape.

2 **Principles are contingent and have to be considered within management goals, 'type of landscape', and spatial and temporal scale.** No one set of 'rules' applies everywhere. This is because of the almost infinite variation that exists, because of differences in environmental context, the range of potential goals for management and the array of spatial and temporal scales that have to be considered. There will be few overarching principles, but a large subset of contingent (more specific) principles dependent on context, conditions, species assemblages, processes and other factors. No set of general principles will substitute for a deep knowledge and understanding of a given landscape.

3 **Beware of general principles that assert black-and-white answers (e.g. edges are bad, corridors are good).** Because of the contingency mentioned in Principle 2, black-and-white answers are rarely appropriate, despite their appeal from a management perspective. What applies in one landscape or type of system may not be immediately transferable to another without careful initial assessment of the similarities and differences between the two situations.

4 **Manage the whole mosaic – not just pieces.** The whole is more than the sum of the parts. Patch-based management is still the norm, but it is increasingly clear that this in itself is not sufficient because it ignores the broader context and does not deal with emergent landscape properties such as flows of biota, water and nutrients, and interactions among mosaic elements. A single patch can be subjected to state-of-the-art management,

but that management can still fail to achieve conservation objectives if the surrounding landscape continues to degrade and lead to adverse impacts on the patch. Hence, patches can only be assessed and managed within the context of the whole landscape.

5 **Manage for change.** Landscapes and their components are dynamic. Change is a natural element of most ecosystems, and often changes can be non-linear and related to threshold phenomena. Superimposed on natural dynamics in response to disturbance at multiple scales are the additional and often novel dynamics initiated by human intervention. Conservation management often aims at stasis and assumes an equilibrium state for natural systems. Failure to acknowledge the dynamic nature of systems will inevitably result in increasing levels of unexpected change and unachieved conservation goals.

6 **Ecological systems are embedded in linked social-ecological systems.** Conservation outcomes are dependent as much on socioeconomic and political issues as on scientific understanding.

7 **Time lags between events and consequences are almost inevitable.** This applies to both the adverse effects of human activities and attempts to restore damaged systems. For instance, the impacts of landscape restoration may not be evident in terms of biotic change for many decades if the vegetation is slow growing. This is particularly important when considering how to assess success of restoration programmes (e.g. to report against milestones for government-funded programmes).

8 **Expect the unexpected.** Given the inherent unpredictability of complex ecological systems and the uncharted territory we are moving into in the face of global change, we need to build flexibility into our management and anticipate surprises rather than simply let them roll over us.

9 **Maintain resilience.** Maintain the ability of a system/landscape to recover from disturbance and adapt to changing circumstances. This includes maintaining processes and flows and the ability of the system to cope with extreme events such as droughts. However, this also involves a recognition that 'resilience costs': that is, it is impossible to manage for resilience and at the same time accomplish single-target management goals such as maximal production.

10 **Manage in an experimental framework.** If we treat management actions as an experiment, we can continuously improve our understanding of the system we are managing. This involves some careful consideration of design and monitoring to ensure that the power of the results is maximized.

11 **Don't put all our eggs in one basket (don't do the same thing everywhere).** Because of contingency, lack of knowledge of biotic responses and complex system dynamics, we don't always really know what we are doing. Hence, it makes sense not to do the same thing everywhere so that there is some degree of insurance (i.e. risk spreading) in the event that one set of actions causes the system to crash and burn: at least then, it won't crash and burn everywhere. And if we treat the variety of management options as experiments (Principle 10), we can actually learn from the experience and aim continually to improve what we are doing.

12 **Amount of native vegetation cover is of key importance (but . . .).** The amount of native vegetation cover remaining in an area is often the key factor in determining persistence of the biota. Sometimes this is regardless of the configuration of that cover. However, configuration is still important in some instances. A key principle will be to avoid low levels of native vegetation cover. Many factors will assume increasing importance when levels of native vegetation cover are low. Threshold effects, regimes shifts, and so forth are more likely to occur under these conditions.

13 **Not all patches are created equal.** Patches vary in their size, shape and relative habitat value, depending on the amount and quality of the habitat elements that they contain for particular species. Some patches may be disproportionately important because of their provision of key scarce resources such as water or nutrients.

14 **Start with the best and most 'intact'.** Any landscape conservation strategy should start by identifying and protecting the most important patches in the landscape in terms of their habitat value and current condition. It is much more cost effective to protect existing patches than to try to recreate them.

15 **Identify disproportionately important species, processes and landscape elements.** Following on from Principle 14, as well as particularly important patches, there may also be species that are of particular concern either because of their relative scarcity or because of their disproportionate impact on the ecosystem (e.g. ecosystem engineers, keystone species). Similarly, it is important to recognize key processes that may be controlling what happens in the landscape (e.g. hydrological flows causing floods, rising water tables). Complexity is constant, and management cannot deal with everything – hence it is imperative to focus on key drivers/threats/species/habitat features.

16 **Integrate aquatic and terrestrial.** The artificial divide between aquatic and terrestrial systems is counterproductive in terms of developing integrated management strategies and understanding the totality of landscape

dynamics. Landscape management is as much about water flows as it is about biotic flows.

17 **Landscapes are not always black-and-white habitat/non-habitat.** The patch-based model of landscapes serves a useful purpose in understanding how species might respond to landscape change, but may be overly simplistic in some cases, particularly when the altered matrix is not entirely inhospitable but offers some habitat value. The matrix matters in many cases, and improving the habitat value of the matrix may often be a profitable management goal to pursue.

18 **Don't forget the ducks – but ducks ain't everything.** Single species are still often the main focus of conservation efforts. Ultimately, conservation actions boil down to protecting key species or sets of species, even in an ecosystem management context. However, all species exist in an ecosystem/landscape context, and ultimately consideration of species, ecosystems and landscapes is required for a comprehensive conservation management strategy. Single-species conservation and ecosystem management are complementary approaches.

19 **Detailed contingent principles can be developed for a range of different landscapes.** Based on our knowledge of how particular landscapes work, we can at least make a start on developing more informed principles that can translate into management guidelines. We can also start to determine what management options are best for a stated goal – that is, when and where to do particular things (e.g. for connectivity, when are corridors important and when is managing the matrix more likely to yield better results?).

Contingent principles: What works where?

Many of the chapters in this book presented arguments in favour of particular models and ways of viewing the landscape. For instance, numerous authors suggested that the patch model of landscapes (or more explicitly the patch-corridor-matrix model) was too simplistic because it assumed a binary relationship between habitat and non-habitat; they asserted that gradient models provide a more realistic representation of how the landscape actually functions for particular species. At the final discussion session at the Bowral workshop, the proposition was put that the patch-corridor-matrix model was 'dead'. However, Woinarski (Chapter 10) discussed the role of fire in creating a mosaic of patches with different characteristics, and Noss (Chapter 23) suggested that urban development creates a clearly differentiated patch-corridor-matrix arrangement. Hence, it is clear that a patch model is appropriate in some situations whereas a gradient model will be more appropriate in others. Similarly, the importance of edges will be high in some contexts but not necessarily in others.

Therefore, we need to aim for a pluralistic approach in our thinking about landscape structure and function, and to ensure that we do not glibly replace one model with another and in so doing also throw the baby out with the bathwater. Cohen and Stewart's (1994) description of models as 'bridges of snow spanning crevasses of ignorance' did not explicitly refer to ecological or landscape models, but nicely captures the perspective that all models are works in progress and should be open to continual improvement and modification as we learn more about the systems under study. The question should not be 'Do we use a patch model or a gradient model to describe landscapes?' but rather 'In which situations is a patch model appropriate, and in which situations does a gradient model work better?' Akin to using different statistical methods, there are likely to be many ways of approaching the same problem or issue, and the task is to identify the most appropriate way while remaining aware of the assumptions and limitations of the different approaches.

Where to from here?

The purpose of this book was to gather different perspectives on landscape ecology, conservation biology and to a lesser extent restoration ecology and try to mould these into a set of principles that are generally applicable in most situations. The idea was to produce principles that could be converted into useful guidelines for practical landscape management, planning and design (landscape design for our purposes meaning the planning and implementation of landscape management and restoration based on sound ecological ideas). Clearly we got to the stage of amassing some principles, but we certainly did not move far down the track of testing them against the real world. Some of the chapters considered practical applications to varying extents, but we have by no means attempted a comprehensive 'road test' of the material. That was not the purpose of the workshop and is largely beyond the scope of this book, which the publishers already found difficult to digest because of its size.

Should we expect the principles listed here and throughout the book to be useful and relevant to landscape managers and planners? The answer to that surely must be that they damned well should be, otherwise we've been wasting our time and simply contributing to the oversupply of unused scientific bullshit littering library shelves and contributing little to the real world. We thus have to take the next step and try the principles out in a real-world situation. However, as Possingham and Nicholson (Chapter 45) suggested, the principles in themselves are pretty useless as management guidelines. Telling a manager that 'The whole is more than the sum of the parts' and that she or he should 'Expect the unexpected' is hardly likely to increase the manager's confidence that science can deliver useful outcomes. Hence, we should not pretend that the principles provide valuable nuggets of advice that are directly translatable into on-ground action. They indicate a set of key issues that need to be considered when developing such guidelines for managing landscapes – and, as discussed earlier, the guidelines will need to consider the issues in the context of the local situation.

Hence, a key next step is trying to do this for a set of differing contexts and seeing whether the principles can be translated into sensible on-ground decisions and actions. An important point is that principles alone won't work – there is an ongoing need for better communication, using a range of tools, including stories, models, checklists, etc. The most appropriate tools might depend on the scale at which one is working or managing. We do, however, need to be confident of at least some science before embarking on major communication efforts. Care is also required to ensure that principles are not applied out of context by managers, and a key communication theme must be to highlight the non-static nature of landscapes (i.e. things change).

Undoubtedly, some principles may prove useful, and others can be quietly dropped before they cause further embarrassment. But at least in this way we can aim to make progress in amassing a set of principles that have broad scientific agreement **and** make sense in the real world. If we can focus our scientific endeavours on researching and discussing this subset of issues, we may be able to make real progress in developing sound scientifically based landscape management. Hopefully, this in turn can contribute positively to meeting the challenge of biodiversity conservation in the face of rapidly changing global, regional and local environments.

Reference

Cohen, J. & Stewart, I. (1994) *The Collapse of Chaos. Discovering Simplicity in a Complex World*, 1st edn. Viking Penguin, London.

Index

Note: Page numbers in **bold** refer to tables, and in *italic* refer to figures.